# MANAGEMENT OF TECHNOLOGY

## THE KEY TO COMPETITIVENESS AND WEALTH CREATION

# MANAGEMENT OF TECHNOLOGY

## THE KEY TO COMPETITIVENESS AND WEALTH CREATION

Tarek M. Khalil
University of Miami

McGraw Hill

Boston Burr Ridge, IL Dubuque, IA Madison, WI
New York San Francisco St. Louis
Bangkok Bogotá Caracas Lisbon London Madrid Mexico City
Milan New Delhi Seoul Singapore Sydney Taipei Toronto

*McGraw-Hill Higher Education*

*A Division of The **McGraw-Hill** Companies*

MANAGEMENT OF TECHNOLOGY
THE KEY TO COMPETITIVENESS AND WEALTH CREATION



This book is printed on acid-free paper.

1 2 3 4 5 6 7 8 9 0 FGR/FGR 0 9 8 7 6 5 4 3 2 1 0

ISBN 0–07–366149–X

Publisher: *Thomas Casson*
Executive editor: *Eric M. Munson*
Editorial coordinator: *Michael Jones*
Senior marketing manager: *John T. Wannemacher*
Project manager: *Paula M. Krauza*
Production supervisor: *Michael R. McCormick*
Freelance design coordinator: *Michelle D. Whitaker*
Cover design: *Sara Sinnard*
Compositor: *Interactive Composition Corporation*
Typeface: *10/12 Times Roman*
Printer: *Quebecor Printing Book Group/Fairfield*

**Library of Congress Cataloging-in-Publication Data**

Khalil, Tarek M.
Management of technology : the key to competitiveness and wealth creation / Tarek Khalil.
p. cm.
ISBN 0–07–336149–X (alk. paper)
1. Technology—Management. I. Title.
T49.5.K45 2000
658.5'21—dc21 99-026735
CIP

www.mhhe.com

To Abla, Basil,
and Ronnie

# CONTENTS

# PREFACE

Technology has always been intertwined with society's progress but never before in history has technology been so visibly linked to improvements in standards of living. The human aspiration for a better life increasingly depends upon technology and its effects on all aspects of life. Because of technology, our world is developing at a phenomenal speed. Technology's pace and scope of change are having profound effects on every human institution.

The enhancement of economic prosperity for countries, industries, and businesses depends upon the effective management of technology. Technology creates wealth. The proper exploitation of technology strongly influences business competitiveness, which is no longer a matter of choice but a matter of survival in the marketplace. At the macro and micro levels, nations and individual firms and organizations are acknowledging the link between innovation and economic success. The development of technology provides an innovator with a leading edge. Clearly, the application of technology, not just its development, is a key to success in the competitive global economy.

Factors that have been associated with improvement in technological competitiveness include long-term planning horizons, research, innovation, quality of products and services, productivity, free trade, and regulatory and social factors. Scholars have been debating the relative importance of these factors for years, and the debate is expected to continue. One factor of undeniable value is the education and training of human resources needed for the ever-changing technological organization. Leading authorities in technology, business, and government recognize that technology can flow across organizational boundaries as well as across borders of countries. Production facilities can also be moved from one location to another in search of the optimal combination of resources. It is people's knowledge and managerial skills that will continue to be the most valuable resource for the success of organizations. The success of organizations is increasingly dependent upon their leaders' ability to properly manage resources in a world of rapidly changing technology. Business enterprises must also be able to compete in a dynamic global market. Engineers, managers, scientists, and policy makers should be aware of the issues associated with the management of technology (MOT). The economic present and future of their businesses, as well as their countries, will depend on it.

This book has evolved from my ten years of teaching MOT, and it responds to engineering and business schools' demand for a textbook that would introduce MOT in their curricula. The book's objectives are:

**1** To stress technology's crucial role in creating wealth and achieving competitiveness.
**2** To introduce the main factors leading to the competitiveness of manufacturing and service enterprises in an increasingly global marketplace.
**3** To emphasize the importance of considering both the speed and the scope of change in technological development and the consequential paradigm shift in the industrial and business enterprise system.
**4** To emphasize the importance of integrating technology planning and business planning.
**5** To introduce the process of technological innovation.
**6** To present the concepts of technology and product life cycles.
**7** To examine the challenges in managing the product life cycle from concept to market.
**8** To stress the importance of research and development management, technology transfer, organizational structures, project management, and third-party influence in achieving and maintaining a competitive edge.
**9** To explore human, social, and environmental concerns associated with technological change.
**10** To link all concepts to the goal of industrial/business development for economic growth and the creation of wealth.

Although many of us have been educators of technology for many years, it was not until the late 1970s and early 1980s that we realized the need for education in MOT. The apparent loss of competitive advantage by many traditionally dominant American industries during that time period heightened the need for education in MOT. The research conducted by U.S. scientists and the technology created by the nation's engineers were still dominant, yet the American competitive advantage in the marketplace declined. Japan and Germany reemerged as economic powers, and several Asian countries became fierce industrial competitors. American industry had a rude awakening to the global competition in technology and markets and was forced to change its business paradigms. Academic institutions have attempted to respond by introducing courses, as well as new programs, in MOT, through either their engineering or their business schools. The material presented in this book is based on the courses that I teach in both the industrial engineering department and the interdisciplinary M.S. program in management of technology, jointly sponsored by the College of Engineering and the School of Business Administration.

The book's fifteen chapters constitute the major components of MOT:

• Chapter 1 introduces the definition of technology and its fundamental role in the development of society. It is important to start with a complete understanding of what we mean by "technology" in the context of an MOT course. I have observed, over the many years that I worked as an engineer involved in technology, and as I taught this course, that even people working in the technology arena have difficulty defining technology or appreciating its impact. This chapter clarifies what we mean by technology and its associated terms. It also introduces the framework for MOT as an interdisciplinary field combining science, engineering, and business administration.

- Chapter 2 introduces the basis for the MOT field and asserts that it is not only technology but the management of technology that creates wealth. The chapter also briefly reviews the evolution of product and production technology over the past century.
- Chapter 3 presents the critical factors in managing technology, with emphasis on the changing environmental conditions in the world.
- Chapter 4 discusses the new paradigms of business. It examines changes in the external environment of business, organizational structure, project management, and human resource development and utilization.
- Chapter 5 introduces the principles of life cycles, including the technology life cycle, the product life cycle, and the market response at various phases.
- Chapter 6 discusses the process of technological innovation and the role of entrepreneurs, and it analyzes the factors contributing to successful innovations in the British Midlands and in Silicon Valley.
- Chapter 7 explores the issues of competitiveness on the global and corporate levels, summarizes data published on U.S. economic performance over the past two decades, and delineates important factors and policies that must be addressed to improve competitive positions at the national (macro) and firm (micro) levels.
- Chapter 8 presents fundamental concepts in strategizing and introduces methodologies used in strategic analysis and decision making. It outlines the concepts of strategic technology management, and it presents the elements of business strategy and technology strategy and stresses the importance of linking the two.
- Chapter 9 discusses technology planning.
- Chapter 10 explores methods of acquiring and exploiting technology, including the R&D mechanism of technology creation. It links the notions of creativity and innovation and discusses technology and human concerns.
- Chapter 11 focuses on technology transfer and its channels of flow across boundaries of countries and industries.
- Chapter 12 introduces and discusses technology management issues in creating world-class manufacturing and service organizations. It stresses the importance of the service sector in the economy.
- Chapter 13 presents traditional contemporary methods of organizational design. It analyzes the problems of the traditional vertical organization and introduces the concept of the horizontal organization. It points out the need for organizations to adjust their structures in order to harness the fruits of the technology revolution and avoid the pitfalls that may hinder their competitive positions.
- Chapter 14 briefly reviews management practices in the twentieth century. It explores the concept of organization reengineering, which became a buzzword for restructuring companies in the 1980s and 1990s.
- Chapter 15, "How America Does It," gives case studies of successful corporations that manage technology to remain competitive. It presents important lessons

from the experiences of some of the most admired corporations and managers. It draws heavily on the high-tech companies of the PC industry to illustrate many of the concepts discussed throughout the book. Lessons are extracted on the basis of real case studies.

The project of writing this textbook has been a challenging one. I wanted to provide comprehensive coverage of as many aspects of technology management as one can put in a book of manageable size for students taking an introductory or survey course. I did not want a handbook. However, I wanted to cover enough topics in sufficient detail for each to stand alone. Topics had to be presented in a concise and clear manner.

Many of the topics covered in this text have the potential of being expanded into a full course with a separate textbook. In fact, at the University of Miami we offer a 12-course master's program in management of technology. Each course is a three-credit-hour semester course. These courses delve into rigorous details of some of the subjects discussed in this book. However, the objective of this text is to tie these subjects together and show how all of them relate to one core and contribute to a main objective. The book concentrates on combining topics traditionally discussed either in engineering or in business administration curricula. It demonstrates the diversity and interdisciplinary nature of the issues associated with technology management. Yet the core objective is creation of wealth for nations, industries, and individuals.

Several relevant and important topics are mentioned only briefly in this text for the sake of completeness. These include general management, the role of capital, managing people, technology protection, third-party influence on technological change, information technology, finance, marketing, diffusion, and effects of technology on the environment. It is felt that specific details of these topics are better dealt with in specialized courses than in a survey course on management of technology.

Management of technology is a very dynamic subject. It is crucial for managing in a changing world environment. Technology is continuously progressing, and world markets are continuously shifting. The success or failure of organizations depends on the organizations' ability to ride the wave of change and still emerge as winners. In this text, several case studies are used to illustrate important concepts. Any example used will probably be dated, for tomorrow will bring a change in world conditions. I have elected to focus on classical case studies of companies that are recognized by most readers. I have extracted fundamental lessons to be learned from these cases that have withstood the test of time. Thus they can be applied, with appropriate modifications, in a generic way to other cases.

This book is an excellent textbook for introductory or survey courses in technology management. It is also suitable for an advanced undergraduate or graduate course. It is recommended for advanced courses if supplemented by additional readings of recently published journal articles, conference proceedings, and case studies. Additional readings and references are given at the end of each chapter from which supplemental material can be drawn. The material presented here has been successfully tested with both university students and mature professional audiences.

This textbook reviews the history and socio-political-economic implications and philosophical importance of technology as a means of creating economic wealth, as well as its importance in ensuring competitiveness in the global economy. The book is valuable reading for government science and technology planners and forecasters, public policy makers, and macro- as well as microeconomic advisors to such individuals.

It is also of value to CEOs and directors of industrial corporations, government officials, investors and businesspeople, entrepreneurs and small-business owners, engineers, managers at all levels, and university faculty involved in technology and business education.

*Tarek Khalil*
Coral Gables, Florida

# ACKNOWLEDGMENTS

I am very grateful to many of my colleagues and students who have helped me shape my thoughts about this field. The many discussions and conferences that we attended or organized over the past decade at the National Academy of Engineers, the National Science Foundation, and the University of Miami have given shape to what now has become the core of the management-of-technology field. Sincere thanks are due to my colleague Dr. Bulent Bayraktar, who helped me introduce the MOT program at the University of Miami and coedited with me the first four books of proceedings for the International Conference on Management of Technology from 1988 to 1994. Thanks also are due to the 20 participants of the NSF/UM workshop on challenges and opportunities for research in the management-of-technology field, held in Miami, Florida, in February 1988. Their insights on the paradigmatic shifts in industry and education helped shape many of my activities in the ensuing years. I am sure that many of the ideas and illustrations in this book have been inspired by the several hundred students who have attended my courses at the University of Miami as well as at sites in major industrial corporations, including IBM, McDonnell-Douglas, Rockwell International, UNISYS, and AT&T Paradyne. In addition, the hundreds of presentations and papers presented by international experts participating in the series of international conferences on management of technology that I have helped organize and chair over the last decade have contributed immeasurably to the development of these ideas. Special thanks go to my graduate student assistant Javier Garcia-Arreola, who has provided significant help in putting this manuscript together. Also sincere thanks are due to Victoria Varela, who helped type the manuscript, and to the reviewers of this manuscript for their constructive comments.

# 1

# INTRODUCTION

## TECHNOLOGY AND SOCIETY

Throughout human history, technology has had a profound effect on human development and on the progress of civilization. It took humans about 2 million years to develop from nomads foraging for food in east central Africa to agricultural settlers capable of augmenting their power with tools and domestic animals. This significant development ushered in the dawn of civilization. The next several thousand years witnessed the development of the wheel, the chariot, the water wheel, and mechanical implements. It was only less than two centuries ago that the steam engine and the factory system ushered in the start of the Industrial Revolution. Energy generated from water and from mechanical, electrical, and nuclear sources enabled humans to achieve unprecedented change in their way of life. Yet in no other time in history has technology been as pervasive in human lives as it is today. To an ever-increasing extent, it has invaded every aspect of human endeavor. Government operations, global corporations, private enterprises, and individuals are highly dependent on technology for their success. The rate of technological progress and society's dependence on technology only promises to intensify as the world moves into the twenty-first century.

In spite of the fact that technology is the most pervasive force influencing human lives today, it remains mysterious to many people, and its exact definition eludes most of the public and many professionals. It is therefore important to start with a clear definition of technology. This will permit a common understanding upon which one can build the arguments expressed in this text.

## DEFINITION OF TECHNOLOGY

*Technology* can be defined as all the knowledge, products, processes, tools, methods, and systems employed in the creation of goods or in providing services. In simple terms, technology is the way we do things. It is the means by which we accomplish objectives.

Technology is the practical implementation of knowledge, a means of aiding human endeavor.

It is common to think of technology in terms of hardware, such as machines, computers, or highly advanced electronic gadgets. However, technology embraces a lot more than just machines. There are several technological entities besides hardware, including software and human skills. Zeleny (1986) highlighted this point by proposing that any technology consists of three interdependent, codetermining, and equally important components:

*Hardware:* The physical structure and logical layout of the equipment or machinery that is to be used to carry out the required tasks.

*Software:* The knowledge of how to use the hardware in order to carry out the required tasks.

*Brainware:* The reasons for using the technology in a particular way. This may also be referred to as the know-why.

In addition to the above three components, a fourth one must be considered independently, for it encompasses all levels of technological achievements:

*Know-how:* The learned or acquired knowledge of or technical skill regarding how to do things well. Know-how may be a result of experience, transfer of knowledge, or hands-on practice. People acquire technical know-how by receiving formal or informal education or training or by working closely with an expert in a certain field. Know-how can also be acquired through a recognized method of technology transfer, a topic described in a later chapter of this book.

### Knowledge and Technology

We currently live in what truly can be termed the "knowledge age." Technology is knowledge applied to the creation of goods, provision of services, and improvement of our stewardship of precious and finite resources; on a negative note, it can also be applied for destructive purposes. Whatever the application may be, advancement of technology follows expansion of knowledge. Knowledge is not information, but it is based on the amount of information available. Knowledge is all that has been perceived or grasped by the mind from the range of information available. Human beings have been able to sort information accumulated throughout their environment into a body of facts, principles, and theories that form a basis for human enlightenment and learning. It is only when knowledge is practically implemented to create new things, operate a system, or provide a service that we enter the realm of technology.

The advances in information technology in the second half of the twentieth century have expanded the amount of information available in the world. This has created an explosion of knowledge and brought about further dramatic progress in technology. The invention of the transistor at Bell Laboratories in 1947 and the subsequent development of integrated circuits have driven the development of computers and the information revolution. The capacity for information processing has been continuously increasing.

Pritchett (1994) reported that there has been more information produced in the last 30 years than during the previous 5,000 years and that the information supply available to us doubles every 5 years. Gordon Moore (Isaacson, 1997), a cofounder of Intel corporation, has predicted that microchips will double in power and halve in price every 18 months ("Moore's law"). Intel has been able to deliver on this prediction, doubling the computing power and the amount of information processed every 18 months or so. Intel and other companies have been finding many new applications for microchips that were scarcely imaginable a year or two earlier. *Time* magazine (Dec. 29, 1997) chose Andrew Grove, CEO of Intel, as its 1997 Man of the Year for his contributions to technology and his impact on the new wave of economic growth in the United States. The advances in microchips and in computer technology in general have led to a new wave of expansion in business expenditure and practice. According to Pritchett, "In 1991, for the first time, companies have spent more money on computing and communications gear than the combined moneys spent on industrial, mining, farm, and construction equipment."

The information age of the late twentieth century has created an explosion of knowledge and has had a major impact on the rate of technological change. The accelerated rate of technological change is having a profound impact on society and the standard of living. As we move into the twenty-first century, the Industrial Revolution has given way to the "technology revolution."

### Technology and Business

An organization can be thought of as a vehicle for introducing one or multiple technologies to society. The goal is to achieve a set of objectives. The technologies used can vary in level, from being very basic, on one end of the scale, to being super-high technologies, on the other end. The organization can be nonprofit or for-profit. Both types are considered production organizations, with *production* being any activity that results in the conversion of resources into goods or services. The resources include natural resources, human resources, and other resources. Business enterprises are formed to make profits. The pool of knowledge available to society is a major source for creating business enterprises. Technology's contribution is not only in how goods and services can be produced but also in what can be or even has the potential to be produced. Technology converts the realm of possibilities into realities.

The technologies that exist in a business are the technological assets of that business. These assets may therefore include hardware, software, brainware, and know-how. They constitute the collective knowledge and technical capabilities of the organization, including its people, equipment, and systems.

In the past, the value of a company was assessed largely on the basis of its capital and physical assets such as land, buildings, equipment, and inventory. Today, the real value of a company is much more than the value of its physical assets or its simple accounting net worth. Technology adds value to the assets of a company. The technology resides *in* the company's people and its technological systems. A case in point is Netscape, a company that is heavily involved in developing Internet browsers/navigators, server software, and applications. Netscape, listed on the NASDAQ stock exchange, started with a few knowledgeable principals and an early lead on Internet technology and software products but very little in capital assets or holdings. Its stock was at $23 in September 1995. Two months later it was valued at more than $47, and the following month it reached more than $85. This is a good example of a company's value residing in its technology and people. Netscape had 71.5 percent of the Internet browser market in August 1996 (Ellis, 1996); its physical assets were relatively insignificant, yet the company's real value, its technology, was huge. The question is, Can such a company maintain its technological edge over competitors and remain financially successful? Only time can tell. However, Netscape's chances of sustaining success depend on the skill of its management team in managing the technology-based business.

According to Quinn et al. (1996), the intellect of contemporary organizations operates at four levels. These levels are presented below in order of increasing importance:

**1** *Cognitive knowledge, or know-what:* This is the basic mastery of the discipline. It is essential but insufficient for commercial success.

**2** *Advanced skill, or know-how:* This is the translation of book learning into effective execution.

**3** *System understanding, or know-why:* This is knowledge of the cause-and-effect relationship underlying a discipline.

**4** *Self-motivated creativity, or care-why:* This is the will, motivation, and adaptability needed for success. Creative and motivated groups outperform groups with greater financial or physical resources.

## CLASSIFICATION OF TECHNOLOGY

Technology can be classified in several ways. The following classifications are important in establishing a common vocabulary for the ensuing discussion in this book.

**New Technology** A *new technology* is any newly introduced or implemented technology that has an explicit impact on the way a company produces products or provides services. One example is new computer software introduced to develop engineering drawings and thus replace manual drafting. Another example is an Internet web site designed to market the company's products. The technology does not have to be new to the world, only to the company. It could have been developed years ago and used by others, but it is classified as new whenever introduced for the first time in a new situation. New technology has a profound effect on improving productivity and maintaining a competitive business enterprise.

**Emerging Technology** An *emerging technology* is any technology that is not yet fully commercialized but will become so within about five years. It may be currently in limited use but is expected to evolve significantly. Examples of emerging technology include genetic engineering, nanotechnology, superconductivity, and the Internet as a replacement for the personal computer. Emerging technologies create new industries and may make existing ones obsolete. They have the potential of triggering large changes in institutions and in society itself.

**High Technology** The term *high technology* (high tech) refers to advanced or sophisticated technologies. High technologies are utilized by a wide variety of industries having certain characteristics. A company is classified as high-tech if it fits the following description (Larsen and Rogers, 1988; Mohrman and Von Ginlow, 1990):

- It employs highly educated people. A large number of the employees are scientists and engineers.
- Its technology is changing at a faster rate than that of other industries.
- It competes with technological innovation.
- It has high levels of research-and-development expenditure. (A general guide is that the ratio of R&D expenditures to sales is 1 to 10 or twice the average for the industry.)
- It has the potential to use technology for rapid growth, and its survival is threatened by the emergence of competing technology.

Some high-tech companies may be working with technologies that are "pushing the envelope." These technologies may be referred to as "super-high technologies."

**Low Technology** The term *low technology* refers to technologies that have permeated large segments of human society. Low technologies are utilized by a wide variety of industries having the following characteristics:

- They employ people with relatively low levels of education or skill.
- They use manual or semiautomatic operations.
- They have low levels of research expenditure (below industry average).
- The technology base used is stable with little change.
- The products produced are mostly of the type that satisfy basic human needs such as food, shelter, clothing, and basic human services.

**Medium Technology** As used in this text, the term *medium technology* comprises a wide set of technologies that fall between high and low technologies. It usually refers to mature technologies that are more amenable than others to technology transfer. Examples of industries in this category are consumer products and the automotive industry.

**Appropriate Technology** The term *appropriate technology* is used to indicate a good match between the technology utilized and the resources required for its optimal use. The technology could be of any level—low, medium, or high. It does not make sense, for example, to use high technology when there is a lack of necessary infrastructure or skilled personnel. This is a dilemma faced by many developing countries that want to transfer technology used in more industrialized countries. They may push for the acquisition of high technology in cases where a medium-level technology would be more effective. Utilizing the appropriate level of technology results in better use of labor resources and better production efficiency.

**Codified versus Tacit Technology** Technology can be preserved and effectively transferred among users if it is expressed in a coded form. An engineering drawing is a coded form expressing shape, dimension, and tolerances about a product. A computer program of an optimization algorithm is a codified form that preserves and transmits knowledge about that algorithm.

*Tacit technology* is nonarticulated knowledge. There is no uniformity in the way it is presented or expressed to a large group of people. It is usually based on experiences and therefore remains within the minds of its developers. The technology developers are the ones who have the know-how in question. Tacit knowledge is transmitted by demonstration or observation, followed by assimilation by those who seek the knowledge. Transfer of tacit technology occurs by close contact and interaction between the source and the host. Apprenticeship programs may serve as a vehicle for transferring the tacit nuances of specific professions or fields.

*Codified technology,* on the other hand, allows people to know how technology works but not necessarily why it works in a certain way. The brainware may be part of the tacit knowledge kept in the minds of developers and shaped by their experiences during the development process. Transfer of technology is easier when the technology is in a codified form. It is harder, less precise, and more time-consuming to transfer tacit technology. A complete mastery of the technology requires an understanding of both the explicit codified knowledge and the nonexplicit tacit knowledge.

## DEFINITION OF MANAGEMENT

Management is an art and to some extent a technology. It is the art of carrying on business. It involves directing and controlling an organization and steering it toward achieving its objectives. It draws on knowledge, experience, and an understanding of human and organizational behavior.

The term "directing" means providing a direction and establishing the path that an organization follows to fulfill its mission. Directing implies a one-way flow of information, usually top-down. It is therefore important to highlight the word "controlling" in the definition of management. Control implies a feedback from the system to the control agent to rectify actions and keep the system on the right track. Management control of corporations is essential to keep the organization on course and prevent problems. Management is not and should not be an open-loop endeavor; it is a closed-loop system. It involves planning and coordination on a continuous basis, and ideally it should permit the flow of information in more than one direction—top down, bottom up, and across the organizational structure.

Management is also a technology, as it is the means by which the desired goals of an enterprise are achieved. Management functions in an organization include planning, organizing, staffing, motivating, and controlling activities of the organization. Management, as a field, has a knowledge base and guiding principles. Frederick Taylor, the "father of scientific management and industrial engineering," applied the scientific method to the art of management. After Taylor, many management gurus contributed their valuable knowledge to create "management technology." The term *management technology* implies technology used to manage organizations or certain functions. It should not be confused with the management *of* technology, which is defined below.

## MANAGEMENT OF TECHNOLOGY

Management of technology (MOT) is an interdisciplinary field that integrates science, engineering, and management knowledge and practice (Figure 1-1). The focus is on technology as the primary factor in wealth creation. Wealth creation involves more than just money; it may encompass factors such as enhancement of knowledge, intellectual capital, effective exploitation of resources, preservation of the natural environment, and other factors that may contribute to raising the standard of living and quality of life. Managing technology implies managing the systems that enable the creation, acquisition and exploitation of technology. It involes assuming responsibility for creating, acquiring, and spinning out technology to aid human endeavors and satisfy customers' needs. Research, inventions, and development are essential components in technology creation and the enhancement of technological progress. However, more important for the creation of wealth is the exploitation or commercialization of technology. It is only when technology is connected with a customer that its benefits are realized (Figure 1-2). A customer is a beneficiary and could be an individual, a corporation, or a government entity such as a defense establishment. An invention made and put on a shelf is not contributing to wealth. An idea that emerges and is not exploited, even if it was patented, does not bring monetary returns. Technology generates wealth when it is commercialized or used to achieve a desired strategic or operational objective for an organization.

While the underlying premise for the MOT field is that technology is the most influential factor in a wealth-creation system, there are other factors that contribute to the system (Figure 1-3). For example, capital formation and investment make significant contributions to economic growth. Labor is another factor in economic growth. Social, political, and environmental considerations facilitate or hinder the wealth-creating process. MOT treats technology as the seed of the wealth-creation system. With proper nourishment and a good environment, a seed grows to become a healthy tree. Other

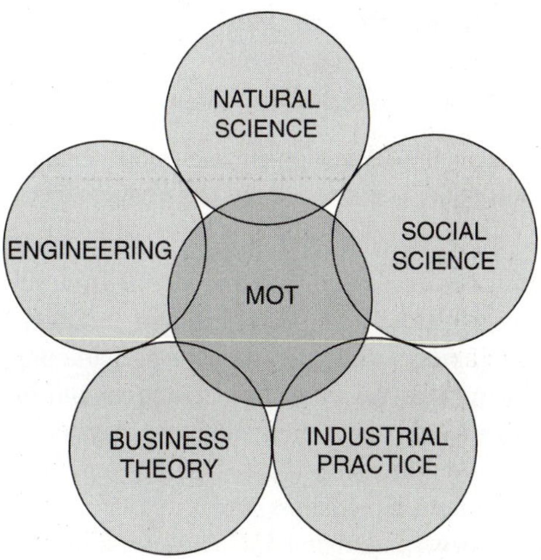


**FIGURE 1-1**
THE INTERDISCIPLINARY NATURE OF MOT

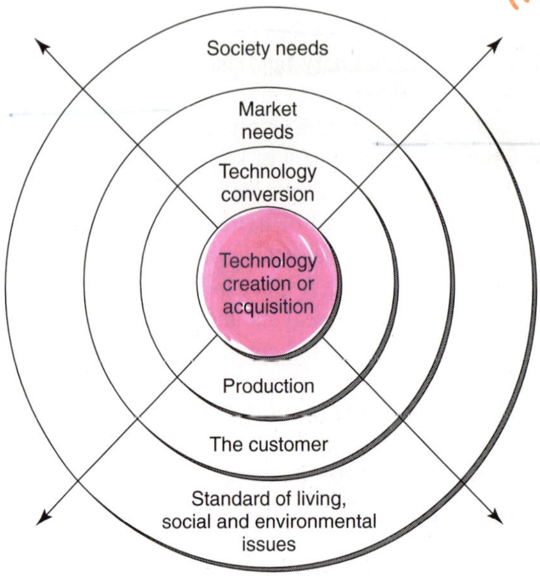


**FIGURE 1-2**
SPINNING OUT TECHNOLOGY
Technology must connect with customer needs to satisfy those needs and achieve societal goals. Technology is the engine of economic growth.

factors contributing to wealth creation—including capital, labor, natural resources, public policy, and so on—provide the fertile land, environment, and nourishment needed for growth. Each one of these factors has its own disciplinary field of study and research. MOT, as an interdisciplinary field, combines knowledge from these disciplines. A comprehensive MOT degree program requires in-depth studies of all these factors.

MOT has national, organizational, and individual dimensions. At the national/government level (macro level) it contributes to shaping public policy. At the firm level (micro level) it contributes to the creation and sustainability of competitive enterprises. At the individual level it contributes to the enhancement of one's worth in society.

### MOT at the Firm Level

A National Research Council report (1987) on management of technology defined it as

> an interdisciplinary field concerned with the planning, development and implementation of technological capabilities to shape and accomplish the operational and strategic objectives of an organization.

MOT is an interdisciplinary field because it involves combined knowledge from science, engineering, and business administration fields. It impacts different functional entities of the corporation: research and development, design, production, marketing, finance, personnel, and information. Its domain involves both the operational and the strategic interests of organizations. The operational aspect deals with the day-to-day activities of the organization, while the strategic dimension focuses on the long-term issues. The organization must take into account both dimensions.

Studies have shown that engineers and managers of many U.S. corporations tend to concentrate their efforts on the operational aspect, focusing on short-term results

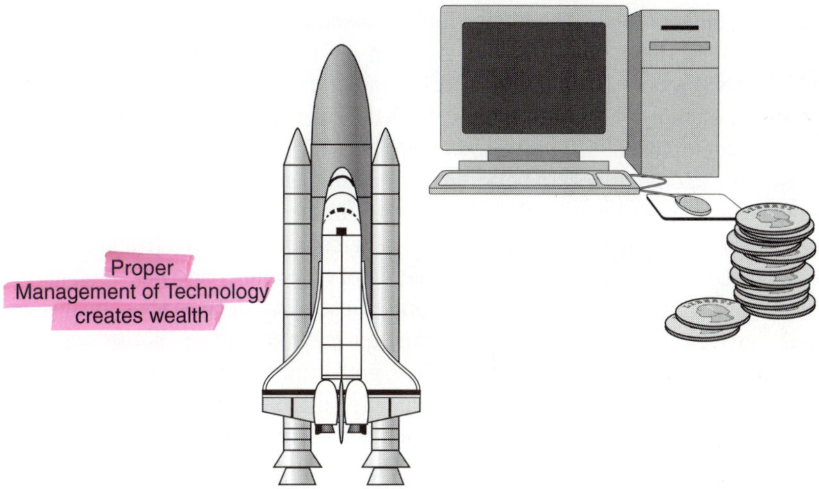


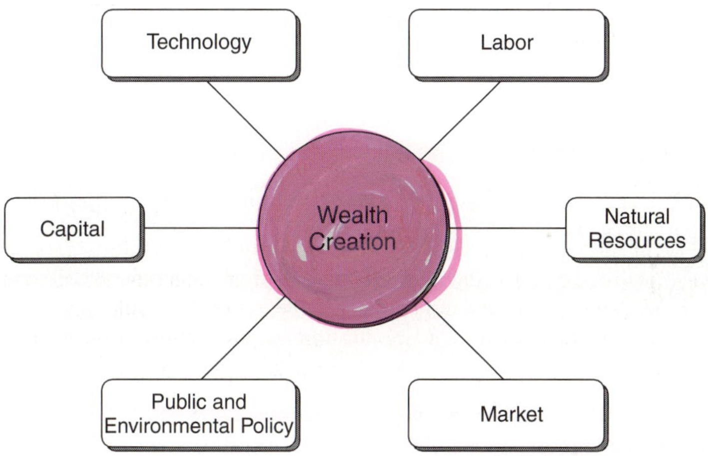


**FIGURE 1-3**
FACTORS CONTRIBUTING TO WEALTH CREATION

and giving less attention to strategic outlook (Berman and Khalil, 1992; Heim and Compton, 1992). This short-term horizon of management underestimates the consequences of today's actions on the future health of the enterprise. MOT addresses these potential weak spots in management systems by putting emphasis on the strategic objectives of any organization. It guides management in its efforts to improve productivity, increase effectiveness, and strengthen the competitive position of the enterprise.

In the 1970s and 1980s U.S. industries started to lose their competitive advantage, mainly to Japanese products but also to products produced in a slew of other Asian and European countries. The decline in competitive position prompted many organizations to express their concern and examine various approaches that might help U.S. industries restore their competitiveness. Organizations involved in this endeavor were the National Research Council (NRC), the National Academy of Engineering, the National Science Foundation (NSF), many industrial organizations, and educational institutions. Their efforts brought attention to the importance of MOT in restoring the United States to its position of economic leadership in the world. The NRC report (1987) proposed bridging the knowledge and practice gap between engineering and science on one side and business and management on the other. It was the link between technology creation and business exploitation that was perceived as inefficient in many private and government-owned organizations. The effective combination of technology and business, and bringing technology to the marketplace in the form of products and services, creates wealth. The link between the technology side and the business side of the house seemed to be missing from paradigms of industrial practice, government policies, and educational institutions. It was realized that the existing educational programs in engineering and business schools, combined with rigid institutional structure, may have become outmoded and deserved a serious review. It was also realized that first-rate educational programs are needed to develop the engineers and managers that can manage technological change and expand global markets. New ways of thinking about these programs and the content of curricula were indicated. The emergence of new specialized programs in management of technology was one of the outcomes of this self-examination exercise.

### MOT at the National/Government Level

From a macro-level perspective a more general definition may be appropriate for MOT. It can be defined as

> a field of knowledge concerned with the setting and implementation of policies to deal with technological development and utilization, and the impact of technology on society, organizations, individuals and nature. It aims to stimulate innovation, create economic growth, and to foster responsible use of technology for the benefit of humankind. (Khalil, 1993)

At the national level, more focus is placed on the role of public policy as it applies to the advancement of science and technology. The overall impact of technology on society is explored, particularly its role in developing sustainable economic growth. The effect of technological change on people, the type of education and training they need, and effects on health and safety as well as on the environment are considered. Government and organizational policies are developed to embrace technological change for the benefit of their constituencies.

## THE CONCEPTUAL FRAMEWORK FOR MOT

Figure 1-4 shows the basic concept of MOT as an interdisciplinary field of study and application. It illustrates how MOT creates a linkage among science, engineering, and management disciplines. From an academic point of view, this conceptual figure indicates that traditional fields in science and engineering contribute to scientific discovery and to technology creation. There are also traditional fields in business administration that contribute to the management of enterprises, economics, finance, marketing, and public policy.

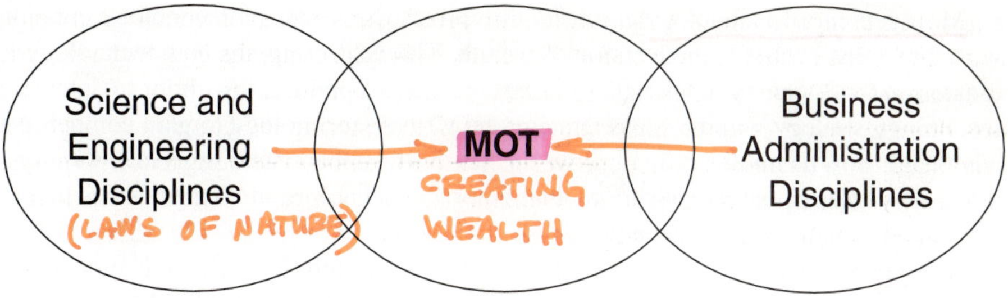


Discipline-based Knowledge
Science Disciplines
Material Technology
Product Technology
Production/Process Technology
Information Technology
Environmental Technology

Strategic/Long-Term Issues Relating Technology
Science and Technology Policy
Process of Technological Innovations
R&D Management
R&D Infrastructure and Technological Change
Technological Entrepreneurship and New Ventures
Product/Process Life Cycles
Technological Forecasting and Planning
Technological Innovations and Strategic Planning
Technological Transfer
International Technology Transfer and the Role of Multinational Corporations
Technological Risk Analysis and Assessment
Technology and Economic Analysis
Technology and Human, Social and Cultural Issues
Training and Education Issues in Mngmt. of Technology
Mngmt. of Technology in Manufacturing Industries
Mngmt. of Technology in Service Industries
Information Technology and Other Emerging Technologies
Manufacturing Marketing and After Market Interface
Technological Change and Organizational Structure
Management of Technical Projects
Financing Technology and Financial Decision Making
Quality and Productivity Issues
Methodologies in Mngmt. of Technology
Eco-efficiency and Environmental Sustainability

Discipline-based Knowledge
Accounting
Finance
Management
Marketing
Economics
Business Law

**FIGURE 1-4**
CORE KNOWLEDGE NEEDED FOR MOT

Management of technology connects disciplines that focus on technology creation with those that enable its conversion to wealth. The field examines how technology is created; how it can be exploited to create business opportunities; how to integrate technology strategy with business strategy; how to use technology to gain competitive advantage; how technology can improve the flexibility of manufacturing and service systems; how to structure organizations that embrace technological change; and when to enter and when to abandon technology.

A series of meetings sponsored by the National Research Council, the National Science Foundation, the University of Miami (Khalil and Bayraktar, 1988), the Oak Ridge Associated Universities, and several professional organizations in the United States addressed these issues. Based on discussions at those meetings a set of topics emerged that were of specific concern in MOT (Khalil, 1991). They are listed under the MOT circle in Figure 1-4.

Science and technology policies have a major impact on a nation's competitiveness; therefore, the process of technological innovation must be well understood. Infrastructure and management of R&D is a key aspect, and entrepreneurship is vital for the development of new technologies. The technology life cycle as well as product and process life cycles are becoming much shorter; technological forecasting is assuming a much greater importance in planning for technology. Other pertinent subjects include technology transfer; the role of multinational corporations; the risks associated with technology; economic analysis; human, social, and cultural issues; education and training aspects; productivity and quality; organizational structure; management of technological projects; the boom in information technology; the marketing of technologies; financial issues related to technological development; and environmental sustainability and eco-efficiency.

All these topics are interwoven to form the fabric of the MOT field. The issues covered have implications for engineers and managers. Engineers deal with the physical components of technology. They need to relate that technology to markets and economic systems. Managers must anticipate implications of technology on business. It is imperative for all concerned to understand the basic concept of connecting technology with the marketplace to create wealth.

## WHY MANAGEMENT OF TECHNOLOGY NOW?

The world is changing. As we move into the twenty-first century, the pace of change continues to increase. New technologies emerge and the dynamics of trade shifts. Management systems must also shift to cope with the change. These shifts create a totally new paradigm for business. Let's examine the nature of this change since the end of World War II.

### The World: Post World War II

World War II created a new order in the world. The industrialized countries dominated the scene. They had the major share of the world's productive capacity. Technology-based products were sold at home and all over the planet, creating great wealth for countries such as the United States, Britain, and France. Many industries flourished and companies such as Ford, General Motors, General Electric, AT&T, IBM, and Westinghouse became conglomerates. The years after the war were the golden years of Western industry.

The postwar era was a period of relative technological stability. The concept of the factory, which emerged in the 1800s, had become well established. Labor was divided mainly into blue-collar and white-collar workers. Industry owners and managers created management systems with well-defined functions such as design, manufacturing, finance, accounting, marketing, and distribution.

U.S. and Western industry in general felt a sense of stability. The general premise was that the future would be similar to the present. Management theories and guiding principles were designed and applied in this stable environment. Increasing productivity through the fine-tuning of operations was adequate. For example, a small improvement in labor efficiency would be reflected in bottom-line profit and would be seen as satisfactory progress by managers and stockholders. Likewise, a new product offering such as a new car model in the automobile industry would be considered a major change.

U.S. industries became accustomed to a relatively predictable competitive environment with a large home market and a stable international market. The same products were sold at home and all over the world, and global competition was insignificant. U.S. management became complacent, and U.S. industry, as one of my former students expressed it, grew "fat, dumb, and happy."

### The World Today

The most pronounced difference between the world today and the world of yesteryear is the rapid pace of technological change. This pace is combined with variation in the scope of technology deployment. Global competition is also relatively new. Competition among nations has intensified in the 1980s and 1990s compounded by the emergence of new countries on the "playing field." This contributes to a continuous shift in the balance of economic power. With the end of the Cold War, a new world order has emerged. Most countries now pursue free global trade, and trade blocs are becoming a defining feature of the time. Let us explore each one of these factors in greater detail.

**The Pace of Technological Change** The magnitude and speed of technological change in recent years have been phenomenal. A very rapid rate of technological innovation is making it imperative to consider technology as the primary factor influencing economic growth and prosperity. The U.S. National Science and Technology Council has noted that "technology is reshaping our world at a speed unimaginable just a few decades ago" (NSTC, 1996).

Technological changes have been of such magnitude that it is difficult for individuals, and often for institutions, to follow them. In several technological sectors, such as the information sector, more changes have occurred in the last few decades than in the previous few thousands years (Pritchett, 1994). The rate of change is increasing exponentially (see Figure 1-5). Nations, industries, and individuals must develop their capabilities to keep abreast of technological changes and to harness technology.

Technological changes in the twentieth century have significantly influenced employment patterns and societal change. During the early 1900s, 85 percent of the U.S. workforce was involved in agriculture. This figure now stands at less than 3 percent. However, the United States is still the world's largest producer of agricultural products. This production efficiency is due in great part to technology. In the 1950s, a large

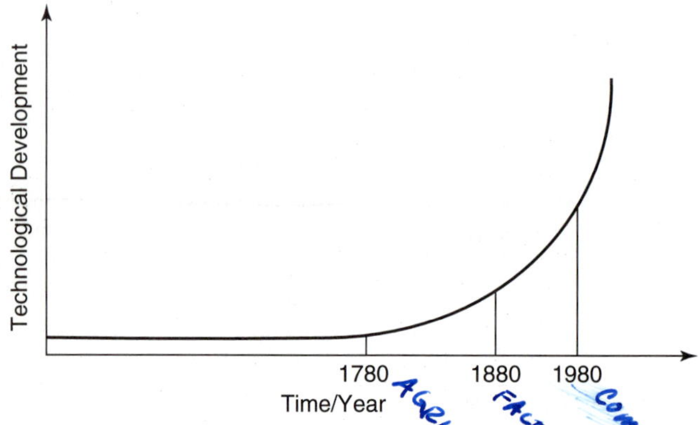


**FIGURE 1-5**
RATE OF TECHNOLOGICAL CHANGE

proportion of the U.S. workforce shifted to manufacturing; production technology had expanded job opportunities and absorbed about 73 percent of the workforce. Now this sector accounts for only about 15 percent of U.S. employment. Still the net output of the manufacturing sector increased due to improvement in technology. In the 1990s, information technology, leading the technology revolution, contributed to another shift in employment—this time toward the service sector, which now accounts for about 80 percent of the workforce. Such changes will continue to happen but at a faster rate as the world moves into the twenty-first century. Technology will continue to be the base for economic growth. At the national or the company level, competing by means of technology is no longer a matter of choice but a matter of survival in the global marketplace.

With the high rate of change in technology, successful managers are those who embrace change for the benefit of their organization. A manager's role, in a predictable, relatively stable technological environment, is to optimize the use of available resources. In a dynamic environment of fast-paced technological change, a manager's job expands beyond the traditional role of managing available resources to a new role of managing change. A manager should be able to manage with technology as well as manage the innovation process.

**The Change in Scope** In conjunction with the rapid pace of technological change, there has been a change in market behavior. Customers are now demanding choice and expect high-quality products. In the early twentieth century, Henry Ford introduced his Model T. To drive the cost down and make the automobile affordable to the masses, he introduced the concept of the assembly line, and mass production was born. Ford is credited with saying that he would produce any color car a customer wanted as long as it was black. Ford's thinking would not be as popular with today's consumers. Today's goods are produced to meet customers' particular needs and demands. Such luxury has fortunately become affordable because of advances in technology.

The rapid change in technology combined with consumers' new attitude has forced a move away from the use of fixed production lines. Flexibility and the ability to quickly respond to change is the name of the game. The use of modern computers and software makes possible such flexibility. Deployment of technology has shifted from focusing on

**EXHIBIT 1-1**
ECONOMIES OF SCALE, SCOPE, AND INTEGRATION

| Characteristics | Scale (volume) | Scope (variety) | Integration (variety & volume) |
|---|---|---|---|
| Process | Continuous flow<br>Special-purpose machinery | Jumbled-flow batch<br>General-purpose machinery | Continuous flow<br>Specialized software, computer integrated multipurpose machinery |
| Product | Standardized commodity | Customized multiproducts | Customized commodity |
| Facility | Centralized | Decentralized | Moderately decentralized |
| | Large (in size) | Small (in size) | Medium (in size) |
| Level of automation | Low (hard to programmable automation) | High | Highest (flexible automation) |
| Total added flexibility | Low | High | Highest |
| Relative unit costs: | | | |
| Fixed | Low | High | Highest |
| Variable | High | Low | Lowest |
| Experience curve | Not too flat | Flat | Flatter and lower |
| Organization | Process focus | Product focus | Product focus |
| Managerial characteristics | Technical | Entrepreneurial | Entrepreneurial, technical |
| Marketing | Low cost, dependability | Flexibility, product innovation | Low cost, dependable product innovation, flexibility |

*Source:* Noori, 1990. Reprinted by permission of Prentice-Hall, Inc., Upper Saddle River, NJ.

economies of scale to focusing on economies of scope or of integration. Noori (1990) reviewed the characteristics of the three economies and contrasted the differences among them, as shown in Exhibit 1-1. Modern corporations must be able to change from one type of economy to another in a timely manner. Modern technology enables them to do so. The emergence of Internet technology has permitted the growth of mass customization where a customer selects the features desired in a product on-line. The producer develops the system capable of meeting the customer's demands.

**Changes in Competition** Relatively new global competition is intensifying each day. The economies of Germany and Japan have been revitalized since World War II. New countries are entering the game: Taiwan, Korea, Singapore, and, more recently, China have emerged as strong competitors. The map of the world's industrial production has shifted, and there are new rules for the new competitors.

**Trade Blocs** Countries are entering into agreements that create trade blocs. It is no longer good enough for the United States to be able to compete with one country, such as France or England or Germany; today's new competitor may be a large bloc of nations, such as the European Union (EU) or the Pacific Rim countries. The international trade blocs are changing the way the game is played and won. Commerce is a global activity that requires global understanding; it is influenced by new agreements and rules, such as the General Agreement on Tariffs and Trade (GATT) and the North American Free Trade Agreement (NAFTA). This trend in the global marketplace is creating a new model for competition among nations and industries.

The global changes have forced a new attitude in government and corporate policies. The U.S. Office of Technology Policy indicates such a change: The federal government now views technology as an integral part of its global strategy for economic growth. The three prongs of that strategy are economics, trade, and technology. National as well as industry competitiveness depends on the integration of these three areas of activity (Mitchell, 1995).

National competitiveness requires the establishment of a sound economic system, strong technological capabilities, and the ability to trade with other nations. The economic system consists of many components, including banking and financial institutions, the stock market, and regulatory agencies. Technological capabilities are based on many factors, including education, research and development, and technology transfer. Global trade is governed by international agreements, negotiated under the auspices of the World Trade Organization (WTO) or among groups of nations. In addition, trade is dependent on market dynamics and the laws of supply and demand.

Creating a national competitive strategy depends on the harmonious integration of economics, technology, and trade systems. Industry competitiveness is dependent on the intersection of these three major systems as well.

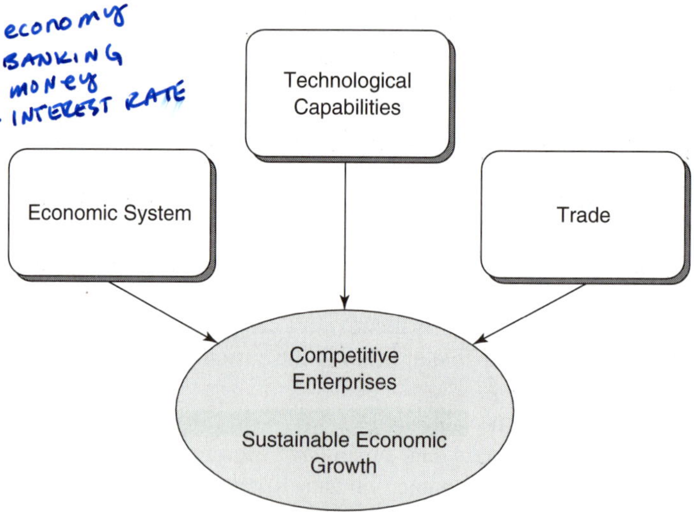


**FIGURE 1-6**
TECHNOLOGY AND SUSTAINABLE GROWTH
The integration of the economic system, technological capabilities, and trade fosters sustainable economic growth.

Effective public policy that addresses these issues is critical for the creation of sustainable economic growth and an improved standard of living. Figure 1-6 illustrates these concepts.

## DISCUSSION QUESTIONS

1 Using the classification of technology presented in this chapter, provide some examples of each type of technology.

2 Compare an old (10 years or more) Fortune 500 or Global 500 list with the most recent one. What movements have occurred in the top 100 companies? Why?

3 Look at an even older (30 years or more) Fortune 500 list. How many technical companies can you find? Compare it with a recent list. What do you find?

4 In a business magazine look for an article about a company at which technology and management have played a key role. What has been the most relevant factor—managerial or technical—in the firm's success or failure? Discuss your thoughts.

## ADDITIONAL READINGS

Michael E. Porter. "Capital Disadvantage: America's Failing Capital Investment System." *Harvard Business Review,* September–October 1992.

Porter recognizes that the American financial system supports emerging fields and has advantages over Japanese and German approaches. However, it also prevents established business from obtaining a sustained competitive advantage. This point of view argues that the configuration of boards of directors, managers' compensation systems, and stockholder's composition creates lack of commitment for long-lasting investments.

Christopher A. Bartlett & Sumantra Ghoshal. "What Is a Global Manager?" *Harvard Business Review,* September–October 1992.

Bartlett and Ghoshal classify international managers as either business, country, functional, or corporate. They provide the characteristics each type should develop to successfully lead in global markets.

## SUGGESTED ASSIGNMENT

For each chapter in this book select a recently published journal, conference proceedings, or magazine article on MOT. Read the article and provide a one-page analysis in which you:

- Briefly summarize the main thoughts of the author.
- List the lessons learned from the article.
- Critically evaluate the lines of thought expressed by the author and indicate your agreement or disagreement with them. Rely on your own experience or previous readings.

## REFERENCES

Berman, E. M., & Khalil, T. 1992. "Technological Competitiveness in the Global Economy: A Survey." *International Journal of Technology Management,* vol. 7, nos. 4/5, pp. 347–358.

Heim, Joseph A., & Compton, W. Dale. 1992. "Operating Principles of World-Class Manufacturing Organizations." In Khalil, T., & Bayraktar, B. (eds.), *Management of Technology III,* pp. 765–776. Industrial Engineering and Management Press, Norcross, GA.

Isaacson, Walter. 1997. "Driven by the Passion of Intel's," *Time,* Dec. 29, 1997–Jan. 5, 1998.

Khalil, T. M. 1991. "Current Issues in Management of Technology." Proceedings of the Third International Congress of Industrial Engineering, Tours, France.

Khalil, T. M. 1993. "Management of Technology Education for the 21st Century." In Sumanth, D., et al. (eds.), *Productivity and Quality Frontiers,* 2d ed. Industrial Engineering and Management Press, Norcross, GA.

Khalil, T. M., & Bayraktar, B. 1988. *Challenges and Opportunities for Research in the Management of Technology*. UM/NSF workshop report, University of Miami, Miami, FL.

Larsen, J., Rogers, E. 1988. Silicone Valley: The Rise and Falling of Entrepreneurial Fever, Chapter 7 in Smilor, R., Kozmetsky, G., and Gibson, D. (eds.), *Creating the Technopolis:* Linking Technology Commercialization and Economic Development, Ballinger, Cambridge, MA.

Mitchell, G. R. 1995. "Technology—Business Strategy—Government Policy." Lecture notes, University of Miami, Mar. 3–4 .

Mohrman, Susan A., & Von Glinow, Mary Ann. 1990. "Beyond the Clash: Managing High Technology Professionals." In Von Glinow, M., and Mohrman, S. (eds.), *Managing Complexity in High Technology Organizations,* pp. 3–14. Oxford University Press.

National Research Council. 1987. *Management of Technology: The Hidden Competitive Advantage*. National Academy Press, Washington, DC. Report No. CETS-CROSS-6.

National Science and Technology Council. 1996. *Technology in the National Interest,* Office of Technology Policy, U.S. Dept. of Commerce, Washington, DC.

Noori, Hamid. 1990. *Managing the Dynamics of New Technology*. Prentice-Hall, Englewood Cliffs, NJ.

Pritchett, Price, 1994. *New Habits for a Radically Changing World.* Pritchett & Associates, Dallas.

Quinn, J. B., Anderson, P., and Finklestein, S. 1996. "Managing Professional Intellect: Making the Most of the Best," *Harvard Business Review,* March–April.

Zeleny, M. 1986. "High Technology Management." *Human Systems Management,* vol. 6, pp. 109–120.

# 2

# THE ROLE OF TECHNOLOGY IN THE CREATION OF WEALTH

## HISTORICAL PERSPECTIVE

Technology has always played a major role in creating the wealth of nations and influencing standards of living and quality of life. When primitive human beings used a tree branch to reach for fruit on trees or shaped a rock to form a spear, they were developing technology necessary for survival. It took years of trial and error, creativity and persistence to develop tools, acquire the knowledge to build cities and monuments, obtain know-how to farm the land, develop methods to move materials and products, and implement procedures to run governments. In short, it took technology to achieve progress. Technology has had such a profound effect on human lives that progress of civilization is frequently identified by the dominant technology of the age, as shown in Figure 2-1.

Civilizations were built around the use of innovative technology, and some vanished when they lagged in technology. The ancient Egyptians gained great wealth using technology in farming the land and storing crops. They used technology and ingenuity to build impressive cities, huge pyramids, and magnificent temples. The Chinese made pottery, forged armor, and built massive walls to preserve and defend their great civilization. Kings, pharaohs, emperors, and sultans used the labor and ingenuity of their people to amass wealth and gain power. The axe, the wheel, the bow and arrow, the sword, and the chariot were technologies that helped communities harness the resources of nature to satisfy human wants and needs. Simultaneously, these tools were used to defend territories and property from intruders, as well as to attain more wealth and power. Technology empowers people in both war and peace. It can be used for the good of humanity or for its destruction. People have a choice between good and evil uses of technology. Decisions to use destructive technology expose the frailty of the human

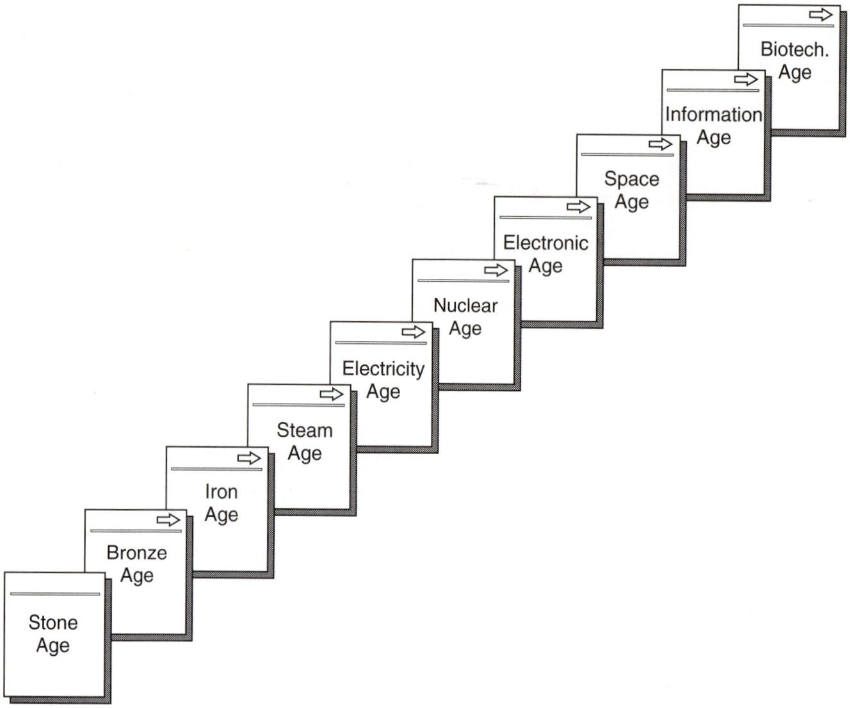


**FIGURE 2-1**
EVOLUTION BY AGE OF TECHNOLOGY

conscience. Yet in both types of endeavors a major motivation is the eternal human desire to create wealth and improve living conditions.

When basic needs for survival were met, artisans developed handicrafts and simple products and traded them in open markets. Societies developed loose business practices—someone had goods or services that someone else needed and wanted; everyone was a customer of someone else. A system of bartering arose and simple methods of accounting emerged. All products were initially handmade, and simple tools were devised to assist workers and improve the accuracy and efficiency of their work. This practice lasted for millennia, with innovations occurring sporadically and infrequently.

In primitive societies, production was achieved by manual labor and the economy was agricultural. It was not until the eighteenth century and the Industrial Revolution, about a mere 200 years ago, that radical technological innovations created a major transformation in the way people live and do business. The factory system was born and mass production became dominant. Technology transformed many national economies from agricultural-based to industrial-based economies. More wealth was created in industrialized countries by utilizing the latest developments in technology and diligently converting resources to marketable products. The result was an improvement in the quality of life and an increase in the products and services available to citizens of industrialized countries. Technology has always been the force that influences production and effects an increase in the standard of living.

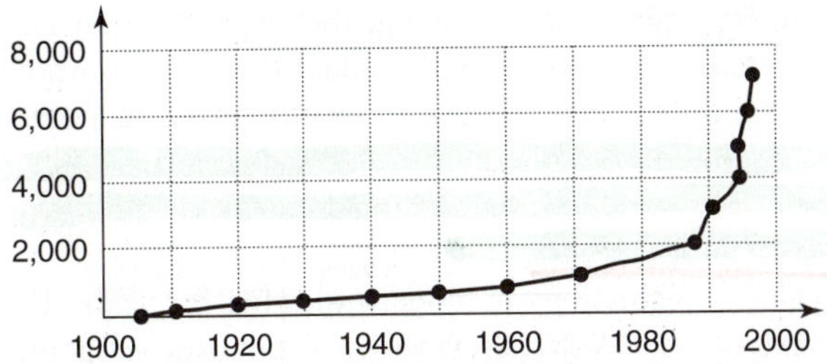


**FIGURE 2-2**
DOW JONES INDUSTRIAL AVERAGE
The Dow Jones Industrial Average has broken symbolic barriers with increasing frequency. This graph shows the thousand-point milestones.
*Source: The Miami Herald,* Feb. 14, 1997.

In the late nineteenth century and early twentieth century, science and technology became increasingly intertwined. Scientific discoveries and methodologies triggered technological breakthroughs, while technological devices and know-how helped advance science. This closer linkage created an explosion in technological development. The war effort during World Wars I and II forced faster technological development in products, manufacturing, quality, logistics, material handling, operations research, human factors, and many other areas. Technology helped win the war and eventually bring peace. Technology also helped build the Western industrialized countries' strong postwar economies and propel their standards of living to new heights.

Figure 2-2 shows growth in the U.S. stock market from 1900 to 2000 as indicated by the Dow Jones Industrial Average. The similarities between this chart and the conceptual chart shown in Figure 1-5 highlight an interesting point: While many factors may have contributed to the strength of the stock market, technological advancement is a primary one of them. American industry started to effectively utilize information technology and to better manage technological resources in the 1990s. The technology sector has led the market in growth over the past several years. This helped the United States regain its competitive edge in the global marketplace and has reflected on the financial market and the economy in general.

The U.S. National Science and Technology Council (NSTC), in its report "Technology in the National Interest" (1996), indicated that technical progress is the single most important determining factor in the nation's sustained economic growth. As much as half the nation's long-term economic growth over the past 50 years was credited to technology.

New and emerging technologies are taking the U.S. and world economies into new frontiers. *Time* magazine credited the "digital revolution" with transforming the end of this century the way the Industrial Revolution transformed the end of the last one. It dubbed the microchip the "dynamo of a new economy" marked by seven years of growth, low unemployment, negligible inflation, and a rational, exuberant stock market (Isaacson, 1997). Communication technology, the Internet, genetic engineering and cloning, logistics, nanotechnology, and a host of other technologies are expected to have an even greater impact in the years to come.

## THE CREATION OF WEALTH

Adam Smith, the influential British philosopher and economist of the eighteenth century, wrote extensively on the nature and causes of nations' wealth. In his 1776 book *The Wealth of Nations* he introduced a powerful analysis of the process whereby economic wealth is produced and distributed. He argued the following: that capital is best employed for the production of wealth; that each nation should produce the goods in which it has absolute advantage—that is, the ones it can produce more efficiently than other countries can; and that market forces and free trade, not government controls, should determine the direction, volume, and composition of international trade and the distribution of wealth—the so-called laissez-faire doctrine.

Economists have debated the value of technology in modern societies for years. The great German economist Joseph Schumpeter (1928) described the capitalistic economic system as being characterized by "private property (private initiative), by production for a market and by the phenomenon of credit." He eloquently argued the case of innovation in creating economic progress. A prevailing thought among many economists of his time was that industrial expansion is incidental to and modulated by general social growth, of which the most important economic forces are growth of population and growth of savings—the so-called received doctrine. David Ricardo's and John Stuart Mill's discussions of progress focus mainly on the relative growth of population or capital. Schumpeter showed that industrial expansion is also the result of economic forces. He wrote:

> It is by means of new combinations of existing factors of production, embodied in new plants and, typically, new firms producing either new commodities, or by a new, i.e. united method, or for a new market, or by means of production in a new market. What we, unscientifically, call economic progress means essentially putting productive resources to uses *hitherto untried in practice,* and withdrawing them from the uses they have served so far. This is what we call "innovation." (Schumpeter, 1928)

Schumpeter argued that innovation in competitive capitalism is typically embodied in the foundation of new firms—the main lever. In fact, regarding the rise of industrial families, he held that entrepreneurs' profit is the primary source of industrial fortunes and that the process of innovation in industry, as engaged in by entrepreneurs, supplies the key to all phenomena of capital and credit creation.

Attempts to define the different sources of economic growth and quantify their relative contributions have been pursued by many economists, including Abramovitz (1956), Solow (1956–1957), Dennison (1962, 1967, 1979, 1985), Kuznets (1971), Kendrick (1973), Jorgenson, et al (1987), and Boskin and Lau (1992).

Economic growth is determined by the rate of change in per capita real gross domestic product (inflation-adjusted GDP). When a nation grows economically, its citizens must be better off in some way. Growth in real GDP per capita reflects improvement in citizens' material well-being.

An embodied assumption in many studies of economic growth is that society has scarce resources and a limited choice of production alternatives and, therefore, that tradeoffs must be made. Production possibilities are considered under the assumption that technology does not change radically over the time period under study. Under these conditions and for any given rate of saving and investment, living standards can rise only through increased worker productivity. Productivity growth means that it is possible to obtain more output from the same amount of input.

Boskin and Lau (1992) indicate that the three principal sources of nations' economic growth are enhanced capital, labor, and technical progress (or, equivalently, *total factor productivity*). *Capital* can be defined as the goods and services used to produce other goods and services. It includes machines, buildings, tools, and improvements to natural resources. *Labor* represents the productive contributions made by all people who work. This is the human resource or human capital. Improvement to labor occurs with education, training, and acquisition of new skills.

The growth rate of physical and human capital, combined with technical progress, accounts for a significant portion of the economic growth of nations. The Nobel Foundation (1997) notes:

> Increased per capita production in a country may be the result of more machines and more factories (a greater stock of real capital). But this increased production may also be due to improved machines and more efficient production methods (which may be termed technical development). In addition, better education and training, and improved methods of organizing production may also give rise to increased productivity.

Robert Solow, a Massachusetts Institute of Technology professor, won the 1987 Noble Prize in Economics for creating a theoretical framework that can be used in discussing the factors that lie behind economic growth. Solow argued that technical progress (the change in production techniques) is built into machines and other capital goods and that this must be taken into account when making empirical measurements of the role played by capital. According to Solow:

> It is easy to list the things that might contribute to economic growth. The problem is, as we say, to make a model understand how these things interact, and to do it in such a way that may have a prayer of measuring it . . . The surprising conclusion was that technological change looms much larger than capital investment . . . Silicon Valley is the sort of thing I am talking about. (quoted in *The Boston Globe,* 1997)

Solow's empirical results show that technical progress accounted for more than half the economic growth in the United States between 1909 and 1949. His work indicates that technological development will be the motor for economic growth in the long run.

Boskin and Lau (1992) estimated the relative contributions of three sources of economic growth—capital, labor, and technical progress—for the United States, France, West Germany, Japan, and the United Kingdom. They showed that

> over the period under study, technical progress is by far the most important source of economic growth, accounting for half or more (three quarters for the European countries), and capital is the second most important source of economic growth (except for the U.S.). Capital and technical progress accounted for more than 95 percent of the economic growth of France, West Germany, Japan, and the United Kingdom. In the U.S. where labor grew more rapidly than in other countries during this period, they still account for 70 percent of economic growth. (Boskin and Lau, 1992)

The U.S. National Science and Technology Council, in its report "Technology in the National Interest" (1996) emphasized that technology is the engine of economic growth. It reported that "performance of individual companies—the agent through which economic growth occurs—is strongly linked to their use of technology." The use of technology proved to enhance manufacturing in every performance category.

## THE LONG-WAVE CYCLE

There is no doubt that improvement in productivity is vitally important to an economic system. It provides relief from inflationary pressures and permits real improvement in the standard of living. Technology is the driver for such improvement. Technology also triggers another mechanism for economic growth that is yet to be fully appreciated, one whose effect has not been quantitatively measured. Through this mechanism, emerging and new technology spurs economic expansion. In traditional economic literature it is known as the *long-wave* or *long economic cycle*. After the Industrial Revolution, economies of Western countries went through major economic expansion followed by a depression. In 1930, the Soviet economist Kondratieff observed that fluctuations occurred in Western economies every 30 years and he attributed them to the long-wave effect.

Mensch (1979) studied this phenomenon and suggested that basic new technology began the economic expansion in each long wave. Graham and Senge (1980) concurred with the view that inventions and innovations trigger economic long cycles.

Betz (1987) suggested that the process behind a long wave is an interaction between new technology, business opportunities the new technology creates, and an eventual overbuilding of capital after the technology ages. He suggested the following sequence of events for the long-wave process:

**1** Discoveries in science create a phenomenal base for technological innovation.

**2** Radical and basic technological innovation creates new products.

**3** These products create new markets and new industries.

**4** The new industries continue to innovate in products and processes, expanding markets.

**5** As the technology matures, many competitors enter internationally, eventually creating excess production capacity.

**6** Excess capacity decreases profitability and increases business failures and unemployment.

**7** Subsequent economic turmoil in financial markets may lead to depressions.

**8** New science and new technology may provide the basis for new economic expansion.

Betz eloquently argued that the long-wave hypothesis merely describes past connections among pervasive basic innovation, long-term economic expansion, and excess capital formation in technology-mature industries: "It does not determine anything in the future." He made the following pertinent observations:

**1** Cutting-edge technology is behind the long waves of economic activity.

**2** High-technology products displace old technology when there is a justification for performance over cost.

**3** Technology life cycles of industries affect long cycles in the national economy.

**4** New technology comes from science, and science comes from new discoveries in nature.

**5** A new technology, when created, will begin a new wave.

In an age characterized by fast-paced technological change, the long wave, as argued in this book, is likely to be much shorter. Many industries are increasingly getting involved in high technology. Scientific discovery and knowledge are at an all-time high. Emerging technologies are promising new and uncharted areas for products and production. All indications support the view that technology will have greater impact on economic growth in the future. Writing about the digital age, Isaacson (1997) stated:

> The outputs of the old economy were simpler to measure: steel, cars, and widgets are easily totted up. But the new economy defies compartmentalized measurement. Corporate software purchases, for instance, are not counted as economic investment. What is the value of cellular phones that keep getting cheaper, or e-mail? By traditional measures banking is contracting, yet there has been explosive growth in automated banking and credit card transactions; the same for the way health care is delivered.

Isaacson also indicated that traditional statistics are increasingly likely to understate productivity and growth. It is likely that economic growth incidental to technological change will continue to be the leading factor in wealth creation.

## THE EVOLUTION OF PRODUCTION TECHNOLOGY

In the first millennium of the Christian calendar, before the Age of Enlightenment, there were very few prevalent technologies. Society was predominantly agricultural. Products were developed by artisans, and manual labor dominated. The Industrial Revolution and the introduction of steam power late in the second millennium changed everything. The concept of the factory was born: Workers now assembled in one place to develop products rather than working individually in small shops.

In the late 1800s, Frederick Taylor introduced the scientific method into factory administration. Taylor's approach led to improved efficiency through the concept of analyzing and designing work. He introduced the idea that management should design the work and workers execute it. The concept of standard time was created to enable management to exercise control over production. Problem areas pertaining to work were identified and scientifically analyzed; alternative solutions were evaluated; and the best alternative was selected and implemented. Incentive systems were also introduced to raise productivity. Taylor's approach ushered in the scientific management process (Barnes, 1967).

Henry Ford introduced the assembly line and specialization of workers in the early 1900s. He believed that producing just a few standardized products would increase efficiency in the system. Thus, the concepts of the assembly line and economy of scale were born. Although these concepts are still in use in modern industry, advances in technology and in the marketplace are creating changes in what was once a dominant model. Flexibility, agility, and economy of scope are the trends in modern manufacturing.

Motion study was introduced in the early 1900s by Frank and Lillian Gilbreth (Gilbreth, 1911; Niebel, 1988). They advocated reducing wasted movements and finding the best way to do a job. Their methods were implemented in the construction industry and later found their way into the factories. Taylor's and the Gilbreths' work formed the basis of the discipline of methods engineering, which profoundly improved the efficiency of operation.

Greater diversification of shops and factories occurred as the industrial age progressed. Many different products were produced in the same facility, requiring scheduling activities. Disciplines such as planning, inventory control, queuing, and modeling were developed. The pace of development increased during World War I, following an observed pattern in which technological developments are expedited in times of war.

During the 1920s and 1930s, concepts of statistical quality control were introduced. Shewhart created control charts and was followed by Deming and Juran, who became active in statistical quality control and quality theory.

World War II brought the necessity of efficient material-handling systems, improvements in productivity, wider applications of statistics, and operations research.

In the 1950s the digital computer emerged, allowing simulation and better efficiency in the solution of complex numerical problems. The computer innovation ushered in the information age and created a revolution in science and technology development. We are still attempting to cope with the growth of this technology and its associated explosion of knowledge as the second millennium ends and the world moves into the third millennium.

Today, developments in material technology, information technology, genetic engineering, bioengineering, communications, robotics, manufacturing processes, and organizational theory are but some of the emerging areas in the technological arena. Their impact on society and on our way of life is expected to be immense.

Figure 2-3 traces the evolution of production technology, based on a review by Turner et al. (1993).

## THE EVOLUTION OF PRODUCT TECHNOLOGY

Technology can be associated with products, production, service, or marketing. Product innovations of the past two centuries have had a profound effect on the world, transforming the way people live and work. Technologies converted into products have contributed to economic growth and prosperity and improved the overall quality of life for many of the world's people. Figure 2-4 outlines the evolution of product technology over the past two centuries.

Innovation in power generation toward the end of the eighteenth century brought with it products such as the steamboat and the steam-powered locomotive. Innovations in radio signal and electric power generation and transmission in the nineteenth century brought applications in communication products such as the radio, the telephone, and the telegraph. The twentieth century was a watershed in the introduction of new products. The air conditioner, the automobile, and the airplane developed in the early part of the century and changed our way and quality of life forever. The electronic age and the introduction of computers in the middle part of the twentieth century vastly expanded the knowledge base. Many innovative products were introduced in that time period, including the television, transistor radio, jet engine, copy machine, and mainframe computer.

From the 1950s to the 1970s we witnessed the production of integrated circuit boards, spaceships, laser products, satellites, composite-materials and fiber optic communication equipment. Technologies of the 1980s brought MRI scanning equipment for

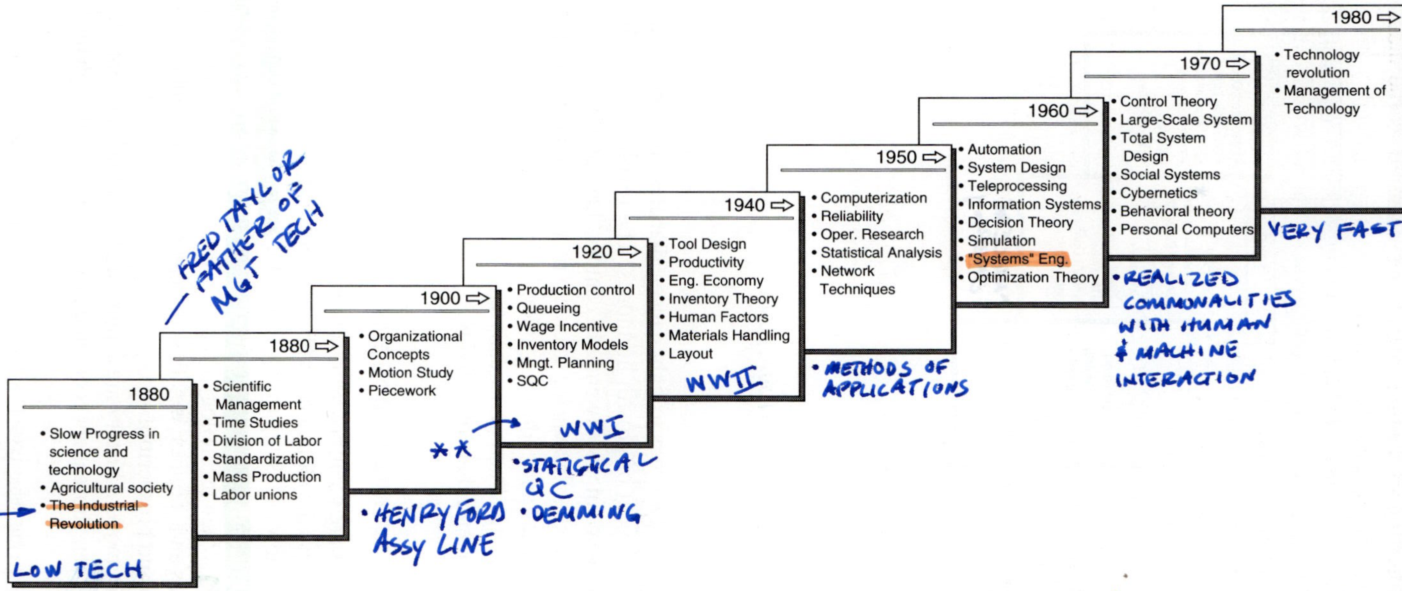


**FIGURE 2-3**
EVOLUTION OF PRODUCTION TECHNOLOGY

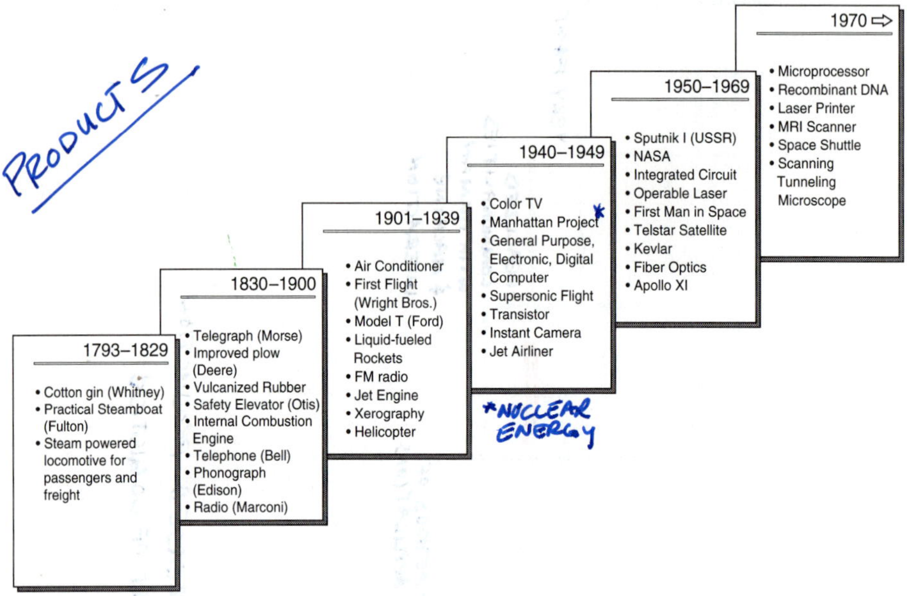


**FIGURE 2-4**
IMPORTANT TECHNOLOGICAL INNOVATIONS
*Source:* U.S. Dept. of Commerce, 1996.

the health care sector; new drugs to fight cancer, diabetes, and genetic disorders; vaccines; and the development of growth hormones. The revolution in information technology toward the end of the twentieth century has produced an explosion of knowledge. Harnessing this knowledge to create products and services is the challenge currently facing countries and industries alike. Software products, telecommunication equipment, semiconductors, and genetically engineered products are but a few of the exciting new developments that are evolving.*

## TECHNOLOGY AND NATIONAL ECONOMY

Developed economies are identified with countries that properly use technology for the creation of wealth. Less developed economies are identified with countries lacking the technological know-how necessary to create wealth. To reiterate, it is not the technology itself that creates the wealth; rather, it is the appropriate and effective use of such technology that does so. When technology is applied to add value to resources and provide marketable products and services, wealth is increased (i.e., the management of technology creates wealth and prosperity).

Two examples in modern history are Japan's and Germany's recent successes in the world markets and their subsequent economic prosperity. Even though the United States

*For further reading on this subject, see U.S. National Science and Technology Council, 1996.

still has the greatest technological edge in the world, Japan and Germany have managed their resources and their technological systems better in recent years and thus have achieved economic advantages. It is important here to note that proper management of technology encompasses all levels of technology, from low-tech to super-high technologies. Certainly managing technology well at the high end may create a greater reward in terms of improvements in the standard of living. However, proper management of low- or medium-level technologies can still create a certain competitive advantage and be effectively used for wealth creation (Khalil, 1993). In support of this point, one can observe the economic growth of newly industrialized countries (NICs), particularly those called the tigers of Asia such as Taiwan, Korea, Singapore, and Malaysia. These countries have succeeded in acquiring low- or medium-level technologies and have done a credible job in managing technological resources. They have gained competitive advantage over many other countries in recent years, including industrialized countries of the North. They have been aided by cheaper labor and fewer regulatory restrictions than exist in highly industrialized countries such as the United States, yet these factors are recognized advantages that must be mitigated in a fundamental strategy of global management of technology.

The U.S. economy went astray in the 1970s and early 1980s when the focus shifted from efficient production and global marketing for wealth creation to a rash of unjustifiable financial transactions and unworkable costing and accounting schemes. Speculators got involved in mergers and acquisitions that neglected technological rationality. Corporations permitted a relative slide in quality and productivity in comparison with international competitors. It was a time of crises in managing technological resources. Several banks and financial institutions failed. One cannot expect to continue to improve economic conditions on the basis of paper transactions on Wall Street or money exchanges in banks, savings and loans associations, or board rooms. Wealth is created on the basis of technology, production, and smart work. These are some of the important factors that sparked the turn-around in the U.S. economy of the 1990s.

## DISCUSSION QUESTIONS

1 Look for a book on the history of the Ford Motor Company. What was the most important factor in Henry Ford's success?
2 Look for a book on the history of Toyota. How did this company dominate the U.S. car industry in the 1980s?
3 Find an article on an "Asian Tiger" in a business magazine. What are the points discussed? What strategy was followed by that country to become a competitor?

## ADDITIONAL READINGS

J. Schumpeter. "The Instability of Capitalism." *Economic Journal,* September 1928, pp. 361–386.

Schumpeter was the first economist to directly address the concept of innovation as a determining factor in the economic progress of society. His paper provides insight on why innovation creates wealth.

F. M. Scherer. *Innovation and Growth,* Part II. MIT Press, Cambridge, MA, 1984.
Scherer, an economist from Stanford, builds on Schumpeter's theory. This book provides a more quantitative economic perspective on the role of innovation as wealth creator. The text provides some economics theory, making it an illustrative reading.

P. K. De & Bernd Huefner. "Technological Competitiveness of Germany: A Post-Second World War Review." *Technology Management,* vol. 2, 1995, pp. 262–274.
In this paper, the authors suggest how to rejuvenate German technological competitiveness after analyzing its rise and subsequent stagnation in the past 50 years.

## REFERENCES

Abramowitz, M. 1956. "Resource and Output Trends in the United States since 1879." *American Economic Review,* vol. 46, pp. 5–23.

Barnes, R. 1967. *Motion and Time Study: Design and Measurement of Work,* 5th ed. Wiley, New York.

Betz, F. 1987. *Managing Technology: Competing through New Ventures, Innovation and Corporate Research,* Prentice Hall, Englewood Cliffs, NJ.

Boskin, M. J., & Lau, J. 1992. "Capital, Technology, and Economic Growth." In Rosenberg, N., Landau, R., & Mowery, D. (eds.), *Technology and the Wealth of Nations,* Stanford University Press, Stanford, CA.

*The Boston Globe,* 1997. http://www.boston.com/globe/search/stories/nobel/1987/1987g.html.

Dennison, E. 1962. "United States Economic Growth," *Journal* of *Business,* 35, pp. 109–121.

Dennison, E. 1967. *Why Growth Rates Differ: Post-War Experiences in Nine Western Countries,* Brookings Institution, Washington, DC.

Dennison, E. 1979. *Accounting for Slower Economic Growth: The United States in the 1970s.* Washington, DC: Brookings Institution.

Dennison, E. 1985. *Trends in American Economic Growth, 1929–1982.* Brookings Institution, Washington, DC.

Gilbreth, F. B. 1911. *Motion Study.* Van Nostrand, Princeton, NJ.

Graham, A., & Senge, P. 1980. "A Long-Wave Hypothesis of Innovation," *Technological Forecasting and Social Change,* vol. 17, August, pp. 283–312.

Isaacson, Walter, 1997. "Driven by the Passion of Intel's," *Time,* Dec. 29, 1997–Jan. 5, 1998.

Jorgenson, D., Gollop, F., & Fraumeni, B. 1987. *Productivity and U.S. Economic Growth.* Harvard University Press, Cambridge, MA.

Kendrick, J. W. 1973. *Postwar Productivity Trends in the United States, 1948–1969.* Columbia University Press, New York.

Khalil, T. M. 1993. "Management of Technology and the Creation of Wealth," *Industrial Engineering,* vol. 25, No. 9, September, pp. 16–17.

Kuznets, S. S. 1971. *Economic Growth of Nations.* Harvard University Press, Cambridge, MA.

Mensch, G. 1979. *Stalemate in Technology.* Ballinger, Cambridge, MA.

Niebel, B. 1988. *Motion and Time Study,* 8th ed. Irwin, Burr Ridge, IL.

The Nobel Foundation, 1997. http://nobel.sdsc.edu/laureates/economy-1987-press.html.

Schumpeter, J. 1928. "The Justability of Capitalism," *Economic Journal,* September, pp. 361–386.

Turner, N., Mize, J., Case, K., & Nazmetz, J. 1993. *Introduction to Industrial and Systems Engineering*. Prentice-Hall, Englewood Cliffs, NJ.

U.S. Department of Commerce. 1996. "Chart on Evolution of Product Technologies." Office of Technology Policy, Washington, DC.

U.S. National Science and Technology Council. 1996. *Technology in the National Interest.* Office of Technology Policy, U.S. Dept. of Commerce, Washington, DC.

# 3

# CRITICAL FACTORS IN MANAGING TECHNOLOGY

## THE CREATIVITY FACTOR

Technology is an expression of human creativity. Managing technology involves continuous effort in creating technology, developing novel products and services, and successfully marketing them. This requires great creativity along with a system designed to exploit it. It also requires an investment in research and development. R&D is a costly endeavor. It is a risky investment and therefore needs to be well managed. However, it is an investment in the future that cannot be neglected, nor can its value be underestimated. Technology creation and exploitation require a chain of events, starting with inventions and ending at the marketplace.

### Invention

When the subject of creativity is discussed in the context of technological change, two closely associated terms are frequently used: "invention" and "innovation." *Invention* is either a concept or the creation of a novel technology. It could be a product, a process, or a previously unknown system. The steam engine, the transistor, and the Xerox machine are examples of important inventions. A new composite material, a new manufactured product, and a new process constitute inventions. The word "new" here implies new to the world. Inventions occur as a result of human ingenuity and imagination. They occur only sporadically, sometimes happening by chance or through trial and error to satisfy a need. In modern times most inventions have followed scientific discoveries. For example, inventions in the nuclear field followed Einstein's discovery of the relationship between mass and energy in the early 1900s. There is usually a time lag between scientific discoveries and inventions. It may take years to convert science into technology. It may take more years to move an invention to the market as a product or service. Even though many inventions are generated by creative people, and many of them are patented, only few reach the marketplace.

### Innovation

*Innovation* involves the creation of a product, service, or process that is new to an organization. It is the introduction into the marketplace, either by utilization or by commercialization, of a new product, service, or process. It does not have to be new to the world; rather, it is viewed as the first use of an idea within an organization (Aiken and Hage, 1979), whether or not the idea has been adopted by other organizations already (Nord and Tucker, 1987). The technology (or the product) need not be novel or groundbreaking. An innovation may be a change in industrial practice, which improves productivity. Schumpeter (1928) defined successful innovation as "a task *sui generis*," a feat not of intellect but of will. The innovation process involves integration of existing technology and inventions to create a new or improved product, process, or system (Jain and Triandis, 1990).

Inventions and innovations are intimately related; however, they are not the same. An invention can be thought of as an event, while innovation can be thought of as a process. Inventions are not as common, and one invention usually precedes a number of innovations. Innovation represents the important connection between an idea and its exploitation or commercialization. Bright (1969) indicated that technological innovation includes "the initiation of the technical idea, the acquisition of the necessary knowledge, its transformation into usable hardware or procedure, and its introduction into society and its diffusion and adoption to the point where its impact is significant." The bottom line of innovation is the market, which will buy it or ignore it, thereby dictating success or failure. MOT encourages invention and the management of innovation. Both are creative processes representing essential components of any technology-creation and application system.

## THE LINK BETWEEN SCIENCE AND TECHNOLOGY

Science deals with understanding the laws of nature. This leads to the discovery of fundamental knowledge about the world, the universe, and all living things. Scientific knowledge focused on natural phenomena is neutral on the question of how this knowledge may be used. It is when scientific knowledge is applied to the things we do in life that knowledge enters the realm of technology.

Brooks (1965) argues that science and technology grew up relatively independent of each other until the nineteenth century. However, Braun (1984) observes that the separate and parallel development paths followed by science and technology have now intersected. Bayraktar (1990) notes that in the twentieth century, most technologies owe their origin to scientific discoveries. Historically speaking, scientific knowledge and technology progressed slowly until very recently. Only when science and technology started to interact and enforce one another did the real explosion in knowledge and technological development occur.

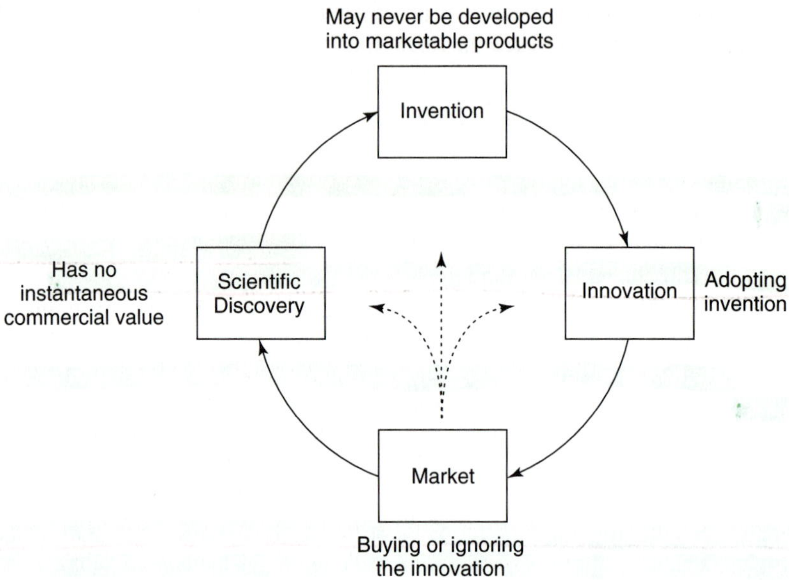


**FIGURE 3-1**
COMPONENTS OF AN INNOVATION CYCLE
Science and technology are linked through the discovery-invention-innovation-market sequence.

When we discuss science, we mention *scientific discoveries;* but when we talk about technology, we mention *technological innovations.* The two are interconnected in that each influences the other. Scientific discoveries can lead to inventions and innovations. New technology permits new scientific discoveries. When science and technology connect with the market, they influence human lives. The market may buy or ignore an innovation. It may also stimulate new innovation and request the initiation of new scientific discoveries to satisfy market needs, such as the need for a vaccine to prevent the spread of a disease. Figure 3-1 depicts the interaction between scientific discovery, invention, innovation, and the market.

## TYPES OF INNOVATION

Innovations can influence a product, a process, a service, or a system. Technological innovation in service is as important as it is in products. It includes enhancing existing service by introducing a good idea or developing a new technological agent such as software. System innovations may involve the development of different components that are integrated into a system. An example of a system innovation is a communication network, in which different components (terminals, computers, fiber optics, satellites, communication protocols, software, etc.) are developed and integrated to attain a desired performance.

To gain market acceptance, an innovation must contribute to the creation of value. Successful innovations are those perceived by customers to add value. For example, information technology developments introduced in the new Xerox copiers have allowed the use of artificial intelligence to predict when the machine's next breakdown will occur. A message is sent to the branch office, where a computer does further analysis and schedules a repairperson to visit the office before the failure actually happens (Brown, 1991). Customers appreciate this innovation, and they are willing to pay for it.

Innovations can be classified either as *radical*, or revolutionary, innovations or as *incremental*, or evolutionary, innovations. Radical breakthrough innovations are usually based on an invention. They change or create new industries. They are relatively rare and typically start outside the boundaries of a firm. When they are developed within the boundaries of a firm, they signify the introduction of something that is not only new to the organization but drastically different from its existing practices. An invention such as the transistor, which was invented at Bell Laboratories, was the starting point of a phenomenal development in the electronics industry, triggering radical innovations in many companies. The development of xerography by Chester Carlson and his collaborators triggered radical innovations in the photocopying industry and created a market of more than $20 billion. (The case of xerography is discussed in greater detail later in this book.)

The other category of innovation comprises the incremental, or evolutionary, innovations. These are small but important improvements in a product, process, or service. They are relatively common and are created within the firms of an industry. They help companies maintain a competitive position in the marketplace. Japan's *Kaizen* philosophy, a process of continuous improvement, is applicable to this type of innovation, which can bring a significant improvement in the operations of existing enterprises. The creation of the portable personal computer (PC) in 1981 was an incremental technological innovation, since the PC already existed. The innovation consisted of putting all the components of a computer together in a way that made portability possible. This is a product concept that is difficult to protect by a patent or other means. The development of a flat screen represents a more radical innovation in the portable PC market.

*Routine innovation* is another term sometimes used to refer to the introduction of something that is new to an organization but very similar to what it had in the past (Nord and Tucker, 1987).

## CREATIVITY AND INNOVATION

Innovation is associated with the creation of value or the satisfaction of a customer need. *Creativity* is the engine of innovation. The essence of creativity is combining two or more ideas to arrive at an entirely new one. For example, Henry Ford's assembly line was based on combining the production of standardized parts, a concept that had been introduced a century earlier by Eli Whitney, with the idea of bringing the parts to the worker rather than moving the worker to the parts. Johannes Gutenberg invented the printing press by combining three existing ideas: (1) a press, as was used in wine making, (2) movable type, as was used in minting coins, and (3) wood blocks, as was used in printing calendars and playing cards.

Creativity favors the prepared mind and is often inspired by a sense of dissatisfaction with existing practice. Creativity depends on both people and environment. According to Jain and Triandis (1990), a creative environment has the following characteristics:

1. Permits people to work in areas of their greatest interest.
2. Encourages employees to have broad contact with stimulating colleagues.
3. Allows taking moderate risks.
4. Tolerates some failures and nonconformity.
5. Provides appropriate rewards and recognition.

Barron (1969) found that the following characteristics exist in creative people who can convert their thoughts into innovation:

- Conceptual fluency (i.e., being able to express ideas well and formulate the ideas as one proceeds).
- The ability to produce a large number of ideas quickly.
- The ability to generate original and unusual ideas.
- The ability to separate source (who said it) from content (what was said) in evaluating information.
- The ability to stand out and be a little deviant from others.
- Interest in the problem one faces.
- Perseverance in following problems wherever they lead.
- Suspension of judgment and no early commitment.
- The willingness to spend time analyzing and exploring.
- A genuine regard for intellectual and cognitive matters.

## BRINGING INNOVATION TO MARKET

As Figure 3-1 indicates, there are time lags between the different stages of the innovation cycle's sequence of events—science, invention, innovation, and market. The manipulation of these time spans is an important and effective competitive weapon. For example, a company may benefit significantly from reducing the time lag between scientific discovery, invention, and introduction to the marketplace. The sooner an innovation reaches the marketplace, the sooner a company can reap its rewards. However, a company that owns a technological innovation may want either to delay or to speed up the diffusion of technology in the marketplace in order to fully exploit its benefits. For example, if the diffusion of technology creates more benefit to a company by developing customer commitment to the technology, the firm should speed up its diffusion. If protection of company's technology is assured, the company may wish to delay its diffusion into the marketplace in order to exploit the technology in its own products. This exclusivity of technology provides higher marginal returns.

One example of speeding diffusion is Microsoft's strategy of licensing its DOS operating system for use by many computer companies and then making its Windows 95 program available on each PC, thus creating customer commitment to the product. The profit Microsoft makes on each unit of software is minimal; however, the market penetration permits the cornering of the software market for many years.

Apple Computer Company, on the other hand, kept its Macintosh user-friendly operating-system software proprietary for many years and refused to license it to other PC makers. Apple was able to use its proprietary operating system to its advantage in marketing its own Macintosh PC. However, Apple's superior technology was not diffused adequately to permit domination of the software market in the long run.

Microsoft's strategy had a long-term technology domination effect; Apple's had a short-term large-profit-margin effect. Microsoft seems to have won the strategy game. Apple could have changed its strategy and diffused its operating system technology, but the company elected not to do so at the opportune time. It has been estimated that Apple chairman John Scully's decision not to license the Macintosh operating system in the mid-1980s cost Apple Computer \$20 to \$40 billion (Byrne, 1996).

### TECHNOLOGY-PRICE RELATIONSHIP

When an entity such as a company has a technological advantage, it is able to command a premium price for its technology (Figure 3-2). The magnitude of this premium is dependent on the value of the technology to customers. If the knowledge gap between the company (as the owner of the technology) and the customer is high, the owner can command a high price for it. However, as the customer gains experience with the technology, the knowledge gap shrinks. The value of the technology, as well as the commanded price, will decline and eventually vanish. It may even reverse and become negative if the knowledge is transferred to a customer who can improve on the initial technology. Strategies for and appropriate rates of technology diffusion should be based on exploiting the price advantage of the difference in knowledge. They should also guard against shrinking the gap in knowledge. One approach is to continue building a technological lead over time.

**FIGURE 3-2**
TECHNOLOGY GAP/PRICE RELATIONSHIP

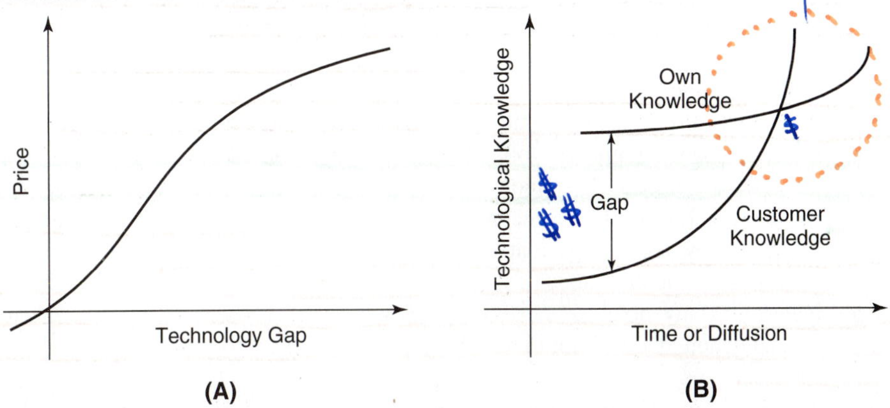

## THE TIMING FACTOR

One of the factors of vital concern in proper management of technology is the timely creation and introduction of technology into the marketplace. Equally important is the timing of the introduction of follow-up technology that will improve performance. Continuous improvement of products and the production capability of the corporation are vital to the firm's health and survival. "Timely" is the key word in this discussion. Actions must be taken at the right time if an enterprise is to succeed in a competitive marketplace. The Osborne Computer Company case study cited by Betz (1987) provides an excellent illustration of the importance of timing. The case occurred in the early 1980s, at the beginning of a new information age that triggered the technological revolution. This was the time that MOT began to assume its importance as an independent area of study, research, and practice.

### The Osborne Computer Company

The personal computer market began in the mid-1970s. Adam Osborne, the founder of a start-up company, decided to package all the PC components together (computer, monitor, and software) as a portable computer, as shown in Figure 3-3. This was the first truly portable personal computer, an innovation that permitted researchers and businesspeople to take their office work home and vice versa. It became an immensely popular idea with the computer-literate population. Osborne shipped his first computer in July 1981. In two months the company had its first $1 million in sales, and by the second year its net revenues reached $100 million. Six months later the company was bankrupt.

What went wrong? There were several major problems associated with the way this innovation was managed: Osborne used a 5-inch display screen to keep the computer compact, but this size allowed only 60 print columns instead of the 80 commonly used by people migrating from punched-card mainframe computers to PCs. It is important to remember, however, that product design inevitably requires a tradeoff among desirable features. To achieve portability of the product, Osborne settled on the technology available to the firm at the time and introduced the small screen. Consumers were interested in a larger, more standard, 80-character display. Osborne lagged in its R&D efforts and was late in introducing a larger screen that would meet consumer demands. This delay eventually proved fatal to the company. Osborne also delayed the generation of needed capital through public offering from the summer of 1982 to early 1983. While this was only about six months, in the dynamic market it proved to be a significant delay.

Kaypro, a competitor company, took advantage of Osborne's weak points and delays. Kaypro introduced improved technology—a portable computer with an 80-character screen. To counter Kaypro's new technological advantage, Osborne announced that it would introduce new technology that would meet consumers' demand for a better display. The timing of the announcement was another major mistake, for Osborne had a large inventory. Many potential customers, knowing that a newer, better model was about to be launched, decided not to buy from the current inventory but to wait and buy the new model.

What was the main reason for Osborne's spectacular success and then sudden demise? Was it the design of the display screen? Was it the problem with R&D? Was it

**FIGURE 3-3**
THE OSBORNE COMPUTER
The Osborne portable computer, open and ready for use. The screen unit fits into slots on the keyboard unit. The 5-inch diagonal screen (that's 3 × 4 inches!) displayed some 60 print columns and the software allowed left-right shifts to see the entire 80-column line. This model had dual floppy drives, and an "internal modem" in the left-hand disk storage slot. The modem is connected by a short jumper cable to the cabinet and has a conventional modular RJ-11 telephone plug. The socket on the right could provide (with an adapter) an NTSC or standard TV monochrome signal to drive a (larger) external TV monitor. A serial port connector is found at the extreme right, and a printer connector is located to the right of the modem connector. This early entry into the area of "portable" computing weighed in at a spritely 25.2 pounds and was powered by the popular Z80A processor, running at 4MHz. For comparison, a variant of the Z80 is currently used in many graphing calculators.

*Source.* Image used with permission from the Department of Computer Science Webmasters at the University of Virginia. http://www.cs.virginia.edu/brochure/museum.html.

the delay in raising the necessary financial resources? Was it the timing of the announcement? Was it the disregard for competitors? Any one of these events could have had a major impact on a company's ability to sustain success. Let's extract some pertinent lessons from this case:

1 Entering the market with a new innovation gives a company an early advantage in sales. Osborne's sales expanded dramatically over a very short period of time.

2 If a company competes with innovation, it should plan to continue competing with innovation. Osborne seems to have underestimated the importance of continued

innovation when it was late with the introduction of an improved screen. Kaypro's introduction of a new display changed Osborne's apparently solid market.

**3** A new product concept creates new markets or alters old markets. The introduction of a portable PC changed people's concepts and habits in working with computers.

**4** All products have a life span. A product's life cycle is greatly defined by the competition and the market. Timely product innovation should be part of every management's technology strategy. Companies must listen to consumers and be positioned to react faster than competitors do. Osborne did not listen or respond in a timely manner to its customers when they expressed the desire for a bigger screen. Maybe management's excitement about the new innovation muffled the voices of the customers. A vigilant organization must be able to devise listening systems in order to get closer to its customers. A procedure for getting feedback from sales representatives is one example of such a system.

**5** The timing of announcements is very important. Companies should not announce new improvements while holding large inventories. Customers are not likely to purchase an old model when they know that a new and improved model will go on the market soon.

**6** Capital formation, financing sources, and cash flow are very important for a growing business. Timing a public offering to take place when the company has technological and market leadership increases the value of its stock and brings new resources for capital expenditure. This is necessary for the expansion of production capability and for funding R&D projects.

**7** Successful new ventures must be able to cope with rapid growth and with increasing competition. Timing innovation is important in three phases of the product life cycle:

- Lead time with innovation provides initial market position.
- Product improvement keeps products technologically competitive.
- Improving the process creates lower cost and keeps the price competitive.

For manufacturing and service organizations, *time-based competition (TBC)* is an important competitive weapon for achieving world-class status (Blackburn, 1991). TBC focuses on the entire value-delivery system to reduce the time required to deliver a product or service. Fast-food firms use time as their competitive advantage in the service sector. Manufacturing organizations have switched to *just-in-time (JIT)* systems to compress time wasted in production and facilitate quick response to customer demands.

Focusing on time has traditionally been the hallmark of success in industry. Henry Ford compressed the time required for producing automobiles and created a very efficient manufacturing system. He achieved great success for his company. In today's world, compressing the time required for innovation provides a major competitive advantage.

## THE VISION TO CHANGE STRATEGY

When a company has a strong market and its revenues are good, management tends to lose sight of environmental changes that may affect the company's competitive position and sometimes even its survival. It is very common for managers to be drawn into the routine day-to-day problems of running a business. This can make them lose their vision for change and enthusiasm for innovation. Short-term success can mask the need for change. The old adage "If it's not broken, don't fix it" may lead management to maintain the status quo. In today's world of fast-paced technological change, this can be a very dangerous posture to take, an attitude that can lead to loss of competitiveness.

### IBM and the Development of the PC

International Business Machines (IBM) is one of the world's largest corporations, having acquired and maintained a leadership position in mainframe computer technology and its marketing. The company's spectacular success and domination of the mainframe market created a culture in its management ranks that was characterized by confidence, invincibility, and resistance to change to newer bottom-up ideas. When the personal computer market developed in the late 1970s and early 1980s, IBM did not have the technology necessary to be the leader in the new market. Its success with the mainframe, combined with the internal inertia, may have interfered with its ability to predict the ensuing huge demand for the PC. It seems as if success in one technology rendered the firm unable to see the need to invest in an emerging competing technology. The result was huge success for start-up PC companies such as Apple, a situation that endangered IBM's historically dominant position. IBM was not about to lose its position, however. It was able to recover from its initial leadership lag in PC technology development and get back into the PC market by relying on its name and other complementary assets, such as its manufacturing capabilities and strong financial position. IBM used existing technologies that were developed elsewhere in order to get back into the race for the expanding PC market. However, it never owned the PC technology or fully controlled it. Its failure to change company strategy—to invest in creating and owning the emerging PC technology in a timely manner—still haunts IBM today. It has been estimated that failing to move expeditiously into PC technology cost IBM as much as $90 billion in lost-market capitalization. This failure is considered one of the biggest strategic blunders of the 1980s (Byrne, 1996).

The IBM case illustrates important principles pertaining to vision and keeping abreast of technological and market changes. The lessons include the following:

**1** When a new technology threatens an older one, it is better to take the risk of investing in the new technology rather than stay with the certainty of the declining one. It is management's job to keep competitors at bay and protect their firm's business and core technologies.

**2** Management must always be on the lookout for new or emerging technologies that can either become a threat or open new opportunities for the business.

**3** A company does not need to invent technological change to implement innovations. What is needed is to embrace change and strategize to acquire the technology. (Technology acquisition is discussed in a later chapter.) IBM was able to get back into the PC production race by using off-the-shelf technologies.

**4** When technology is developed outside the firm, or is not well protected, it is difficult to sustain a leadership position in the market.

### Microsoft and the Internet

In 1996, the Microsoft Corporation, a giant in the software industry, was facing the same dilemma that IBM faced in the late 1970s. Microsoft has been dominant in PC software. It developed the highly successful and useful Windows 95, a version of its Windows-based software that contributed to a significant increase in Microsoft's profits and helped the company expand its domination of the PC market. Yet there is an emerging technology looming on the horizon—the Internet. The growth of net technology has been exploited by Netscape, Yahoo, and Sun Microsystems. Netscape developed a browser that is very popular among Internet users. Yahoo's software allows users to search for topics on the World Wide Web (WWW); its initial public offering of stock was made at $12 per share and went up to more than $40 per share the same day the stock was traded on NASDAQ. Sun Microsystems introduced Java software, which quickly became one of the standard languages of the web in 1996.

The question is, Could Bill Gates, founder and CEO of Microsoft and champion of its PC-related business success, have the vision to change the direction of his firm and embrace the emerging Internet technology? Could he forget all the profits he made in the PC-related business and make potentially risky investments in the new technology? Apparently he made the difficult choice. It was reported that "Gates agreed that the Internet was going to drive demand for PCs and software. He opted to play by the Internet's rules, not the PC rules Microsoft had written with Windows. Gates ordered that Microsoft become web-centric, dumping that which didn't fit and reshaping everything else. In Silicon Valley parlance, Gates showed he was willing to 'eat his young' to stay on top" (*Business Week,* July 15, 1996, p. 98).

Proper management of technology requires making tough decisions and being willing to accept change, move where new technologies are headed, and invest in the future. These are the marks of successful managers.

## MANAGING CHANGE

Thirty years ago was the era of managing production. Engineers and managers concentrated on improving their firm's productivity. Through the fine-tuning of operational efficiency they were able to reduce cost and improve profit margins. The focus was on how to manage available resources.

Today the approach must be very different and much more comprehensive. Management must be able to deal with both stability and change. Management no longer manages in a static or stable technological environment. Rather, it manages in a highly dynamic and frequently turbulent environment. The only thing certain is change. The tasks associated with generating new ideas, creating new products, controlling production, and dealing with a new breed of competitors and with demanding customers are some of the challenges faced by today's managers. The key issue for management in the current environment is how to utilize the existing capabilities of the organization to take advantage of the possibilities. The solution lies in a creating a flexible, highly competitive organization capable of coping with the state of the external world. Whether it involves technology, markets, suppliers, or competitors, successful managers of technology must be able to recognize and react to external change as early as possible.

Survival is at risk if a company cannot forecast or foresee the changes in the external environment. It is recognized that 60 to 80 percent of small businesses fail within five years of starting up. Even large, established firms face the problem of staying power. Of the top 100 firms on the Fortune 500 list in 1956, only 29 were still in the top 100 in 1992 (*Fortune,* April 4, 1992). In one year between 1998 and 1999, 47 firms were displaced from the Fortune 500 list. Many factors could lead to the change of fortunes. They may include mergers, acquisitions or changes in the business environment.

## PRODUCTIVITY, EFFECTIVENESS, AND COMPETITIVENESS

Differences between productivity and effectiveness and their relationships to competitiveness need to be understood in order to optimize the performance of production systems (manufacturing and service). ***Productivity*** is the ratio of the output to input resources (see Figure 3-4). For a manufacturing firm, input resources include capital, material, labor, and energy. The output is what the firm produces. Both input and output should be expressed in similar units, such as dollar value. Productivity is a good measure of efficiency.

On a national level, productivity is a very important factor in raising the standard of living. National productivity can be expressed as output in terms of gross domestic product (GDP) over input in terms of total hours worked:

$$\text{Productivity} = \frac{\text{output (GDP)}}{\text{input (total hours worked)}} = \text{\$/hour worked}$$

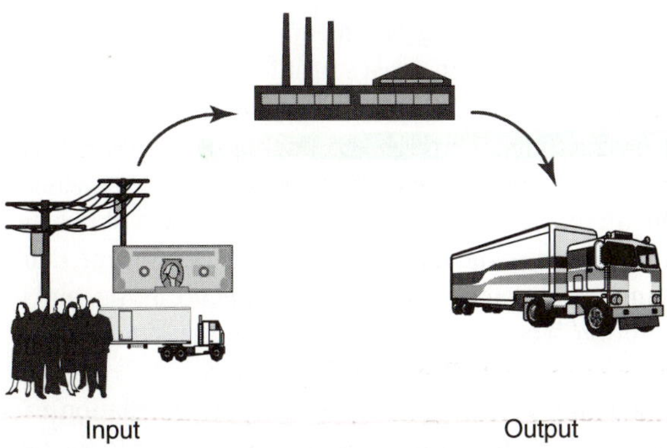


$$\text{Productivity} = \frac{\text{Output}}{\text{Input}}$$

Input: The amount of resources consumed to produce the firm's output

Resources: Capital, Material, Labor, Energy

**FIGURE 3-4**
PRODUCTIVITY

Standard of living is often associated with a population's per capita income. This relationship can be expressed as

$$\text{Per capita income} = \frac{\text{GDP}}{\text{total population}} = \$/\text{person}$$

A rise in the productivity of workers in a population contributes to the growth of GDP and increases per capita income.

Productivity is highly dependent on technology. Denison (1985) estimated that two-thirds, and perhaps as much as 80 percent, of U.S. productivity growth since the Great Depression can be directly or indirectly attributed to technological innovation.

Productivity or efficiency in operation implies doing things right. However, there is an abysmal difference between doing things right and doing the right things. Paying attention to productivity helps to do things right. Doing the right things is an indication of effectiveness. The productivity of a firm is an internal standard of the performance of the organization; it implies minimizing the loss or waste of resources in producing output. It is a necessary but not adequate condition for ensuring the organization's success or continued survival in the marketplace.

*Effectiveness* implies the ability to achieve desired goals, such as increasing the market share of the company or attaining an acceptable level of profit. Effectiveness can be understood as the degree to which an organization accomplishes its objectives.

Effectiveness is about producing or being capable of producing results. It implies producing outcomes that meet the demands or expectations of stakeholders. The stakeholders could be customers, shareholders, owners, employees, suppliers, or the community. Effectiveness reflects the external standards of the performance of an organization. It is affected by and related to the business and social environment of the organization. For example, a charitable organization is effective if it offers services needed by a disadvantaged group. A manufacturing organization is effective if it increases sales, achieves a large market share, and increases profitability. These are factors influenced by the environment of the firm.

The terms "productivity" and "effectiveness" are not synonymous. A company can have very high productivity indexes, yet its management may not be effective in achieving desired goals. To remain competitive, a company must attain high levels of productivity and be effective in its own markets. A problematic scenario may occur if the company's products are based on aging technology, its market share is declining, and its technology is threatened by substitution. This is a recipe for disaster that could lead to the company's demise.

*Competitiveness* indicates the standing of a country or a company in relation to a known group. The competitiveness of a company compares the firm's output to the outputs of its competitors and indicates its standing in the marketplace. When competition is high, only the fit survive. A company is competitive when it maintains a profitable status while producing a product or service that meets the tests of the marketplace. MOT is concerned with achieving higher levels of efficiency and productivity while rendering the organization more effective in achieving its desired goals. It is also concerned with being competitive in an increasingly global marketplace to ensure survival. All of this

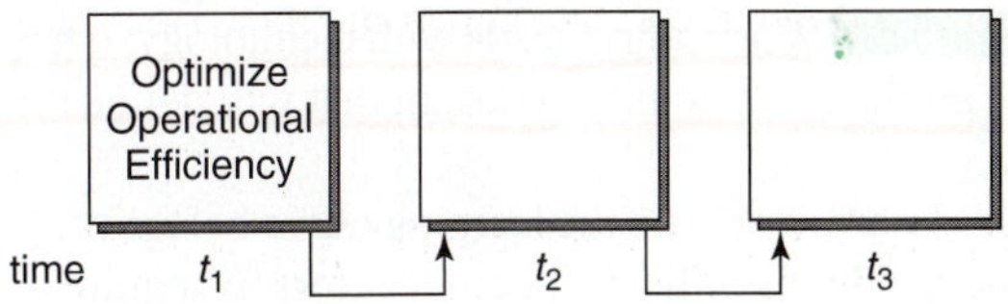


**FIGURE 3-5**
MOT WORKING AT TIME $t_1$ AND PLANNING FOR TIME $t_2$ AND $t_3$

requires a continuous process of assessment and planning that must be repeated with each one of the technologies existent in the company. The ability to identify, plan, and implement technological capabilities in a timely manner is the key to success.

MOT encourages managers to keep an eye on the future without neglecting the need to continue optimizing current operations. Therefore, a manager of a system at time $t_1$ may be working to implement systems for improving operational efficiency, yet he or she must be able to predict technological and environmental changes that will have an impact on the organization in the future, at time $t_2$ or $t_3$ (see Figure 3-5). This will render the organization more effective with tactical and strategic implications.

## LEADERS VERSUS FOLLOWERS

Winners are those who can bring technology to the market. In terms of technological innovation, a firm can be one of three types:

- *A leader:* A leader is a firm that is the first to market an innovation.
- *A follower:* This firm misses the initial wave of capitalization on the technology but recognizes the technology's impact on its business. Such firms follow closely behind the leader. They may be able to catch up or surpass the leader if they can capitalize on their own strengths. For example, complementary assets such as manufacturing capability, marketing, knowledge, or service outlets may help a follower overcome the consequences of the initial shock.
- *A laggard:* This type of firm realizes a potential for profit on technology but seldom influences technology's use. Laggards engage in "me-too" behavior, according to Holt (1990). Often, their survival may depend on adopting new technology.

The advantages of being a leader in innovation are:

**1** *Name recognition:* The names of leaders with innovative products come to be well known by the public. If a brand or product is associated with the corporate name, that name may become synonymous with the item in general. Consider, for example, the public use of the following names:

- "Kleenex" instead of tissue
- "Xerox" instead of photocopy
- "Coke" instead of soft drink or cola
- "Frigidaire" instead of refrigerator

Such strong name recognition can be converted into profitable long-term sustainability.

**2** *Better market position:* Being first to market gives a firm an opportunity to capture a large market share.

**3** *A chance to define the industry standard:* When a firm leads with technology, it has an opportunity to establish a dominant product or a dominant design that will define the industry standard.

**4** *A head start on the learning curve:* Leaders start on the learning curve before competitors. They also develop tacit knowledge that is difficult for competitors to acquire or transfer.

**5** *Protective barriers:* Leaders can protect their technology through patents and other means to prevent late entrants from competing. They also have better opportunities to exploit their technology.

**6** *High profit:* Leaders command the market. They establish a technology gap between their products and their customers or competitors. Thus, they are able to obtain a high price for their products, thereby earning large profits.

**7** *Delayed customer switching:* Leaders establish special relations with their customers. Brand loyalty and the cost of switching cause customers to delay switching to a competitor's product.

**8** *Favorable response by outsiders:* Leaders have better potential for getting support from government, venture capitalists, and other industries interested in their innovations.

There are several disadvantages to being a leader in innovation:

**1** The leader assumes the large cost associated with research, prototyping, testing, and overall development.

**2** The leader must be able to sustain the lead. There are costs associated with updating the technology.

**3** The initial investment in design, tooling, and production may create difficulty in reversing the course of action should a competitor introduce a better technology or an improved design.

**4** There is market uncertainty associated with the introduction of new technology. It is difficult to predict demand and to set an optimal price.

**5** The leader is a target for competition.

Innovation leaders can sustain their leadership through a combination of the following strategies: They invest in R&D to continue the development of technology. They rely on developing the technology within, as opposed to outside, the firm (Porter, 1985). They build their technological resources and assemble a workforce with the highest technical skills. They work to diffuse their own technology while delaying competitors from getting into that technology.

A follower firm does have some advantages over a leader in that it is not committed to any special design, process, or technology. It has a chance to examine a leader's product and attack it at its weakest point. The chances are that it will have a lower initial investment in R&D. Thus, it can leverage its investment to experiment and improve on what already exists.

A laggard has the least chance of being effective in influencing the directions of the technology or the standards of the industry. This is a situation in which some small firms with low R&D budgets become entrenched. They have an opportunity, however, to find niches or to create alliances with others that will help propel their technology position forward.

Teece (1987) introduced a taxonomy of outcomes from the innovation process and identified some winners and some losers (see Figure 3-6). In the taxonomy he considered innovators as those who were first to the market with a product or a process. Winning implies getting a new earning stream, enhancing an existing one, or sustaining a competitive advantage. Losing implies the opposite. Leading with innovation may give a company a head start, but it does not guarantee a win over the long stretch. As in a track race, a follower may come from behind and overtake the leader. In a competitive arena with many participants, proper management of technology can provide a company with the winning edge.

As can be seen in Figure 3-6, innovators that were winners included:

- *Pilkington,* with its Float Glass process.
- *Searle,* with the introduction of its NutraSweet product.
- *Du Pont,* with the commercialization of Teflon.

According to Teece, innovators that were losers included:

• *RC Cola,* which was the first to introduce diet cola and cola in a can. Coca-Cola and Pepsi-Cola followed almost immediately and deprived RC Cola's leading-edge advantage.

• *EMI,* which introduced computer axial tomography (CAT) scanners but was later overtaken by General Electric (a good review of this case is in Martin, 1994).

• *Bowmar Instruments Corporation,* which introduced the pocket calculator and was later overshadowed by Texas Instruments and Hewlett-Packard.

| | INNOVATOR | IMITATOR–FOLLOWER |
|---|---|---|
| WIN | 1<br>• Pilkington (Float Glass)<br>• G.D. Searle (NutraSweet)<br>• Du Pont (Teflon) | 2<br>• IBM (PC)<br>• Matsushita (VHS format for CR)<br>• Seiko (Quartz watch) |
| LOSE | 3<br>• RC Cola (Diet Cola)<br>• EMI (Scanner)<br>• Bowmar (calculator)<br>• Xerox ("Star")<br>• DeHavilland (Comet) | 4<br>• Kodak (Instant photography)<br>• Northrup (F20)<br>• DEC (PC) |

**FIGURE 3-6**
OUTCOME FROM THE INNOVATION PROCESS
*Source:* Teece, 1987.

- *Xerox,* which introduced many of the desirable ideas in an office computer (see the Xerox-PARC case later in this book) yet was overtaken by Apple and other PC makers.
- *DeHavilland,* which introduced the first jet airliner—the Comet—but Boeing was able to win with its 707 and other 700-series airplanes.

Imitators that won included:

- *IBM,* which missed out on the original phase of PC development but was able to get back into the game with its own PC development at its Boca Raton, Florida, site, using a venture team led by Philip Estridge. IBM's success in the PC market is largely attributed to its name and complementary assets. As discussed elsewhere in this book, IBM dominance in the PC market did not last because it did not own or control the PC technology.
- *Matsushita,* which defined the industry standards with its VHS video recorders. Although Sony had invented and controlled the Beta-Max, which has technology superior to that of the VHS, it was unable to diffuse its technology fast enough to define the industry standards. Other companies were able to break Sony's monopoly of the technology and define the VHS as the industry standard.
- *Seiko,* which was a follower with the quartz watch but was able to create a sustained competitive advantage for its product.

Imitators and followers that were judged to be losers included:

- *Kodak,* which attempted to compete with Polaroid in instant photography. Polaroid had its technology protected, and Kodak was unable to break Polaroid's hold on it.
- *Northrup,* whose development of the F20 was an unsuccessful venture over time.
- *Digital Equipment Corporation (DEC),* whose entry into the PC market was an unsustainable venture.

## DISCUSSION QUESTIONS

1. Look in an entrepreneurial or business magazine for an article about a technical venture that has succeeded (*Business Week* publishes special issues on growing companies). What was the entrepreneur's background? Who were her or his partners? How long did it take to have the company running?
2. Do some research on the origins of Apple Computer. Do you think Steve Jobs would succeed today? What kind of computer company is likely to be created nowadays? How have the opportunities changed over time?
3. Identify a company that changed from being a follower to being a leader. What was its approach? How long did it take?

## ADDITIONAL READINGS

W. Chan Kim & Renee Mauborgne. "Value Innovation: The Strategic Logic of High Growth." *Harvard Business Review,* January–February 1997.

The authors analyze a basic managerial problem: What is the difference between regular firms and high-growth competitors? They argue that it is the way managers approach their strategies that makes the difference. Managers should use their portfolio of business to create value through innovation.

T. M. Nevens, G. L. Summe, & B. Uttal. "Commercializing Technology: What the Best Companies Do." *Harvard Business Review,* May–June 1990.
The authors of the paper make the point that "Just as quality and manufacturing were key to competitiveness in the 1980s, businesses will rise and fall depending on whether they discipline their commercialization efforts." This article provides guidance on the basic skills that companies must develop in order to commercialize technology.

F. M. Scherer. *Innovation and Growth.* MIT Press, Cambridge, MA, 1984.
The concepts of invention, innovation, and entrepreneurship are discussed in Part I of Scherer's book. He uses the Watt-Boulton steam engine venture as a case study.

## REFERENCES

Aiken, M. & Hage, J. 1979. "The Organic Organization and Innovation." In Zey-Ferrel, M. (ed.), *Readings on Dimensions of Organizations,* pp. 263–279. Goodyear, Santa Monica, CA.

Barron, F. 1969. *Creative Person and Creative Process.* Holt, Reinhart & Winston, New York.

Bayraktar, B. 1990. "On the Concept of Technology and Management of Technology." In Khalil, T. and Bayraktar, B. (eds.) *Management of Technology II: The Key to Global Competitiveness.* Industrial Engineering and Management Press, Norcross, GA.

Betz, Frederick. 1987. *Managing Technology. Competing through New Ventures, Innovation, and Corporate Research.* Prentice-Hall, Englewood Cliffs, NJ.

Blackburn, Joseph D. 1991. *Time-Based Competition: The Next Battleground in American Manufacturing.* Business One Irwin, Homewood, IL.

Braun, E. 1984. *Wayward Technology.* Greenwood Press, Westport, Conn.

Bright, J. R. 1969. "Some Management Lessons from Technological Innovation Research." *Long Range Planning,* vol. 2, no. 1, pp. 36–41.

Brooks, H. 1965. "The Interaction of Science and Technology: Another View," in Wanner, W., Morse, D. and Eicher, A. (eds.) *The Impact of Science and Technology.* Columbia University Press, New York.

Brown, John S. 1991. "Research that Reinvents the Corporation." *Harvard Business Review,* January–February, pp. 102–111.

Byrne, J. A. 1996. "Strategic Planning." *Business Week,* Aug. 26.

Denison, Edward F. 1985. *Trends in American Economic Growth 1929–1982: The United States in the 1970's.* Brookings Institute, Washington, DC.

Holt, K. 1990. "Technology Strategy—Is There a Need for It?" In Khalil, T., & Bayraktar, B. (eds.), *Management of Technology II,* Industrial Engineering and Management Press, Norcross, GA.

Jain, R. K., & Triandis, H. C. 1990. *Management of R&D Organizations.* Wiley Interscience, New York.

Martin, Michael J. C. 1994. *Managing Innovation and Entrepreneurship in Technology Firms.* Wiley Interscience, New York.

Nord, W., & Tucker, S. 1987. *Implementing Routine and Radical Innovations.* Lexington Books, Lexington, MA.

Porter, Michael E. 1985. *Competitive Advantage.* Free Press, New York.

Schumpeter, J. 1928. "The Instability of Capitalism," *Economic Journal,* Sept., pp. 361–386.

Teece, David J. 1987. "Capturing Value from Technological Innovation: Integration, Strategic Partnering, and Licensing Decisions." In Guile, B., & Brooks, H. (eds.), *Technology and Global Industry,* pp. 65–95. National Academy of Engineering, National Academy Press, Washington, DC.

# 4

# MANAGEMENT OF TECHNOLOGY: THE NEW PARADIGMS

The decline of U.S. industrial competitiveness in the 1970s and 1980s became a subject of increasing concern to people in education, industry, and government. Several initiatives were mounted to determine the sources of this decline and to formulate a successful response to the challenge it presented.

Very quickly a consensus was reached on the idea that a significant amount of this effort had to be directed toward improving the management of technology. The decline of U.S. industry during those two decades is widely perceived to have resulted not from an inability to develop new technologies but from the failure to manage available and emerging technologies in an effective and timely manner.

Technology plays a pivotal role in the interactions among the individual, society, and nature. Technological advances have major effects on each of these entities and are, in turn, influenced by them. Management of technology involves developing an understanding of these relationships and dealing with them in a rational and effective manner.

Because the management-of-technology field is a critical component of the effort to address questions of American competitiveness in the marketplace, a 1986 workshop sponsored by the National Research Council (1987) focused attention on it as a hidden competitive advantage. It was recommended that significant measures be taken to build national awareness of the strategic importance of MOT and to support research in this area. The NRC workshop report identified industry needs in MOT to be as follows:

**1** How to integrate technology into the overall strategic objectives of the firm.
**2** How to get into and out of technologies faster and more efficiently.
**3** How to assess/evaluate technology more effectively.
**4** How best to accomplish technology transfer.
**5** How to reduce new product development time.

**6** How to manage large, complex, and interdisciplinary or interorganizational projects/systems.
**7** How to manage the organization's internal use of technology.
**8** How to leverage the effectiveness of technical professionals.

In 1987 a workshop organized by the Public Affairs Council of the American Association of Engineering Societies (1988) sought to build greater understanding and awareness of MOT issues. The report of this workshop contrasted the widespread belief in the importance of technology management for U.S. industrial competitiveness with the corresponding absence of concrete steps to pursue relevant avenues of research and application. There was a definite need for a paradigm shift in how to manage in the new environment created by the emerging technology revolution.

Paralleling these efforts, the University of Miami (UM) Department of Industrial Engineering organized two meetings in 1988. The First International Conference on Technology Management was held in Miami, Florida, in February 1988 (Khalil et al., 1988). More than 300 scientists, engineers, and managers from 30 countries attended. A workshop, jointly sponsored by the National Science Foundation's Division of Cross-Disciplinary Research and the University of Miami, followed. This series of meetings resulted in a clearer formulation of management of technology as a new field of study and research.

The ideas that arose from the various meetings were summarized by Khalil and Bayraktar (1988, 1990) and are presented here. They focus on the scope of MOT and the issues that need to be addressed in a modern organization. The issues are presented with a number of stipulations that should be observed. These stipulations are:

**1** Topics identified are viewed only as a representative sample of important issues in MOT.

**2** While the discussions and examples may sometimes be drawn from a manufacturing environment, the issues raised have equal relevance for service industries. Although a sound manufacturing base represents a major wellspring of technology and innovation, the service industries now constitute the largest sector of the U.S. economy. Service industries are the locus of the greatest economic growth in the United States. Any agenda on technology management must address the needs of this expanding sector, or it will fail to address the needs of the future.

**3** A fundamental axiom in discussing issues of managing technology is that they must be examined increasingly in a global context. In an era marked by the rise of multinational corporations and the rapid emergence of strong international competitors, the high cost of technological innovation and the cyclical nature of business will demand greater dependence on global alliances as a strategy.

**4** The current revolution in information technology promises only to intensify in the years to come. As the information available to decision makers increases exponentially, the critical issue will be how to absorb and manage this information and evaluate its impact on individual organizations and society at large.

**5** MOT is frequently mistaken as being limited to the management of the creation of technology, that is, limited to R&D. Research and development pushes technological advances, but it is not the only means of establishing a needed technological base.

Economic considerations frequently dictate that more effective and efficient technologies be adopted; these may already be a part of the worldwide technological inventory, obtainable through any of several transfer mechanisms. The purchase and sale of technologies are now common events; technology has become a marketable commodity transcending national boundaries. Consideration of any array of possible technologies should thus be embedded in the strategic planning of organizations.

**6** Technology transcends manufacturing and service organizations through a set of enablers. These include technical and financial resources, environmental factors influencing the business, organizational structure, projects, and people (see Figure 4-1). Therefore, issues falling under the scope of management of technology can be explored in their relation to one of the following five categories:

- Methods and tools for effective management of resources.
- The business environment and the ability to manage the interface between the organization and the external environment.
- The structure and management of organizations.
- Management of R&D and engineering projects.
- Management of human resources under conditions of rapid technological and social change.

**7** Another dimension of the matrix of technology enablers shown in Figure 4-1 is the *technology life cycle*. This is the development of technology from the concept stage to

**FIGURE 4-1**
TECHNOLOGY AND ITS ENABLERS

| Enablers / Technologies | Resources (Technical & Financial) | Business Environment | Structure and Management of Organization | Project Planning & Management | Human Resources |
|---|---|---|---|---|---|
| Natural Resources (Materials) | | | | | |
| Product Technology (Concepts & Design) | | | | | |
| Production Technology (Processes & Operations) | | | | | |
| Information Technology | | | | | |
| Marketing Technology (Traditional & Innovative) | | | | | |
| Service and Customer Satisfaction Technology | | | | | |
| Safety & Environmental Technology | | | | | |

| Idea Generation and Concept Definition | Market Analysis | Technical Analysis | Business Plan and Approval | Developing and Testing | Production | Commer-cialization | Disposal or Recycling |
|---|---|---|---|---|---|---|---|

**FIGURE 4-2**
STAGES IN NEW PRODUCT/TECHNOLOGY LIFE CYCLE

the prototype, production, market, and after-market stages, as shown in Figure 4-2. This cycle determines the birth, life, and death of technology.

## ESSENTIAL ISSUES IN MANAGING TECHNOLOGY

In the third millennium, we will witness drastic changes in the business environment. There is a search for new paradigms that are suitable for this new environment. A *paradigm* is a framework of ideas that establishes the general context of analysis. The unifying theme for the entire field of MOT is technology as a creator of wealth. As an interdisciplinary field it draws upon knowledge from existing fields such as engineering, management, accounting, finance, economics, production, and political science. The search is for ways to argue the case of technology as drawn from all these areas of knowledge.

An overriding concern is the highly dynamic conditions that exist for manufacturing and service organizations, conditions dictated by changes in technology and by the global business environment. Exhibit 4-1 lists several of the changing trends in industry during recent years. To embrace change, managers of organizations should consider a number of essential issues. These are classified below according to the technology enablers' categories. The issues discussed have been recommended to the National Science Foundation (NSF) as deserving of additional research and to industry as worthy of intensified attention (Khalil and Bayraktar, 1990).

### Resources

The efficient utilization of technological resources is a critical aspect of the management of techno-economic enterprises. In a way of life based on technology, the rational and productive use of available instruments, equipment, tools, materials, methods, software, skilled workers, information, intellectual assets, and financial resources is crucial in providing a competitive posture for corporations.

In a highly competitive environment where error tolerance is limited, managers must be equipped with predictive methodologies and decision tools that are reliable, flexible, practical, and fast. There is a need for new ideas, imaginative methodologies, and performance criteria that have been tested in real-life situations.

The following areas are considered critical for industrial competitiveness. Their continued research, validation, refinement, and application deserve a high priority. A brief rationale for each is given.

**EXHIBIT 4-1**
CHANGING TRENDS IN INDUSTRY

| Factor | Traditional | New |
|---|---|---|
| Life cycle | Long life cycles | Short life cycles |
| Innovation | Few innovations | Continuous innovations |
| Competition | Expected competition<br>Competitors are the enemy<br>Cooperation not allowed | Stronger competition<br>Alliance with competitors accepted |
| Market | Expected market<br>Local market | Uncertain market<br>Global market |
| Quality | Quality is desirable | Quality is imperative (a hygiene factor, a survival factor) |
| Production | Mass production<br>Produce in large lots<br>No commitment to suppliers<br>Large inventories<br>Fixed manufacturing | Customized production<br>Produce in small lots<br>Suppliers are partners<br>Reduce inventories (JIT)<br>Flexible manufacturing |
| Organization | Large corporations vertically<br>Integrated companies<br>Bureaucratic organizations<br>Financial methods control the organization | Smaller plants; companies rely on outsourcing<br>Nimble organizations<br>Financial methods to serve the organization's objective |

**Methods of Performance Assessment** Traditional measures of performance are sterile and unimaginative. Methods of accounting and financial assessment are biased against technological innovation and underestimate the risk of maintaining the status quo. There is a need to know the extent to which the specific characteristics of a new technology influence methods of general and financial assessment of performance, that is, the risks, the quality of the process and products, market entry, competitive positioning, and short- and long-term profitability. In recent years, many firms have attempted to address this issue by introducing new costing and accounting techniques such as activity-based costing (ABC).

**The Measure of Performance of a Technology** Reliable methods of measuring the performance and competitiveness of specific technologies in the marketplace should be developed and used. The problem is more acute for new technologies that are untested in a competitive environment. Different criteria may be needed for different circumstances.

**The Measure of Benefits from R&D Activities** Benefits from R&D activities may be manifold. A rational approach to identifying potential benefits and a set of measurement criteria for comparing the outcomes of R&D projects should be developed and used systematically.

**New Tools for Optimizing Decisions** Since resources are scarce and time is limited, optimal allocation of financial, material, and human resources is critical. Areas of interest include:

**1** *Improving methodologies of technology forecasting and of integrating technology forecasting within the planning and decision-making processes.* Because of technological discontinuities, adaptive models providing for the continuous evaluation of prior assumptions are required in lieu of methods based on the extrapolation of past data.

**2** *Developing a new and more representative set of criteria to optimize the performance of high-technology firms.* Classical optimization methods rely on performance measures such as return on investment (ROI), return on sales (ROS), and price/earnings ratio (PER). In recent years, corporate managers have become better sensitized to the limitations of these measures and management should be more cautious in using strict, tangible, financial justification models in evaluating technological projects. Both tangible and intangible criteria are needed for the optimization of decisions.

**3** *Determining the optimum mix of high-technology versus traditional products in a large business.* New tools are needed to assist managers in making correct resource allocation decisions. In the meantime at what level of decision should particular criteria and attributes of decision be taken into account?

**Alliances as Alternatives to Rivalry** Fierce competition in the form of market-share shifts and takeover bids is not necessarily the most efficient way of using scarce human, financial, and material resources. It causes an excessive level of insecurity and disorganizes industrial teams. It will be useful to know:

**1** Would alliances on specific projects among national and international competitors offer successful alternatives to conducting in-house R&D? If so, how to structure an agreement to facilitate negotiations on new products and technologies?

**2** How effective are government initiatives (e.g., incubators and technopolis) in introducing new technologies and opening up large industrial firms to outside suggestions? Would joint ventures and the sharing of innovation between a large firm and a small firm take advantage of the strengths and supplement the weaknesses of each?

**3** When can research results and/or generic technologies be shared through alliances? To what extent do patent laws assist or prohibit cooperation among companies?

### The Business Environment

Organizations operate in a socio-techno-economic environment and interact with it. Within the context of technology management, interest is primarily focused on technological factors, activities, and plans. How do external factors affect the creation and introduction of technological change within an organization, and how do technological changes that take place within an organization influence the environment?

The introduction of a technological innovation in the marketplace, particularly when it has been widely adopted through the processes of transfer and diffusion, impacts a society, its economy, and the natural environment to varying degrees. The effects may have varying levels of acceptance and/or desirability based on the prevailing value systems of the society.

### ALLIANCES AND THE RACE FOR TECHNOLOGY

IBM, Apple, and Motorola have been traditional rivals in the manufacturing and selling of personal computers and their components. When Apple introduced its Apple II PC in the early 1980s, the computer met with spectacular success. IBM, which initially resisted PC technology, had to respond quickly to meet the challenge from Apple in the expanding PC market. However, IBM did not have the necessary technology and had to rely on other companies, such as Intel (for the microprocessor) and Microsoft (for the operating system). By 1990, Intel had a strong hold on the microprocessor technology and Microsoft on the operating systems and software technology. The dominance of Intel and Microsoft in these two technologies, critical for the production of PCs, threatened the interests of IBM and Apple. In 1991, IBM, Apple, and Motorola established a joint venture known as Somerset. Its aim was to break Intel's hold on the microprocessor technology and its dominance of that market. Somerset was a technological alliance forged to take advantage of the design and manufacturing powers of Motorola and IBM and the purchasing-volume potential of IBM and Apple. It would also permit sharing resources and defraying the cost of R&D among the three giant companies. Somerset was to create a new Power PC chip with higher performance than that of chips provided by Intel. The Power PC would also give IBM and Apple proprietary technology that they never owned before.

Besides having traditional problems that face R&D projects in general, Somerset was plagued by many of the problems that face new alliances for producing technology, including technical difficulties, business bickering, management upheavals, and clashes of culture among workers from different organizations. These difficulties delayed the production of the Power PC and its more powerful successor chips ("Time May Have Passed the Power PC," *Business Week,* Mar. 4, 1996). The project ran behind schedule, missing windows of opportunity that are essential for exploiting new technology and making it a commercial success. Because of this, Somerset had to develop a superior technology, bring it to the market in a timely manner and diffuse it enough to create market dominance, and establish new industry standards. This proved to be a monumental and illusive goal. In the meantime, Intel was not standing idle waiting for the Power PC technology to overtake its own. Intel was busy developing its own microchip technology with higher performance capabilities. Intel was able to stay ahead in this race for technology.

Any technological change that affects the public at large and/or the natural environment becomes an issue for the potential exercise of public power at different institutional levels. Legislative, administrative, and regulatory machinery can be set in motion to counteract perceived or potentially undesirable consequences or to facilitate the widespread adoption of changes expected to promote the public welfare.

The involvement of public power entails the careful assessment of technologies that are of societal concern and the evaluation of risks associated with public exposure to them. Organizations themselves must be aware of potential public concerns regarding the product that they are about to market and the processes they plan to introduce in their production and operation systems.

The ultimate measure of the success or survival of a corporation is the market performance of its products and services. It is incumbent upon organizations to translate market indicators to strategic decisions and operational plans. Another important environmental factor that influences business strategy is competition in the marketplace. No corporation can afford to ignore what its competitors are doing, especially with regard to technological opportunities. In order to remain competitive, a firm must anticipate and evaluate technological opportunities before other firms attain an insurmountable competitive edge. These considerations are reflected in technological plans that must be incorporated into the strategies and plans of the business firm. The following are considered priority issues.

**The Integration of Technological and Strategic Plans** Technological planning involves decisions affecting the selection of R&D projects, the allocation of resources, and timetables for successful implementation. It also involves choosing the technologies to be incorporated into the production process and evaluating whether they should be produced internally or purchased. Each of these options must be addressed in the strategic plan. Does methodological guidance for managers and planners regarding rational and efficient means of discharging this responsibility exist in the corporation? If not, it should be developed and applied.

**The Impact of Third Parties on Technological Change** What are the effects of third-party regulations (e.g., judicial decisions, legislative and regulatory actions, decisions regarding insurance risks and liability) on the firm's decisions to pursue and implement particular technologies? While there is general acceptance of the proposition that each of these factors affects a firm's utilization of some technologies and its market policy toward some questionable products, there is not enough understanding of the underlying relationships to provide management with guidance on anticipating and taking timely actions.

**Increasing the User's Influence in the Selection and Application of Technologies** There is a need for more understanding of the feedback mechanism between the users and the producers of a technology and for means of strengthening user influence in the selection and application of technology. Businesses may then address the actual needs of the marketplace so that products having little or no potential market are not produced under the push of science and technology. The consumers and the public at large have a stake in newly developed technology and they influence its acceptance in the marketplace.

**Decreasing Social Resistance to the Introduction and Adoption of Technology in the Workplace** Given the prevailing tendency to resist changes that influence work rules and organizational structure, plans to introduce new technological systems to the workplace may cause apprehension and opposition. To overcome these difficulties, management must develop adequate insights as to the factors underlying such opposition and must develop strategies for dealing with it constructively.

**Distributing the Benefits from New Technologies to Gain Acceptance** Adoption of new technologies is easier to accept if their benefits are shared or rationally explained. For example, a technological solution such as automation may lead to a reduction in the labor share of the total manufacturing cost of a product. Rationalization may emphasize that the lower production cost reduces the incentive for offshore facilities and thereby increases the availability of manufacturing jobs in the United States.

**Other Areas of Concern to the Firm** Other issues worthy of consideration in developing firm-specific methodologies to guide managers in dealing with technology include:

- Potential obstacles to and benefits of interfirm cooperation.
- Appropriate strategies and time points for the transition from cooperation to competition in interfirm technological alliances.
- The impact of technology on the quality of life, health, and safety.

### The Structure and Management of Organizations

Rapid technological change accompanied by intense global competition creates considerable problems in structuring and managing organizations in every sector of the economy. In industries such as manufacturing, where installations of highly sophisticated information and communication systems, computer-based manufacturing, and direct linkage to customers are widespread, it is necessary to staff these systems with highly skilled personnel. Empowering employees for decision making improves productivity and reduces the time needed to respond to markets or customers' demands. These are factors that tend to decrease the need for hierarchical organizational structures and support shallower, or "flat," structures. Computer-integrated information and manufacturing systems allow close, real-time coordination and cooperation among departments charged with separate functions. However, new opportunities for cooperation raise questions regarding organizational structures designed along functional lines. The existence of in-house R&D activities and the necessity of coördinating them with the production and marketing functions at early points in the design and development stages make it imperative that the organizational structure permit effective integration of these activities. Some organizations may opt for the use of outside sources for their R&D and may outsource many components of their products.

In technologically dynamic companies, the installation of technological gatekeepers, the encouragement of internal entrepreneurship, and the increase of joint ventures in both R&D and production have major consequences for organizational structure, all of which need to be addressed rationally. Reexamination of the effect of organizational change on technological creativity and the internal dynamics of the organization is also needed.

In sum, organizational structure interacts intimately with the technological posture of an organization and provides an array of topics for reflection by modern managers. The following topics are viewed as priority issues in the area of organizational structure.

**Factors Leading to Reorganization of Technological Activities in Firms** The question of reorganization is usually considered in terms of centralization versus decentralization. Dynamic technological imperatives necessitate the investigation of other issues and forms of reorganization. Shuffling organizational arrangements may be an inadequate way of dealing with more fundamental technical or managerial problems. Tradeoffs may exist between organizational structures that are efficient in motivating and carrying out technological advances and those that favor current production modes and activities.

Organizational structures that are responsive to industrial and technological requirements may be industry-dependent and therefore must be examined in this context. Proposing improvements to organizational structures requires a better understanding of existing motives and practices in restructuring organizations.

**Evaluating the Impacts of Reorganization on Technical Activities** Reorganization may directly impact the technical activities (e.g., research, development, and manufacturing engineering) of a firm and also the interaction of these activities with production and marketing. The issues raised here address the set of possible tradeoffs incurred when reorganization of the firm occurs. Evaluation of the perceptions and expectations regarding the benefits and costs of reorganization, and an objective *ex post* evaluation of the results of the reorganization, should be a part of any reorganization effort.

**The Effects of Different Organizational Structures on the Efficiency of the Product Development Cycle** The key issue here relates to the timing and organization of R&D, design engineering, and the various groups involved in the product development process. Coordination among R&D, design engineering, manufacturing engineering, operations, and marketing groups is a critical ingredient for success. Some of the issues to consider are:

**1** Determining which interfaces are helpful.
**2** Identifying what kinds of interfaces impede higher levels of innovation.
**3** Considering differences in arrangements among basic technologies or industries.

**Facilitators and Inhibitors of Technological Innovations and Transfers within Organizations** There are two issues to consider. The first is the organizational arrangements and incentives that facilitate the transfer of technology within organizations, as opposed to those that seem to encourage the secrecy and insularity of teams, groups, and divisions. For example, internal entrepreneurship (intrapreneurship) is a two-edged sword: Technological advances may well be achieved, but the strong "ownership" of the information developed may prevent the rest of the organization from benefiting from it.

The second issue to consider is insight into the organizational factors that encourage and support, as well as those that hinder, the effective performance of technological gatekeepers, internal entrepreneurs, and others who are committed to bringing about innovations.

**Documentation of the Decision Processes Leading to Organizational Changes** Keeping a record of the decision-making process itself is a good practice. It can be used in guiding future actions.

### Project Planning and Management

Complex R&D projects require the mobilization of substantial resources and the coordination of activities in different laboratories and sometimes in different countries. The management of such projects constitutes a formidable task that demands considerable skill.

Comparable challenges exist when a new industrial product is designed, developed, and marketed. The entire process involves people from several departments and disciplines in one or more laboratories, firms, and/or institutions. Scientists, specialists, and engineers must collaborate to carry the product from concept to "launch," yet organizationally they belong to separate parts of the enterprise or even to another organization. In large projects, where several firms and institutions are involved, the project manager's difficulties are obviously compounded.

R&D and engineering projects have other characteristics. Projects are carried out by a highly trained, highly motivated professional staff; mostly, they are one-of-a-kind undertakings involving considerable uncertainty and risk. Project managers therefore need tools and techniques that will help them better grasp the intricacies of the relationships among various components and will also equip them with capabilities for skillfully handling human problems. The mechanics of project management involving the scheduling of tasks and the allocation of resources are well explored; many software systems are available to help project managers. However, there should be a mechanism that provides them with rapid feedback about the process, particularly about early warning signals regarding potential failures.

There is also a need for improved understanding of the human components of the project management process: how to select people with different skills and training and how to enable them to operate in a multidisciplinary and multicultural environment. These are issues requiring special skills. The majority of project managers are engineers promoted to managerial positions, and they need training in people-related skills. Such managers primarily rely on their personal aptitudes and the skills they acquire on the job, with little or no formal training. Engineers and scientists must be encouraged to learn and develop "people" skills. They should be given professional training in this area for later use as either a team member or a team leader. Such investment in employees is beneficial to any organization in the long run.

One of the important tasks of organizations is to select only R&D projects that may have some potential for future exploitation consistent with the firm's growth strategy. There is a need for practical and powerful selection methods. R&D projects that are part of the innovation process pose a challenge to management, which must reconcile the visions and ideas originating among the organization's scientists and engineers with the views and plans of upper management and then translate it all into a workable program. Understanding the internal dynamics of these relationships is important. Attention to the following items deserve special attention.

**Project Portfolio Selection** Organizations frequently have several, and occasionally a large number, of ongoing projects. Not only do these projects need to be constantly monitored and evaluated for their potential usefulness to the organization, but their utility must also be compared with new opportunities. The reexamination and reevaluation of priorities are thus an important part of the management of R&D. There is a concomitant need to develop easy-to-use decision-support systems for managers so that they can focus on the content rather than the mechanical aspects of project management.

**Initiation of Innovative Ideas in Organizations: Top Down or Bottom Up?** In organizations with long-term business strategies, it is expected that strategic decisions and operational plans, when communicated from upper management, will be translated into focused programs and projects. However, such instructions may not fit well with the ideas and aspirations of scientists and engineers in the innovation chain. Consideration must be given to:

**1** Creating a balance between these divergent views.
**2** Understanding the dynamics of this balance.
**3** Finding and evaluating existing patterns and identifying their impact on organizational performance.

**Human Problems in Project Management** Project management continues to grow in importance, domestically and internationally, in public and private organizations. Although many capabilities have been developed for scheduling and monitoring, much needs to be done to help project managers in selecting personnel and in coping with the problems created by the multidisciplinary and multicultural background of the professional project staff.

**Postmortem Analyses of Projects** It is important to conduct postmortem analyses of projects, both those with successful and those with unsuccessful outcomes. Understanding the commonalities and divergences among projects with similar outcomes can be helpful to decision makers. It is also necessary to understand the interfaces between a project and the rest of the organization, and between the project and the external environment, and their effects on the success or failure of the project.

### Management of Human Resources

Recent advances in communications technology, transportation systems, and computer-based information systems and new developments in computer-integrated manufacturing and office automation have drastically altered the character of modern manufacturing and service enterprises. The spatial, as well as temporal, characteristics of the workplace have been undergoing significant changes. Work locations, the scale of operations, the critical time factor, skill requirements, and operational parameters represent a core of issues associated with the management of future organizations.

The incremental technical obsolescence of the professional staff, insufficient past training, and the inexperience of even skilled labor in handling newly implemented tools and equipment create continuing problems for management and workers alike. Professional staff must keep abreast of recent developments in scientific knowledge and technological innovation through a variety of means, such as books and journals, professional meetings, and continuing education programs. Management must anticipate the skill requirements of new technologies scheduled for implementation and try to match existing skills to them or seek, through retraining and relocating measures, to minimize operational disruption and redundancy. Cost accounting methods, productivity evaluations, and operational procedures must be revised to meet the needs of new and changing situations. The reexamination of the performance characteristics of human resources is also required.

In technologically dynamic corporations, highly trained professionals are needed to evaluate newly implemented technological advances and those that could become a reality in the near future. The motivational reward systems of corporations should be designed to encourage and support the activities of technological gatekeepers and internal entrepreneurs who will take the lead in stimulating corporate awareness of new technological opportunities and their efficient applications.

The biggest challenge for organizations is harnessing and fully utilizing the capability of employees. Therefore, recruitment, selection, training, proper placement, teaming, and motivation of employees have been priority issues for organizations. These activities assume even greater importance in the face of the continuously changing business environment. Alarmists tend to cast a shadow on the value of the fast pace of technological change by claiming that technology may replace people and threaten the social structure of society. The fact is that technological advances have always contributed to economic growth and the betterment of human lives. Technology may create a temporary displacement of a certain type of labor, such as manual labor, but people can be retrained to assume higher-level tasks, such as those requiring mental or service labor. What is needed to effect a smooth transition is a commitment by organizations to their employees. The U.S. *Economic Report to the President,* released in February 1994, asserted this fact:

> Since the dawn of the industrial revolution, alarmists have argued that technology and automation threaten jobs. Such claims are still heard today. But history shows that they have never been right in the past and suggests that they are wrong again. Time after time, in epoch after epoch and country after country, technological advance has produced higher wages and living standards, not mass unemployment. This is exactly what we expect to happen again in the 21st century. And the government should be helping this process along, facilitating growth and change, not impeding it.

This quotation underlines the overall benefit of technological progress from a national perspective. From an organization's perspective, special attention should be accorded to the following points.

**The Effects of Technological Change on the Skill Requirements of the Workforce** The introduction of new and advanced technologies into the workplace immediately results in different skill requirements. The magnitude and nature of the changes will be influenced by the economic sector and the type of industry involved.

**Matching and Training the Skilled Workforce to Meet the Requirements of New Technologies** Once the decision to adopt a new technology has been made, management must determine, before implementation, the skills necessary to run the new installations efficiently and effectively. Management must also develop operational plans to accomplish the transition with minimal disruption to operations and minimal adverse effects on the existing workforce. Reliable, perhaps industry-dependent, data can guide management in its decisions about how much change the workforce can handle and the types of reeducation, retraining, and relocation that are needed.

**Obsolescence of Professional Staff and the Continuing Need for Professional Development Activities** The growth in scientific knowledge and the escalating rate of technological change renders obsolete the training that professional staff acquired during their formal education or prior work experience. There is a growing need for continuing education for the professional staff. Reliable data and necessary strategies must be developed to determine how to meet the needs of the organization under various circumstances.

**The Role of Technological Gatekeepers and Internal Entrepreneurs** In view of the rapid nature of technological change, organizations must find ways to determine, choose, adopt, and implement appropriate technologies. The role of an organization's technological gatekeepers and internal entrepreneurs in the successful identification, implementation, and utilization of new technologies is essential and must be thoroughly understood.

**Social Consequences of Technological Change** Technology is the most important source of change in human experience. Its impact on our daily lives, socioeconomic structure, political system, and employment necessitates a thorough understanding of its implications and the development of reliable predictive models. Industry should determine what social-support structures within organizations, particularly high-technology organizations, exist or should exist to assist the following groups in coping with the demands of new or changing technologies:

- Working couples, single parents, or individuals with extended family obligations.
- Workers and professionals with changing or interrupted careers.
- Workers and professionals displaced by technology.

**Other Areas of Importance** Management should consider the implementation of:

**1** Reward and incentive systems for engineers, scientists, and internal entrepreneurs in corporations (e.g., evaluation and use of a "dual-ladder" reward system).

**2** Measures to facilitate the transition from technical specialist to technical manager.

**3** Measurement methodologies related to professional, human, and worker-machine interactions.

### Uses in the Analysis of Case Studies

The issues raised above can be used in the analysis and discussion of case studies of individual companies in order to extract lessons learned. In such analyses the steps listed here may be followed:

**1** Check the applicability of the issues to the case under study.

**2** Enumerate the issues that must be considered and prioritize them.

3 Review whether the company has a plan to deal with these issues.

4 Examine the methodology currently used in dealing with the pertinent issues.

5 Critically evaluate the existing methodologies.

6 If plans and methodologies do not exist, develop a plan of action and propose appropriate methodologies.

7 Propose an implementation procedure.

8 Indicate how you intend to measure results.

## MOT GUIDING PRINCIPLES FOR MANAGING ENTERPRISES

Recognizing the new paradigms for managing technology, Betz et al. (1995; see also Reading 4.1) presents eight guiding principles for the management of the modern enterprise. These are:

**1** *Value creation:* Value added constitutes the basic social responsibility of the enterprise. It is the key to long-term survivability of the enterprise.

**2** *Quality:* Quality is a fundamental requirement influencing competitiveness. Quality need not be thought of as a tradeoff with cost but as a hygiene factor. Organizations cannot sustain success without offering quality products or services.

**3** *Responsiveness:* An enterprise must manage not only for stability but also for change. It must be able to manage short cycles and respond to external environmental changes and customer demands promptly.

**4** *Agility:* A production facility must be flexible enough to (1) produce a variety of product lines and (2) facilitate communication and operation between suppliers, production, and customers. This may require changes within an organization's structures to meet changing demands.

**5** *Innovation:* A firm must be able to improve its ability to innovate and to use innovation to gain competitive advantage. Innovation may be relevant in a number of categories, including products, production, and services. Competing through technology is a fact of life today.

**6** *Integration:* A modern firm must be able to acquire and integrate a portfolio of technologies that will give it a unique and defined advantage over its competitors. The portfolio may include more than one generation of product or process technologies. Integrating all resources including technology, people, energy, information, and capital is essential for improving productivity and increasing effectiveness.

**7** *Teaming:* The complexity of integrating mixed technologies with varying life cycles requires a workforce with high levels of training. Workers must be able to work together in interdisciplinary teams to carry out and coordinate the operations of the enterprise.

**8** *Fairness:* A firm must develop a fair way to distribute to all its stakeholders the wealth created by a successful production operation. Fairness reduces conflicts among managers, labor, government, and the public. It leads to long-term survival of the enterprise.

The following chapters of this book address many, but certainly not all, of the issues emerging from the new MOT paradigms. The objective is to present underlying foundations that will help the reader establish a framework of thinking and will permit analysis of critical issues and synthesis of solutions. The concepts and methodologies presented are those with proven value. Case studies are used as a source of lessons and guiding principles. It is suggested that the reader supplement the material presented here with up-to-date articles published in professional journals and in business-related magazines.

READING 4.1

## Management Paradigms and the Technology Factor

**Frederick Betz, Kenneth Keys, Tarek Khalil, and Richard Smith**

Major changes in the world's economy now include: rapid technological change and diffusion, increasing technological complexity, new computer-based service technologies, and globalization of competition and markets. These require changes of the dominant paradigm by which the productive enterprise is managed. The topic of the management of technology (MOT) focuses upon how technological change can be managed to improve the competitiveness of the business enterprise. This focus is resulting in some major alterations in the management paradigm by which competitive enterprises should be managed. We will discuss how such paradigms have been altered by attention to the technological factor of the enterprise, and how a new paradigm for management is emerging.

### BACKGROUND: FACTORS IN THE BUSINESS ENTERPRISE

We recall that any productive enterprise involves the six principal economic factors: capital, labor, management, products, resources, and technology. Capital provides the financial capability of assembling the necessary factors of an enterprise into productive operations. Labor and management provide organizational capability and skills and knowledge to carry out productive operations. Products provide the outputs in the forms of goods, processes, and services that, through marketing and sales, replenish and grow the capital of the productive operations. Resources provide the materials and energy required to produce the products of the operations. Technology represents the knowledge, skills, techniques, and tools required for the transformation of the resources into the products of the operations.

We also recall that all these economic factors of the enterprise need to be managed, which requires attention to detail, responsibility, planning, implementation, supervision and discipline, and evaluation. For example, the management of capital requires raising capital, deciding on and controlling capital expenditure, and distributing the results of capital growth. Labor and management also require management, including organizing, staffing, evaluating, rewarding, and teaming. Products require management of design, production, marketing, and service. Resources require management, including acquisition, processing, assembling, safety, and conservation. So too does technology require management, including technology strategy and planning, research and development, and innovation in products, processes, and services.

And we recall that management problems on the economic factors of the enterprise arise from both complexity and integration. Each economic factor has its own complexity. For example, the forms of capital include cash, assets, inventory, debt, and equity. The forms of technology include technologies within products, technologies within production, technologies within services, and technologies within management activities. Integration of the activities concerning the different economic factors also creates problems. For example, decisions on the use of capital must be made on every other economic factor, e.g., capital decisions on organization, products, resources, and technology. As we approach the twenty-first century, we are continuing to witness very rapid and increasingly complex technological change. This pace and complexity have made technology a critical and central factor for being competitive in the global marketplace of interlinked economies.

What we need to consider is how complexity and integration of the technology factor have been affecting the changes in the paradigms of what we regard as "good" management practice.

## MANAGEMENT PARADIGMS AND TECHNOLOGY

We recall that in the industrial revolution in England during the eighteenth century, the technological innovation of steam-powered machinery resulted in altering the management paradigm of the textile industry. The industry changed from a productive cottage industry into a factory organization of production. This paradigm then involved owner-operator and labor in a factory situation (with labor directly supervised and paid on a time basis).

A second major shift in the paradigm of good management practice next occurred in the middle of the nineteenth century with the concept of interchangeable parts in product production.

A third major shift in the paradigm of good management practice occurred in the nineteenth century when Frank Church innovated the accounting practice for the costs of production (based upon direct labor and materials, and all else as overhead proportioned to the direct costs). The technology of the production operations provided the bases of the costing system.

A fourth major paradigm shift also occurred during that century with Frederick Taylor's concept of scientific management. His central idea was that all productive operations could be scientifically analyzed and optimized (into units of actions and sequences of unit actions). Management would analyze operations and specify procedures to be executed by labor. Here again, the technologies of production provided the bases for specifying organizational procedures.

However, one of the unfortunate consequences of Taylor's conception of the appropriate roles of management and labor was the loss of the contribution that labor can and should make to decisions on productive operations. A too-narrow concept of scientific management was used and contributed to poor management-labor relations in America and in England by encouraging managers to define themselves as the sole "brains" of the outfit and labor as only the "brawn." Even after Elton Mayo's studies demonstrated the inevitability of labor participation in decisions on production, the idea that workers' knowledge and cooperation were essential to production efficiency and quality was ignored by American management. (Only after the Second World War did Japanese management demonstrate the costliness of ignoring labor's necessary contributions to continually improving quality.)

A fifth major paradigm shift occurred in the twentieth century when in the 1920s Henry Ford introduced the concept of the assembly-line organization of production.

After the Second World War, a sixth major paradigm shift occurred from Japanese management's focus on quality of production. In addition, Toyota management led in a new attitude toward competition in a crowded market through improving profitability in low volumes of production. New techniques for flexible production, just-in-time production, continuous improvement of production, and rapid response were introduced.

Also after that war, another major paradigm shift has been occurring in management from the several applications of computers and communications technologies as service technologies. Some have called this shift "managing with technology." These computer-based service technologies are both external and internal. The external service technologies have altered the quality and quantity of services in banking, transportation, communications, and so on. The internal service technologies have affected accounting and control, engineering and manufacturing, and enterprise organization. For example, internally distributed computational and communications systems for management information and control are altering the hierarchical structuring of large organizations, leading to flatter structures and eliminating layers of middle management. In engineering and manufacturing, computers and communications are leading to virtual product development techniques and computer-integrated and agile manufacturing techniques.

In addition to these shifts, an eighth paradigm shift has been occurring in the second half of the twentieth century that is focused upon the deliberate creation of technological change and innovation. Previously, the economic benefits of technological innovation could be captured nationally as the new technology more slowly diffused throughout the world. After the Second World War, national policies have turned to the deliberate borrowing and developing of critical technologies on a national scale. The United States focused upon defense technologies and funded these with large-scale federal investments in R&D. Japan and Germany reconstructed their economies with national efforts focused on civilian technologies. Companies in these nations aggressively borrowed new technologies from other countries, and improved and implemented these in new industries. As a result, the pace of technology diffusion in the world has increased dramatically, resulting in a globalization of technology, production, and markets.

Now, management practices are directing more systematic attention to the pervasive effects of rapid technological change in mixed core-technology-based products, processes, and services. All modern products involve several technologies, both in the design of the product and in the production of the product. A critical technology is one which paces the rate of obsolescence of either the product or its production. Traditionally (even under all the above management paradigms), products and their production were managed by the pacing of a single critical technology (either in product or production).

For example, machine tools have traditionally depended upon two critical core technologies, metallurgy and mechanical machinery. However, with the advent of computer-controlled machinery, a third critical technology has been added, electronic control. The rates of change of technology in metallurgy and in mechanical machinery have been relatively slow and incremental. But, the rate of change of technology in electronic control has been fast and discontinuous. Therefore, the pacing critical technology for the recent and coming generations of machine tool products has been in electronic control. Moreover, in the next generations of machinery, another critical technology is being added in sensors for more feedback types of electronic control. Although metallurgy and mechanical machinery will continue to be critical technologies for the machine tool industry, changes in product lifetimes of machine tools will be paced by the rates of change in the other critical technologies of sensing and control.

Thus, another new shift in management paradigm is to the management of product change due to the complexities of multi-critical technology product lines. These require two important changes in management style:

1 Time regarded as a resource; and
2 Product-development interdisciplinary teaming.

## CONDITIONS FOR A NEW MANAGEMENT PARADIGM

Taken together, these paradigm shifts add up to the need for the articulation of a new paradigm for the enterprise, differing substantially from the older Church-Taylor-Ford paradigm, which emphasized:

**1** Reducing the direct costs of production should be the primary focus of management concern;

**2** Management should be primarily regarded as decision makers, and labor as passive followers of instructions;

**3** The operations of an enterprise could be analyzed as a stable set of unit operations;

**4** Production economies require large volumes of standardized products on assembly lines with fixed automation;

**5** Single critical-technology-based product lines will have long product lifetimes, providing long periods of stability in the organization; and

**6** World markets can be divided on a national basis, with national firms dominant in domestic markets.

These paradigm assumptions about management have now been negated from the combined impacts of the recent changes in the technology factor of the enterprise:

**1** Japanese management approaches to quality (*kaizen*),
**2** Computerization of the technologies for managing the enterprise,

**3** Product lines based upon multiple core technologies, and
**4** Global access to markets and rapid diffusion of technological progress.

A new paradigm would require: (1) managing the technological factors of the enterprise (managing technology and managing with technology), and (2) managing the enterprise for controlled change as well as stability. Together, these may be simply called MOT.

A MOT paradigm for management should require:

**1** The indirect costs of the enterprise should be reduced while improving competitiveness is viewed as a major challenge and opportunity;
**2** Decentralized, multi-functional, and multi-disciplinary enterprise teams should decide and operate the productive activities of the enterprise;
**3** The operations of an enterprise should be flexible, agile, and continuously improvable in successive states of quasi-stable production conditions;
**4** Production economies of scope are equally important with economies of scale, and production automation should be appropriately balanced between hard and soft automation (depending upon product volumes and product lifetimes);
**5** Multi-core-technology product lines will have shorter product lifetimes and should be planned as generations of products (paced by the most rapidly changing critical technology), and the organization must be flexibly organized for rapid and correct response; and
**6** World markets and technology are now global, and enterprises should be globally based to "think globally and act locally."

## A MOT PARADIGM FOR THE MANAGEMENT OF THE ENTERPRISE

There are several principles that can help focus management in a new paradigm to deal with both change and stability:

**1** Value creation,
**2** Quality,
**3** Responsiveness,
**4** Agility,
**5** Innovation,
**6** Integration,
**7** Teaming, and
**8** Fairness.

The basic social responsibility of the enterprise is to provide value-creating activities for society. The difference between legal and illegal enterprises is the concept that the products produced for society add value to society and do not subtract value. The failure in the twentieth century of the communist approach to economy was its inability to foster value-creating activities in the economy. The focus upon the nature of the enterprise's products (goods, processes, or services), as to how they add value for the customer and how that value may be increased, is the key to the long-term survivability of the enterprise.

In a highly competitive situation, the *quality* of product and production must be at least equivalent to (or better than) competitors at the same price. Quality and cost are not necessarily tradeoffs. There are several kinds of quality: quality of performance, quality of safety, and quality of production. A properly designed and produced product should be able to provide quality of safety and production at the same cost as lower-quality products. Both quality and cost leadership are necessary for long-term competitive success.

Responsiveness is necessary to the enterprise in a world of fast-changing conditions, and certainly the twentieth century was fast changing and the twenty-first century promises to be. There were rapid changes in technology, markets, competition, communications and transportation, national economies, resources, environment, and in global interrelationships. A modern enterprise must be able to manage not only for stability but also for change. To do this, the enterprise needs to manage short cycles in all of its operations, from product development and production to distribution and marketing.

To gain profitability from quick response to changing opportunities in the market, agility in production capability is necessary. One of the major costs of producing a product is investment in the production facilities, which then must be amortized over the product lifetime. Computer-based technologies in manufacturing are making possible (1) production facilities that are flexible enough to produce a variety of product lines, and (2) improved utilization of suppliers' production facilities and capability to produce the variety of parts required for agile production.

Because all technological capability eventually diffuses and all new technologies mature, no firm can continue to gain a technological competitive advantage without the ability to innovate. Otherwise, a firm has no advantage over competitors in differentiating its product or in having lower production costs and higher quality. Innovation produces the distinct competencies that provide products, production, and services that are superior to those of competitors. Innovation also provides the capability for managing change in an organization to complement the stability. For long-term survival, firms must manage both for stability and for change.

Most modern products, production, and services require several technologies as core to their design and production. A modern firm must be able to acquire and integrate technologies using different skill and knowledge bases in engineering and science. The pace of change in product lifetimes is dominated by the most rapidly changing of the critical technologies in the design or production of a particular generation of that product. Without the capability of continually integrating technologies, a modern firm cannot dominate competition products having rapid product cycles. Two kinds of technologies must be integrated: the technologies of products and production of the enterprise, and the technologies by which the enterprise is managed.

The combination of (1) the rapidity of technological change, (2) the complexity of integrating mixed technologies, and (3) creating high-quality, low-cost production requires that the firm hire and encourage a work force with high levels of training and continual upgrading of skills. Workers must maintain a professional attitude toward work and be able to work together as interdisciplinary teams in carrying out and coordinating the operations of the enterprise. The old antagonisms between management and labor are too inefficient and costly for a modern firm to be competitive. The flatter nature of organizations, the higher levels of skills required for production and marketing, and the need for cross-functional teaming for decisions and operations: all these require managers to develop a professional workplace.

Finally, the firm surviving over the long term must now create conditions of fairness in the type and distribution of the wealth created by the success of productive operations. The stakeholder theory of the firm as a successor to the older paradigm of owner/operator and labor is no longer a nice thing but a necessity. Executive teams, who have excessively controlled corporate boards to provide themselves with very large and disproportionate rewards from profits, have consistently over the long term been shown to have serious problems with: (1) maintaining a cooperative and efficient work force, (2) maximizing revenue and profit margin per employee, (3) maintaining share price to prevent hostile takeovers, (4) maintaining public credibility to prevent punitive government regulation, and (5) preserving the environment for future generations.

### SUMMARY

The major changes taking place in the world's economy require a new modification of the dominant paradigm by which the productive enterprise is managed. This modification must focus upon several issues, including the basic value-creating mission of business: producing high quality products with a fair and responsible distribution of the created wealth. It must include developing a professionalized work force, working as interdisciplinary teams, integrating complex technologies, and providing a responsive and agile production capability.

### THE TWENTY-FIRST CENTURY

More than a decade has passed since the publication of the NRC report, "Management of Technology: The Hidden Competitive Advantage" and the UM-NSF report on challenges and opportunities for research in the management of technology. These two reports focused on the perceived decline of U.S. industrial competitiveness in domestic and overseas markets. The decline was attributed largely to the inability of U.S. organizations to manage the available technology in an effective and timely manner. Participants in the debates of the 1980s predicted a paradigm shift in management and demanded a change in the way organizations manage their resources. The new field of management of technology was born and has grown steadily ever since. U.S. industry by and large adapted the changes necessary to restore its competitive advantage. A shift in management style is already under way in many successful organizations.

While the new paradigms are expected to continue to hold for the near future, the one thing that is certain is continuous change. What are the drivers for change in the field of technology management as the world enters a new century? Answering this question entails gazing into the future. It is important to evaluate the changing business environment in order to devise necessary responses that permit the sustainability of growth and prosperity. Therefore, in September 1998, under the auspices of the National Science Foundation, the University of Miami organized a two-day workshop to examine the industry and business-community demand for education in the emerging field of management of technology. The workshop was held at the NSF headquarters in Arlington, Virginia.

The workshop participants convened to examine new directions and future needs in MOT. Participants shared the common belief that technology has become the most pervasive factor influencing business competitiveness and societal changes.

The following issues were explored:

**1** What are the critical issues influencing the direction of MOT in the twenty-first century?
**2** What are the market needs for MOT education?
**3** What should be the content of MOT?
**4** Are existing programs meeting the demand?
**5** What needs to be done, if anything?

The backdrop for the discussion is a common recognition that the fast pace of technological change is forcing industry to change. It is also forcing a change in human resource development practices. Moreover, the globalization of industry and markets, as well as free-trade agreements, is creating major economic, cultural, and societal changes.

The educational system still distinguishes between technical education and management education. Educational institutions seem to have responded to the need for MOT education by continuing to compartmentalize education either in business or in engineering. Barriers to interdisciplinary education still exist in academic institutions. Applied and qualitative research seem to be less rewarded than basic and quantitative research, and cooperation with industry, though improved in recent years, is still quite marginal.

The workshop was intended to solicit industry's input and to clarify demand for MOT education. The approach is based on listening to the stakeholders. Attendance at the workshop was limited to 20 invited participants representing business and industry, educational institutions, and public and private groups that have direct interest or involvement in the topic.

The participants identified several factors that are perceived to be the drivers of change in the twenty-first century. Each of these factors is discussed below (Khalil, 1999).

## Technology

The rapid change in technology promises to intensify in the new century. The following changes are expected to occur:

**1** Technological complexity is expected to increase. This will require a higher level of human knowledge and skill to deal with complex technology, putting additional demands on human resource development. Technological complexity will also require multidisciplinary involvement. This, in turn, will necessitate cross-disciplinary, cross-cultural training with an emphasis on teaming, interpersonal relationships, and sensitivity to both technological and human issues.

**2** Technology fusion will be more pronounced. Technologies from one discipline will cross-fertilize technologies of other disciplines, thus enhancing the level of performance of technology.

**3** A diffusion of information and communication technology is under way and will continue into the future. The Internet is becoming the most influential medium for providing both raw and processed information. It is permitting two-way and multiway interaction and will foster collaboration between departments within an organization as well as between organizations.

Mergers and the acquisition of information- and communication-technology firms are expected to increase.

**4** The emerging technologies, particularly molecular biology and computer information technology, will have a major impact on industry and on all walks of life. Applications of genetic engineering, biotechnology, and nanotechnology are expected to change existing industries and create new markets. These emerging technologies are expected to find many applications in the health care industry, in agriculture, and in human and animal genetics.

Computer information technology has already shown its potency in increasing the productivity and effectiveness of organizations. When properly used, it can influence the competitive position of industry in all sectors of the economy. The use of computer information technology is fundamental to linking organizational functions. Information technology forms the base for managing in a technological environment.

The Internet is considered a watershed in the way organizations will conduct their business in the future. Not only has information become available on demand, but also the ways of using this new technology can change industrial practices globally.

**5** It used to be that technology life cycles were much longer than business life cycles. This created difficulty in synchronizing business and technology strategies. Today, however, technology life cycles for high-tech industries are becoming much shorter. This trend will continue in the future, enabling closer linkage and better harmonization of business and technology strategies.

### Changes in the Business Environment

The business environment of the twenty-first century will be more complex than it is today. The following characteristics are expected to emerge:

**1** There will be new types of stakeholders with whom management will need to contend. Global owners, other corporations with vested interests in specific technology, environmentalists, and global customers are but a few examples. Managers have to be ready to coordinate the involvement and differing interests of such stakeholders.

**2** Global alliances are expected to increase. Management, therefore, will need greater knowledge of international business practices. A more dynamic global marketplace will also be an important factor to contend with. Global competition and changing national fortunes will necessitate greater attention to international economic and political structures. In addition, because of the emergence of many small industries worldwide, organizations will need to conduct better monitoring of the environment to fend against competition and take advantage of opportunities through joint ventures, mergers, acquisitions, and outsourcing practices.

**3** The protection of intellectual property rights (IPRs) in a global environment with many participants and multiple partners will pose a challenge to corporate managers. Multinational, multicorporate teams will be involved in various phases of the innovation process. Structuring fair ownership agreements and enforcing IPRs constitute a new area of responsibility for businesses.

**4** Multiple parties will be involved in innovation, and multiple sources of financing will be arranged. Shared technologies and multiple-structured financial deals will assume added importance in a global environment.

**5** Technology will be a part of every aspect of business practice. "Technology imperialism" will invade all business disciplines. Finance, marketing, innovation, engineering, public relations, and business law will all use technology as a common basis of competition.

**6** The growth of entrepreneurship and of small and medium-size enterprises will be a hallmark of the new century. The consolidation of mature technology firms into huge international conglomerates will also intensify.

### Communication, Integration, and Collaboration

Organizations will work to improve communication, strive for integration, and move toward more collaboration than ever before. Expected changes include the following:

**1** There will be a change in organizations' cultures. Organizations that have not yet done so will move toward introducing a technology culture. Organizations will also become more sensitive to global needs.

**2** Closer interaction and partnering with customers, suppliers, and distributors will be needed. Mass customization will replace mass production. Consumers are becoming more sophisticated, and they demand more attention and better-quality products and services. Companies must integrate customer needs and demands into the products and services they offer.

**3** Greater collaboration among companies is expected. This implies more global alliances, more joint R&D and production projects, and more use of supply-chain management where suppliers, production, and distributors are better-integrated. Outsourcing instead of vertical integration will increase. Information technology and logistics will permit these changes to happen more smoothly.

**4** Collaborative efforts between business and government is expected to increase. Cooperation will replace the hands-off, adversarial, or policing role of government in relation to industry.

### Strategic Directions of Industry

Companies are likely to appreciate the role of strategic planning in gaining a technological edge. The strategic planning process must recognize intellectual capital and tacit knowledge. The following projections are made:

**1** Corporations will need more sophisticated strategic thinking. Both long-range and short-range planning efforts will be of utmost importance in order for organizations to succeed.

**2** Linking technology and business strategy will assume greater importance.

**3** The old management paradigms are collapsing. The concept that management knows what needs to be done will not be as applicable in the future as it was during the last century. Learning by doing is more applicable in view of the changes seen in technology.

**4** Industry will lean toward distributing its effort into smaller entrepreneurial ventures to take advantage of technological innovation without jeopardizing existing lines of business.

### Changes in Organizational Structure

The structure of organizations is expected to change. The following points are pertinent:

**1** Temporary organizations are expected to increase as permanent organizations decrease. A *temporary organization* is created to take advantage of a specific need or special technology and then disbanded when the life cycle of the technology ends.

**2** Virtual corporations will increase to take advantage of technology in logistics, transportation, and globalization to satisfy customers' needs. A virtual corporation requires no factory, few employees, small capital and minimal physical assets. It relies on networking suppliers with customers using information technology and electronic media.

**3** The vertical, hierarchical organizational structure will yield to horizontal and matrix organizational structures. Shallower organizations with fewer layers of management will be more popular than multilayered ones. The line and functional-staff organization, known as the "citadel model," will yield to a more interactive network model.

**4** Greater integration within the organization will be established. The use of intranet technology will help organizations integrate better.

**5** Knowledge will drive innovations in products, production, distribution, and marketing, as well as in the creation of a new supply-chain strategy. New types of organizations which value change will be formed. They will depend on the use of computers, information technology, and logistics to integrate their business. They will offer services varying from knowledge generation to offering consultations in total integrated solutions.

**6** There will emerge a number of organizations that concentrate their business on generating and distributing knowledge. Nonphysical production will assume greater importance than physical production.

**7** National cultures will dictate the presence of culture-specific organizations to design for and serve the needs of specific cultures.

### Financial Sector Structure

There are several issues facing the financial community that are related to investment in the technology sector. The following points are relevant:

**1** New and emerging technologies need financial resources. The financial sector must devise new innovations to fund venture projects.

**2** Venture capitalists will require better methodologies to assess the merits and potential of technology projects.

**3** Multiple parties will be involved in innovation projects. They may include several companies from various countries. Structuring deals for multiparty financing of innovation will be needed. The sharing of costs and benefits must govern such deals' structure.

**4** The valuation of intellectual capital and the wealth-creating capacity of high-tech workers and emerging information industry companies will pose serious problems for financial analysts and economists. New methods of economic forecasting and valuation are needed.

**5** National funding of projects, as well as public policy, are currently very ethnocentric and reflect global parochialism. This is particularly true in regard to the funding of projects of global interest and implications, such as global warming or earthquake disaster mitigation. This nationalistic attitude needs to be tempered so that policy and funding favor the global benefits expected from a project.

**VALUATION OF INTERNET COMPANIES**

The stock market performances of many emerging Internet companies have been the subject of heated debate in financial and academic circles. A case in point is Priceline's price.com, a new web-site venture that lets consumers bid, via the Internet, on products or services such as a car or an airline ticket. Transactions occur when a match is found between a seller and a bidder's price. After 11 months in existence, Priceline lost $114 million selling $35 million of airline tickets. Despite this performance, Priceline's shares jumped from the initial offering of $16 a share to a high of $85 on the same day. Priceline's market value reached $9.8 billion, surpassing the combined worth of United Airlines, Continental Airlines, and Northwest Airlines. This is an astounding value when analyzed by conventional financial and economic measures.

In other, similar valuations, relatively new technology-based companies have surpassed well-established powerhouses. For example, America Online is worth more than ABC, CBS, and NBC combined. Yahoo is worth more than the *New York Times,* and Amazon.com is worth more than Barnes & Noble and Borders combined.

### Education and Training

The current educational system structure, particularly at the university level, is viewed as rigid and unable to change to meet the demands of the changing world environment. The following points deserve attention:

**1** The existing system of education is seen as hindering the rate of progress achieved in technology and economic growth. A change in the educational model is needed.

**2** A change in formal education style is required. A style that fosters free thinking, creativity, innovation, and interdisciplinary flow, rather than compartmentalization of thoughts, is needed.

**3** A change in the mode of education delivery is also needed. Advances in communication and multimedia technology, including the Internet, offer new ways of teaching and delivering course materials to students.

**4** New disciplines, such as knowledge infrastructure engineering, should emerge. Transcultural engineering, which involves designing for cultural acceptance, should also be stressed.

**5** Multiple degree seekers will emerge, where no manager will have only a single degree in the future. The explosion of knowledge will require life-long learning experience. There will be a need for more technological and managerial education for all managers.

**6** Higher education is lagging behind industry in making the changes necessary to meet the challenges of the twenty-first century. This presents a current challenge to policy makers and knowledge drivers. Institutions of higher education need champions to implement changes. Virtual universities and electronic teaching methods will increase in number, size, and popularity. Virtual laboratories will be created. The challenge is how to harness the new technology in education to optimize learning and deliver quality, cost-effective education to the public at large, as well as how to satisfy industry's needs for a highly skilled, technologically literate labor force.

## CONCLUDING REMARKS

The past two decades have witnessed major changes in the business environment. These changes have influenced the standing of individual firms as well as the national economies of many countries. Competition intensified and a paradigm shift in the management of companies and of technology emerged. Concerned groups of stakeholders in the United States and throughout the world started to focus on issues that deserve attention by business managers, public policy makers, and educators. A number of workshops and conferences were held to debate the issues. They concluded that a significant amount of effort must be devoted to improving our ability to manage business enterprises in a world experiencing a technological revolution.

This revolution has an impact on every aspect of life. The continuous change of technology is forcing changes in business practices and organizational structures. Therefore, drivers and motivators for change must be understood. Strategies to deal with the impact of changes need to be developed. In this chapter we clarified the new paradigms of business. In subsequent chapters we provide information on technology management approaches and practices commonly used to achieve business success.

## DISCUSSION QUESTIONS

1 Discuss the effect of the fast pace of technological change on human resources.
2 There are several emerging industrial practices that have been gaining wide acceptance in recent years. One example is the virtual corporation; another is supply-chain management. Find a recent article on each subject and discuss the concept.
3 Read a book chapter or an article on the accounting methods of return on investment (ROI) and activity-based costing (ABC). Discuss the uses and limitations of each in regard to a technology development project.

## ADDITIONAL READING

National Research Council. 1987. *Management of Technology: The Hidden Competitive Advantage.* National Academy Press, Washington, DC. Report No. CETS-CROSS-6.

## REFERENCES

American Association of Engineering Societies. 1988. *Management of Technology: The Key to America's Competitive Future.* AAES, Washington, DC.

Betz, Frederick, & Keys, Kenneth. 1995. "Management Paradigms and the Technology Factor." *Technology Management,* vol. 1, pp. 242–246.

Khalil, Tarek M. 1999. *Management of Technology: Future Directions and Needs of the New Century,* UM/NSF workshop report, University of Miami, Miami, FL.

Khalil, Tarek M., & Bayraktar, Bulent A. 1988. *Challenges and Opportunities for Research in the Management of Technology.* UM/NSF workshop report, University of Miami, Miami, FL.

Khalil, T., & Bayraktar, B. (eds.). 1990. *Management of Technology II: The Key to Global Competitiveness.* Industrial Engineering and Management Press, Norcross, GA.

National Research Council. 1987. *Management of Technology: The Hidden Competitive Advantage,* National Academy Press, Washington, DC. Report No. CETS-CROSS-6.

# 5

# TECHNOLOGY LIFE CYCLES

The performance of a technology has a recognized pattern over time that, if properly understood, can be of great use in strategic planning. As a matter of fact, neglecting this pattern as a key factor in the planning process may prove very costly to the competitive position of a corporation. Managing technology requires deep understanding of the life cycles of the technology, product, process, and system.

## THE S-CURVE OF TECHNOLOGICAL PROGRESS

A technology's improvement of performance follows the S-curve. When a technology-performance parameter ($y$ axis) is plotted against time ($x$ axis), the result resembles an s-shaped diagram called the *S-curve.* Technological performance can be expressed in terms of any attribute, such as density in the electronics industry (number of transistors per chip) or aircraft speed in miles per hour.

As can be seen in Figure 5-1, technology progresses through a three-stage *technology life cycle (TLC):* (1) the new invention period, also known as the embryonic stage; (2) the technology improvement period, also known as the growth stage; and (3) the mature-technology period. The technology becomes vulnerable to substitution or obsolescence when a new or better-performing technology emerges.

The new invention period is characterized by a period of slow initial growth. This is the time when experimentation and initial bugs are worked out of the system. The technology improvement period is characterized by rapid and sustained growth.

The mature-technology period starts when the upper limit of the technology is approached and progress in performance slows down. This is when the technology reaches its natural limits as dictated by factors such as physical limits. For example, the vacuum-tube technology was limited by the tube's size and the power consumption of the heated

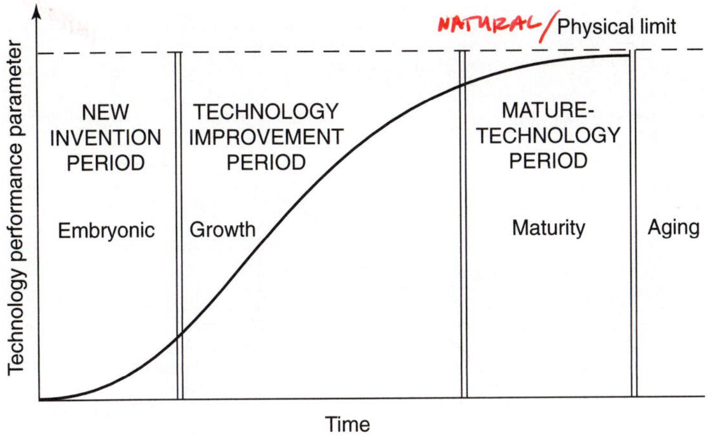


**FIGURE 5-1**
THE S-CURVE OF TECHNOLOGICAL PROGRESS

filament. Both of these factors were natural barriers to electron conduction in a vacuum tube. Electronic engineers could not overcome these limitations. The arrival of the solid-state technology, or transistor, which permitted electron conduction in solid material, changed the physical barriers of size and power. The transistor technology started a new technology life cycle and rendered the vacuum-tube technology obsolete.

This example illustrates a very important concept in MOT: *When a technology reaches its natural limits it becomes a mature technology vulnerable to substitution or obsolescence.*

The S-curve of technology progress is a very useful model in technological forecasting, as shown by Fisher and Pry (1971). A more detailed discussion of technological forecasting is presented in Chapter 9.

A technology's rate of performance improvement is dependent on the effort devoted to its development. As shown in Figure 5-2, a technology may progress on curve A or A′, depending on a number of factors, including the type of the technology itself and the cost and time devoted to its development. A newer technology (B) has a higher limit of performance for the same parameter. It may progress at a faster rate and will influence the progression of the older technology. At a certain point in time it will replace the earlier technology (A). An example is ceramics, which have higher operating temperatures and substitute for metals used in internal combustion engines; the newer technology permits better performance of the engines. The performance of the engines can continue to improve as a result of a sequence of newer technologies, each with a higher limit of the performance parameter of interest.

### The Technology Life Cycle and Market Growth

When technology reaches the market, it generates income. Technology under development has no real income-producing value. Technology on the shelf (i.e., not being

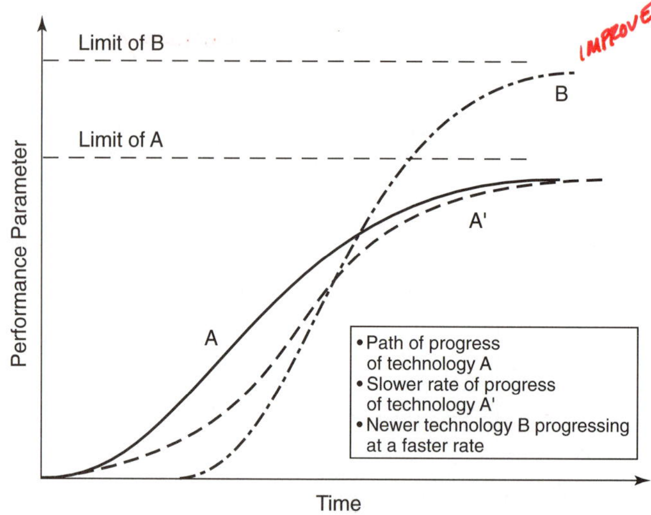


**FIGURE 5-2**
CHANGES IN NATURAL LIMITS OF TECHNOLOGY
The rate of performance improvement is shown for two technologies.

marketed) provides no return. As technology develops, following the recognized technology life cycle, market penetration occurs and so does market growth, expressed as market volume. Figure 5-3 shows the market-growth pattern at different phases of the technology life cycle. The *x* axis represents time and the *y* axis represents the market volume expected at six technology phases: (1) technology development phase, (2) application launch phase, (3) application growth phase, (4) mature-technology phase, (5) technology substitution phase, and (6) technology obsolescence phase.

During the technology development phase the market does not recognize the technology at all; it has zero response. However, this is the important period in which scientists and engineers are spending significant amounts of effort and money to create the technology, develop prototypes, and test the new technology. The goal of any R&D manager should be to reduce this time period as much as possible, since it is very expensive and does not produce revenue.

Once the first wave of the new technology application is launched into the market, the market volume follows the path of technological progress. This is characterized by slow initial growth during the launching period, followed by rapid growth.

During the growth phase of the technology, penetration into the market will depend on the rate of innovation and the market needs for the new technology. The growth rate slows down as the technology approaches its maturity. At some point, the market volume will peak and then start to decline. This will happen when the technology matures and enters its substitution phase. Companies that continue to use the old technology in this phase will be faced with a shrinking market share and a fall in revenues. The final phase is technology obsolescence, during which the technology has little or no value.

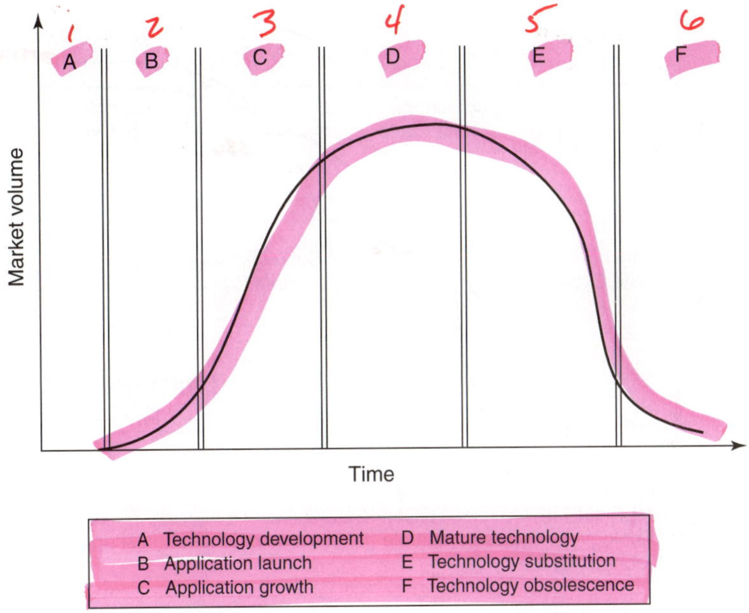


**FIGURE 5-3**
MARKET GROWTH AT DIFFERENT STAGES OF THE TECHNOLOGY LIFE CYCLE

## MULTIPLE-GENERATION TECHNOLOGIES

Technology, like all systems, has a hierarchy. A system can consist of a number of subsystems, and each subsystem may have a number of components.

Technology need not consist of a single component or derive from a single innovation. Technology can consist of multiple technologies and derive from different generations of innovation. The personal computer is a technology and has a technology life cycle. It consists of several subtechnologies. One such subtechnology is the microprocessor, which can also be defined as a technology with a technology life cycle all its own. In turn the microprocessor has its own multiple-generation technologies or subtechnologies. For example, the microprocessor technology developed by a company such as Intel has undergone several generations of changes (8088, 286, 386, 486, Pentium). Each of these generations of innovation helped boost the technology life cycle of the microprocessor and, in turn, that of the PC. (See Figure 5-4.)

The same concept applies to software technology. Any software developed for a major application undergoes several generations of change. The changes improve the software and extend its useful life. If a company developing software stops its development after one generation and another company continues to develop new generations, the former will find itself unable to compete with the latter's newer-generation technology. Another situation occurs for a company that is making an acquisition investment in the software. If it buys one generation of software and an update is introduced, the new version has

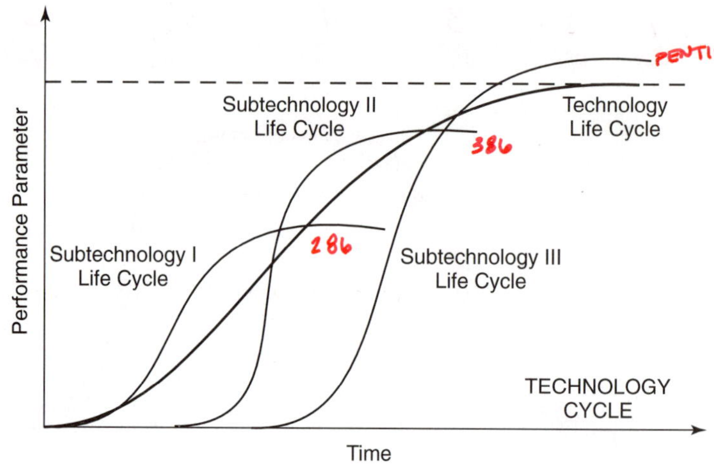


**FIGURE 5-4**
MULTIPLE-GENERATION TECHNOLOGIES
Subtechnology life cycles in multiple generations of innovation shape the overall technology life cycle.

more capabilities and extends the application of the software. The company may have to invest to update its software in order to extend the life cycle of its software technology.

## TECHNOLOGY AND MARKET INTERACTION

A very strong dynamic relationship exists between technological innovation and the marketplace. The presence of a market or the creation of a new market represents the reward for technological development. Breakthroughs in technology open new vistas for industrial developments and economic growth. However, it is only when technological developments find a market that scientific research pays off and the development cost is reimbursed in economic or social terms.

### Science-Technology Push

Events of the nineteenth and twentieth centuries provide many illustrations of how science and technology have become closely intertwined. Most of the recent technological breakthroughs are based on earlier scientific discoveries. Science provides the base for technological development, which in turn creates new markets. Bayraktar (1990) cites several examples of technologies that owe their bases to scientific discoveries, as shown in Exhibit 5-1: The field of electronics is based on Maxwell's theory of electromagnetism; nuclear energy is based on Einstein's 1905 paper, which established the famous $E = mc^2$ equation; the transistors are based on A. H. Wilson's 1931 paper on the theory of semiconductors; and genetic engineering followed the discovery of the structure of DNA by Watson and Crick in 1952. In this sense, we can say that science provides the base for the technological push. Figure 5-5 illustrates this concept. Innovations that ensued from the technologies in Exhibit 5-1 caused major industry upheavals and totally

**EXHIBIT 5-1**
TWENTIETH CENTURY TECHNOLOGIES AND THEIR SCIENTIFIC BASE

| Technology | Scientific discovery |
|---|---|
| Nuclear energy | Based on Einstein's 1905 paper, which established the equivalence of mass and energy |
| Transistors | Based on A. H. Wilson's 1931 theory of semiconductors |
| Electronics | Based on Maxwell's theory of electromagnetism, developed in the 1880s |
| Genetic engineering | Followed Watson and Crick's 1952 discovery of the structure of DNA |

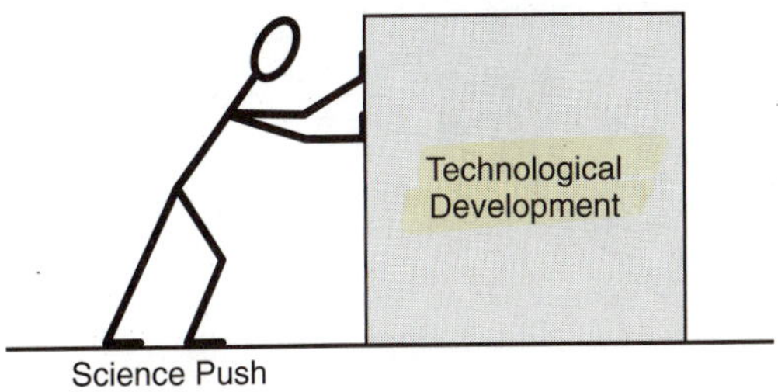


**FIGURE 5-5**
SCIENCE/TECHNOLOGY PUSH
The push opens new vistas for industrial development and economic growth.

changed the markets. They brought major economic growth. Radical innovations of products within a technology area create similar effects. An example of a radical innovation that created a major change in the way we do business is xerography. When the Xerox machine was developed, it was dubbed an invention with little promise and a product concept without a market (Mort, 1990). Observe where this copying industry is today. Radical innovations create new markets and expand existing markets.

### Market Pull

Technological development is also stimulated by market pull. Technology is often developed to meet a market need or demand. This is the most effective way to connect technology with the market. However, in the majority of cases, market pull is stimulated by consumers. Consumers may or may not know whether a new technology exists or is being developed, or if they do, they may not understand the technology. Most of the technological developments stimulated by market pull are of an incremental nature, or represent improvements to existing technologies. Incremental technological improvements have a cumulative effect, and they can have a tremendous impact on productivity and competitiveness. When there is a strong collective demand for a solution to a specific problem (such as a vaccine for AIDS), market pull may provoke major breakthroughs. Figure 5-6 illustrates this concept. Both mechanisms, push and pull, contribute to stimulating innovation and technological change. Integrating them accelerates the change (Figure 5-7). Munro and Noori (1988) proposed that commitment to technology adoption is dependent on an integrative approach to technology push and market pull combined with management's attitude toward technology and the firm's technical and

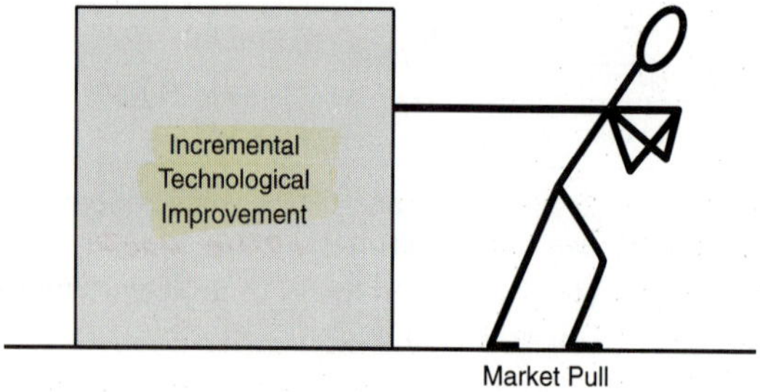


**FIGURE 5-6**
MARKET PULL
The pull stimulates incremental technological improvements.

**FIGURE 5-7**
COMBINED EFFECT OF TECHNOLOGY PUSH AND MARKET PULL

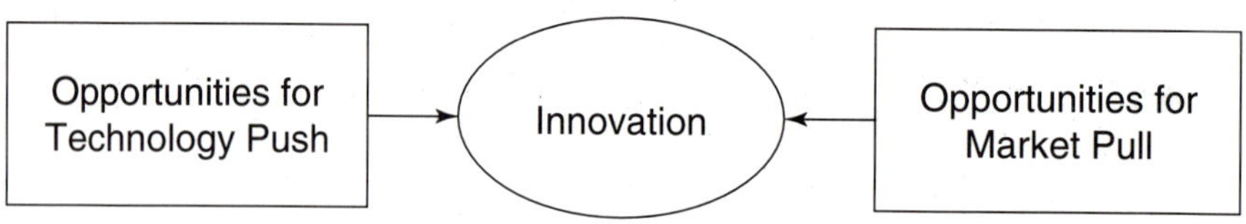


- Scientific discoveries
- Applied knowledge
- Recognized needs
- Intellectual capital (scientists and engineers)

- Market demand
- Proliferation of application areas
- Recognized needs
- Opportunities for increased: profitability, quality, productivity
- Entrepreneurs

**FIGURE 5-8**
INTEGRATING TECHNOLOGY PUSH AND MARKET PULL TO STIMULATE INNOVATION

financial resources. Figure 5-8 depicts how opportunities for technology push and market pull can be integrated to stimulate innovation.

### THE PRODUCT LIFE CYCLE

A product life cycle closely resembles the profile of the technology life cycle and its associated market-growth profile (Figure 5-9). A product emerges from a concept, which is translated into an engineering design and usually illustrated through an engineering drawing. A prototype is developed and tested to make sure that the product specifications are met and the performance parameters achieved. In this initial design-and-prototype-development phase, the product has not yet met the market and has no wealth value to the company.

The second phase is the product-launching phase, followed by the growth phase, whose profile depends on the market response to the product. Typically, sales start slow and then accelerate as the product becomes known and accepted in the marketplace. As the product is diffused in the market and the market becomes saturated with a well-established mature-technology product, the growth rate is likely to slow down. New products threaten mature-technology products and may substitute for them and eventually render them obsolete. Obsolete products have little or no monetary value. They may be recycled, placed in museums, or kept as collection items if they possess aesthetic or appealing characteristics.

When scientific and engineering advances lead to the introduction of new technology, turbulence is created in existing systems. New products emerge in the embryonic phase of a technology and many product innovations occur. As the rate of product

**FIGURE 5-9**
PRODUCT-MARKET LIFE CYCLE

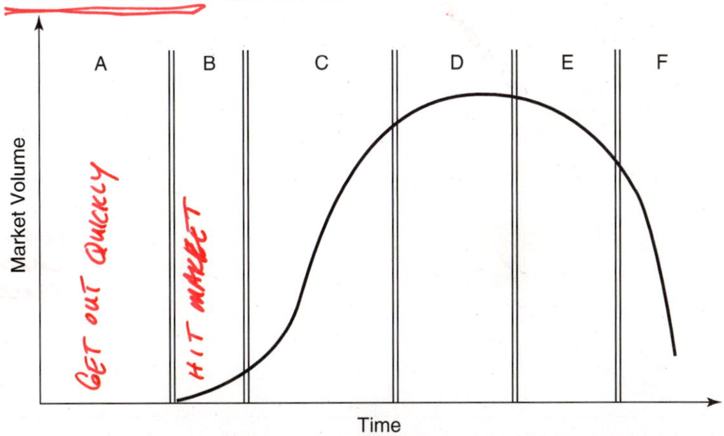


| | |
|---|---|
| A Concept Design Prototype | D Mature Stage |
| B Product Launch | E Substitution Products |
| C Product Growth | F Product Obsolescence |

innovation reaches its peak and starts to decline, a dominant product design emerges and the industry standard is defined accordingly (Figure 5-10). Process innovation follows new product designs. It continues throughout the technology life cycle in support of both radical and incremental product innovations. Process innovations are important for the different generations of products. Process innovations increase a product's life cycle and help maintain competitiveness until a substitute technology creates a discontinuity in the system and a new life cycle emerges. For example, switching from steam-powered engines to diesel-powered engines creates turbulence in the diesel technology and discontinuity of the steam technology. The diesel technology will have its own products, which go through different designs until an industry standard emerges and dominates the market. Process innovation continues to create improvements in the performance of the dominant design until a new technological discontinuity occurs, such as an electric-powered engine. The electric technology may render the diesel technology obsolete. The product and process innovation of electric products will run their cycles until another discontinuity occurs, perhaps hydrogen-powered engines.

For a single product, the technology life cycle and the product life cycle coincide. Technological discontinuity ends one product's life cycle and starts a new product life cycle. Technological discontinuities used to be few and far between. In the technology age this is no longer the case. The digital age, for example, has created very rapid rates of innovation for components and products. A microprocessor's design and manufacturing process change almost on a yearly basis. Software is changing at a faster pace. The product life cycle is certainly much shorter than it was in the nineteenth and twentieth centuries.

**FIGURE 5-10**
TECHNOLOGICAL PROGRESS
The progress of technology is shown in relation to product and process innovation.

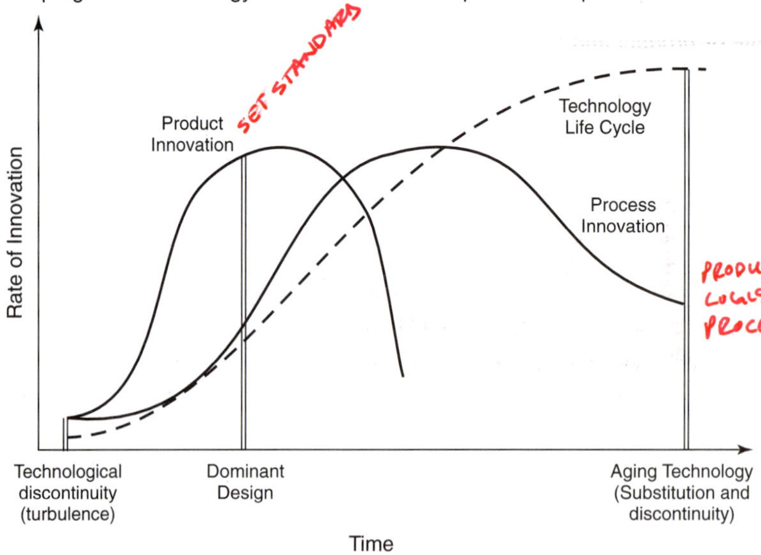

## COMPETITION AT DIFFERENT PHASES OF THE TECHNOLOGY LIFE CYCLE

In the early stage of the technology life cycle, also known as the embryonic or emerging-technology stage, competition is based on innovation. In this stage, the technology is still developing and has not been fully accepted. Companies depend on their innovation to add value to products and services they bring to their customers. The introduced technology has not yet demonstrated its potential for changing the basis of competition.

In the early phase of the growth stage of the technology life cycle, the introduced technology helps expand the market size for the product or service offered. The technology becomes a pacing technology in that it has the potential for changing the basis of the competition. In this stage a company must be able to balance its growth strategies with its marketing strategies. Attention to growth must not distract the company from continuing innovation. The Osborne Computer Company case discussed in Chapter 3 illustrated this fact.

Once the innovation has proved itself in the market, it permits its owner to take a patented position or to define the industry standard. A dominant design of the product emerges, and the technology has a major impact on the value-added stream of performance, cost, and quality. Technology in this phase of the growth stage is known as *key technology,* and a company should increase its capabilities in this area to compete.

When the technology reaches a stage of maturity and the rate of innovation declines, it becomes a commodity, available to all competitors. Technologies in this category are also recognized as *base technologies* and have little ability to give a company a strong competitive edge.

### Competition with Product and Process Innovation

The rate of product and process innovations follow a general pattern, as shown in Figure 5-10. This pattern can be used to formulate policies and procedures to better manage the process of technological innovation.

When a new product or process is introduced to the market, it creates certain energy within the innovation community, triggering a series of changes to the product or process. Over time, the rate of innovation of new products or processes increases, reaches a plateau, and then decreases, creating the inverted U-shaped curve shown in Figure 5-10. At the early stages of product development, competition in innovation and improvement delays agreement on a standard design. A leader in innovation has the opportunity to set the standard. A company should strive to be in such a position because once a dominant design is established in the market by another company, it will be too late for the company to set a different industry standard based on its own product. It may have to settle for being a follower, in which case it will have to develop another strategy to obtain a leading position in the marketplace. One approach is to rely on process innovation to reduce cost. Another is to rely on complementary assets, such as name recognition, to increase market share. Yet another is to use marketing innovation and improve customer service to lure customers away from competitors.

### Competition in Mature Technology

As the technology approaches the maturity stage, the rules for competition change, as follows:

**1** The competition switches from being based on innovation to being based on price and quality.

**2** Process innovations tend to dominate, and they assume greater importance in achieving a competitive edge.

**3** Companies compete by introducing product lines into segmented markets.

**4** Companies rely on economy of scale to reduce price.

**5** Specialization and production efficiency within companies assume greater importance.

**6** Only firms with dominant markets tend to survive. This favors large companies. Mergers and acquisitions of companies assume greater importance in companies' strategies.

**7** Large organizations with mature technology tend to be rigid, bureaucratic, and multilayered. Such a structure often impedes innovation and is a threat to sustainable success.

**8** Companies with mature technology become subject to increased competition by those who have lower production costs, lower labor rates, or lower overheads. This introduces international competition as a major factor.

**9** Mature technology is continuously threatened by substitution of newer technology. Management must be alert to emerging or competing technologies.

A company's success in introducing a product innovation gives it a leading edge but does not guarantee sustained competitive advantage. A company that leads with product innovation, establishes the industry standards, and follows through with incremental and process innovation can sustain success. It is important to maintain control over products and their domination of the market throughout the product life cycle. It is also important to take a proactive approach to developing or dealing with technological disturbances. Migrating to the emerging technology in a timely manner keeps a company's products competitive.

Managing technological innovation requires that an organization continue to introduce incremental innovations and forecast future changes in order to ensure continued existence in the face of discontinuous innovation. Companies that have been able to do this successfully are 3M, General Electric, Sony, and Microsoft. These companies compete with innovation and work hard to be leaders in technology (case studies are presented in later chapters).

## DIFFUSION OF TECHNOLOGY

A technological innovation, a new idea, or a new system is considered to be successful when it is adopted by users and diffused through the user population. *Diffusion* is the process by which an innovation is communicated, over time, through certain channels to members of a social system (Rogers, 1995). The term "innovation" is frequently used in the diffusion literature as being synonymous with "technology." Adoption of a certain type of technology is usually based on the possible efficacy of that technology in solving a perceived problem. Information about an innovation reaches a potential adopter through communication channels. There are many channels for communicating new ideas to potential users, including interpersonal channels and mass media. The rate of adoption of an innovation by members of a social system is dependent on the following factors:

**1** *The degree to which the innovation is perceived to be offering better advantage than does existing practice:* An example is an innovation that offers a less expensive method of producing a product.

**2** *The degree to which the innovation is compatible with the values and needs of the users:* An example of an incompatible innovation is a new product that may produce pollution in an environmentally sensitive community.

**3** *The degree to which the innovation is considered complex and difficult to use:* An example is a new process that requires a great deal of effort in retraining employees and has a high cost of implementation.

**4** *The degree to which the innovation can be introduced on a trial basis before users must fully commit to its adoption:* An example is a new drug that physicians can use on a limited trial basis before prescribing it to all patients. Free samples of drugs given to physicians permit them to do so.

**5** *The degree to which the innovation is seen, and its results are observed, by potential adopters:* An example is a small satellite dish for television viewing. As people see it in use and observe their neighbors' satisfaction with its performance, they are more likely to be willing to use it.

Innovations that are perceived by individuals as having greater relative advantage, compatibility, and less complexity and that can be tried and observed will be adopted more rapidly than other innovations (Rogers, 1995). An example of the diffusion curves of two technologies is shown in Figure 5-11.

**FIGURE 5-11**
DIFFUSION CURVES

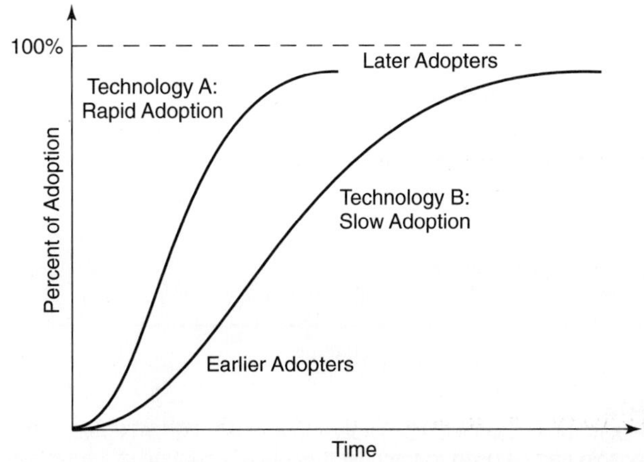

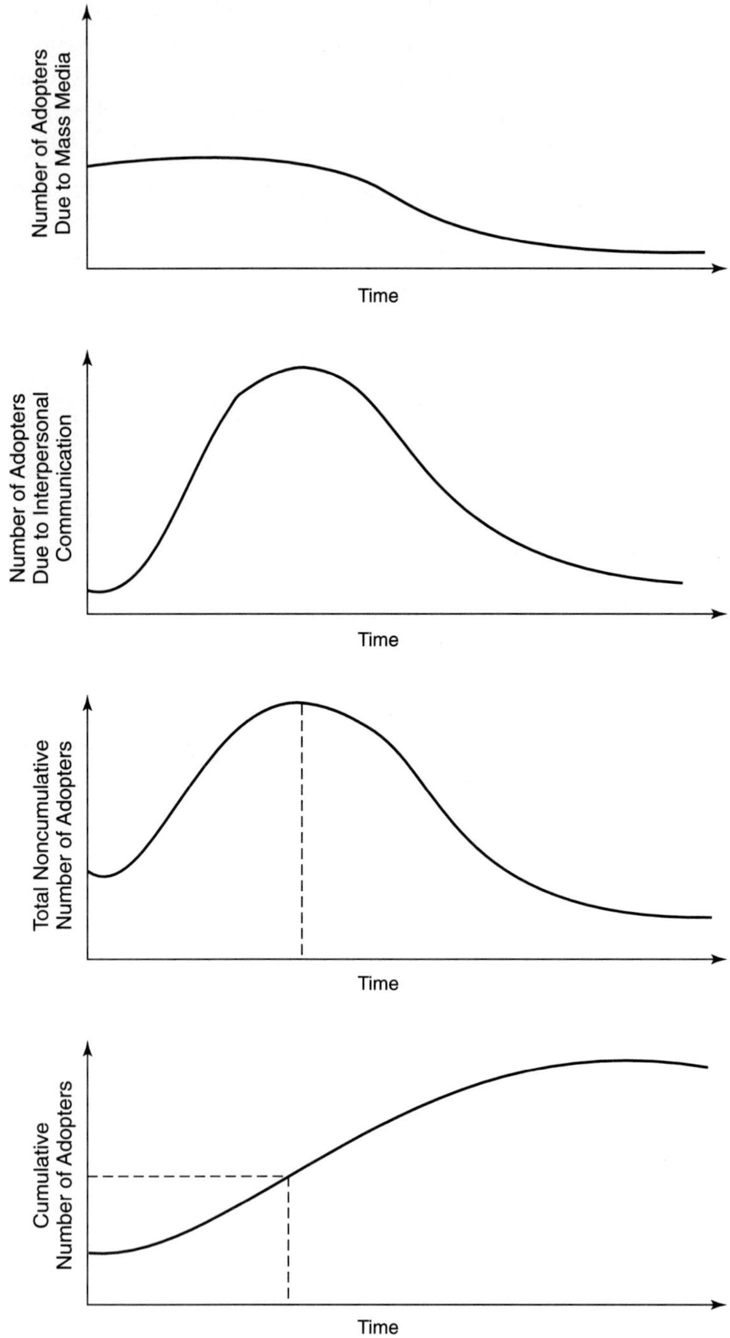


**FIGURE 5-12**
THE DIFFUSION–COMMUNICATION-CHANNEL RELATIONSHIP
This relationship can be used to forecast the rate of adoption of innovation.

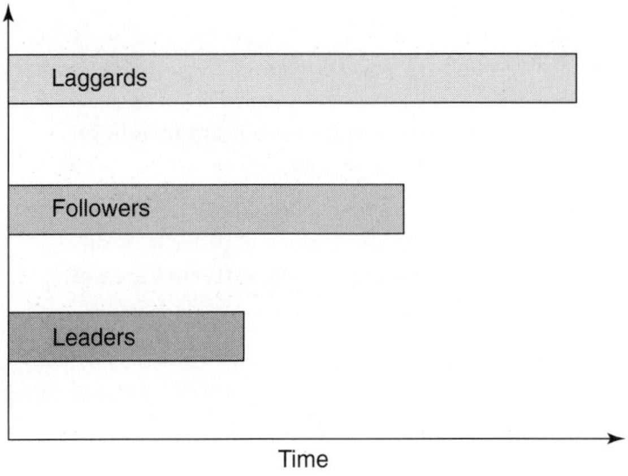


**FIGURE 5-13**
INNOVATION DECISION TIME

## THE DIFFUSION–COMMUNICATION-CHANNEL RELATIONSHIP

Mahajan et al. (1990) suggest that adopters of an innovation are influenced by two types of communication channels: interpersonal word of mouth and mass media channels. Mass media influence is greatest in the early phase of diffusion but occurs continually throughout the diffusion process. In contrast, the number of users who adopt a new innovation as a result of interpersonal communication expands during the early phase of the diffusion process and declines during the second half of the process. This behavior results in a cumulative S-shaped diffusion curve (Figure 5-12).

The decision to adopt an innovation by an individual or an organization takes a certain period of time and consists of several stages. It starts with gaining knowledge of the innovation, forming a favorable opinion about it, making the decision to adopt it, implementing the innovation, and following up on its performance. Innovative organizations that are considered technology leaders require a shorter time period than others to go through the innovation-decision process. Followers take longer to effect the same process, and laggards take much longer to make a decision for technology adoption (Figure 5-13).

## DISCUSSION QUESTIONS

1 Select a technology you are interested in (car or plane speed, computer power, screen resolution, etc.). Look for performance parameters and plot them against the years they were launched. Use the data to obtain a regression equation. (You don't need to be a statistics expert; MS Excel can do that with a click of the mouse.) Can you predict future technological developments?

2 Select a company you want to study. Read articles about the company in business magazines, trying to understand its origins, latest achievements, and problems. (*Fortune* magazine and *Business Week* usually provide good background information; corporate Internet sites are also a good source for this purpose.) How have the elements in Figure 5-7 contributed to the firm's performance? Explain how Figure 5-11 applies to the case of the company you selected.

## ADDITIONAL READINGS

Theodore Levitt. "Exploit the Product Life Cycle." *Harvard Business Review,* November–December 1965.

This apparently old-fashioned but classic article provides very good insight on strategic actions to be taken on the basis of market maturity.

Everett M. Rogers. *Diffusion of Innovation,* 4th ed. Free Press, New York, 1995.

This book provides a comprehensive review of the concepts and process of diffusion. It has an extensive reference list of publications in the diffusion area of research.

## REFERENCES

Bayraktar, B. 1990. "On Technology and the Management of Technology." In Khalil, T., and Bayraktar, B. (eds.), *Management of Technology II: The Key to Global Competitiveness,* Industrial Engineering and Management Press, Norcross, GA.

Fisher, J. C., & Pry, R. H. 1971. "A Simple Substitution Model of Technical Change." *Technological Forecasting and Social Change,* vol. 3, pp. 75–88.

Mahajan, V., Eitan, M., & Bass, F. 1990. "New Product Diffusion Models in Marketing—A Review and Directions for Research." *Journal of Marketing,* vol. 54, pp. 1–26.

Mort, J. 1990. "Xerography—50 Years of Technological Innovation." In Khalil, T., and Bayraktar, B. (eds.), *Management of Technology II: The Key to Global Competitiveness,* Industrial Engineering and Management Press, Norcross, GA.

Munro, H., & Noori, H. 1988. "Measuring Commitment to New Manufacturing Technology: Integrating Push and Pull Concepts." *IEEE Transactions on Engineering Management,* vol. 2, pp. 63–70.

Rogers, E. M. 1995. *Diffusion of Innovation,* 4th ed., Free Press, New York.

# 6

# THE PROCESS OF TECHNOLOGICAL INNOVATION

The process of technological innovation is a complex set of activities that transforms ideas and scientific knowledge into physical reality and real-world applications. It is a process that converts knowledge into useful products and services that have socioeconomic impact. It requires the integration of inventions and existing technologies to bring innovations to the marketplace.

There are eight stages in the process of technological innovation, as shown in Figure 6-1. Some activities within those stages may overlap with each other. The stages of technological innovation are:

**1** *Basic research:* This is research for the sake of increasing our general understanding of the laws of nature. It is a process of generating knowledge over a long period of time. It may or may not result in specific application.

**2** *Applied research:* This is research directed toward solving one or more of society's problems. An example is research conducted to develop a drug for treating a known disease. Basic and applied research advance science by systematically building knowledge on previous knowledge. Successful applied research results in technology development and implementation.

**3** *Technology development:* This is a human activity that converts knowledge and ideas into physical hardware, software, or service. It may involve demonstrating the feasibility of an idea, verifying a design concept, or building and testing a prototype.

**4** *Technology implementation:* This is the set of activities associated with introducing a product into the marketplace. Technology implementation involves the first operational use of an idea or a product by society. It entails the activities associated with ensuring the successful commercial introduction of the product or service, such as cost, safety, and environmental considerations.

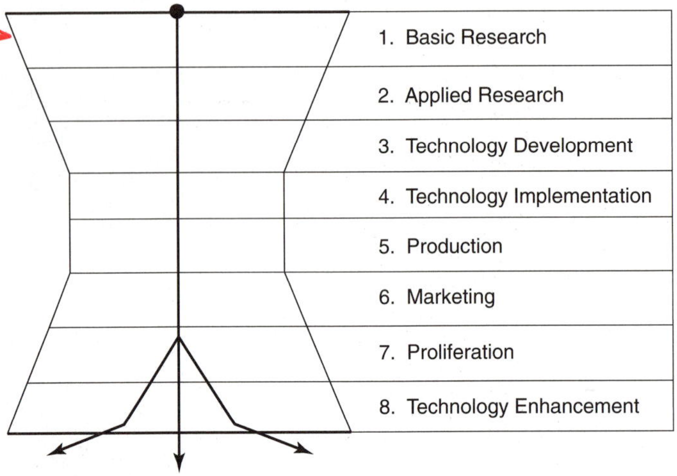


**FIGURE 6-1**
THE EIGHT STAGES OF TECHNOLOGICAL INNOVATION

**5** *Production:* This is the set of activities associated with the widespread conversion of design concepts or ideas into products and services. Production involves manufacturing, production control, logistics, and distribution.

**6** *Marketing:* This is the set of activities that ensures that consumers embrace the technology. It entails market assessment, distribution strategy, promotion, and the gauging of consumers' behavior.

**7** *Proliferation:* This is the strategy and associated activities that ensure the widespread use of the technology and its dominance in the marketplace. Proliferation depends on methods of exploiting the technology and on the practice used for marketing the technology. For example, Microsoft spreads the use of its Internet browser technology by including the browser with its popular Windows software.

**8** *Technology enhancement:* This is the set of activities associated with maintaining a competitive edge for the technology. It entails improving the technology, developing new generations or new applications for the technology, improving quality, reducing cost, and meeting customers' special needs. Technology enhancement increases the life cycle of the technology.

A generic model of the process of technological innovation in a company is shown in Figure 6-2. The innovation could be a new product in a manufacturing organization or a new service in a service organization.

Regardless of the nature of the innovation, be it a product, service, or system, the fundamental components of the model are the same, but the nature of the business may dictate variations in implementation techniques. For example, in service innovation, there may not be a need for a physical prototype or for specific equipment and tooling, as in the case of product innovation.

Using a chemical analogy, Martin (1994) illustrates the technological innovation process as a chain equation (Figure 6-3). A commercially successful innovation is the

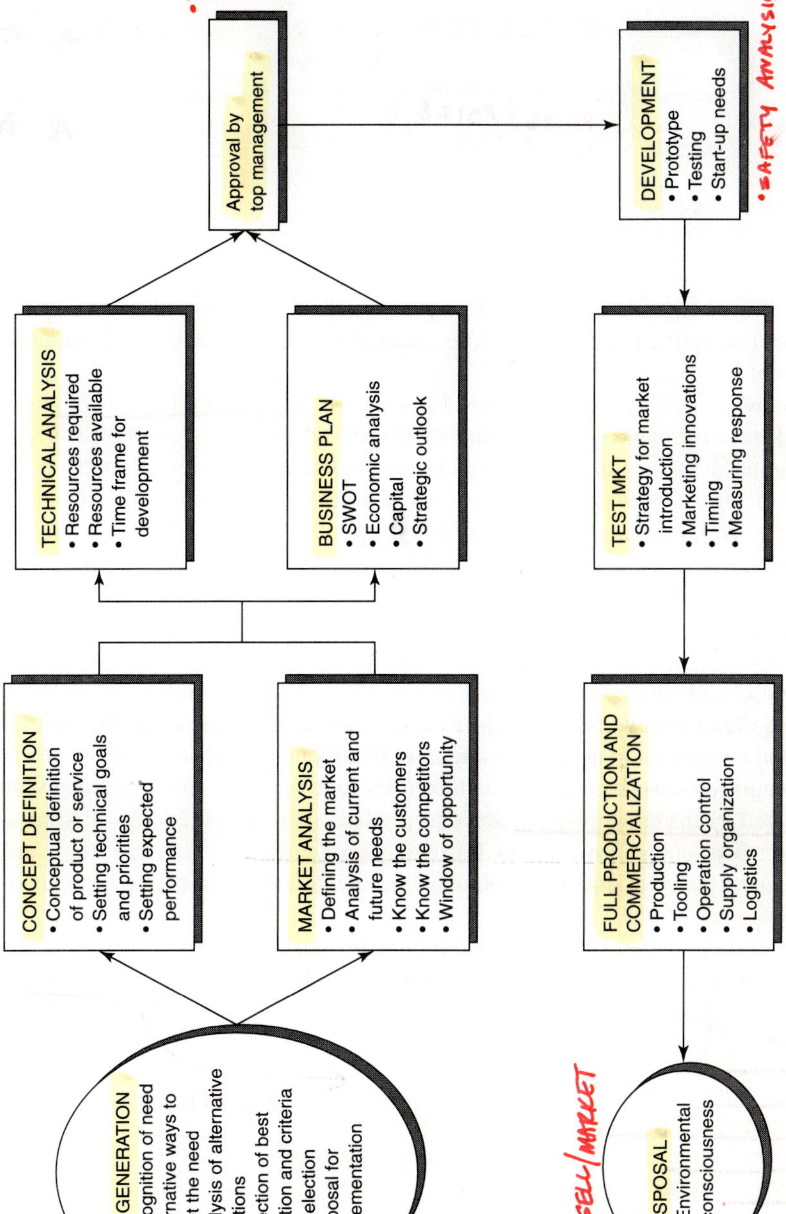


**FIGURE 6-2**
THE PROCESS OF TECHNOLOGICAL INNOVATION

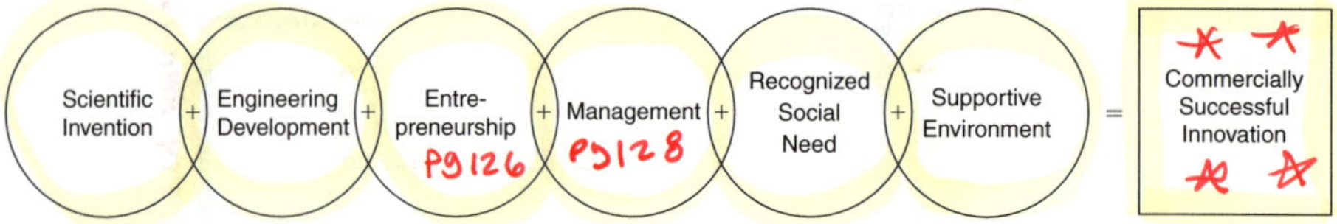


**FIGURE 6-3**
THE INNOVATION CHAIN EQUATION
*Source:* Martin, 1994. Reprinted by permission of John Wiley & Sons, Inc.

product of a sustained chain reaction. It requires synthesis of knowledge and expertise, entrepreneurial spirit, management skills, recognized social need, and a supportive environment.

The innovation chain equation reveals the important role of entrepreneurship in connecting ideas to the marketplace. When a technological turbulence takes place because of a scientific discovery or an engineering development, it creates a loosely defined product structure. It takes entrepreneurial and technical competencies, experimentation, and creative work processes to solidly define the product. It takes persistence and willingness to take risk to bring the product to the marketplace. These are the conditions of entrepreneurial culture, young mindset, and heterogeneous population—the type of conditions that lead to radical innovation.

The management role in the innovation chain emphasizes the need for stability and control at a certain phase of the innovation process. It permits running operations efficiently in order to achieve commercial success. Formalized management takes a defined product and connects it to a market using a set of structured functions, formalized procedures, engineered work processes, defined roles for employees, and safer capabilities. A formalized managerial structure tends to produce incremental and process innovations more than radical innovations. Entrepreneurial culture is different from formal management (stewardship) culture and may create conflicting forces in a firm pursuing innovations.

Mills (1996) provided very simplified but interesting definitions for several components of the process of technological innovation. He called them minimalist definitions:

*Science:* How things are.
*Technology:* How to do things.
*Management:* How to get things done.
*Technology management:* Doing things.
*Entrepreneurship:* Doing things to make money.
*Innovation:* Doing entrepreneurship.

Bordogna (1997; see also Reading 6.1) presents a contemporary model of technological innovation that challenges the traditional linear-path process characterized by new scientific knowledge created at the front of the path and new products and services garnered at the path's end. He defines innovation as

> a concurrent, interactive, and nonlinear activity. It includes not only science, engineering, and technology, but social, political, and economic interactions as well . . . and the public policy that either enables or mutes the whole wealth creation process.

Bordogna illustrates the key elements of the innovation process as shown in Figure 1 of Reading 6.1. This view elevates the innovation process to a new level of total system integration, taking into consideration social, political, and economic issues. The management of technological innovation is vital to the creation of wealth.

READING 6.1

## Innovation and Creative Transformation in the Knowledge Age: Critical Trajectories

**Joseph Bordogna**

Acting Deputy Director, Chief Operating Officer,
National Science Foundation

### INTRODUCTION

We live in an era of breathtaking change and complexity. Twenty-five years ago, typewriters were one of the top products; now it's PCs. A few decades back, I worked my way through college creating India ink drawings by hand for RCA Corporation. I made a name, and not incidentally, a living, for myself by demonstrating expertise with a simple set of instruments known as French curves. Now mouse-clicking a sketch on a PC screen not only produces a better drawing in less time—I also don't get indelible India ink all over my fingers.

Although we are invigorated by change, many of us have difficulty grasping the full potential of the advances at our fingertips. Today we are experiencing great economic strength while many people feel insecure about their jobs. Indeed, there are no more "safe" careers. We are witnessing the era of "commodity" workers—whose contemporary skills are ubiquitous and thus easily garnered at minimum cost in a global market. This is no way to make a living: rewarding careers should be rewarded.

While inexorable technological change challenges our current ethical, social, and economic systems, we are also presented with opportunities to improve our lot. Let's consider, for a moment, the wealth creation process, which enables our welfare, quality of life, and even our quest for knowledge.

### THE INNOVATION/WEALTH CREATION PROCESS

A number of reliable studies indicate that, during the past fifty years, *industrial innovation* has been responsible for about 40 percent of the productivity gain in the United States. However, the old model of innovation as a linear path process, with new scientific knowledge created at the front of the path and new products, services, and markets garnered at the path's "end," is increasingly challenged. Recent economic history has

*Source:* Paper presented at the Plenary Session, Portland International Conference on Management of Engineering and Technology, July 29, 1997.

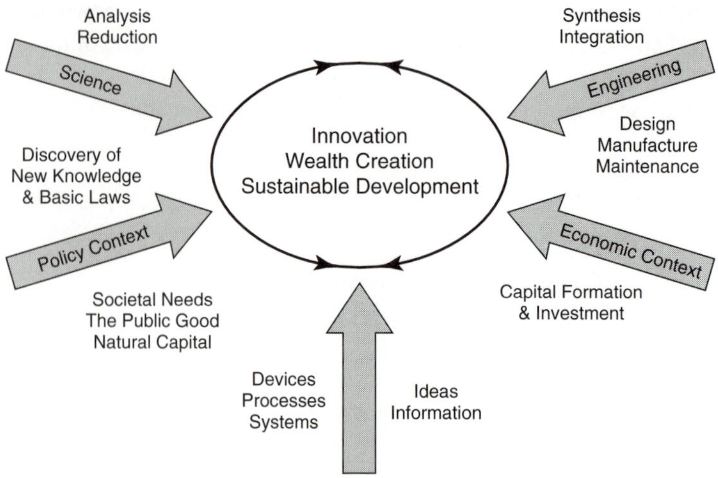


**FIGURE 1**
INNOVATION-CONCURRENT INTEGRATION
*Source:* Bordogna, 1997.

made it all too clear that research leadership does not translate automatically into economic success: physical capital (including information and databases), enriched human capital, and technological capital are needed as well. To better illustrate this, let's for a moment examine the key elements of the *innovation process.*

As portrayed in Figure 1, *innovation* is a concurrent, interactive, and nonlinear activity. It includes not only science, engineering, and technology, but social, political, and economic interactions as well and the public policy that either enables or mutes the whole wealth creation process.

A critical element in the innovation process is scientific inquiry, an analytic, reductionist process which involves delving into the secrets of the universe to discover new knowledge. Those who excel at this paradigm sustain and nurture the world's rich intellectual infrastructure.

The essence of engineering, on the other hand, is the process of integrating all knowledge to some purpose. In a poetic sense, paraphrasing the words of Italo Calvino, the engineer must be adept at "correlating exactitude with chaos to bring visions into focus."

The "stuff" of technology underpins enablement but the whole process is muted if the public policy and economic context are awry.

## THREE CRITICAL TRAJECTORIES IMPACTING THE INNOVATION PROCESS

With these thoughts in mind, I would like to examine three critical trajectories that are strategically impacting the innovation process. (See Figure 2.) These evolve from trends that are inventing, and being invented by, each other. I use the word "trajectory" here because it conveys a useful idea—something moving along a path with some "oomph" behind it.

- Border Crossings (National and Sectorial)
- Emergence of Complex Technologies
- Knowledge and Distributed Intelligence

**FIGURE 2**
CRITICAL TRAJECTORIES IMPACTING THE INNOVATION PROCESS
*Source:* Bordogna, 1997.

### Border Crossings (National and Sectorial)

The first trajectory comes under the heading of "border crossings." It refers to the growth in both scale and importance of cooperative approaches to scientific and technological research.

As Figure 3 shows, we have seen a marked increase in recent years in research collaborations that span international boundaries. The number of internationally co-authored articles increased by 150 percent from 1981 to 1993. The share of all published articles with co-authors from two or more different nations has more than doubled over the past decade.

The uninhibited flow of fundamental knowledge in science and engineering through publication and peer review remains a defining characteristic of our global enterprise. These data make clear that this tradition remains indispensable to the progress of research in all scientific and technological fields.

Another type of border crossing has only recently begun to occur with regularity. Cooperative activities that cross sectorial boundaries—notably industry-university

**FIGURE 3**
INTERNATIONAL CO-AUTHORSHIP
The percentage of science-technology articles with international co-authorship has been steadily increasing in recent years.
*Source:* Bordogna, 1997.

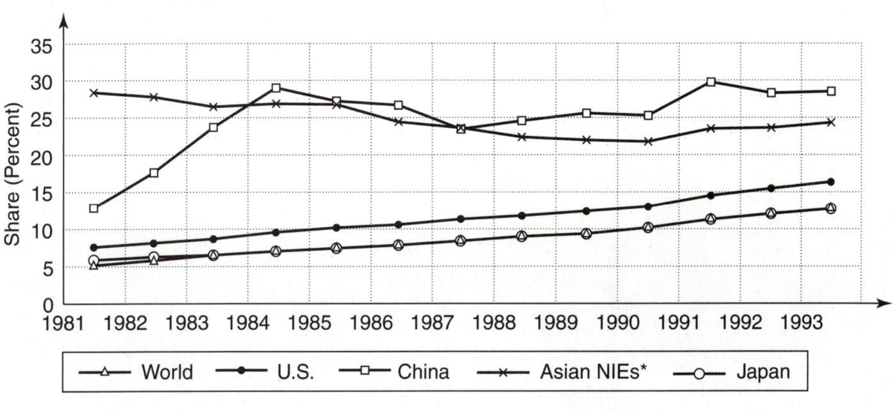

partnerships—are a relatively new addition to the strategic intent of research investment in contemporary universities, but they too show signs of proceeding at an accelerating pace.

In the U.S., this trend is most pronounced when examined from the perspective of the industrial researcher, as is shown in Figure 4. Cross-sectorial co-authorship has grown steadily in the U.S. since the early 1980s. A large share, nearly 40 percent, of journal articles published by researchers based in private industry now include a co-author from a university or government laboratory. In 1981, this share was hovering at just over 20 percent, so we have seen it roughly double over the past dozen years or so.

Only very recently have we begun to see concrete evidence that highlights the importance of university-based research in determining a nation's capacity to innovate and compete economically. A just-completed study of citations from U.S. patents to the scientific literature has documented the linkage between university research and industrial innovation. (See Figure 5.) This study, developed by Dr. Francis Narin and several colleagues, has already been featured prominently in a number of major news outlets, including the *New York Times*.

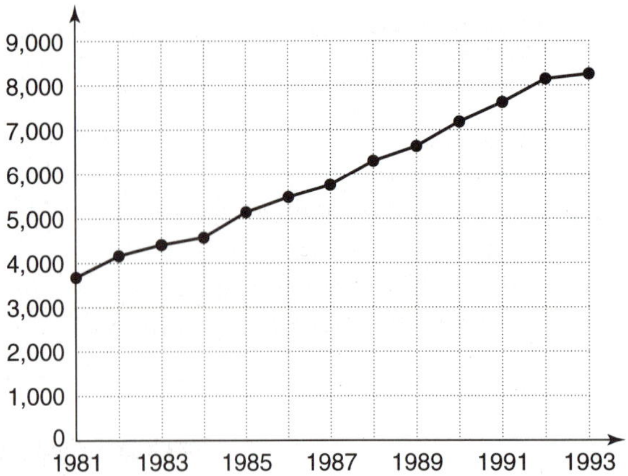


**FIGURE 4**
CO-AUTHORSHIP ACROSS SECTORS
The number of industry articles with co-authors from academe and government has increased dramatically since 1981.
*Source: S/E Indicators,* Chap. 5; used in Bordogna, 1997.

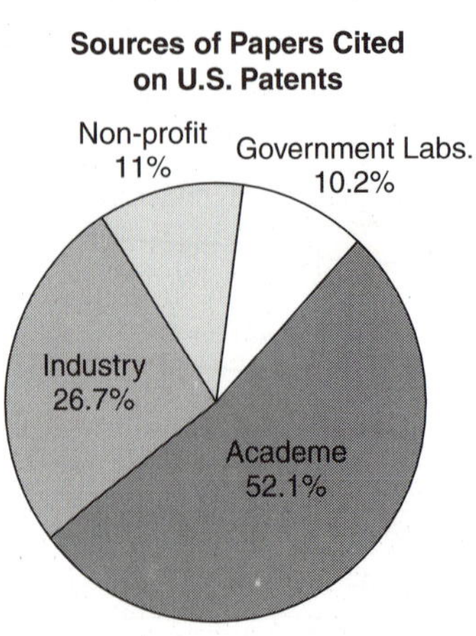


**FIGURE 5**
PATENT CITES TO "PUBLIC" SCIENCE
*Source:* CHI Research (Narin, featured in the *New York Times,* May 13, 1997); used in Bordogna, 1997.

The Times article ran under the headline, "Study Finds Public Science Is a Pillar of Industry." The study found that 73 percent of recent patents awarded in the U.S. cite research from public and non-profit organizations. The academic sector was found to be the principal source of key findings, as it proved to be the source of just over half of the articles cited.

These findings, coupled with today's constrained U.S. Federal budget environment, make this an especially crucial period for industry-university linkages. Over the last two decades, we have seen partnerships between academe and industry grow into a bountiful landscape of innovative endeavors.

### Emergence of Complex Technologies

This brings me to the second trajectory—the changing nature of the products and processes demanded by today's global marketplace.

In a study presented this February at the annual meeting of the American Association for the Advancement of Science, Donald Kash and Robert Rycroft found that the most successful commercial technologies have changed in one basic way over the past quarter century: they have become more complex.

Kash and Rycroft analyzed the 30 most valuable exports in the global market in the years 1970 and 1994. They divided them into the categories shown in Figure 6. The boxes on the matrix are determined by whether the products themselves can be considered simple or complex, and whether they require simple or complex manufacturing processes.

Kash and Rycroft's key finding is quite striking. In 1970, a quarter century ago, nearly 60 percent of the world's top exports were essentially simple products that could be manufactured through simple processes. Today, that same percentage (60 percent) of the world's top exports are complex products that require complex manufacturing processes.

Kash and Rycroft write that "economic well-being in the future will likely go to those who are successful in innovating complex technologies." Put simply, the future belongs to those who can make sense of the complex, to those who can integrate diverse knowledge located in many different organizations to produce previously non-existent capabilities. (See Figure 7.)

**FIGURE 6**

THIRTY MOST VALUABLE EXPORTS, 1970 AND 1994—IN BILLIONS

*Source:* Bordogna, 1997; based on data from Kash and Rycroft, "Technology Policy in the 21st Century."

| Simple Process/Simple Product | Simple Process/Complex Product |
|---|---|
| 1970 = 58% (US$87)<br>1994 = 8% (US$347) | 1970 = 0%<br>1994 = 0% |
| **Complex Process/Simple Product** | **Complex Process/Complex Product** |
| 1970 = 12% (US$35)<br>1994 = 25% (US$435) | 1970 = 31% (US$46)<br>1994 = 59% (US$1128) |

"Economic well being in the future will likely go to those who are successful in innovating complex technologies."

**Innovating Complex Technologies**

"The innovation of complex technologies is distinguished by *synthesis,* the capability to integrate diverse knowledge located in many different organizations to produce previously non-existent capabilities."

"Diversity is integral to complexity. The innovation of complex technologies is normally accomplished by accessing or creating new knowledge, decoupling from existing knowledge, and/or reconfiguring knowledge."

**FIGURE 7**
INNOVATING COMPLEX TECHNOLOGIES
*Source:* Bordogna, 1997, based on Kash and Rycroft, "Technology Policy in the 21st Century."

Diversity is a must—diversity in views, in approaches, and in backgrounds. Without it, we will never see beyond the limits of our individual perspectives and achieve the breakthroughs that occur only through the synthesis of widely different skills and perspectives.

### Age of Knowledge and Distributed Intelligence (KDI)

The third trajectory I'll examine today is the impact of advanced information technologies on society—what my colleagues and I at NSF describe as Knowledge and Distributed Intelligence or simply "KDI."

When recently asked about the future of the Internet, Bob Lucky, vice president at Bellcore, said: "There are two things I know about the future. First, after the turn of the century there will be one billion people using the Internet. The second thing I know is that I haven't the foggiest idea of what they are going to be using it for."

The Internet is indeed a tremendous breakthrough (which NSF helped to enable) in that it cobbles together millions of computers, servers, all kinds of software and databases, and documents—and makes huge amounts of "stuff" available to millions of people. The next revolution, however, will be making the Internet "intelligent"—a "place" where people and machines collaborate.

What we are seeing today is only the beginning for forging connections to learning and creativity. We are moving from the *Internet Decade* to the *Information Everywhere Decade*. Will we develop new ways to express and unleash our creative talents—talents that are now limited by our ability to interface via a keyboard and mouse? What tools will enable us to control and master this ultra-rapid flow of information? Will having the proverbial Library of Congress in your pocket be a blessing or a burden?

The answers to these questions are being pursued on many different fronts from many different directions. Our efforts and our leadership can transform this immense, unprecedented, and somewhat intimidating potential into true progress, economic opportunity, social gain, and rising living standards for human civilization.

At NSF we have developed a theme that we refer to as *Knowledge and Distributed Intelligence* (or simply "*KDI*"). KDI is perhaps the most encompassing venture NSF has

| • **Knowledge Networks** | • **New Challenges for Computation** |
|---|---|
| Multi-Media Environments | Data Mining |
| Resource Sharing Technologies | Visualization |
| Digital Libraries | Pattern Recognition |
| Collaboratories | Partnerships for Advanced Computational Infrastructure |

• **Learning and Intelligent Systems**
Learning Technologies (based on insights into learning and cognitive functioning)
Collaborative Learning Across Physical and Virtual Communities
Knowledge-on-Demand Pedagogies
Fresh Creativity-Enabling Infrastructure

**FIGURE 8**
KNOWLEDGE AND DISTRIBUTED INTELLIGENCE
*Source:* Bordogna, 1997.

ever pursued. It cuts across all fields of research and touches education at all levels. And it is inseparable from the trends and technologies that are driving growth and opportunity in our economy and society—from networks to sensors to virtual reality systems.

In the next few years, KDI research will help us take the next quantum leap forward in terms of both technological progress and societal benefit. It is impossible to predict the next level of tools and capabilities. But, we can be confident they will be spectacular!

For fiscal year 1998, NSF's KDI investment falls into three basic categories (see Figure 8):

**1** *Knowledge Networking* focuses on the integration of knowledge from different sources and domains across space and time.

**2** *Learning and Intelligent Systems* seeks to unify experimentally and theoretically derived concepts relating to how humans learn and create, in collaboration with machines.

**3** *New Challenges in Computation* focuses on research and tools needed to model, simulate, analyze, display, and understand complicated phenomena, to control resources and deal with massive volumes of data in real time, and to predict the behavior of complex systems.

Cutting across these three activities is the *Next Generation Internet.* NSF's role in this multi-agency effort is intended to keep academic science and engineering at the cutting edge of computing and networking technologies.

## CREATIVE TRANSFORMATIONS

Joseph Schumpeter introduced the concepts of creative destruction and creative transformations over half a century ago. He admonishes that unless an entity continually transforms itself, it will ultimately be destroyed by market competition. (See Figure 9.)

> Business leaders usually visualize a market economy in the context of how capitalism administers existing structures, whereas the wiser approach is to understand how it creates and destroys them.
>
> Paraphrased from Joseph Schumpeter
> *Capitalism, Socialism and Democracy*
> Chapter VII, page 84, 1942.

**FIGURE 9**
CREATIVE TRANSFORMATIONS: THE SCHUMPETERIAN FACTOR
*Source:* Bordogna, 1997.

As you are well aware, Corporate America has been going through a period of restructuring. The three integrated trajectories that I described have driven the shift in corporate focus away from the individual and toward the group. Indeed, products and processes have become so complex that no one individual can bring all the needed skills to the table.

Today, a new model for the successful corporation has emerged. It's epitomized by high-tech firms like Sun Microsystems and Netscape. Robert Keidel dubbed this "the cooperation-driven" corporation—and it is different in a number of ways from what we've been used to in the past. Its overarching purpose is to enhance group performance, celebrate teamwork and flexibility, and create a tempo that is electric. Cooperation is now a key to continually re-creating a corporate entity and remaining competitive in the global economy. Not only is this true for corporations; it is true for universities and other institutions as well.

Consider that over 2,000 years ago a well-to-do citizen of ancient Greece offered some of his real estate, a grove, to a thoughtful fellow citizen—to be used as a place where fellow thinkers could gather for hearty discussions on matters of common and uncommon interest. The grove became Plato's Academy, and the generous benefactor's name was Academus—the name from which our higher education enterprise derives its own name.

In those days, a physical place was needed in order to build connections to learning and creativity. Today, knowledge is becoming available to anyone, anywhere, anytime, and power, information, and responsibility are moving away from centralized control to the individual. At many universities and elsewhere, books are already being published and courses taught on the World Wide Web.

The noted guru of artificial intelligence, Edward Feigenbaum, states: "The library of the future will be a network of knowledge systems in which people and machines collaborate." We can only speculate on the enormous impact this will have on what we now call a "university."

The dynamics that underlie the process of creative transformation are poorly understood. At NSF, we have begun to address the principles underlying creative transformations by bringing together research in two areas—the Management of Technological Innovation (MOTI) and research on Transformations to Quality Organizations (TQO). Currently, the first program is administered by our Engineering Directorate, while the TQO program resides in our Directorate for Social, Behavioral, and Economic Sciences.

- How do organizations create, develop, & implement new technologies, processes, and structures?
- How do organizations come to understand the need for innovation and change?
- How can products and processes be most effectively designed to meet customer needs?
- How does technological change affect organizational change?
- How do transformations affect performance?

**FIGURE 10**
THE STUDY OF CREATIVE TRANSFORMATIONS: A MERGING OF TECHNOLOGY AND THE SOCIAL SCIENCES
*Source:* Bordogna, 1997.

These organizational details are important, because we have learned that we must draw upon work in both engineering and the social and behavioral sciences to address the fundamental questions that hold the key to progress (Figure 10), such as:

- How do organizations come to understand the need for innovation and change?
- How does technological change affect organizational change?
- How do transformations affect performance?
- How can organizations effectively create, develop, and implement new technologies, processes, and structures to meet customer needs?

It is clear that these questions and many others like them cannot be addressed by relying exclusively on either the so-called "hard" or "soft" sciences. Addressing them requires that we develop new approaches to research that are highly integrative across all fields of science and engineering.

## CONCLUSION

In closing, I should like to speculate about what the three trajectories I have highlighted hold for the future:

- International and inter-sectorial cooperation is likely to continue growing at an accelerating pace—to the benefit of all of us.
- The emergence of complex technologies and their impact on wealth creation will increase the need for integration across all fields and sectors.
- The arrival of the era of knowledge and distributed intelligence will enable us to pursue previously unimaginable avenues of technology, and it also will restore and reinvigorate the natural linkages between research and learning.

Let me add that these trajectories are much more likely to change us than we are to change them. That may cause some discomfort in our ranks, but we should keep in mind something Douglas MacArthur once said, "There is no security in life, only opportunity."

In the final analysis, I believe these trends bode very well for economic and social progress, and our ability to innovate.

### CASE STUDIES IN INNOVATION

Technological innovation is not only about creativity and new ideas. Of course, they represent the first step toward innovation; however, there are many other factors that are important and must be considered and managed.

The following two case studies illustrate the dynamics of the technological innovation process in reality. The first case (Reading 6.2) describes the invention of xerography and explains how an invention that was deemed by some to be of little technological promise and by others a product concept without a market became a huge success. The second case (Reading 6.3) illustrates the sequence of activities in the innovation of a biomedical device. It shows the peculiarities as well as the similarities existing in the innovation of an industry-specific technology.

## CASE 1: Xerox—The Beginning

#### READING 6.2

## Xerography—50 Years of Technological Innovation

**J. Mort**

Xerox Corporation, NY, USA

In the fifty years since the first xerographic copy, "*10.- 22.- 38 Astoria,*" was made by Chester Carlson in Astoria, New York, xerography has established itself as one of the major successful technological innovations in history (Mort, 1989). Today, this technology generates billions of dollars in revenues and more than one trillion documents worldwide annually. This fact alone establishes it as a paradigm of successful innovation. Xerography also affords an opportunity to examine how an innovation can evolve or renew itself; although originally used almost totally for document copying, xerography has now emerged as a major component in the burgeoning fields of document creation, processing, and desktop publishing.

Xerography involves the confluence and interplay of a number of disparate developments in science, technology, and business. It employs a broad range of materials and physical phenomena, ranging from non-crystalline or amorphous solids to polymers, photoconductivity, and triboelectricity, for which little scientific understanding initially existed. The subsequent evolution of xerography depended on scientific and technological progress, not only in these diverse areas but also in originally unrelated fields, such as digital solid-state electronics, word-processing computers, and lasers. Of equal importance in the initial and later innovations was the role of entrepreneurial businesspeople, innovative marketing strategies, and the final arbiters of any innovation—the customers.

At the outset, it is helpful to define terminology, since terms like invention, innovation, technology, and science are subject to a variety of interpretations. Invention is nothing more nor less than an idea for a better way of doing or using something, although it often involves a rudimentary feasibility demonstration. Technology development is concerned with transforming the original invention into materials and with the device specifications sufficient for product conceptualization and the definition of the manufacturing processes. Technological innovation, on the other hand, means the profitable marketing of a product. The roles of science and technology in innovation are the subjects of much discussion and research activities are often classified, as if neatly dissectible, using terms such as "basic or pure" science, applied science, and technology. Such terms can be misleading if the inference is drawn that fundamental differences exist in the methodologies involved. In this context, science has to do with the acquisition of understanding as an end in itself, whereas technology is the acquisition and application of knowledge for the production of a product; the methods may be the same but the goals are different.

Scientific understanding can be the primary catalyst for a technological innovation, as was the case for solid-state electronics (Braun and MacDonald, 1978), although it is a misconception that science is an essential precursor to technology. More often than not, purely empirical approaches without a clear understanding of the basic principles involved lead to innovation; such was the case with xerography. Once an innovation is successful, however, empiricism can be inefficient and costly, particularly in a competitive environment. In such a situation, the establishment of a body of understanding leading to quantitative specification of the critical parameters for materials, processes, device operation, or markets is vital for further improvements or maintenance of market leadership. This brief paper attempts to delineate the roles that each of these elements played in the story of xerography and presents a discussion of the generic lessons that may be drawn for the successful management of future technological innovation.

## INVENTION: THE CREATION OF XEROGRAPHY

The most important determinant of a successful innovation is that it satisfy a discernible need or create a demand. Although in the case of xerography a market-pull was identified, the enormous commercial success came about without its full market potential being initially recognized. Thus, improved or more efficient ways of making copies of documents were generally seen as more of a convenience than a necessity, carbon copies being of sufficient quality to fulfill the need for cheap, convenient, archival copies. Carlson, more acutely aware of the market need due to his work as a patent attorney, thought differently.

Like most inventions, xerography did not materialize out of thin air. Various products, for example, were available in the 1930s for producing copies, but were complex, time-consuming, and expensive. Technically, the production of images based on the ability of electrostatically charged insulators to attract triboelectrically charged powders was well-known; indeed, triboelectricity, charging produced by friction, was first recorded by the Greeks hundreds of years before Christ. Such phenomena, however, were largely viewed as scientific curiosities until the early twentieth century when serious efforts began to develop what became known as electrography for reproducing images. Selenyi and others investigated the concept of charging an insulator with an image-wise pattern of charged ions. This required the scanning of an original document, line by line, with a light-sensitive scanner to produce an analog voltage then used to modulate the output from an ion source. Copies could thus be produced by displacing of a document-size, insulating layer beneath a stationary ion source and rendering the resultant charged image visible or developed by dusting with a triboelectrically charged powder.

Carlson's invention was to combine such electrostatic charging and development with the phenomenon of photoconductivity, first discovered in selenium crystals in the 1870s. Photoconductivity, or the enhancement of the electrical conductivity of a material by illumination with light, has the major advantage that the image of a complete document to be copied can be simply projected onto a page-sized photoconductive layer, the photoreceptor, uniformly charged with ions. The reflected light from the document then produces selective photodischarge proportional to the incident light intensity in the photoreceptor. The resultant image, consisting of the remaining surface charge, replicates the information content of the document and can be developed by its electrostatic attraction for charged powder. In the original reduction to practice, the powder image was transferred from the surface of the photoreceptor to waxed paper by simple pressure to create the final copy. Carlson called his invention electrophotography (Williams, 1984; Carlson, 1965).

## TECHNOLOGY: THE DEVELOPMENT OF XEROGRAPHY

Carlson's original demonstration was far from being a technology. The photoreceptor used was a sublimed sulfur film with little sensitivity to visible light; the developer was lycopodium powder; and the initial charging of the sulfur involved rubbing with a handkerchief. Given this rather primitive state of affairs, it is not surprising that the quality of the image produced only established the feasibility of the invention and much work remained to transform it into a viable technology. Carlson spent the next six years trying to generate support for the required development work, but despite his contacting many companies, no one was sufficiently impressed to take up the challenge. The first significant break came in 1944 because of a chance encounter with Battelle Memorial Research Institute in Columbus, Ohio, which supplemented its income by soliciting contracts from industry and government agencies. On a visit in 1944 related to his work as a patent attorney, Carlson mentioned his own patents on a new copying process. By good fortune, Battelle was thinking of establishing a research group in graphic arts and, despite the obvious problems with Carlson's invention, felt substantial contributions could be made with their expertise in chemistry and physics. As a result, Battelle acquired exclusive rights to Carlson's patents and agreed to give him a substantial share of any proceeds from subsequent profits. In return, Carlson agreed to commit $3,000 to pursue development of the invention. The first approaches were a search for better photoconductive and developer materials, better conditions for obtaining sharper images, and the best means for rendering these latent electrostatic images visible, consistent with an ability to transfer them to paper.

A 1944 article published in *Radio News* magazine came to the attention of the Haloid Company, a small photographic paper manufacturer in Rochester, New York. The company sold photocopying products based on wet chemistry, directly to the customers, using a large force of salesmen and demonstrators. As a result of reading the article, the president of Haloid, Joseph Wilson, and his research director visited Battelle to look at the new invention that might enlarge their business opportunities. They were sufficiently interested for Haloid to commission market research to see if a market for a new copying device existed. Inevitably, the results were ambiguous because the product was hypothetical and the most basic questions as to advantages, cost, size, and speed could not be answered. Still, Haloid's interest grew, and in 1946 an agreement was signed in which Haloid supported the Battelle research at a level of $25,000 a year, beginning in 1947; this was a major risk considering that in 1947 the company had a net income of only $138,000 on net sales of $7 million. As a result, essentially all the research on Carlson's invention in the period 1944 through 1948 was carried out at Battelle, initially using its own funds but with the later support of Haloid.

During this period, important advances were made that enabled the ultimate commercialization of electrophotography. The highly photosensitive amorphous form of selenium was discovered independently in the mid-1940s by two laboratories searching for large-area thin-layer photoconductors for applications in imaging devices. One of these was the RCA Laboratories, which was exploring vidicon technology involving the use of photosensors in television cameras. The other was at Battelle, where better photoreceptor materials for Carlson's invention were being sought. In fact, the requirements for both technologies are very similar in that the photoconductive layer must have a sufficiently high resistivity to ensure that in the dark the voltage produced across the film by charging with an electron beam (vidicon) or ions (electrophotography) does not discharge until exposed to the optical image. Other improvements involved ion-charging processes, electrostatic transfer, and dry-ink or toner materials and processes for their manufacture.

## INNOVATION: THE MARKETING OF XEROGRAPHY

With these various improvements, commercialization moved closer to reality. This raised the issue of disclosing the technology publicly to protect the significant Haloid investment, avoid pre-emption by any unknown photoprocess, and more significantly, to attract the additional financing required for product development. It was also decided that the name electrophotography did not convey the required aura of a new, unique invention, and the Greek words for dry, "xeros" and writing or drawing, "graphein," were combined to give the name xerography to the invention and Xerox (sic) as a tradename. The public announcement of the new technology was made at the meeting of the Optical Society of America on the tenth anniversary of the first xerographic copy. The invention of the transistor, ultimately to greatly influence the evolution of xerography, had already been announced on June 30th, although neither was a product at the time. In the same year, the U.S. Army Signal Corps awarded a grant of $120,000 to Haloid to develop a dry photographic process. This proved a major stimulus both to the morale of those involved and the future development efforts.

The first product, the Xerox Copier Machine, Model A, announced in 1949, was field tested by loaning units to four large companies who found them of little value, since their manual operation was complex and copies of acceptable quality were operator dependent. The obvious solution was to automate the process, but major technical uncertainties existed as to its cyclic stability. In any case, this required further substantial investment and the experiences with the Model A made external investors even more wary of getting involved. At about this time, however, it was realized that the process could be used to make offset lithographic master plates by transferring the developed image to a specially prepared offset master plate. On small lithographic presses, such offset plates could produce large numbers of copies quickly at low cost. However, the masters produced by conventional methods required an hour to make and cost several dollars each. By contrast, the master made by xerography cost 30–40¢ per plate and turned out copies within minutes of receiving the original. This was an important niche for the embryonic technology, and it also provided cash, knowledge and experience, all of which were needed to achieve the objective of automating xerography.

The next xerographic product, the CopyFlo printer, introduced in 1955, was the first automated xerographic product and enabled the production of copies on a continuous web of ordinary paper. By 1955, Haloid's annual sales had reached $21 million, of which almost 50% came from xerographic products. As recognition of this fact, the name of the company was changed to Haloid-Xerox Inc. The company also bought from Battelle the original Carlson patents for 50,000 shares of Haloid-Xerox stock, worth about $4 million. These transactions had substantial implications for the company's cash-flow position, as did its decision to lease rather than sell its xerographic products, since by leasing Haloid-Xerox retained legal ownership and the significant tax advantages of depreciation. As a result, from 1951 to 1955 the company's cash flow improved from $720,000 to more than $2 million.

In 1958, it was decided to enter the office copying machine market with the Xerox 914, so named because it would make copies on paper up to 9″ × 14″ in size. Significant research and engineering had to be done even to demonstrate the first model, involving, as it did, a coated amorphous selenium drum, optical systems, paper feeder, paper transport, fusing, and more. Based on the results of the first prototype and manufacturing cost estimates, it became obvious that millions of dollars would have to be expended to bring the product to market. Given the risks involved, several large firms were approached to see if they would manufacture the product for Haloid-Xerox in return for a share of the profits. Market surveys made for these companies, confirming similar ones done for Haloid-Xerox, estimated the maximum number of units that would be sold over the life of the 914 at no more than 3,000. These estimates were in part based on the extant market in which automatic copying already existed, including electrophotographic products introduced by RCA using paper coated with dye-sensitized zinc oxide. Such coated papers met market resistance in terms of esthetics and cost, so in 1958 the total market was only about $100 million. Moreover, the 914 represented radical departures from the existing copiers, and despite the touted advantages of xerography, these departures were viewed negatively. First, the projected size of the 914 was considered too bulky and heavy, at 650 pounds, for the office. Second, in terms of a conventional sales approach, the cost of an outright sale was high. Consequently, the 914 generated no euphoria in either market researchers or other companies. Haloid-Xerox was thus faced with the choice of establishing its own manufacturing capability or abandoning the whole venture. They chose the former path and the rest is history.

As great as Carlson's invention and the skills of the Haloid and Battelle technical staff, they were matched by the creative marketing necessary to complete any successful innovation. A major role in the development and implementation of the new marketing approaches was played by C. Peter McColough, a future president of the Xerox Corporation. These approaches had a number of interrelated elements, but a critical feature was the use of a two-tiered pricing structure. This was based on the premise that the ultimate cost to the customer should be a function of the number of copies produced. First, the base price, including a number of "free" copies, was set sufficiently high to ensure an acceptable return, yet sufficiently low, on a per copy basis, that customers would consider the 914 worth the charge per copy. In addition, the base price, ultimately set at $95, had to be low enough and the contract term short enough to reassure any customers concerned about their investment risk. Meter pricing was then used so copies in excess of those included in the base price were charged to the customer. As a result, the higher the utilization of the machine, the higher the revenues and profits. This, in turn, justified the required establishment and maintenance of a nation-wide sales and service force. The latter, essential for a leasing strategy, also provided a mechanism for facilitating the diffusion of the 914. Such marketing innovations, just as the technical aspects of xerography, were in some respects evolutionary in that they built on Haloid's earlier experiences in direct marketing.

The Xerox 914 copier, first shown in September 1959, was a fully automatic machine requiring no special skill or adjustments to operate. Although basic by modern standards, it produced seven copies per minute on plain paper with a reusable photoreceptor. The image, composed of a carbon-black impregnated polymer toner impervious to degradation by light or chemicals, was ideally suited for archival storage. The impact was phenomenal and caused a revolution in both the copier industry and the office. In contrast to the market predictions of 3,000 placements over the life of the product, a total of over 200,000 were sold, with a peak population of approximately 65,000 units reached in six years. A monthly copy volume of 50 million in September 1961 climbed to 490 million by March 1966. The inexorable growth of Xerox's business was slowed by neither economic recession nor seasonal fluctuations, so the financial impact was equally dramatic. The revenues of Haloid-Xerox in 1960 were $37 million, of which only $3 million were from xerography. One year later the revenues of the newly named Xerox Corporation had leapt to $59.5 million, and by 1966 were $428 million. In 1986, the total copier business world-wide for all companies had grown to roughly $20 billion, and the annual total number of copies made approached 1 trillion. Impressive as these figures are, the evolution and growth of xerography should not be described only in terms of financial success. Behind the numbers lies a story of continued research and advances in materials, process, and product concepts which have led to dramatic improvements in xerographic technology and product value to the customer. Likewise, parallel developments in other areas of science occurred during the years in which Carlson, Wilson, and others struggled to make xerography real. These apparently unconnected advances ultimately impacted and enlarged the capability of xerography beyond anyone's expectations.

## SCIENCE: THE IMPROVEMENT OF XEROGRAPHY

The genesis of xerography did not lie in science, except insofar as all progress builds on accumulated knowledge. Hence, although much use was made of electrostatics, photoconductivity, and materials, this was done to a large degree empirically. Moreover, at the time no understanding of the materials or processes associated with amorphous photoconductors or developer materials existed at anything approaching a fundamental level. As a result, early problems in production or performance had to be solved on the run using a combination of limited insight and intuition. While this provided sufficient operational know-how to permit the technology and product development of the Xerox 914 copier, a more systematic approach was clearly essential for future improvement in the technology. Consequently, xerography is an example of a technological innovation that gave birth to new scientific fields of study.

Carlson's invention had provided a new way to create copies using photoconductive, non-crystalline thin films. Glass is the most familiar member of a general class of materials known as amorphous solids. Unlike the atoms of crystals, which are distributed with uniform order in three dimensions, the atoms in amorphous materials are distributed randomly. Freed from the constraint of positional order, amorphous materials can be easily produced in large areas of arbitrary shape, at low cost, and as a result glass is a centuries-old technology. Still, there was nothing to suggest that such materials were even remotely interesting from the perspective of photoelectronic properties; indeed, glasses were viewed as quintessential high-resistivity insulators. Even more striking was the fact that the scientific community had initiated no coherent or sustained research on amorphous materials as of 1960. Undoubtedly a number of reasons accounted for this. First, the field of solid-state physics in the 1940s and 1950s focused almost exclusively on crystalline solids. Not surprisingly, these were viewed as being the most fruitful area of endeavor, given the fact that for scientists they appeared the most tractable. The dramatic discovery of the transistor and research in crystalline silicon only heightened this focus as more and more laboratories, particularly those in industry, competed to find another proverbial pot of technological gold in the world of crystals. By contrast, the attraction of research in amorphous materials and phenomena like triboelectricity was not widely appreciated or pursued. Therefore, despite the successful innovation of the Xerox 914 and the critical role of photogeneration and the movement of charge in amorphous selenium, the first direct, unambiguous measurement of these properties was not made until the late 1950s when a tentative start was made to understand the materials and phenomena involved in xerography in a more fundamental way (Mort, 1983). This required not only new interpretations of the electrical, structural, and thermomechanical properties of solids and the phenomena of electrostatics and triboelectricity, but even the development of appropriate measurement techniques to systematically characterize these properties.

Amorphous selenium played a central role in the phenomenal success of xerography. Despite this, selenium had some definite shortcomings in terms of the future evolution of the technology. First, it exhibited a propensity to crystallize, although solutions to this particular problem were found. Less easily remedied shortcomings were its lack of photosensitivity to red light and intrinsic lack of flexibility. From an engineering perspective, considerable virtue was ascribed to having flexible photoreceptors and, by using a materials engineering approach, amorphous selenium belts were produced by Xerox in the 1970s for use in the high speed Xerox 9200 machines. Another impetus, arising from business considerations, was the need for cheaper photoreceptors to reduce costs and make the products more affordable for the customer. The combined needs for extended spectral response, flexibility, and low cost were, therefore, major driving forces to identify alternative photoreceptor materials and devices during the 1960s and 1970s. The advances capitalized on the accumulated insights on the first generation of photoreceptors, an improved understanding of the amorphous state in the wider scientific community, and the identification of its relevance to actual device performance. Alternatives to amorphous selenium and its alloys in terms of flexible thin-films exist in organic polymers, and it was thus natural that attention of research laboratories turned to these materials in the 1960s.

Polymers are a particular class of organic compounds which have distinctive characteristics, including outstanding flexibility, transparency, impact strength, and chemical inertness. They consist of molecular units connected together, much like a string of pearls in a necklace, to produce chains which may contain millions of these sub-units. Many polymers are glass-like in that the chains do not line up or chemically bond to each other and so can be produced cheaply in large sheets. Despite these desirable mechanical and chemical properties, polymers, like inorganic glasses, generally constitute some of the best electrical insulators known. Initial studies on the electrical properties of organic materials in general and polymers in particular, however, created the hope that coupling creative chemical synthesis with an understanding of the organic solid-state might yield a combination of electronic function and desirable mechanical features. Today, polymeric-based photoreceptors using polymers molecularly doped to make them electronically active are widely employed by the xerographic industry. Since the polymers themselves typically do not absorb visible light, they are combined with a second thinner photogenerator layer whose function is to act as the photoconductive element. Usually this photogenerator layer is overcoated with the thicker, doped-charge transport layer in order to protect it. Comparable advances in the understanding of polymer chemistry and physics have led to the formulation of toners with improved characteristics for fusing properties, life, and color in developer applications.

## EVOLUTION: THE EXTENSION OF XEROGRAPHY

The evolution of xerography, as with other innovations, has been profoundly influenced by developments in other fields, although when such advances occurred their relevance was initially less obvious than it seems with hindsight. In itself, the notion of using xerography in electronic publishing is less than a giant step, since Selenyi's work in the 1930s and a xerographic computer printer of the late 1950s pointed in that direction. However, two critical elements were missing: the capability to effectively exploit the idea and, just as important, the need and motivation to do so. The first required a means to easily and efficiently write in faster fashion with a light beam. This had to await the laser, although nothing could have been further from the thoughts of the scientists responsible for its discovery in 1960. This discovery was followed by the necessary development phase that made possible a practical, affordable, and reliable device, so it was only in the mid-1970s that xerographic laser printers became a reality (Starkweather, 1980). The emergence of a major market for such printers, however, was also predicated by increased use of electronic data processing in the world of text and graphics. First and foremost, this required the infiltration of computer technology into the office. The time constant for this to occur in a substantial fashion was essentially a decade, despite the absence of intrinsic technical limitations. The primary roadblock was cost, so the rate of decline in the price of semiconductors due to advances in integrated circuitry and the increased computing power per dollar it enabled were of greater importance. Xerographic printing and its role in desktop publishing, therefore, was a consequence of the maturing of both computer and xerographic technology, leading to their eventual coexistence in the office, together with the rise of digital data and network transmission technology.

## CONCLUSIONS: THE LESSONS OF XEROGRAPHY

Xerography is a classic example of a successful technological innovation and therefore worthy of emulation. The difficulty exists in delineating what lessons one should draw. By any criteria, the innovation of xerography was an exceptional event, and it is hard to imagine that the unique collection of circumstances surrounding its birth can be preordained. The best that can be done is to highlight the individual ingredients that synergistically proved so successful. It is logical to start not at the beginning but at the end of the process, for no matter how exciting an invention or impressive a technology, unless there is success in the marketplace no innovation has occurred. The ultimate determining factor, therefore, was the customer. Even in this respect, no easy answers exist, since repeated market surveys did not reveal the latent market potential of xerography. This is not to denigrate the use of marketing research, where for evolutionary or commodity products numerous successes can be documented. Rather, it suggests its limitations when applied to revolutionary technological innovations. Here, both researcher and customer must grapple with the implications of a product that enables capabilities beyond experience or imagination, and it is extraordinarily difficult to ascertain with any degree of confidence the needs of the marketplace (Braun and MacDonald, 1978). On the other hand, the innovative marketing strategy, employing the leasing approach, proved central to facilitating customer acceptance and the subsequent diffusion of the 914 copier.

It has been said that xerography was an "invention nobody wanted" with the clear implication that it should have been. Retrospectively, this is a tautology since technologies, still less innovations, do not materialize in ready-made or recognizable form. Thus the rejection of xerography twice, first on the grounds that it was an invention with little technological promise, and second, because it was a product concept without a market, occurred for totally rational technical and business reasons, *on the basis of the facts known at the time.* Therefore, for a technological innovation to have a chance of success it must have one or more champions who *believe;* xerography had two, first and always

Carlson, and later Wilson. Individually or collectively, such champion(s), must have, in addition to faith, an unusual combination of technical and business judgment and, given a lack of complete information or control over events, a good intuitive sense. Even these are no guarantors of success, however, and the value of luck cannot be denied. In the case of xerography, all were involved. The technological development of xerography initially occurred in a phenomenological fashion, although no less impressive for that. In fact, given the melange of interdependent phenomena and materials involved, it is hard to imagine that it could have evolved effectively through a reductionist approach in which the various technology parts were independently investigated for later integration. For research, whether internal long-range research targeted at a technology, or discoveries generated in the external scientific community, it has two general characteristics. The first is that, because it is knowledge-limited, there is inevitably a phase lag with the immediate needs of developing technology. Technologists and engineers struggling with the everyday necessity to converge on a set of materials and process specifications have totally different time constants for integrating new ideas into technology. Second, scientific discoveries in the widest sense, including, for example, new manufacturing processes, may occur with no evident relevance but which ultimately, sometimes much later, become enabling. Obviously, new technologies should not be incorrectly identified as successful innovations, no matter how impressive they may appear, but neither should they be prematurely evaluated; as with many endeavors, timing can be everything. Examples from xerography of the first characteristic is the study and understanding of charge transport in amorphous materials, including polymers, which over a decade of research led to the materials engineering of organic photoreceptors. Examples of the second include the development of integrated circuit processing technology leading to the desktop word-processor, or the discovery in 1960 of the gas laser, followed by the solid state laser of the late 1970s which now enables desktop laser printers.

As a coda, the question has been raised as to whether innovation is a game of skill played with chance, or a game of chance played with skill (Braun and MacDonald, 1978). The story of xerography, which, while differing in details, has remarkable parallels with other innovations, suggests that the answer is both.

### Reading 6.2 References

Braun, E., and MacDonald, S. (1978). *Revolution in Miniature,* Cambridge, Cambridge University Press.

Carlson, C. F. (1965). *Xerography and Related Processes,* edited by J. H. Dessauer and H. E. Clark. Chapter 1: "History of Electrostatic Recording," New York, Focal Press.

Mort, J. (1983). "Amorphous Semiconductors: From Selenium to Silicon," *Journal of Materials Education,* vol. 5, 197.

Mort, J. (1989). *The Anatomy of Xerography,* and references therein, Jefferson, NC, McFarland & Company Inc.

Starkweather, G. K. (1980). "High-Speed Laser Printing Systems," *Laser Applications,* vol. 4, 125.

Williams, E. M. (1984). *The Physics and Technology of Xerographic Processes,* John Wiley & Sons, Inc., New York.

### Lessons from Xerox: The Beginning

The case study about the invention of xerography has been presented here for many reasons. First, it demonstrates the anatomy of the innovation process and illustrates the complexity of such a process along with the many factors that lead to innovation. It also demonstrates the differences between scientific discoveries and technological innovation, showing how a purely empirical approach without a clear understanding of the basic principles involved can lead to innovation. Yet once the innovation is headed toward success, pure empiricism can be insufficient.

The inventor was a patent attorney, an entrepreneur, not an engineer or scientist but a man with a vision, an idea, a passion for this idea, and persistence. Carlson spent six years trying to generate support for his idea. His first break came when Battelle became interested in his idea. Battelle brought to the process important knowledge, talent, and scientific approaches. Haloid's Wilson found out about the idea by chance. However, being a visionary, entrepreneur, and venture capitalist in his own right, he moved quickly to support Battelle's research (with $25,000). Being a businessman, Wilson, and Haloid, moved to protect the technology by giving it a trade name.

The U.S. Army, a frequent supporter of basic and applied research of interest to its operation, moved to support Haloid's development effort with a $120,000 grant. Financial support by third parties and by government bodies is frequently the catalyst that propels new ideas toward development and marketing.

The first Xerox machine was announced in 1949. Yet it took several innovations to improve it and bring its technology to a level acceptable by the marketplace. The initial predicted market size for the early-generation Xerox 914 copy machine was 3,000 placements over the life of the product. By any traditional accounting method, such as return on investment (ROI) or rate of return (RR), this was a volume that would not justify further development of the product. In reality, more than 200,000 machines were sold. A revolutionary innovation that was touted as a product concept without a market indeed created an unexpected and phenomenal change in the marketplace.

When dealing with revolutionary innovations, market surveys and predictions cannot be trusted to account for the expected change. Meanwhile, it was a marketing innovation that facilitated customer acceptance of the product. A leasing and pricing structure introduced by Peter McColough helped to overcome the obstacle of the machine's initial cost of an outright sale. This strategy certainly helped the diffusion of the 914 machine and retained legal ownership of the product with significant tax advantages of depreciation. Once the technology was well developed, accepted by the market, and diffused, a huge explosion in the volume of copies produced ensued. The company's revenues multiplied.

The following MOT lessons can be drawn from this case:

**1** The success of an invention is dependent on its marketability.

**2** Multiple generations of innovations may be required to facilitate customer acceptability of the product.

**3** Market surveys have limitations when used to predict the impact of revolutionary technological innovations.

**4** Inventions and innovations need one or more champions to create them and introduce them to the marketplace.

**5** Successful innovations require a combination of factors, including vision, persistence, technical and business skill, good intuitive sense, and good luck.

READING 6.3

# A Model for Technological Innovation in Biomedical Devices

**Ahmed Zaki and Tarek Khalil**

## INTRODUCTION

Innovation and the issue of competitiveness are of prime concern in the United States due to their economic impact. Foreign competitors are challenging American industries in many sectors of the economy. The automobile industry and home electronics are examples of industries facing this challenge. Lack of competitiveness can be attributed to several factors, including: management's short-term vision, delays in technology transfer, investment policies, trade policies, quality issues, labor cost, and lack of interaction between organizations (Berman, 1990; Berman and Khalil, 1992; Dertouzos et al., 1989). These factors can result in the failure to translate new ideas into successful marketable products or in delays in the delivery of products to the marketplace. Much of the information presented in this paper is based on the authors' own experience with the development of biomedical devices from concept to market.

## A TECHNOLOGICAL INNOVATION MODEL

The process of technological innovation in biomedical devices follows the same generic path as other technological innovations (Utterback, 1971). However, the timing and methodology of each stage of the innovation process could differ due to the nature of the products. The model shown in Figure 1 illustrates the important components of this process with emphasis on the special characteristics of the biomedical devices industry. The details of each component in the process are described and illustrated in subsequent figures.

### Generation of Ideas

As can be seen in Figure 1, the generation of ideas is only the starting point of a complex process which includes two types of dimensions: internal, involving all the departments

*Source:* Modified from T. Khalil and B. Bayraktar (eds.), *Management of Technology III,* Industrial Engineering and Management Press, Atlanta/Norcross, GA, 1992. © Institute of Industrial Engineers. Used with permission.

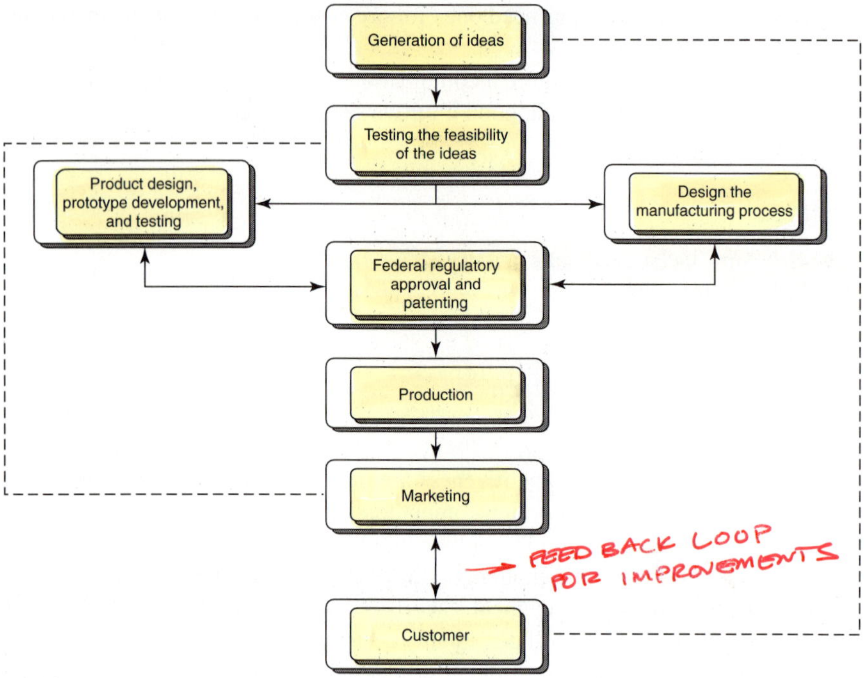


**FIGURE 1**
AN OVERALL MODEL FOR TECHNOLOGICAL INNOVATION IN BIOMEDICAL DEVICES
*Source:* Zaki and Khalil, 1992, © Institute of Industrial Engineers.

in the company, and external, involving customers, government agencies, as well as social and legal regulations. It is particularly interesting to note that in this industry most good ideas are based on customer needs even though many are still advanced by technological breakthroughs. Figure 2 presents in detail the factors that influence the generation of ideas. Innovative ideas are generated by technology push and market pull. If a company is to develop successful products, it must be connected to these two sources.

The market pull usually results from the special needs of physicians, healthcare professionals, and patients. Crises in healthcare and frustration with existing devices can also create the market pull. The needs of physicians and healthcare professionals are voiced and assessed through interaction with the sales force in biomedical equipment companies and through marketing analysis and surveys conducted by marketing specialists. Physicians' needs could be translated into new products or into improvements of existing products through the individual efforts of entrepreneurial physicians or through the interaction between a healthcare provider and a technically inclined counterpart. The technology push results from advancements in technology which stimulate researchers to generate new ideas and find cheaper and more efficient ways to solve problems. Biomedical engineers play a very important role in creating the technology push due to their close interaction with healthcare professionals in combination with their technical background.

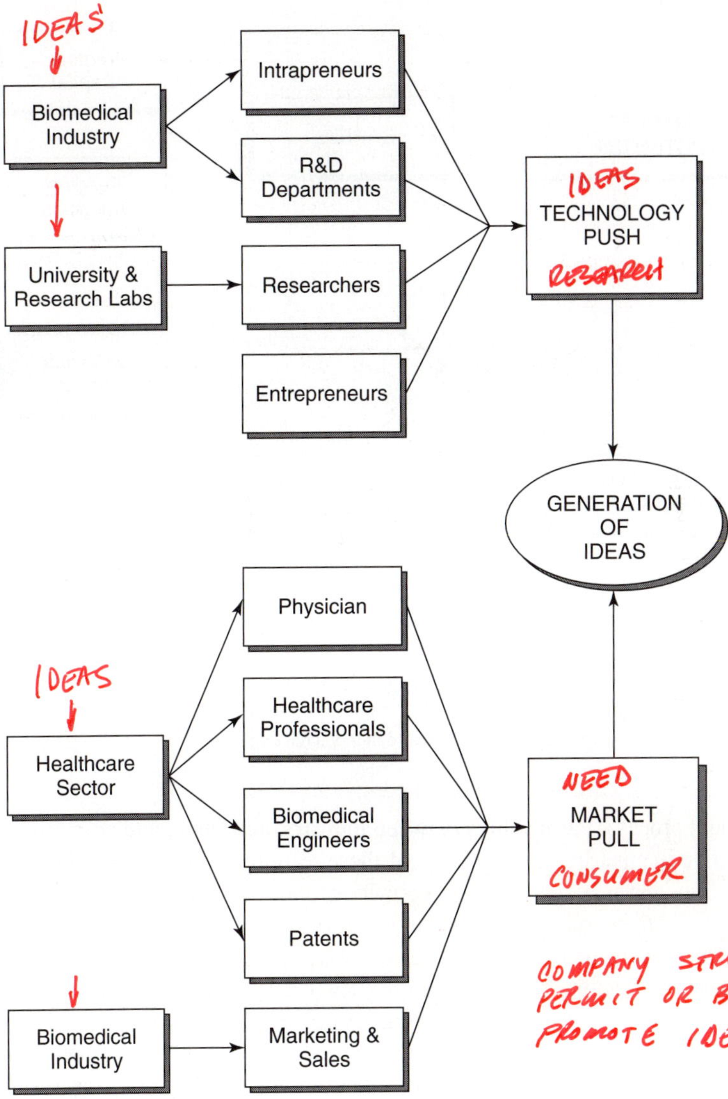


**FIGURE 2**
GENERATION OF IDEAS
*Source:* Zaki and Khalil, 1992, © Institute of Industrial Engineers.

### Testing the Feasibility of Ideas

The generation of an innovative idea is the starting point toward implementation. The feasibility of the idea has to be investigated. The model for this phase is illustrated in Figure 3. Market potential for products has to be assessed through market analysis. The nature of some specialized biomedical technologies is limited by the number of market users. This may place constraints on the market size. Resources (funds, qualified personnel, and professional advice) have to be readily available before proceeding with the

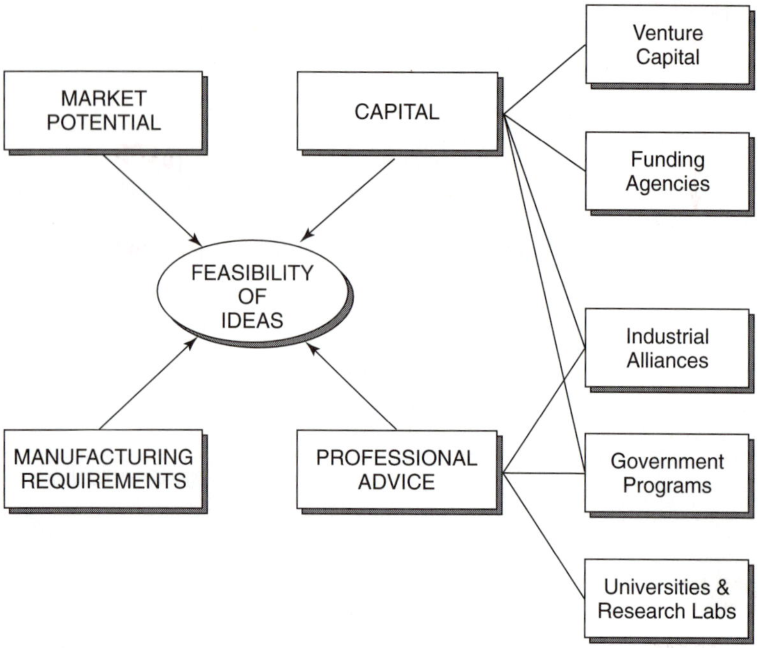


**FIGURE 3**
FEASIBILITY OF IDEAS
*Source:* Zaki and Khalil, 1992, © Institute of Industrial Engineers.

development process. Partnerships between industry, academia, and government can be very effective in ensuring the existence of these resources (Port, 1989). The cost of developing innovations, expertise, and knowledge can be shared through these partnerships. Funds can be obtained through venture capital, funding agencies, industrial alliances, and government programs. The involvement of the government in the innovation process can affect the outcome significantly (Carey, 1989). Testing the feasibility of ideas for biomedical devices is a more complex process than innovation, and requires additional steps and time due to the nature of the products. The complexity of investigating the feasibility of ideas arises from the need to use humans or animals for testing, which requires a stringent adherence to government, medical, and ethical standards.

The availability of or access to appropriate testing facilities contributes to the complexity of this process. The availability of qualified physicians, healthcare professionals, researchers, and engineers to conduct the testing process is an important consideration that should thoroughly be investigated at this stage.

This feasibility phase is very important in determining if the green light will be given for the innovation process to continue.

### Product Design, Prototype Development, and Testing

This stage is the beginning of the development process of the innovation. It is a very critical stage in the life of the project, as it will determine the future success and competitiveness of the innovation. Several factors in the design of the product should be

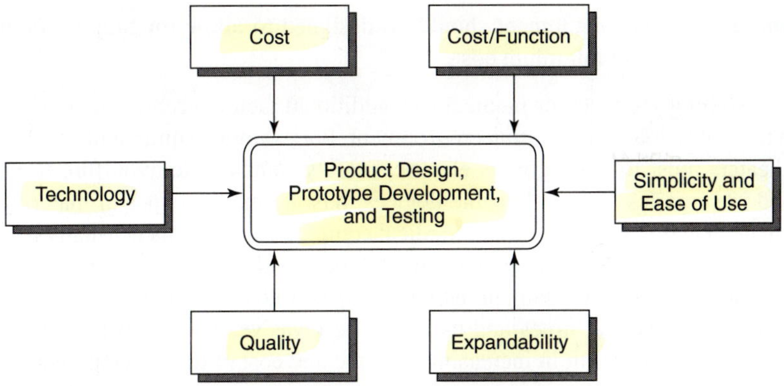


**FIGURE 4**
PRODUCT DESIGN, DEVELOPMENT, AND TESTING
*Source:* Zaki and Khalil, 1992, © Institute of Industrial Engineers.

seriously considered (Figure 4). Prominent among these are:

**Technology** Making use of available technology is usually the most expedient way in developing the biomedical innovation product or process. If this is not possible, researchers should investigate new technology to achieve their goals. It should be clear that using new or emerging technology can be expensive and time consuming. If an emerging technology is used, it must be protected by the innovators in order to maximize its benefit to them.

**Quality** The quality of the product or process innovation is of utmost importance in this industry. Any slippage in quality exposes the firm to very expensive retrofit problems as well as liability and legal ramifications.

**Cost** The cost of equipment should be kept to a minimum without sacrificing quality. The use of fancy components that perform more than the required functions should be avoided (e.g., using sophisticated microprocessors that have more power and capability than the application requires). This consideration is sometimes neglected by biomedical device producers who may use extras as a marketing scheme when dealing with novice users of the technology.

**Cost-function relationship** A balance between the cost of the device and the features included should be maintained. It is essential to design and evaluate the equipment based on the equipment's user-related features rather than its technological features. This consideration is unfortunately ignored by some innovators in the industry, particularly if they feel that the buyer is not cost-conscious.

**Simplicity and ease of use** The equipment should be simple in design, user-friendly, and easy to use. Ergonomic concepts for human-machine interaction should be considered. This is a relatively new but important addition into the biomedical devices industry. If the equipment has global-market potential, proper design for ease of use becomes more important (Khalil and Waly, 1988).

**Expandability** The equipment should be designed to allow for future modifications and development at minimum cost.

When designing products for medical use, additional factors need to be considered, based on their function and operating environment. For example, equipment used in operating theaters must follow rigorous electrical safety standards to avoid fire hazards. Tools used in surgeries and implantable devices must be made of special materials to avoid rust and reaction with live tissue. Medical equipment and products usually require extensive testing since they are used in the treatment of patients. Testing medical equipment is a more complex process than the testing of non-medical equipment, because it requires human subjects at times and can even be invasive. The extensive testing of medical equipment can result in increasing the time and cost of the development.

### Design of the Manufacturing Process

Concepts used in the design of the manufacturing process in the biomedical devices industry are not significantly different from those used in other products. Different manufacturing techniques are investigated and the one that ensures better price and quality is implemented. The major distinction is that in the manufacturing process for medical devices and disposables, the quality of the process should be closely monitored and rigorous testing and quality control stressed. At times the process will require special environments (e.g., clean rooms) or additional processes (e.g., sterilization) or special packaging.

### Federal Regulatory Requirements

Medical devices and implantables require approvals from the Food and Drug Administration (FDA) before being used by or marketed to customers. A premarket notification, 510(k), must be submitted to the FDA at least 90 days before the introduction of the invention to the marketplace. Testing of prototypes also requires FDA approvals. Prior to testing, protocols are approved by the FDA and the Human-Subjects Committee of the institution involved in testing. The time-consuming application and approval process requires paper work and professional expertise. Applying for patents is usually done during this stage.

### Production

In this phase, the product is manufactured and produced in its final form. Again process and product quality control is an important component of this stage.

### Marketing of the Innovation

The marketing process involves the following classical components:

Market analysis (opportunities, threats, and competition)
Market Segmentation
Pricing
Distribution
Promotion

Marketing specialists usually conduct studies to identify the correct markets, determine the equipment price, and determine the methods by which to effectively reach the end-user. Marketing effort of medical equipment and disposables is similar to that of non-medical equipment except in channels of distribution and promotion. Medical equipment usually has limited channels of distribution, either from the manufacturer directly to the customer, or through a dealer, or distributor. The promotion strategy is mostly dependent on direct sales, exhibitions, and conferences. Advertisements are mostly limited to professional journals and magazines or promotional samples given or loaned to physicians or hospitals. The position of the product in the product life cycle will determine the type of pricing and promotion policies that will be used. If the product is a new innovation with no existing competition, promotion activities will be concentrated on educating the customers about the product, perhaps with penetrating or skimming pricing strategies. If the product has competition in the market, the product is either in the growth, maturity, or decline stage; promotional efforts will be concentrated on highlighting the difference between the product and its competition. In the biomedical devices industry, the rule of thumb in pricing a product is approximately eight times its manufacturing cost. If the market can bear this price, then the product commercialization is quite feasible. Recuperation of development cost usually takes place from the commercialization of the first run of the product if the innovation is judged a success. Innovative marketing strategies, such as leasing of equipment or basing prices on insurance reimbursement policies, are often used to permit successful commercialization of the innovation.

## CONCLUSIONS

The biomedical devices industry has some unique characteristics resulting from the special nature of the healthcare industry which influences the market pull of the innovation. The market size depends on the problem being addressed. Rigorous federal regulation combined with users' expectations for quality render the process more complex. It also contributes to higher cost of development. The potential for success, which is generally low, and the exposure to risk add another dimension to this industry.

Managing the innovation process is a key factor in the success of the innovation. Managing the development includes the planning of the innovation process and coordinating the different activities, individuals, disciplines, and organizations involved. Certainly key factors that can contribute to the success of the innovation are the timely introduction of the innovation to the market and the shortening of innovation process from concept to market. Delays can allow competitors to enter the market first, and have the name lead and strong market share.

### Reading 6.3 References

Berman, E. M. (1990). "R&D Consortia: Impact on Competitiveness." *Technology Transfer,* Summer Issue, pp. 5–12.

Berman, E. M., and Khalil, T. M. (1992). "US Technological Competitiveness in the Global Economy"; a survey. *International Journal of Technology Management,* vol. 7, nos. 4/5, pp. 347–358.

Carey, J. (1989). "Washington Inc.?" *Business Week,* Innovation issue, pp. 40–41.

Dertouzos, M., Nestor, R., and Solow, R. (1989). "Made in America: Regaining the Productive Edge." *M.I.T. Commission of Industrial Productivity,* Massachusetts Institute of Technology Press, Cambridge, MA.

Khalil, T. M., and Waly, S. M. (1988). "Planning for Health-Care Technology Transfer." In Khalil et al., *Technology Management I,* Inderscience Enterprises Ltd., Geneva, Switzerland, pp. 424–431.

Port, O. (1989). "Financing Innovation; Agenda for Change." *Business Week,* Innovation issue, pp. 186–173.

Utterback, J. M. (1971). "The Process of Technological Innovation within the Firm." *Academy of Management Journal,* March, pp. 75–88.

## ENTREPRENEURSHIP

Technological progress is frequently sparked by entrepreneurs and entrepreneurial spirit. Chester Carlson and Joseph Wilson of Xerox, Steve Jobs of Apple, and Bill Gates of Microsoft are but a few examples of successful entrepreneurs. Entrepreneurs are a special breed of people who have the ability to sell or market ideas to others. They possess a particular set of qualities, including vision, courage, initiative, commitment, persistence, independent thinking, drive to succeed, and ambition. Most entrepreneurs have an appreciation for a particular technology branch, good motivational skills, and a commanding personality. They tend to deviate from mainstream thinking, enjoy being the center of attention, and savor recognition. They could be compulsive in seeking their goals or perfection. Entrepreneurs are not necessarily inventors. Their skill usually lies in bringing innovation to the marketplace.

Successful entrepreneurs with business savvy have a staying power unlike those who are only creative and thrive on chaos. An example of a successful entrepreneur with staying power is Bill Gates. He has been able to lead a company that started with two people (Gates and Paul Allen) and guide it through a technology revolution to become the largest software company in the world (Goldblatt, 1995). In 1996, Bill Gates's net worth was greater than the gross national product of Nicaragua (Starling, 1996). Between 1996 and 1997, his net worth increased from about $18 billion to more than $36 billion. In 1999 it was more than $90 billion and with the continued growth of his technology sector company, his net worth today is greater than the GNP of several countries combined.

Another example is Steve Jobs, who teamed with Stephen Wozniak to create Apple Computers. Once an entrepreneur, always an entrepreneur. Jobs was forced out of the company in a power management struggle in the mid-1980s. He moved on to a new venture, buying and leading Pixar, a successful computer-based movie animation company that produced the very successful movie *Toy Story*. He had paid $10 million to buy Pixar from George Lucas. He invested another $50 million in the company. In 1995, Pixar went for public offering and Steve Jobs's stake in it was valued at $1.2 billion (Warshaw, 1996).

Entrepreneurs play a key role in creating and advancing technology and in sparking economic growth and job creation. The entrepreneurial spirit of Silicon Valley helped create thousands of new jobs in that region of the country and elsewhere. Paradi (1994) reported that "from 1978 through 1992, eighty-five percent of the new jobs in Canada were created by small entrepreneurial firms (firms with twenty or fewer employees). By comparison, large corporations (500 employees or more) recorded no growth at all."

Entrepreneurship, however, is not limited to individuals or small companies. It exists and must indeed be encouraged in large corporations if the corporation is to maintain its vitality and ensure its long-term survival. Entrepreneurship within the confines of an organization is frequently referred to as *intrapreneurship.* Bergelman et al. (1996) refer to this concept as *internal entrepreneurship,* in which the entrepreneurial process of combining resources remains nested in the larger resource combination constituted by the firm. Bergelman et al. (1996) distinguish between this mode of entrepreneurship and *external entrepreneurship,* which they define as the individual entrepreneurial process of combining resources discovered in the environment with the entrepreneur's very unique resources to create a new combination that is basically independent of all other resource combinations.

## ENTREPRENEURIAL VERSUS STEWARDSHIP MANAGEMENT

Organizations are created to transfer knowledge or technology to the marketplace. The structure of the organization provides a system to integrate the resources required to exploit the knowledge. The objective of management is to guide the organization toward achieving its mission while optimizing the use of resources. Organizations can be managed in a highly structured, rigidly standardized way or in a loosely defined, less formal, entrepreneurial way. The former is associated with professional management, the latter with the entrepreneurial management style. Personal characteristics of managers represent both ends of the spectrum of organizational leadership. A summary of these characteristics is presented in Exhibit 6-1.

In MOT the style of management adopted by organizations requires adjustment according to the different phases of the technology life cycle. In the early phase of technology development, an entrepreneurial style of management is needed. An entrepreneurial spirit creates disturbance in the system and helps organizations create new technology cycles or discontinue existing ones. However, once the technology life cycle is in the full growth phase or in the maturity phase, stewardship or professional management may be needed to exercise control over operations. Systems need to be developed for production, inventory control, and logistics. Facilities must be built and machines acquired to meet the demand for the product or service. The organization must be prepared for full production to satisfy growth and the market demands. The professional managers become more concerned with running a business day in and day out and preserving it through tightly controlled procedures. A stewardship managing style avoids risk taking. It often leads to a bureaucratic environment. It frequently has a short vision and in many cases fails to recognize the importance of innovation. This may kill new initiatives. An example of a company whose conservative, stewardship management style was unable to take advantage of technological strength within the organization is Xerox in the mid-1970s. Xerox management's inability to recognize and

**EXHIBIT 6-1**
PERSONAL CHARACTERISTICS OF MANAGERS

| Professional manager | Entrepreneur |
|---|---|
| 1. Career-oriented with well-defined goals. | 1. Self-starter; defines goals as he or she goes alone. |
| 2. Accomplishes tasks through people. | 2. Does the important things by himself or herself. |
| 3. A good delegator and motivator. | 3. Not a good delegator; strong need to control. |
| 4. Good leader and people person. | 4. Charismatic leader, but hard to follow. |
| 5. Competitive and politically astute. | 5. Extremely strong drive and capacity for work. |
| 6. Reward-oriented for:<br>• Cash<br>• Visible rewards<br>• Status<br>• Perquisites | 6. Reward oriented for:<br>• Money<br>• Visible rewards (cars, boats, clubs, etc.)<br>• Community admiration for accomplishments<br>• Perquisites |
| 7. Experience, ability, and accomplishments are evident. | 7. Excellent problem-solving abilities. |
| 8. Plays by the rules, not a risk taker. | 8. Innovative thinker. |
| 9. Committed to self more than to company. | 9. Realistic; takes moderate and well-calculated risks. |
| | 10. Committed to the company. |

*Source:* Paradi, 1994.

implement ideas generated in the company's R&D Palo Alto Research Center (PARC) cost Xerox its potential to lead the PC market (see Reading 6.4).

## THE MANAGEMENT RENEWAL CYCLE

Entrepreneurs have difficulty keeping order and conforming to rigid procedures. They also have a strong attachment to their ideas and products. It is difficult for them to let go of something they championed, gave birth to, nurtured to health, and watched grow. However, there is a time when the entrepreneurial management style must give way to a stewardship one. When the latter arrives, it often conflicts with entrepreneurs, forcing them out or stifling their creativity. This need not be the case, yet it often happens. The bureaucracy created by stewardship resists change and gives way to an organized but static system. Entrepreneurial spirit fades. A stable or declining period ensues. To revitalize the organization, entrepreneurial spirit must be brought back and the cycle continued. Figure 6-4 depicts this cycle. Successful organizations must be able to embrace elements of these two managerial styles. The organization must be structured to innovate, respond to and sustain growth, and continue to be in the forefront of technological change.

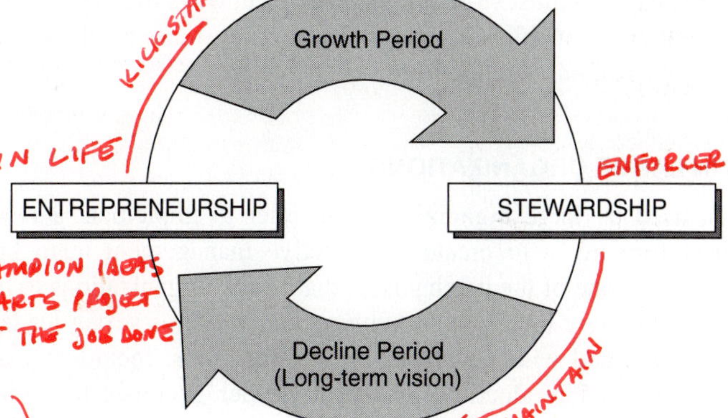


**FIGURE 6-4**
MANAGEMENT RENEWAL CYCLE

Many companies have shown that it is possible to encourage and compete with innovation. The Minnesota Mining and Manufacturing Company (3M) has been successful in competing with innovation (see Chapter 15). The company encourages entrepreneurship among its employees. It requires that its divisions derive 30 percent of their sales from products created in the last four years. It also supports employees in bringing innovations to the market. Other cases also show that stewardship is needed at certain phases of the life of an organization. Apple came to view Steve Jobs, its founder, as too demanding and as spending too much of its resources pursuing new ideas that did not contribute positively to the bottom line in the short run. Apple hired a new CEO, Sculley, who agreed with Apple's board to force Jobs out and to foster a more orderly management style to sustain progress. Sculley's strength was in marketing, having spent two decades in PepsiCo before joining Apple. Sculley emphasized high-profile corporate sales and expanded Apple's operations around the globe. He was successful for about five years, until Microsoft started gaining grounds with its Windows technology. Apple was unable to adequately exploit its lead in technology to stave off Microsoft's advances. It was also unable to renew its entrepreneurial spirit to recapture the lead (Pebello et al., 1996). After several years of decline, Apple turned back to Steve Jobs to save it from demise. He forged an alliance between Apple and Microsoft. Playing the management-of-technology game continues: Sometimes, professional management ignores or is blind to the importance of innovation, as in the case of Xerox (described in Reading 6.4). Other times, hierarchy, conformance, and rigid systems stifle innovation, as in the case of IBM. In the early 1980s, IBM was a follower in the PC market. However, because of its name and complementary assets, IBM was able to catch up with Apple in the PC race and gain a larger market share with its PC manufactured at Boca Raton, Florida. Nevertheless, IBM was not able to sustain that success because it failed to control the PC technology. Its hierarchical organizational structure did not permit a continued success for its Boca Raton venture.

Organizations can and must structure themselves in ways that permit the proper balance between entrepreneurship and stewardship. They must be flexible, or agile, to allow for necessary responses to environmental change.

### VENTURES IN LARGE ORGANIZATIONS

One approach used by large organizations in order to implement a new venture or move a project forward is to create an effective management team known as the *venture team.* This is one of the mechanisms that enable organizations to deal with turbulence in the technology life cycle. A venture team requires strong leadership, excellent technical skills, and wide representation of functions, including manufacturing, marketing, finance, and production. The team leader becomes the *product/process champion* and assumes leadership over technical matters. He or she should possess entrepreneurial skills.

Roberts (1977) explains that any project team working on an innovation must be able to perform five critical functions: idea generation, technological and market gatekeeping, project championing, project managing, and coaching.

Physically locating the team away from the company's main operations is a common practice. It shields the team from being distracted by the day-to-day problems of operating the company, and it allows the team to concentrate on the task at hand. IBM followed this approach when developing the IBM PC: The venture team was located in Boca Raton, Florida, away from IBM headquarters in New York. Xerox also did so when it located a research team at the Palo Alto Research Center (PARC) in Palo Alto, California.

## CASE 2

#### READING 6.4

# Xerox—After the Invention

**Javier Garcia-Arreola and Tarek Khalil**

### XEROX'S EARLY DAYS

Xerox's history encompasses important lessons for people interested in the management-of-technology field. From the early years of Chester Carlson's efforts to introduce this technology (Reading 6.2) to recent restructuring in the 1990s, Xerox presents a wide variety of examples on managerial practices and pitfalls that illustrate important concepts in management of technology (MOT).

Carlson, who became tired of the mistakes and impracticality of retyping documents, envisioned the process of image reproduction using electrophotography. He offered the technology to companies such as RCA, Kodak, General Electric, and IBM, but they turned him down because they believed that carbon paper could produce the same results.

Although Carlson was often frustrated by the lack of interest in his invention, he never gave up, and he and his associates successfully developed a machine which was dubbed by *Fortune* magazine "the most successful product ever marketed in America."

## PARC: SCIENTIFIC ACHIEVEMENT AND BUSINESS FAILURE

In the early 1970s, the newly appointed president, Peter McColough, had the vision to purchase a computer company in order to enter the information industry. This was a brilliant vision; however, vision cannot be realized without the existence of many other supporting components. McColough wanted Xerox to compete directly with IBM. In May 1970 he approved the acquisition of Scientific Data Systems (SDS). Although a small firm—SDS sales during 1968 were only $9.8 million—McColough paid $900 million in Xerox stock (Walker, 1993).

In 1970, Jack Goldman, Xerox research director, proposed, and McColough approved, the creation of a new research center to transform the company into a digital corporation. During a speech before the New York Society of Security Analysis, McColough laid out the vision for the next decade: "The basic purpose of Xerox Corporation is to find the best means to bring greater order and discipline to information. Thus our fundamental thrust, our common denominator, has evolved toward establishing leadership in what we call 'the architecture of information' " (Walker, 1993).

With the vision set and the research budget approved, Goldman started working. He hired George Pake to run a new center, which was located in Palo Alto, close to the emerging semiconductor industry sources, hence the name "Palo Alto Research Center" (PARC). Pake hired Bob Taylor, who started recruiting the best computer minds in the country using an extremely thorough selection process.

PARC was an excellent example of how a research team can be assembled and used to foster technological innovation. The synergy that the researchers achieved in PARC was unparalleled due greatly to the excellent communication among the scientists and the flat organizational chart. All the employees reported directly to Taylor and weekly meetings were held to share their individual progress. During meetings an individual may have led the discussion; however, the objective always was to share models and visions. In the first five years of operation, PARC devised working versions of a personal computer (the Alto), a laser printer, advanced word-processing programs with what-you-see-is-what-you-get (WYSIWYG) technology, multitasking operating systems, the first object-oriented programming language, the mouse control, and an Ethernet. Had Xerox been successful in managing these technologies and swiftly bringing these products to the market, the story of the PC would have been different. As it happened, Xerox was not able to capitalize on its early lead in this technology sector.

## DECLINING YEARS

Studies by Walker (1993) reveal some very interesting facts about the case of Xerox. When Chairman Joe Wilson died in 1971, Archie McCardell was named president of the company. McCardell was a Ford-trained financial executive whose management theory was defined by thorough financial controls and intolerance to any risk. He brought with him a strict "stewardship" management style. Under this regime, Xerox research became highly bureaucratic: dozens of review points were established and the entire process was subject to committee decisions.

Bureaucracy was not the only problem. Quality declined as well. According to some reports, a machine once caught fire in the White House (Walker, 1993). Instead of fixing the quality problems at the source, Xerox hired 700 more service representatives.

In addition to quality, time to market was extremely slow, inhibiting Xerox from introducing new models to the market. The company was organized in such a way that product planning, engineering, and manufacturing did not meet until the projects got to the president's level. During the fatal decade, only three completely new copying machines were introduced in the U.S.

In 1973, McCardell realized the importance of moving the development operation to a new location, away from the Rochester area. He asked the new director of the Office Product Division, Bob Potter, to select the new location. Two cities were candidates, Dallas and San Francisco. Based on complex financial models—which were already common practice of the company—Potter selected Dallas. The distant geographic location from PARC and other internal rivalry factors did not help communication or create needed synergy between research and development. The unfortunate result was a cold war among them, initiating competition instead of cooperation.

In 1976, in another shortsighted financial decision, Xerox's PC development was killed. The engineers at PARC had developed almost every innovation that exists in modern PCs: the Alto, a computer system with unique features such as icons, windows, and mouse control. Researchers at PARC could send electronic mail through an Ethernet network, and they had the capability to print with a Xerox-developed laser printer. The Alto system was the first automated office ever created. In spite of all these pioneering technological developments, managers at the Xerox headquarters were very cautious in decisions regarding the development of computer technologies. All Alto's features were unimportant to an analytical team which contended that the Alto manufacturing costs were not accurate. They suspended its funding and lost the PC race. Instead, Xerox's Office Product Division launched a very expensive workstation and started competing with many technological followers that entered the copying market.

## XEROX'S RECOVERY

The Japanese companies' entry into Xerox's traditional marketplace affected its market dominance and profitability drastically. The corporation introduced a major benchmarking effort in 1979 with the aim of improving performance. It discovered it had huge overhead problems and manufacturing costs when compared with the Japanese. This is an expected situation in a large company that starts competing in mature technology.

In 1982 David Kearns became CEO of Xerox at a time when market share was only 13%. Kearns set up a team of executives to launch a major cultural change program that included quality goals, teamwork, systematic problem solving, and avoidance of internal competition. Ten factors were the main driving forces behind the effort (Jacobson and Hillkirk, 1986):

1. Competitive benchmarking
2. Pushing responsibility down
3. More emphasis on market research
4. A relying point (the 10 Series machines)
5. Driving technology
6. Internationalization
7. Just-in-time manufacturing
8. Automation and computerization
9. Moving faster
10. Emphasis on quality

It should be noted here that all these factors are very important for becoming competitive in a global marketplace. They are particularly critical for an industry competing in a mature technology.

In 1987 the conflict between quality and financial results was the cause of poor quality at Xerox. Managers selected profits over quality in any decision. After a quality evaluation, Kearns made customer satisfaction the sole focus of the company. By 1989, Xerox won the Malcom Baldrige National Quality Award. The quality effort had a positive result in the financial measures and the market share was back to 19% (Walker, 1993).

In 1989, Xerox announced a major layoff (2,000 employees) and a new restructuring effort. John Seely Brown was appointed PARC director. He expressed the following philosophy in driving the research center at Palo Alto (Brown, 1991):

**1** Research on new work practices is as important as research on new products.

**2** Innovation is everywhere; the problem is learning from it.

**3** Research cannot just produce innovation; it must co-produce it. One must rethink the process by which innovation is transmitted throughout the organization.

**4** The research department's ultimate innovation partner is the customer.

## CHANGING THE ORGANIZATIONAL STRUCTURE

In 1992 Xerox announced more restructuring efforts under Paul Allaire which were aimed at making Xerox even more responsive to customer demands. A total redesign of the company structure was sought.

The new structure is aimed at linking technological advancement with market needs. This is done through business divisions which are responsible for specific phases of the product's entire life cycle from concept to market. According to Howard (1992), Allaire explains this organizational structure as follows:

> The centerpiece of our new architecture is a set of nine relatively independent business divisions. These are stand-alone businesses organized around specific products and markets and with profit and loss responsibility. Each division consists of a number of business teams, smaller entities tied to the marketplace by a specific customer need. In fact, the business team leader is the new entry-level general manager job in the company.

> [The divisions] are linked to other parts of the company whose responsibility is to leverage corporate wide resources and relationships in support of the divisions. At one side [of the organization] is technology, and we have retained and integrated corporate research and technology organization. At the other end is the customer. We have organized our sales and service people into three geographic customer-operations divisions. Their purpose is to create some "suctions" on technology and pull into the marketplace. Finally, we have created a new unit called "strategic services" that will provide support to the business divisions in areas such as specialized manufacturing and purchasing, where economies of scale still make a big difference. And we've established a six person corporate office, served by a much leaner corporate staff, to make sure all these units fit together in a strategically coherent way.

Allaire's team worked on a complete change where communication, cultural change, and continuous learning were covered. Allaire calls this process "organizational architecture." It implies not only a change in the structure, but in the culture of the organization.

## CHANGING THE REWARD SYSTEM

Compensation has also changed at Xerox in order to reinforce the desired behavior. For the 50 top positions in the company, managers have share incentives, as do managers in many other companies, but these are only earned if the company meets certain objectives fixed in advance and based on return on assets. On top of this, management must actually purchase one year's salary worth of Xerox stock (Howard, 1992). The bonus policies have also changed. They are now based on corporate performance, division performance, and personal performance.

## PROMOTING ENTREPRENEURSHIP

Entrepreneurship is also promoted within the new architecture in Xerox. Managers of the different divisions ask for money from corporate management in the way an entrepreneur goes to a venture capitalist or banker. In this way, budgets are not set up from the beginning, but as a result of the individual plans of each division. Corporate management's role is to provide coherence to the individual strategies, based on how they fit together for the company as a system.

### Reading 6.4 References

Brown, John S. 1991. "Research That Reinvents the Corporation." *Harvard Business Review,* January–February, 102–111.

Howard, Robert. 1992. "The CEO as Organizational Architect: An Interview with Xerox's Paul Allaire." *Harvard Business Review,* September–October, 107–119.

Jacobson, Gary, and Hillkirk, John. 1986. *Xerox: American Samurai*. New York, Macmillan Publishing Company.

Walker, Wayne G. 1993. *Recovering the Fumbles and Organizing for the Future: Xerox Integrates R&D into Corporate Strategy with Pioneering Research and Restructures to Become a Learning Organization—with Lessons for Military Acquisition.* Rand, Santa Monica, CA.

### Lessons Learned from Xerox—After the Invention

Several lessons can be extracted from the case study presented above, including:

- Research groups with freedom from daily-defined assignments are more creative.
- Locating R&D ventures away from the main corporate operation center gives a research team the freedom to innovate.
- Bureaucratic management can seriously hinder technological innovation.
- Top management's focus on short-term financial objectives may be detrimental to the realization of a long-term objective or vision.
- Synergy and cooperation between the research arm and the product-development arm of a company are essential for the successful push of technology. Synergy and cooperation between R&D, production, and marketing are essential for successful commercialization of technology.

### Comments on Xerography—After the Invention

When Xerox began, Carlson's radical innovation created a completely new market. The new technology was valued by customers and was the base for the unimaginable growth of Xerox. However, profits at that time were mostly based on a single product: the 914 copier. This product became the industry standard and made Xerox the leader in the copying industry. It provided enormous profits to Xerox. Once the technology became recognized, competition shifted from radical innovation to incremental and process innovation. Competitors were growing as the technology matured. Xerox started losing its leadership advantage to more nimble corporations. The company's unbelievable growth required better management controls in order to foster future developments. Thus, the board appointed McCardell to be a steward in the effort to guide Xerox's growth. The decision was apparently correct as related to the printing technology sector of the company. However, the developments in the information technology sector at PARC required a much more entrepreneurial way of thinking. By 1975 Xerox's market share and revenues where lost to new competitors such as Canon, Sharp, Minolta, and Panasonic, whose quality was inching higher and costs were getting lower.

The truly outstanding research in the 1970s at PARC was, without a doubt, the result of state-of-the-art organizational culture of the center itself. The culture at PARC allowed a great degree of coordination and communication and permitted ideas to flow among its engineers. Moreover, the location of the laboratory in the heart of Silicon Valley permitted those engineers to share opinions, meet colleagues, and be aware of the latest discoveries in the field. These are factors that assist in enhancing creativity (Jacobson and Hillkirk, 1986).

But PARC had a serious problem. The technology created at the center was never really appreciated by the bureaucrats at Xerox. The technology, therefore, was not transferred to the rest of the company or to the commercial market. Whereas the communication within the walls of the laboratory was excellent, communication with the rest of the company was extremely poor. There was no linkage between the vision of the center's scientists and engineers and Xerox's strategy. Walker (1993) has established the following as the major causes:

- Cultural separation and technological asymmetry.
- The Xerox hierarchy.
- Mental maps of viable technology.
- Fixation on the bottom line.
- The financial management system brought in from Ford in the late 1960s.

Potter's decision to relocate Xerox's Development Division in Dallas might have been financially accurate, but it was strategically wrong. It had the great disadvantage of undermining the needed synergy between the scientists and engineers at PARC and the Development Division.

PARC's Alto system presented to Xerox the opportunity to be the leader in innovating the PC business. This would have ensured its domination of business practice in the workplace. It seems strange that a company built on risk and innovation could neglect the importance of continuing to compete with new innovations. A main difference between this time and the early days of the company was the lack of a committed champion who pressed for the development of the new technology. Management was concerned with the bottom line and did not have entrepreneurial skills to foster developments in an entirely new arena.

During the 1990s, Xerox introduced drastic changes in the way it does business. CEO and Chairman Allaire, who also served as chairman of the Council on Competitiveness, implemented a radical change in Xerox's core technology. Changes are the result of tough competition, but as Allaire points out, technology and core competencies also require a new approach: Allaire redirected Xerox's strategic vision from being a producer of discrete products to being a more comprehensive "document company." He stated:

> Our traditional business is light-lens copiers and duplicators, fairly sophisticated electro-optical mechanical devices. Increasingly, they incorporate computer systems—to control for quality copy, for instance—but the guts of the machines are electro-optical and mechanical. These are standard alone devices. With the evolution of digital technology, however, and its rapid reduction in cost, the light-lens element in a copier can now be replaced with a scanning device that digitizes the information on the page. Once you've captured that image electronically, you can do all kinds of things with it in addition to just making a copy. In this new technological environment, it's not enough anymore just to make and sell discrete products. Rather, we need to offer our customers distinctive capabilities that will have a major impact on the way they do business. (Howard, 1992)

This is the reason why Xerox has redefined its strategic vision toward creating "the document company."

To restore its competitive edge in the marketplace, Xerox made it known that the new organization will demand new skills from employees. Characteristics such as strategic thinking, results-oriented teamwork, the ability to delegate, and empowerment skills are highly appreciated in the new structure. These are desired characteristics for employees of a company that is competing in a highly volatile technological environment.

## TECHNOLOGICAL INNOVATION—THE MACRO LEVEL

Scientific discoveries and technological breakthroughs have a potential for changing the way people live and do business. They spur major economic growth and may change the entire socioeconomic landscape of regions or nations. Therefore, understanding the dynamics of the process of technological innovation and how to foster it at the macro level of countries or regions, as opposed to the micro level of a company, is essential. Even though similarities may be found among models at both levels, we will focus below on the success stories of two macro-level developments and the lessons to be learned from them.

### The British Midlands

Creative people who tend to depart from the mundane and the status quo usually advance innovations. Entrepreneurs mostly drive innovations with support from skilled technologists. As Cadbury puts it (1995; see Reading 6.5), "Innovators tend to be drawn from the ranks of those with no investment or vested interest in the established order."

The Industrial Revolution was forged by people who moved to the British Midlands primarily to escape religious persecution. The people of Birmingham and the surrounding area were nonconformists, free from established or forced customs and practices. They believed in freedom of worship and freedom of trade, without the burden of the guild restrictions that applied in other cities in England (Cadbury, 1995). They were determined to prove their worth, and they valued independent, imaginative thinking and education and contact with like-minded thinkers. They were excluded from positions of public office and established organizations, so they turned to business and trade.

Nonconformist Quakers in the eighteenth century made significant contributions; for example, Darby founded the steel industry, and Ransome invented the self-sharpening ploughshare and introduced the idea of interchangeable parts in its manufacture in the early 1800s. The Lunar Society of Birmingham was formally founded in 1776. It consisted of 14 members who arranged to meet when the moon was full so that they "could have the benefits of its light in returning home" (Cadbury, 1995). The members of the Lunar Society made major contributions to the Industrial Revolution: Bolton created metal and silverware works; Watt contributed to the power and manufacturing industries; Keir created a chemical industry; and Galton made guns.

The Lunar Society members exchanged ideas, supported each other's activities, and shared information across disciplines and industry boundaries. They pooled their knowledge and experience in the interest of advancing technological innovations. Some of them also had great business savvy. For example, Matthew Bolton understood the value of quality industrial management and worldwide trading. He developed many of the features that have characterized major industries since the time of the Industrial Revolution: He established large manufacturing facilities (Soho Works), with the labor force living in close proximity to the factories; instituted large-scale production; paid attention to good design and quality; and marketed his products worldwide. As Cadbury points out, the British Midlands led the world into the Industrial Revolution and created great wealth and prosperity for Great Britain.

READING 6.5

# The English Midlands: Cradle of Technology

**Sir Adrian Cadbury**

First, may I welcome you all most warmly to the University of Aston and to this important and timely Conference. I would also like to thank and congratulate those responsible for its staging. That in itself represents a considerable feat of management. The title you have chosen is a reminder that there are no geographical barriers to innovation or to the exploitation of innovation, now that markets have become global.

You could not have chosen a more appropriate venue for your Conference than the English Midlands, Birmingham and this University. The Industrial Revolution was forged in the Midlands, Birmingham was where technology was first applied on a factory scale and this university grew out of the Mechanics' institutes which were founded in Birmingham from the 1830s onwards. On its way to becoming a university, Aston was established as one of the first Colleges of Advanced Technology. The themes of technological innovation, its management and global markets are all closely linked with the history of this region and this institution.

These themes have such a broad sweep that we may seem to be discussing impersonal social and economic forces whose forward march is predictable. The past suggests otherwise and points to technological innovation not as a deterministic process, but as one which depends on the individual spark of enterprise combined with seizing market opportunity.

It is relevant that Birmingham's first historian, William Hutton, who came here in 1741, was struck mainly by the people whom he encountered:

> I was surprised at the place, but more so at the people, they presented a vivacity I had never before beheld. I had been among dreamers, but now I saw men awake. Their very step along the street showed alacrity. The town was large and full of inhabitants and these inhabitants full of industry.

Given the origins of Birmingham, that vivacity was more a matter of self-selection than of chance. Birmingham's inhabitants were for the most part drawn here to escape religious persecution. In the 1660s, a variety of restrictions were imposed on those who did not belong to the Church of England, under what was known as the Clarendon Code. Because Birmingham was more than five miles from any City, Town Corporate or Borough, it escaped the Code's restrictions.

Birmingham, therefore, became a center for non-conformists of all persuasions and its inhabitants not only had freedom of worship, but freedom to trade as well, without the burden of the guild restrictions which applied to corporate towns and cities. Birmingham retained this independence from the established civic structure for many years, only becoming a Parliamentary Borough in 1838.

*Source:* Keynote address at the IAMOT European Conference on Management of Technology, Aston University, Birmingham, UK, July 5–7, 1995. Printed with permission.

Clearly the ability of Birmingham citizens to ply their hand at any trade and to train their workpeople to meet their needs, neither of which could be done under the dead hand of the guilds, provided the opportunity to innovate. It was, however, the drive and enterprise of the non-conformists which turned that opportunity to advantage. Those who broke with established religions were men and women who thought for themselves, who took responsibility for their lives and work and who were determined to prove their worth. They were innovatory in belief and therefore in action, since they saw life as a whole. Their innovation in action was assisted by their exclusion from established trades and professions, with all the barnacles of past custom and practice that went with them.

By the same token, although they set great store by education, they were sheltered from the stultifying effects of university education of the day. Their exclusion from the universities continued in this country until 1871, when the Corporation Act was finally repealed. The established universities in the 18th and 19th centuries looked to the past and their approach to education can be somewhat unfairly summed up by this quotation from an Easter sermon by the Rev. Thomas Gaisford, who was a 19th Century Regius Professor of Greek at the University of Oxford:

> The advantages of a classical education are twofold. It enables us to look down with contempt on those who have not shared its advantages and also fits us for places of emolument, not only in this world but in that which is to come.

Non-conformists were spared that deadening influence on their thinking, while being well aware of the importance of education for themselves and for those who worked with them. They favoured, however, the study of practical subjects like arithmetic, geography, natural history and what they referred to as the Modern Tongues. The first adult Sunday school in Britain was started in Birmingham and institutions for training in a range of mechanical skills flourished here in the 19th Century.

We have, therefore, a concentration of people who think for themselves and who value education as a means of applying knowledge to everyday affairs, but why should their energies have been directed so single-mindedly to trade and industry? A basic reason was that non-conformists were excluded from all positions of public office, from most professions, from the larger trade guilds and from academic life, as we have seen. Business was the main outlet for their drive and they brought to its pursuit a questioning approach, a persistent willingness to search all the time for improved methods and to harness knowledge to that search for greater efficiency. They made a major, but not exclusive, contribution to the technological and managerial innovations which brought about the Industrial Revolution.

The Quakers typify the contribution of non-conformist groups to technological innovation. A Quaker, Abraham Darby of Coalbrookdale, was the founding father of the iron and steel industry of today. He and his son were responsible for the two great innovations which transformed the manufacture of iron. Abraham Darby the first devised the process of smelting iron-ore with coke; Abraham Darby the second succeeded in making pig-iron of a quality that could be turned into bar-iron at forges. Another Quaker, though not a Midlander, Benjamin Huntsman, completed the sequence by inventing the process for casting steel.

You can see the contribution of the Darby family to technological innovation at first hand at Ironbridge. It is worth noting that the iron bridge itself, while a remarkable feat of design and engineering, had a practical purpose. It enabled Abraham Darby to bring work people to Coalbrookdale from the other side of the Severn at a time when his demand for labour was increasing rapidly.

The Quaker guiding hand can be seen across the broad spread of British industry from iron and steel, banking, railways, chemicals and engineering to consumer goods of all kinds, with a particular corner in tea, biscuits and chocolate.

Ransome is one of the best examples of a Quaker innovator; he made agricultural machinery; his business continues to this day. In 1803, he patented a ploughshare, which was self-sharpening and the method of its manufacture has not yet been superseded. He went on, however, to make an even more remarkable contribution to manufacturing technology. Ploughs tended to break at the beginning of the season and usually needed to be both dismantled and rebuilt—a lengthy process. To cut down the repair time involved and to deal with the seasonal build-up, Ransome invented a plough built from interchangeable parts. This meant that spares could be supplied and repairs quickly effected, at the start of the ploughing season. Thus the manufacturing approach which Ford was to exploit to such good advantage, in fact dates back to Ransome.

To put a measure on the degree to which this small religious group contributed to technological innovation, during the long history of the Royal Society, Quakers have secured forty times their due proportion of fellowships, in comparison with their numbers in the population.

A vital element in technological innovation is therefore that of individual initiative. It springs from those who, for whatever reason, are dissatisfied with the status quo, with things as they are. It requires a questioning mind, a restless search for improvement and a willingness to draw appropriately on knowledge of all kinds.

I would suggest that this leads to two practical conclusions. One is that innovators tend to be drawn from the ranks of those with no investment or vested interest in the established order. Immigrants are a group who fit that profile. We should recognize the contribution which successive waves of immigrants have made to innovation in this country and be prepared to back newly arrived potential entrepreneurs. The movement of people across Europe at present, from East to West and South to North, should be a spur to innovation, provided it is seen by receiving countries as an opportunity and not a threat.

The other conclusion applies to all of us involved in education. We need to devise courses which encourage inquiry, doubt and dissent. We should beware of specious certainty and the narrow focus of a too precise specialization.

While individuals are the mainsprings of innovation, their creativity and breadth of imagination can be enhanced by contact with like-minded thinkers.

For major forward steps in progress to be made, knowledge normally needs to spread beyond the straightjacket of a single discipline. The free flow of ideas and argument between enterprising people, bringing different skills and experience to bear on common problems, is a potent source of innovation.

An excellent example of this kind of encouragement to imaginative thinking was set by the Lunar Society of Birmingham, which made a specifically Midland contribution to the Industrial Revolution. It was formally founded in 1776, but the remarkable group of men who made it up had been meeting for some time before then. They arranged their meetings to coincide with the full moon so that they "could have the benefit of its light in returning home."

The contacts between them were not limited to these pre-arranged meetings, since they all lived in and around Birmingham and met or wrote to each other regularly. There were only 14 members of the Lunar Society over the years: Matthew Boulton, Erasmus Darwin, Thomas Day, Richard Lovell Edgeworth, Samuel Galton, Robert Augustus Johnson, James Keir, Joseph Priestley, William Small, Jonathan Stokes, James Watt, Josiah Wedgwood, John Whitehurst and William Withering.

It was said of them, "There was not an individual, institution or industry with pretensions of contact with advancing technology throughout the land, but some member of the Lunar Society group had connections with it."

Their interests were astonishingly varied, covering the whole field of human activity, but the special interest they had in common was science and how it could be applied to industry. Their network extended to America and Benjamin Franklin introduced William Small to Matthew Boulton at the same time asking Boulton to let him know "of anything new in Magnetism or Electricity or any other branch of natural knowledge which had occurred to his fruitful genius since last they had met."

The Lunar Society was therefore an outward-looking group whose members pooled their astonishing range of knowledge and experience in the interest of the scientific advance of industry and much else besides.

Four of the members ran their own businesses: Matthew Boulton made a wide range of metal goods and silverware at his Soho Works in Birmingham; James Keir established an alkali works at Tipton and the chemical industry with it; Josiah Wedgwood founded his pottery business at Etruria in Staffordshire; and Samuel Galton made guns in Birmingham—to the considerable concern of his fellow Quakers. As a result of these industrial contacts, a stream of practical problems came before the members of the Society, who contributed advice, ideas and an unbounded enthusiasm for experimentation.

As an example, Wedgwood made the first earthenware drain and water pipes, thereby making a major contribution to public health. The problem was how to make them strong enough and how to make them in sufficient quantity. In this, Wedgwood was helped by both Darwin and Watt to arrive at the right formula and at an appropriate manufacturing process. In the same way, the members between them made considerable advances in metallurgy to assist Matthew Boulton in his production of metal goods and in geology in the course of promoting canals.

The versatility of the Lunar Society members is exemplified by Dr. Erasmus Darwin, Charles Darwin's grandfather. He was a highly respected physician, but his publications span natural history, the classification of flowers and vegetables, "a plan for the conduct of female education in boarding schools" and poetry.

What can we learn from the workings of the Lunar Society? The main conclusion, perhaps, is the value of the flow of knowledge and ideas across the artificial boundaries of subjects, disciplines and professions. Not only were problems in one field solved by insights from another, but the pooling of all these skills and interests encouraged a process of intellectual fermentation which generated new thoughts and approaches with great vitality.

In seeking to stimulate technological innovation, we should be looking out for ways of breaching the compartments so impermeably constructed by specialists around their particular interest or function. Your conference here today is surely building on a similar approach to that of the Lunar Society.

Technological innovation is not, however, an inevitable process. Hero of Alexandria described the power of steam in the first century A.D., but its application awaited the demand for power in the 1790s. Charles Babbage's difference and analytical engines were too far in advance of the supporting technologies of their time for them to be constructed. Babbage also points up one of the vital elements in the process of successful innovation, which is that there has to be a latent demand for the innovation to satisfy, if it is to take off commercially.

There was no effective demand in the early 19th century for speed and accuracy in computing, which was what Babbage had to offer. The same was true of the punched card system, which controlled his engines. Jacquard operated the first fully automatic draw-loom by punched cards in 1801. Babbage, interestingly enough, had a five-feet-square, woven silk portrait of Jacquard; which had been made on a Jacquard loom; 24,000 punched cards each capable of taking 1050 punch holes were used to make it.

The potential of the punched card for the processing of information and for the control of machines was not realised until the demand for information, with the rise of the census, and the demand for automation in this century. Demand-pull and innovatory-push are both essential for the effective diffusion of technological advances, with market demand determining the timing of the advance. This brings me to my last historical example of a Midlands innovator and one who fully understood the global challenge, Matthew Boulton. He did not, as far as I know, belong to a non-conformist sect, but he was certainly not bound by establishment conventions. He came from the landed gentry and yet devoted himself to trade and above all to manufacturing, when he could have lived the life of a country squire. He acquired his capital by marrying the well-endowed daughter of a rich merchant and when she died marrying her sister, so keeping the money in the family—hardly in line with the doctrines of the established Church!

Matthew Boulton is the best example one could find of an entrepreneur who pulled together in his remarkable career all the threads of the Conference theme. He was a technological and managerial innovator, he marketed his goods and his ideas internationally and he understood the importance of quality and of good design.

Born in 1728, he was the first manufacturer to break through the 18th century distribution chain and serve his extensive export trade in decorative metal wares such as steel and silver buttons, buckles, watch chains and guards, seals, snuff boxes, toothpick cases and sword hilts—his pattern books in the Birmingham Reference Library include some 1470 designs for these kinds of objects, referred to as "toys." The greater part of Birmingham's production of toys and jewelry were sold abroad at that time and this overseas trade depended on middlemen, who booked orders from foreign customers and then made them up by buying from a range of small specialist manufacturers who lacked the means to sell direct.

Boulton and his partner Fothergill decided to cut out the middleman. In doing so, Boulton appreciated that he would have to be in a position to meet foreign orders in full.

Accordingly, he established the Soho Manufactory, equipped with a range of water-powered machinery and capable of producing the majority of items in his order book more rapidly and more cheaply than his competitors.

Boulton had in one bold and imaginative stroke combined merchanting and manufacturing, established a new pattern of industrial management (to give some idea of his scale of operation, there was living accommodation for a thousand employees at the Soho Works), mechanized the processes for making toys and jewelry, developed his foreign trade and struck an economic balance between goods for the quality and the mass markets—a truly astonishing record.

On those last two points, Matthew Boulton wrote:

> I have established a correspondence in almost every mercantile town in Europe which regularly supplies me with orders for the grosser and current articles which enables me constantly to employ such a number of hands as yields a choice of artists for the finer branches and am thus enabled to erect and employ a more extensive and more convenient apparatus than would be prudent to erect for the finer articles only.

It is also relevant to add his comment on training:

> I have trained up many and am training up more young plain country lads, all of which that betray any genius are taught to draw, from which I derive many advantages.

Then, in 1775, came his partnership with Watt. This to an extent grew out of the success of the Soho Manufactory, since there was probably no other enterprise that had the depth of skills and the managerial organization to cope with the standards of accuracy and complexity of production, which turning Watt's innovations into commercial propositions demanded. The Soho Foundry was added to the Manufactory and the vital parts for Watt's engines made there, with the remaining parts bought in from outside suppliers. In addition, Boulton and Watt sold the plans for their engines and then supervised their construction.

To keep track of an increasing flow of documents, Watt invented a wet copying press—the first duplicating machine—an excellent example of innovation—responding to demand. The copies, which Watt thereby made of his extensive correspondence, are now safely lodged in Birmingham Library and form a treasured archive. Typically, it was Boulton who saw the duplicator's commercial potential and exploited it.

There are two aspects of Boulton's approach to technological innovation, which are particularly relevant to today's themes. First, he saw that the ability of the Boulton and Watt engines to compete effectively and profitably would depend on the scale on which they were made and sold and therefore on the size of the market he could dominate. Hence his famous statement:

> It would not be worth my while to make for three counties only, but I find it very well worth my while to make for all the world.

Second, Boulton was well aware that he was not in the business of making and selling steam engines, but in that of supplying the means to produce goods relatively cheaply and in quantity. He expressed it simply when he told James Boswell in 1776:

> I sell here, Sir, what all the world desires to have—POWER.

It was because he clearly perceived the scale of demand for what he was selling and its nature—his customers were not buying steam engines but an efficient means of driving their machinery—that he was never satisfied with the then state of the art technology. As a result, for example, he pressed Watt to work on a rotative engine:

> The people in London, Manchester and Birmingham are all steam mill mad . . . I don't mean to hurry you but I think that in the course of a month or two we should determine to take out a patent for certain methods of producing rotative motion.

That purposive dissatisfaction combined with a sense of vision on a world scale are the hallmark of a technological innovator capable of matching up to a global challenge.

### Silicon Valley

The rise of entrepreneurial fever in Silicon Valley in the late 1970s and early 1980s provides excellent examples of how the spirit of entrepreneurship is built, what factors foster creativity and innovation, how these factors can be nourished to create a culture for the growth of high technology, and, last but not least, how all of these developments can transform the economy of an entire region and bring great wealth not only to people but to an entire country.

The story begins with a visionary and technologist, Frederick Terman, engineering professor and dean at Stanford University. Terman understood the value of technology; however, more important, he appreciated that in order to achieve its potential, technology must be exploited commercially. He established contacts with local industry and encouraged his students to push their knowledge and technological innovations into the marketplace. Terman advanced a loan of $538 to two of his students, William Hewlett and David Packard, to start production on a variable-frequency oscillator, the topic of Hewlett's master's thesis (Larson and Rogers, 1988). Hewlett and Packard started production in 1938 in a small garage in Palo Alto, and their operation grew from there, making them both billionaires. They made contributions of more than $300 million to Stanford, including a gift in 1994 of $77.4 million to construct a science and engineering complex and to endow the Terman Fellowship Program in honor of Terman. Packard died in 1996, leaving a charitable foundation with assets of $6.9 billion (*Chronicle of Higher Education,* Apr. 5, 1996).

Terman's major contribution was the idea of establishing a research park, where industry and the university could collaborate to develop and market technology. He encouraged the university to lease some of its land to high-technology companies. Many companies availed themselves of the opportunity to locate near a university where science and technology could be generated or advanced. The university researchers also found outlets for applying and commercializing their ideas. The collaboration was a win-win model that has since been adopted by many regions and countries interested in industrial development or technology transfer.

In addition to the vision and actions of Terman, several other factors contributed to the rise of Silicon Valley as a bed for high technology. (See Exhibit 6-2.) William Shockley, who co-invented the transistor at Bell Labs, moved to the Palo Alto region in

1955. He established the Shockley Semiconductor Laboratory, which was the start of major growth in the semiconductor business in Silicon Valley in the years to come. Shockley was able to recruit brilliant personnel to his operation in Palo Alto, including Robert Noyce, who later, with several of Shockley's recruited employees, formed the Fairchild Semiconductor Company. Shockley's company was not successful. However, it led to an entrepreneurial flood of spin-off companies formed with the help of people Shockley had recruited. These people were helped with loans from venture capitalists who saw the promise of the emerging technology and funded entrepreneurs to start their own operations. In 1968, Noyce left Fairchild to start another company—Intel. The Intel microprocessor now exists inside the great majority of personal computers.

The technological developments of the transistor, the semiconductor chip (several transistors on a silicon chip), and the microprocessor (central function of a computer on a semiconductor chip) all found their way to thousands of applications in industry and led to the creation of completely new industries. The PC industry is one of them.

### The Home Brew Computer Club

An important group that made its mark on the growth of the PC industry was the Home Brew Computer Club. It was an informal group that met regularly at Stanford and exchanged ideas in an open forum. The members came from different backgrounds and shared their ideas freely. The interaction of this group paralleled the interaction of the Lunar Society in the eighteenth century. The 1960s and 1970s were also decades of rebellion in California, a time when people were eager to depart from the normal ways of thinking, to defy the establishment, and to think freely. This antiestablishment mind-set was another fueling factor for entrepreneurial fever and is reminiscent of the era of the rebellious Quakers of the British Midlands. Steve Jobs and Steve Wozniak of Apple Computer benefited from the meetings of the Home Brew Computer Club. They were able to package a computer on a single board and sell it to the public. Jobs, at the age of 21, was able to secure venture capital for his ideas and launch Apple Computer Company. Jobs and Wozniak's success inspired many like-minded people to start their own ventures. By 1983, there were 85 semiconductor firms and more than 3,000 firms in microelectronics in Silicon Valley (Larsen and Rogers, 1984). The Silicon Valley entrepreneurial fever was in full force. It created a significant amount of technological progress and vast wealth.

### Lessons from Silicon Valley

After reviewing the chronology of events that led to the economic growth of Silicon Valley, it is possible to establish some guiding principles that lead to technological development at the regional and national levels:

*Vision:* Professor Terman had the vision to create a system that would commercialize technology. He was able to put together all the important factors for the creation of wealth.

*Technical expertise:* Silicon Valley benefited from the technical expertise of the Stanford College of Engineering and of the Nobel laureates who came to work in the area.

**EXHIBIT 6-2**
CHRONOLOGY OF THE IMPORTANT INVENTIONS, EVENTS, AND PEOPLE IN THE MICROELECTRONICS HIGH TECHNOLOGY INDUSTRY

| Year | Event |
|---|---|
| 1912 | Lee de Forest discovers the amplification qualities of the vacuum tube in Palo Alto, California, thus making possible radio, television, film, and other communication technologies. |
| 1938 | Hewlett-Packard is founded in a garage in Palo Alto by William Hewlett and David Packard, two of the first entrepreneurs in Silicon Valley. |
| 1946 | ENIAC, the first mainframe computer, with 18,000 vacuum tubes, is invented at the University of Pennsylvania. |
| 1947 | William Shockley, John Bardeen, and Walter Brattain invent the transistor at Bell Labs in Murray Hill, New Jersey. The transistor eventually replaces vacuum tubes. |
| 1955 | Shockley leaves Bell Labs to establish Shockley Semiconductor Laboratory in Palo Alto. |
| 1956 | Shockley, Bardeen, and Brattain win the Nobel Prize in physics. |
| 1957 | The entrepreneurial spirit in Silicon Valley gets under way when Robert Noyce and seven other brilliant young engineers quit Shockley Semiconductor Laboratory to launch Fairchild Semiconductor. These cofounders later split to launch over 80 semiconductor firms in Silicon Valley over the next 35 years. |
| 1968 | Noyce leaves Fairchild to start Intel. |
| 1971 | Ted Hoff, of Intel, invents the microprocessor, a computer control unit on a semiconductor chip.<br>Silicon Valley is named by the late Don Hoefler, then editor of a local electronics newsletter.<br>Nolan Bushnell designs Pong and launches Atari, and the video-game industry is begun. |
| 1976 | Steve Jobs and Steve Wozniak build the Apple microcomputer. |
| 1980 | Apple goes public: Art Rock, the venture capitalist who had invested \$57,000, earns \$14 million; Jobs is worth \$165 million. |
| 1982 | About 3,100 microelectronics firms exist in Silicon Valley; two-thirds have less than 10 employees, and only 50 or so have more than 1,000 workers. |
| 1984 | Silicon Valley has 15,000 millionaires and 2 billionaires. |

*Source:* Larsen and Rogers, 1988. Reprinted by permission of HarperCollins Publishers, Inc.

*Infrastructure:* To be able to develop successful new technologies, it is important to have the proper equipment, laboratories, installations, and other major elements of the infrastructure. The creation of the Science Park in Stanford to a great extent provided these elements.

*Venture capital:* The availability of financial resources is an essential ingredient. It is also important to have investors who are willing to take risks and to support new technological ventures.

*Job mobility:* This is a great way of transferring knowledge and technology. Self-motivated people who move among companies or form new companies create an environment of competition, which leads to technological advancements.

*Information-exchange network:* When people with similar professional interests and motivations are brought together as a community, the exchange of knowledge is increased. This flow of information helps advance technology.

*Entrepreneurial learning:* When employees and other members of society see the fruits of entrepreneurship, they learn that it is a good way of making a living. The fever catches on.

Would it be possible to repeat the Silicon Valley model again? Certainly. States such as North Carolina, Massachusetts, and Texas are developing technology zones. In fact, the Silicon Valley model can even be applied to a whole country, as has been done in Taiwan, Korea, and Singapore. The primary requirements are understanding the rules and playing the game.

### Lessons from the British Midlands and Silicon Valley

Two major points emerge from our analysis of the British Midlands and Silicon Valley:

**1** Freedom of thought and entrepreneurial spirit are essential elements in large-scale innovations.

**2** The exchange of information among groups from various disciplines is important.

The formation of the Lunar Society (1776) in the British Midlands helped to:

- Exchange ideas.
- Support member activities.
- Share information across disciplines.
- Share information across industry.
- Exhibit business strategy.
- Create role models.

The formation of the Home Brew Computer Club (1970s) helped to:

- Exchange ideas freely.
- Support member activities.
- Share information across disciplines.
- Exercise technical skills.
- Produce real entrepreneurs.
- Create role models.

History repeats itself when the environmental conditions permit a certain formula for social interaction. As for the process of technological innovation, when conditions are ripe for mental interaction in the existence of infrastructure, innovations flourish.

## FACTORS INFLUENCING TECHNOLOGICAL INNOVATION

Factors that influence technological development and the rate of innovation include:

**1** *The presence of scientific knowledge:* Technological change is dependent on scientific discoveries.

**2** *The level of maturity of the underlying science:* A wide base of knowledge enhances technological development.

**3** *The type of technology and the phase of its life cycle:* The innovation rate is high for emerging and growing technology.

**4** *The level of investment in technology:* Technological development is connected to the level of R&D funding.

**5** *The level of political commitment:* The pace of innovation is sensitive to high-level policy decisions at the company level or at the national level. For example, when the United States made a commitment to put a man on the moon in the 1960s, innovations connected to space flight increased.

**6** *The ability to borrow advances from related technologies:* For example, advances in communication technology are dependent on advances in laser technology or satellite components.

**7** *The diffusion rate and patterns:* A technology that is widely diffused in the market may delay or preclude other technologies from entry.

Several other factors emerging from current world conditions are influencing technological innovation in industry. These are:

**1** *The changing world environment:* Many countries have realized that wealth is created through technology and by adding value to natural resources. These countries are pushing for technology transfer and technology development. Competition within innovation is global.

**2** *Improvements in communication:* The improvement in communication systems worldwide, including satellite technology, news, television, and the Internet, makes information about science and technology available instantly throughout the world. Viewers can tune in to CNN International and receive news at the time it is broadcast, whether they are in New York, Tokyo, Moscow, or Berlin.

**3** *Multiple-site continuous R&D:* Research and development can go on continuously, 24 hours a day. If a company has scientists and engineers in different locations around the world, its R&D can occur interactively and continuously in, for example, Boston, Los Angeles, Tokyo, Kuala Lumpur, Tel Aviv, Rome, and Birmingham. This can speed up the time it takes to bring a new product to the market.

**4** *Time to market:* Time to market is one of the major factors influencing competitiveness in the marketplace. One of the goals of innovative organizations is the reduction of a new technology's development time from concept to market. The time lag from scientific discovery to technological innovation to product concept to marketplace can be significant. For instance, the idea behind the transistor goes back at least to Wilson's discovery of the theory of semiconductors in 1931. The transistor was invented at Bell Laboratories in 1947, more than 15 years later. This new technology allowed companies to develop concepts for new products. An electronic radio of reduced size was developed, "the transistor radio," and introduced into the market in 1955. Many other products ensued.

The sooner the technology gets to the market, the sooner a company can reap its rewards. A short time to market gives a company a competitive advantage. Time to market has been shortened over the years. For example, Hero of Alexandria described the power of steam in the first century A.D., but it was not applied to the steam engine until the late eighteenth century. On the other hand, it took about 20 years to bring the transistor to the market. Today, we speak of the time to market of technological innovations in terms of months, not centuries or years.

**5** *The push for education:* Countries have realized that to increase economic growth, they must establish good educational systems. England, France, Germany, and the United States traditionally had good educational systems that helped them in creating technology and pushing scientific discoveries. More recently, India, Korea, and Taiwan have developed good science and technology educational systems and are producing scores of talented scientists and engineers. Other countries of Southeast Asia and the Middle East have invested heavily in education in recent years and are developing talents that can be influential in technological innovation.

**6** *Changes in institutional interactions:* The speed of technological change and the corresponding high cost associated with technological development are forcing a change in the way organizations interact. Cooperation and alliances are becoming fundamental aspects of technological innovation. These alliances are taking place at different levels in different organizations:

- *Country and country,* such as the cooperation between France and England on R&D to develop the Concorde supersonic airplane;
- *Industry and industry,* such as the alliance between IBM, Motorola, and Apple to develop the power chip for the PC;
- *Government and industry,* such as the strategic alliance of the Japanese Ministry of Trade and Industry (MITI) with Japanese industry.
- *Industry and/or university and scientific agency,* such as the NSF-sponsored centers at U.S. universities that concentrate on specific technologies. An example is the center for net shape manufacturing at Ohio State University.

**7** *Changes in organizational structures:* Technological innovations are enhanced by better collaboration between scientists and technologists. In a company, good communication and teamwork among different groups are essential elements in pushing innovation, forming and empowering teams to carry ideas forward, creating efficiency, and speeding up the process. The structure of an organization must allow for collaboration and take advantage of the strengths and know-how of its people.

**8** *Infusion of resources into technological development and penetration:* This is a requirement of continued progress.

Managing technology is about meeting the challenges of creating an environment that fosters scientific discoveries and technological development. It involves the ability to predict the demands of the marketplace, to respond to these demands with technological solutions, to create an organization that links the customers to the research organization and matches research priorities with actual demands of the marketplace, and to structure an organization that encourages entrepreneurship and efficiently moves ideas from research to manufacturing to the market.

## DISCUSSION QUESTIONS

1 Look in a small-business magazine (*Inc., Success,* etc.) for an article about an entrepreneur. Then look in a large-business magazine (*Fortune, Forbes,* etc.) for an article about a successful CEO. What are their motivations? How do they measure success? What are their problems?

2 Both entrepreneurs and managers confront financial problems. Do you think their problems are different in nature? Do you think financial pressures affect their behavior? Do some research on the case of Steve Jobs and the Macintosh computer. What forced Jobs out of Apple? (You may also use Chapter 15 of this book for your research.)

3 What is innovation's role in the competitiveness of a firm? Can a company survive without innovation? The Coca-Cola Company is one of the most admired corporations on *Fortune*'s list. Its CEO has created enormous wealth for investors (*Fortune,* Feb. 5, 1996). Are there any differences between Coca-Cola and Xerox?

## ADDITIONAL READINGS

Jeffry A. Timmons. *New Venture Creation: Entrepreneurship for the 21st Century.* McGraw-Hill, 1997, New York, NY.

This is a good entrepreneurial guide. It provides the reader with theory, exercises, and self-assessments to determine the success potential of a new venture. It guides the entrepreneur-to-be in the preparation of a business plan.

Russell M. Knight. "Criteria Used by Venture Capitalists: A Cross Cultural Analysis." In Khalil, T., & Bayraktar, B. (eds.), *Management of Technology III: Proceedings of the Third International Conference on Management of Technology,* vol. 1, pp. 574–583. Industrial Engineering and Management Press, Norcross, GA.

This paper describes a cross-cultural survey of international venture capitalists to investigate the criteria they use to evaluate venture proposals. The study includes Asia Pacific, Canada, Europe, and the U.S. In general, venture capitalists share a common criteria on evaluating proposals; however, high-technology investments are not nearly as popular in the rest of the world as they are in the United States.

W. Chan Kim & Renee Mauborgne. "Value Innovation: The Strategic Logic of High Growth." *Harvard Business Review,* January–February 1997.

The authors studied 30 companies for a period of five years to uncover why some companies achieve success. They claim that innovation focused on creating customer value is the key to competitiveness.

G. Hammel & C. K. Prahalad. "Corporate Imagination and Expeditionary Marketing." *Harvard Business Review,* July–August 1991.

New technologies do not have a market. It has to be created. "To realize the potential that core competencies create, management needs the imagination to envision markets that do not yet exist." The authors maintain that Japanese dominance over U.S. firms—mainly in consumer electronics—is due to a combination of competencies and marketing research.

J. Mort. "Innovation as a Percolation Phenomenon." In Khalil, T., & Bayraktar, B. (eds.), *Management of Technology III: Proceedings of the Third International Conference on Management of Technology,* vol. 1, pp. 79–88. Industrial Engineering and Management Press, Norcross, GA.

Understanding the innovation process is a key element in effective MOT. Mort (research fellow/manager at Xerox) uses percolation concepts from mathematical theory to explain innovation. He explains *percolation* as a "special case of diffusion where motion from any site is constrained to a finite number of directions so that intrinsically the motion [of knowledge] is controlled by a medium [individuals, companies, customers, etc.]." The author uses the fax as a case study.

## SUGGESTED CASES*

The following Harvard case studies offer opportunities for MOT analysis on an advanced level:

- "Silicon Graphics." Harvard Business School, Case 9-697-038.
- "Porsche AG." Harvard Business School, Case 9-193-071.
- "Digital Imaging in 1995: Opportunities in the Descent to the Desktop." Harvard Business School, Case 9-796-060.
- "Alpha-Beta Technology, Inc.: Pioneering Carbohydrate Technology." Harvard Business School, Case 9-794-093.
- "Eli Lilly Co.: Innovation in Diabetes Care." Harvard Business School, Case 9-696-077.
- "World VCR Industry." Harvard Business School, Case 9-387-098.
- "Corporate New Ventures at Procter & Gamble." Harvard Business School, Case 9-897-088.

## REFERENCES

Bergelman, R. A., Maidique, M. A., & Wheelright, S. C. 1996. *Strategic Management of Technology and Innovation,* 2nd ed., Irwin, Chicago.

Bordogna, Joseph. 1997. *Innovation and Creative Transformation in the Knowledge Age: Critical Trajectories.* Plenary Session presentation; PICMET. Http://www.nsf.gov/bordogna.

Cadbury, A. 1995. *The English Midlands: Cradle of Technology.* Keynote address, IAMOT. European Conference on Management of Technology, Aston University, Birmingham, UK.

Goldblatt, H. 1995. "Bill Gates and Paul Allen Talk." *Fortune,* Oct. 2, pp. 68–86.

Howard, R. 1992. "The CEO as Organizational Architect: An Interview with Xerox CEO Paul Allaire." *Harvard Business Review,* September–October, pp. 107–119.

Jacobson, Gary, & Hillkirk, John. 1986. *Xerox: American Samurai.* Macmillan, London.

Larsen, J., & Rogers, E. 1988. "Silicon Valley: The Rise and Falling of Entrepreneurial Fever." In Smilor, R., Kozmetsky, G., and Gibson, D. (eds.), *Creating the Technopolis: Linking Technology Commercialization and Economic Development,* Ballinger, Cambridge, MA.

*The President and Fellows of Harvard College hold copyright of the Harvard Case Studies. To order copies or request permission to reproduce materials, call 1-800-545-7685 or write to Harvard Business School Publishing, Boston, MA 02163. Case description and order form can also be found on the web site: www.hbsp.harvard.edu/bin/showbook/

Martin, Michael J. C. 1994. *Managing Innovation and Entrepreneurship in Technology Firms*. Wiley Interscience, New York.

Mills, Robert. 1996. *Minimalist Definitions*. TMI Unit, University of Waikato, New Zealand; *b.mills@waikato.ac.NZ.*

Paradi, J. 1994. "Entrepreneurship." In Pliniussen, J. and Wilson, L. (eds.) *Introduction to Canadian Business Management,* McGraw-Hill Ryerson and Captus Press, Toronto.

Roberts, E. B. 1977. "Generating Effective Corporate Innovation." *Technology Review,* October–November, pp. 26–33.

Walker, Wayne G. 1993. *Recovering the Fumbles and Organizing for the Future: Xerox Integrates R&D into Corporate Strategy with Pioneering Research and Restructures to Become a Learning Organization—with Lessons for Military Applications.* P-7802, Rand, Santa Monica, CA.

Warshaw, Michael (ed.). 1996. "The Billion Dollar Comeback Kid." *Success,* July–August.

Zaki, Ahmed, & Khalil, Tarek. 1992. "A Model for Technological Innovation in Biomedical Devices." In Khalil, T. and Bayraktar, B. (eds.), *Management of Technology III,* Industrial Engineering and Management Press, Norcross, GA.

# 7

# COMPETITIVENESS

"Competitiveness" is one of the terms that has emerged strongly in the new era of globalization. In the last decade it became a key word used to describe the economic strength of countries or the position of a certain company with respect to its competitors in the marketplace. In this chapter, we will define the terms as they apply to the economic race among nations as well as to companies' performances. We will explore the basis upon which a country or a company can compete. We will also introduce a number of indexes used as measures of competitiveness and will illustrate their usage with graphical examples. Recent global changes in competitiveness are described and discussed.

## DEFINITIONS AND INDICATORS OF COMPETITIVENESS

*Competitiveness* is the process by which one entity strives to outperform another. Whether the entity is a person, a corporation, or a country, the goal is to win. Competition between business rivals within and outside the boundaries of a country has intensified in recent times. To be competitive, several factors must exist: ability, the desire to win, commitment or perseverance, and the availability of certain resources. For a company, being competitive means producing or providing, in a timely and cost-effective manner, a product or a service that meets the test of the marketplace and the needs of customers. To maintain its competitive position, the company must continue to outperform its business rivals. In today's global markets, those rivals may be operating within local, regional, national, or global markets.

At the macro level, the competitiveness of nations reflects the standard of living of their citizens. National competitiveness is a consolidation of the micro-level performances of companies and individuals—the true agents of economic growth. Issues of competitiveness have gained prominence in the post-cold war era. The fall of

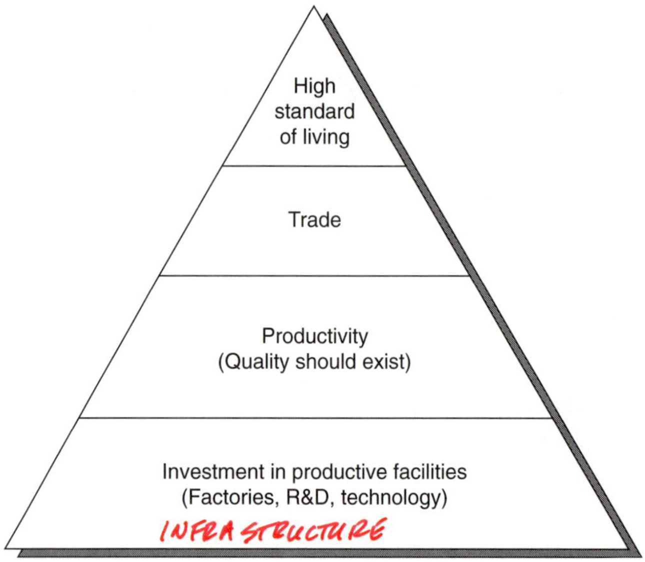


**FIGURE 7-1**
THE COMPETITIVENESS PYRAMID
*Source:* Based on Council on Competitiveness, 1995.

communism, the trend toward democracy, the opening of the market in the eastern bloc, and reduced military spending have created a new environment for business. Countries' objectives have converged on creating sustainable economic growth.

In 1985, the U.S. President's Commission on Industrial Competitiveness (Council on Competitiveness, 1994) defined competitiveness as

> the degree to which a nation can, under free and fair market conditions, produce goods and services that will meet the test of international markets, while simultaneously maintaining or expanding the real income of its citizens.

The Washington-based U.S. Council on Competitiveness adopted this definition and depicted the determining factors of competitiveness as a four-section pyramid, shown in Figure 7-1. An explanation of the indicators in each section is given in Exhibit 7-1.

### Investment

No economy can operate successfully without proper investment. The creation of wealth requires a productive base as the foundation of economic growth. Investments in technology, factories, equipment, infrastructure, and people help create such a foundation.

### Productivity

The next step is improving productivity. Productivity reflects the efficiency with which goods and services are produced. High levels of productivity provide an organization

**EXHIBIT 7-1**
INDICATORS OF COMPETITIVENESS

*Standard of living:* Standard of living crowns the competitiveness pyramid because it is the ultimate objective of a capitalist, free-market economy. Standard of living is, perhaps, the central indicator of national competitiveness. The *Index* measures standard of living as GDP per capita. It tracks both long-term and latest-year growth in standard of living, as well as standard of living levels using purchasing power parities.

*Trade:* Exports are dependent on national rates of productivity and national levels of investment in products and processes of production. Growth rates and levels of exports can be a significant indicator of national competitiveness. Useful indices include growth in exports of manufactured products, merchandise goods and services, as well as trade balances. Exchange rates and their impact on the growth of exports should also be traced.

*Productivity:* Productivity is the efficiency with which goods and services are produced and provided. It is largely determined by previous investments and by the quality and performance of the workforce, technological innovation, the quality of plant and equipment, and the effectiveness with which these factors of production are utilized. As such, productivity is both a determinant and an indicator of national competitiveness.

*Investment:* Investment is the fundamental building block of current and future economic activity and is, therefore, at the base of the Council's competitiveness pyramid. Investment is also the fundamental determinant of national competitiveness. The *Index* defines investment to include both "hard" assets, such as private-sector expenditure on factories and equipment, as well as "soft" assets, such as public- and private-sector spending on R&D and public-sector expenditure on education. Useful indices include information on patents—a key indicator of the success of civilian R&D efforts—and national savings rates—a key determinant of investment.

*Source:* Council on Competitiveness, 1995.

with a distinct advantage over its rivals. Productivity helps drive cost down and improve profitability. Efforts to improve productivity should not, however, sacrifice quality. Gone is the time when quality was considered a luxury; today, it is a minimum requirement—an essential factor. Product quality and performance are determinants of overall competitiveness.

### Trade

Trade connects production with markets. Today's trade is global. Trade operations have become more complex with the creation of trade blocs such as the European Union (EU), the North American Free Trade Agreement (NAFTA), the Asia Pacific Economic Cooperation (APEC), and the Association of Southeastern Asian Nations known as the "ASEAN." Each bloc has been created with the objective of fostering commercial activities within the bloc. These commercial clusters, however, do not operate completely as closed entities; they are still open to free trade with nonbloc nations. This is why treaties such as the General Agreement on Tariffs and Trade (GATT) and the World Trade Organization (WTO) play key roles in the modern world. An in-depth study of the world trend on trade and competition is needed to fully appreciate the effect of the trade factor on competitiveness. Suffice it to say that products or services that cannot be

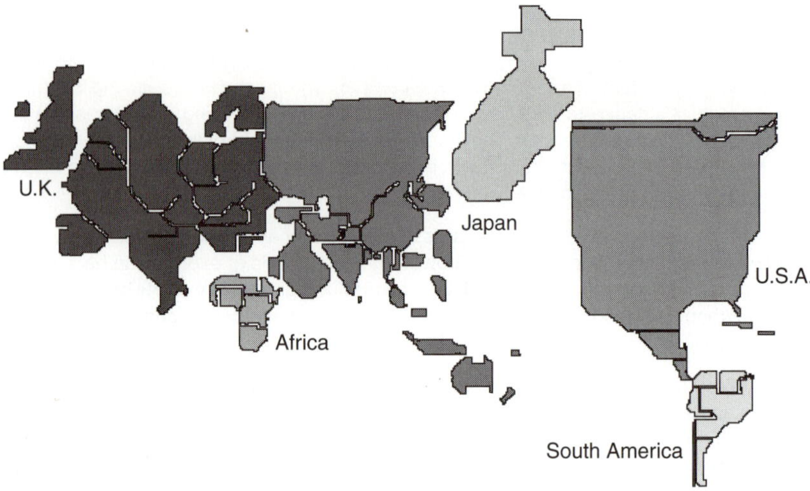


**FIGURE 7-2**
COUNTRIES' RELATIVE SIZES IN TERMS OF GDP

traded in open markets do not produce the economic growth that can trigger a significant improvement in standard of living.

### Standard of Living

Gross domestic product (GDP) and gross national product (GNP) are economic measures of the amount of wealth created in a country. This wealth is passed to the citizens and reflected in their standard of living. It is possible to determine a country's competitiveness on the basis of its citizens' standard of living as defined by GDP per person. A nation's technological advancement is a major contributor to its economic prowess. Figure 7-2 shows a map of the world drawn to the scale of GDP. It is clear that wealth created in the United States or Japan far exceeds wealth created in Africa or South America. This capacity to generate wealth is reflected in the standard of living of people in these regions of the world. Technology is also the major factor of creating companies' wealth. A review of the top 100 economies of the world indicates that about 50 percent of that list include companies and not countries. Technology also permits people to accumulate wealth. In 1999 the net worth of a single individual—Bill Gates, founder of Microsoft—is estimated to exceed $90 billion, more than the combined GDP of several countries; his wealth and power are a direct result of Microsoft's dominance in information technology.

## MANAGEMENT OF TECHNOLOGY AND GLOBAL COMPETITIVENESS

Management of technology plays a major role in creating and maintaining competitiveness in the global arena. MOT activities may be undertaken at the national/international, or macro, level, or at the firm, or micro, level.

At the macro level, countries must be able to:

- Create an economic growth policy, taking into consideration the fact that technology policy is a major contributor to economic strength.
- Provide an infrastructure permitting the support of technological enterprises and the facilitation of commerce and trade. Planning for human resource development must also be an integral part of any technology development strategy.
- Encourage cooperation between government, industry, and education and research institutions.
- Energize and support technological innovation and develop plans to enhance creativity and support R&D activity.
- Promulgate necessary but unburdensome legislation and regulation measures to protect the environment and strengthen social structure.

In the past, national competitive advantage focused on the availability and successful exploitation of raw materials, labor, transportation, and sources of capital. These factors are still important today. However, in today's global economy, multinational corporations have crossed national boundaries to establish their facilities where production cost is lowest. This globalization of production has erased most of the traditional bases upon which the industrialized countries, such as the United States and the United Kingdom, have built their competitive advantage. Industrialized countries are now taking advantage of the explosion of knowledge to create advanced technology that will help them maintain a competitive edge. At the same time, however, improvements in communication and transportation technology have brought the world closer together and facilitated the rapid migration of technology across borders, thereby decreasing the wide technological gap among countries. These changing world conditions and the changing environment of business make it evident that competitive advantage is increasingly dependent on our talent and skill in managing technology and technological enterprises.

The National Academy of Engineering's Committee on Engineering as an International Enterprise concluded that the comparative strength of a nation's technical enterprise depends upon the following factors (Lee and Reid, 1991):

- The strength of the national research enterprise.
- The quality of technical education.
- The presence of a large pool of technical talents.
- The strength of information technology infrastructure.
- The ability to cultivate individual creativity and initiative.
- Synergy between basic research and downstream technical activities such as design and production capabilities.
- The scale of domestic markets and the openness of global markets as engines for innovation and its commercialization.
- The ability to continually modernize plant and equipment in private industry, and the commitment to do so.
- Collaboration between industries and universities and the government.
- National savings and the level of investment in industrial modernization.

- National policy supporting initiatives to enhance adoption, adaptation, and diffusion of technology and related know-how.
- The development of the necessary human, physical, financial, regulatory, and institutional infrastructures to attract individuals, companies, and institutional entities, regardless of national origin, to invest in and conduct technical activities within the boundaries of the country. This ensures the long-term wealth-generating capacity of the economy.
- Public support of generic and domestically developed technologies.

### The Case of Japan

Japan is a classic example of a country that was able to change the map of international competitiveness. Factors contributing to Japan's success include (1) thoughtful strategic planning, (2) the planned transfer of technology, (3) targeting of niche products and markets, (4) teamwork and excellent execution, and (5) commitment and the desire to win. After World War II, Japan competed in mature industries, such as steel, automobiles, and consumer products. This strategy made sense since it was easier for Japan to acquire the technology of mature industries than that of newer industries. In the automotive industry, Japan observed that the quality of American cars could be improved to provide better customer satisfaction. Moreover, the oil embargo, following the 1973 Middle East war, changed the needs of the market, which shifted to smaller, more economic vehicles. This helped the Japanese find a niche within the automobile industry and build an effective strategy to exploit it. Japanese industry, although helped by the global economic situation, did what had to be done: compete in process technologies with techniques such as *kanban,* single method die exchange (SMDE), and statistical process control. Such techniques were under the umbrellas of corporate programs such as total quality management (TQM) and just in time (JIT). Once this process was reliable and flexible, Japan began to introduce innovation into the products, increasing its R&D effort. Japan was able to consolidate several important ingredients for technology development: strategic planning, know-how, and collaboration between institutions. A sound strategy for cooperation was formulated by the Ministry of International Trade and Industry (MITI) (Cheney and Grimes, 1991). The required know-how was provided by cooperating universities and industries. The capital was provided by strong conglomerates known as *kereitsus.*

### The Case of Singapore

Another interesting example is Singapore, now following in Japan's steps to become a global competitor. Poh-Kam Wong (1995) addresses three main problems facing small nations such as Singapore: a small domestic market, limited natural resources, and a limited supply of indigenous human resources. Wong mentions three strategic approaches that Singapore used in overcoming these problems:

1 Serving as a regional business service hub for other nearby nations
2 Engaging in niche specialization
3 Acting as a home base and R&D hub for global firms

All of these approaches are based on acquiring technologies from outside the country. This external transfer of technology boosts the development of internal technologies while allowing the country to join the ranks of those contributing to the value-added chain. After the initial success of Singapore's economic development strategy, another shift in Singapore's plans took place. According to Wong (1995), "Singapore government's economic development strategies in recent years indicate a clear strategic intent to seek to reposition the Singaporean economy from its traditional regional business hub and niche manufacturing role to a regional, if not global, technology hub role. This new emphasis on technological innovation can be seen from the establishment of a new government agency, the National Science and Technology Board." The government is now putting emphasis on promoting innovation and creativity in design and services along the entire business-value chain.

The shift from being a technology user to being a technology innovator is a more advanced step in an economic development strategy. To support the drive toward a high-technology hub, the Singapore government created programs to attract highly qualified foreign scientists and engineers.

Singapore's spectacular success in economic development provides a good model of a successful national strategy in a small, young, yet growing nation.

## A COMPARISON OF INTERNATIONAL COMPETITIVENESS: ECONOMIC INDICATORS

The economic performance of a nation is commonly expressed in terms of its gross domestic product. This index reflects the wealth created within the borders of a nation and represents the output (total market value) produced by people, firms, and governments domestically. Firms owned by other countries and foreign citizens working within the country contribute to this index. In 1993 the GDP of the United States was \$6.5 trillion. In 1998 it was more than \$8.5 trillion. Exhibit 7-2 lists the GDPs of 22 nations, as well as the average annual rate of GDP growth for each of these nations between 1990 and 1994. It also lists the 1998 GDP and percentage real GDP growth in 1998. The figures given in the exhibit show the changing fortunes of nations over time, which directly influence the standard of living.

The GDP index is different from the GNP index, which measures output produced by citizens of a country either within or outside the borders of that country. For a country such as Japan, GNP includes profits made by Japanese firms in the United States, and for the United States, GNP includes profits made by U.S. companies overseas and earnings of U.S. citizens abroad but not profits or earnings of Japanese companies or citizens located in the United States.

The GDP index is becoming a more commonly used index because it correlates well with many other economic indicators, such as industrial production and employment. GDP can be adjusted for inflation to produce another index, called *real GDP.* This is a good index for tracing the actual increases or decreases in a nation's output after adjustments for inflation.

**EXHIBIT 7-2**
GDP AND RATE OF GDP GROWTH, SELECT COUNTRIES

| Country | GDP, 1994 ($, billion) | GDP growth rate, 1990–1994 (%) | GDP, 1998* ($, billion) | Real GDP** growth, 1998 (%) |
|---|---|---|---|---|
| United States | 6,648.0 | 2.5 | 8,508.9 | 3.90 |
| Japan | 4,590.9 | 1.2 | 3,786.2 | −2.84 |
| Germany | 2,045.9 | 1.1 | 2,118.3 | 2.80 |
| France | 1,330.3 | .8 | 1,418.7 | 3.80 |
| United Kingdom | 1,071.3 | 5.7 | 1,377.8 | 2.60 |
| Canada | 639.9 | 5.7 | 595.3 | 3.00 |
| Brazil | 554.6 | 2.2 | 776.8 | .02 |
| China | 522.2 | 12.9 | 960.9 | 9.06 |
| Mexico | 377.1 | 2.5 | 415.0 | 3.90 |
| Korea | 376.5 | 6.6 | 301.6 | −4.26 |
| Australia | 331.9 | 3.4 | 359.8 | 4.00 |
| India | 293.6 | 3.8 | 345.8 | 5.89 |
| Argentina | 281.9 | 7.6 | 336.9 | 4.40 |
| Taiwan | 234.0 | 6.5 | 261.4 | 4.83 |
| Indonesia | 174.6 | 7.6 | 53.3 | −14.50 |
| Thailand | 143.2 | 8.2 | 107.6 | −7.80 |
| South Africa | 121.9 | −.1 | 107.6 | .00 |
| Poland | 92.6 | 1.6 | 145.2 | 6.00 |
| Malaysia | 70.6 | 8.4 | 71.3 | −4.79 |
| Chile | 51.9 | 7.5 | 73.0 | 4.00 |
| Hungary | 41.3 | −2.0 | 47.1 | 5.00 |
| Czech Republic | 36.0 | −4.7 | 49.8 | −1.80 |

*Source:* World Bank, 1996, 1999. OECD,1998 a, b, 1999. IMD. (Figures are actual or estimated.)
*at current prices and exchange rates.
**percentage change, computed on a local currency at constant prices basis.

## THE U.S. COUNCIL ON COMPETITIVENESS

In 1986, a group of chief executives representing a large number of business leaders, education leaders, and labor organizations formed a council to conduct studies aimed at improving U.S. competitiveness in world markets. Known as the Council on Competitiveness and located in Washington, D.C., it publishes an annual competitiveness index. The publication assesses U.S. economic performance relative to that of a group of advanced industrialized nations known as the *Group of Seven (G-7)*. The heads of the G-7 nations meet annually to discuss economic issues affecting their countries. The G-7 countries are the United States, Canada, Japan, France, Germany, Italy, and the United Kingdom; in 1996, Russia joined the group. Much of the information contained in this section is based on information published by the Council on Competitiveness. We are interested here in presenting and defining indexes used to evaluate competitiveness, since these are of universal value. The data change continuously, however, so the reader is advised to periodically consult Council on Competitiveness publications or the IMD-published *World Competitiveness Year Book* (www.imd.ch/wey.html) for details on and updates of the competitiveness indexes.

### Standard-of-Living Indexes

Standard of living is a reflection of how well people live in a certain country or region of the world. It reflects the distribution of a country's wealth among its citizens. The Council on Competitiveness defines standard of living as gross domestic product per person. This index of standard of living assumes that the country's wealth is distributed evenly among the inhabitants regardless of social or political differences. Figure 7-3 shows the growth in standard of living for the G-7 countries between 1973 and 1993. As can be seen in the figure, healthy gains were achieved by Japan, Germany, and Italy during that 20-year period. Of all the G-7 countries, the United States had the lowest gain. Similar trends continued between 1984–1994 (see Figure 7-4). This trend has been reversing since 1992, with the United States emerging as a better competitor. A more representative index for the standard of living is *purchasing-power parity (PPP).* This index measures how much it costs to buy a standard basket of goods and services in a country relative to how much the same basket costs in the United States. It adjusts for current prices and exchange rates between countries. The Organization for Economic Cooperation and Development (OECD) uses the PPP index. The basket of goods and services is selected on the basis of people's purchasing patterns and is updated periodically to account for changes over time.

**FIGURE 7-3**
LONG-TERM REAL GROWTH IN STANDARD OF LIVING, 1973–1993
For this chart, standard of living is based on GDP per person.
*Source:* Council on Competitiveness, 1994.

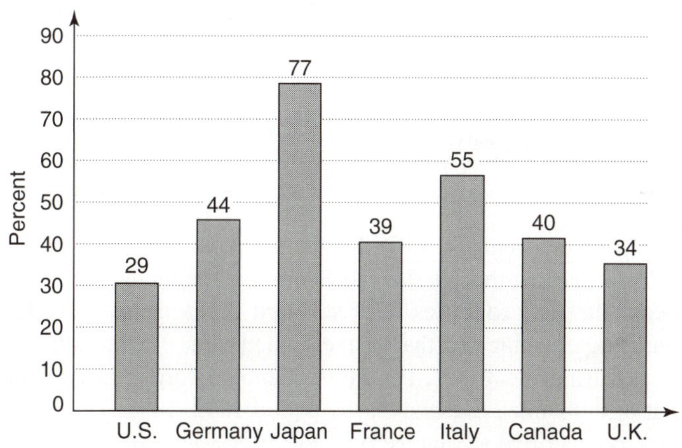

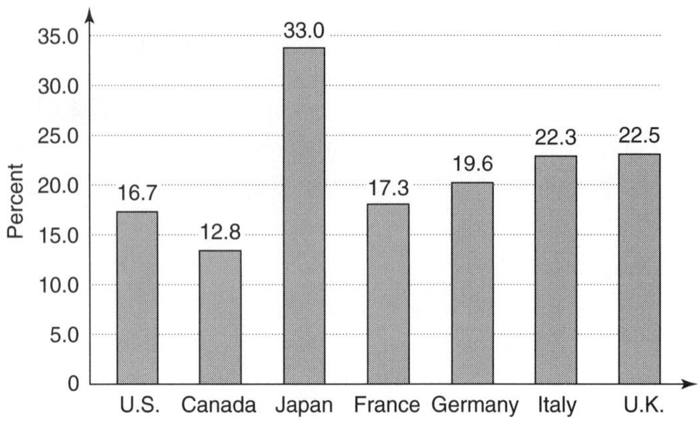


**FIGURE 7-4**
LONG-TERM REAL GROWTH IN STANDARD OF LIVING, 1984–1994
For this chart, standard of living is based on GDP per person.
*Source:* Council on Competitiveness, 1995.

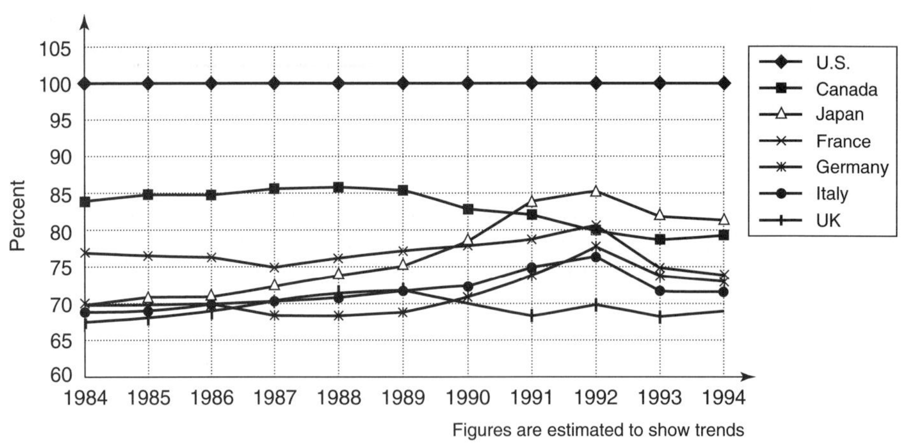


**FIGURE 7-5**
STANDARD OF LIVING, 1984–1994
For this chart, standard of living is based on purchasing-power parity. Purchasing-power parities measure how much it costs to buy a standard basket of goods and services in a foreign country relative to the cost of the same basket in the United States.
*Source:* Council on Competitiveness, 1995.

Figure 7-5 shows the trends in PPPs for the G-7 countries over the 10-year period 1984–1994. It can be seen that the United States has maintained the highest standard of living among the G-7 countries. The standard of living of the other G-7 countries is shown as a percentage of that of the United States. After some losses in U.S. competitiveness during the 1980s, the PPPs of most of the other countries declined relative to that of the United States over the period 1992–1998. This reflects efforts to restore the U.S. competitive advantage over the 1990's decade.

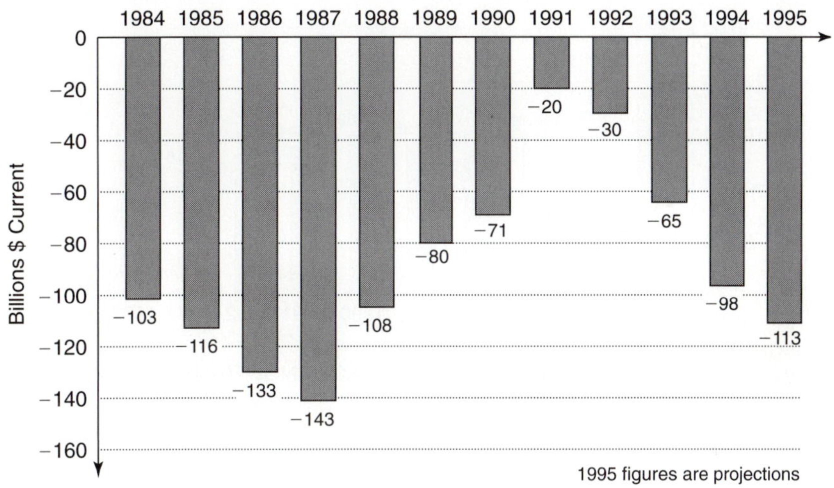


**FIGURE 7-6**
U.S. TRADE BALANCE, 1984–1994
Merchandise goods include manufactured goods, as well as food and live animals, beverages and tobacco, crude materials, and animal and vegetable oils.
*Source:* Council on Competitiveness, 1995.

### Trade Indexes

Figure 7-6 shows the U.S. trade balance in merchandise goods and services from 1984 to 1995. A *trade balance* represents the difference between the total value of merchandise goods and services exported by a country and the total value of merchandise goods and services imported. The *trade deficit* is an index of the relative competitiveness of a country's industry and service organizations. The United States moved from a positive trade balance in the early 1980s to a serious trade deficit as the 1980s progressed. The negative trend continued in the 1990s reaching $210 million in 1997. A great contributor to the U.S. deficit is the loss of a competitive edge to countries such as Japan, Germany, and the Asian Tiger nations. This is particularly true in the manufactured-goods sector. In the service sector, the United States has a positive trade balance.

The U.S. trade deficit is even more serious when one considers the fluctuations in exchange rates between the U.S. dollar and competitors' currencies, such as the Japanese yen or the German mark. Figure 7-7 shows yen/dollar exchange rates for the years 1985 to 1995. As the value of U.S. currency declines, it becomes more expensive for Americans to buy imported products. It also becomes less expensive for people overseas to buy U.S. products, thus increasing U.S. exports. Even though the dollar's value in 1995 was less than half its value in 1985, the U.S. trade deficit with Japan remained huge. In 1996 and 1997, the dollar's value started inching up against the values of the Japanese yen and German mark. In 1999 the exchange rate of the dollar was 121 Japanese yen and about 1.82 German marks, reversing the trend shown in Figure 7-7. The U.S. balance of trade, however, continued to increase in deficit reaching minus $210 billion in 1997.

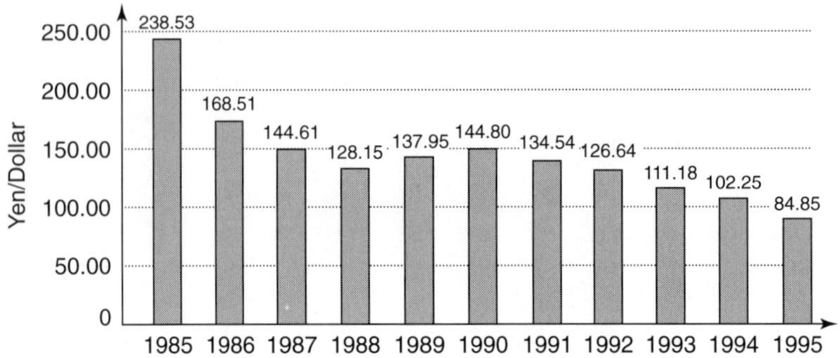


**FIGURE 7-7**
JAPANESE YEN/U.S. DOLLAR MARKET EXCHANGE RATES
*Source:* Council on Competitiveness, 1995.

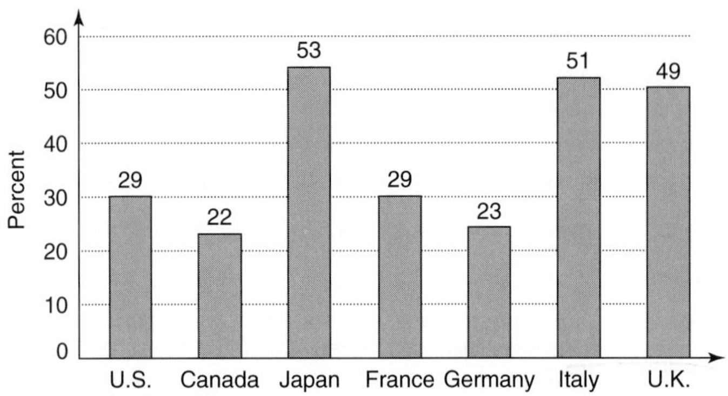


**FIGURE 7-8**
LONG-TERM REAL GROWTH IN MANUFACTURING PRODUCTIVITY, 1983–1993
For this chart, manufacturing productivity growth is based on output per manufacturing hour.
*Source:* Council on Competitiveness, 1995.

### Productivity Indexes

Productivity, as defined earlier, is the ratio of output to input. It reflects the efficiency of an operation. Several indexes can be used to express and track productivity (Sumanth, 1984). The most common index used to track productivity in manufacturing is output per worker-hour input, shown in Figure 7-8. The index commonly used to track national productivity as an indicator of national competitiveness is based on GDP per total employed persons. Figure 7-9 shows the growth of national productivity from 1984 to 1994 for the G-7 countries.

It should be noted that in spite of the relative productivity gains achieved by most G-7 countries over the United States during the past two decades, the United States still has the highest national productivity index. However, as shown in Figure 7-10, the gap

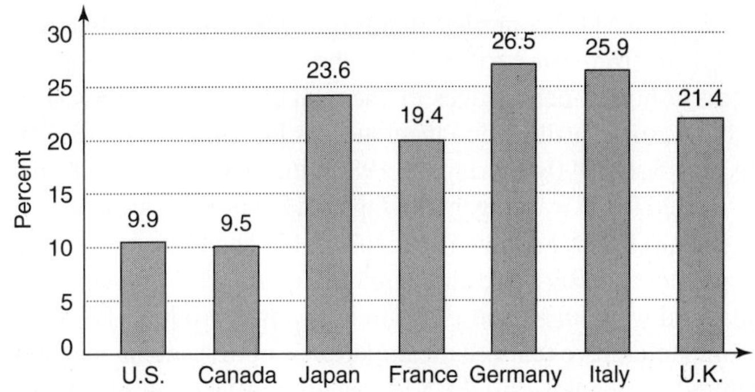


**FIGURE 7-9**
LONG-TERM REAL GROWTH IN NATIONAL PRODUCTIVITY, 1984–1994
For this chart, national productivity is based on GDP per total employed persons.
*Source:* Council on Competitiveness, 1995.

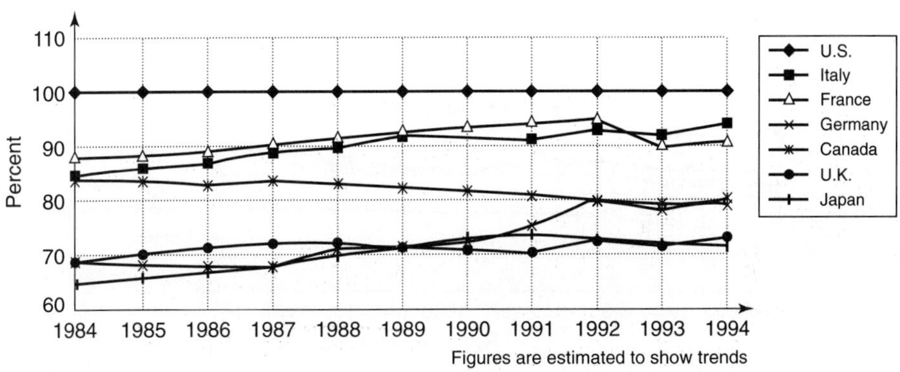


**FIGURE 7-10**
NATIONAL PRODUCTIVITY TRENDS, 1984–1994
*Source:* Council on Competitiveness, 1995.

between the United States and several G-7 countries has narrowed significantly since the early 1980s.

In 1998 overall productivity expressed as GDP per person employed was $64,325 for the United States, $63,271 for France, $59,237 for Germany, $57,828 for Italy, $57,808 for Japan, $50,688 for the United Kingdom, and $40,579 for Canada.

### Investment Indexes

Investment in R&D, plant and equipment (P&E), and education provides a base for long-term economic growth. Therefore, it is very important to track these indicators and to sound the alarm when they take a wrong turn. Savings are another indicator, reflecting the accumulation of resources necessary to unleash investment.

Figure 7-11 presents the G-7 countries' trends of investment in civilian R&D as a percentage of their GDPs from 1982 to 1992. As the figure shows, Japan and Germany invested significantly higher percentages of their GDPs in civilian R&D than did the United States. This factor contributed to their success in commercializing civilian products in the 1980s and early 1990s. During the 1980s, the U.S. overall investment in civilian plus defense-related R&D was larger than Japan's, but more than 50 percent of it went to defense (noncivilian) R&D, as shown in Figure 7-12. The U.S. and Japanese ratios of GNP devoted to nondefense R&D are compared in Figure 7-13. There should be no real surprise that Japan did well during the 1980s in innovating products for the commercial market. Major companies have realized the importance of investment in technology. The top 300 international companies raised R&D spending in 1998 by 11.9 percent to $254 billion following a 12.8 percent increase in 1997 (*Financial Times,* June 25, 1999).

**FIGURE 7-11**
INVESTMENT IN CIVILIAN R&D, 1982–1992
Civilian R&D is defined as nondefense R&D funded and performed by the federal government, industry, universities, and nonprofit organizations.
*Source:* Council on Competitiveness, 1995.

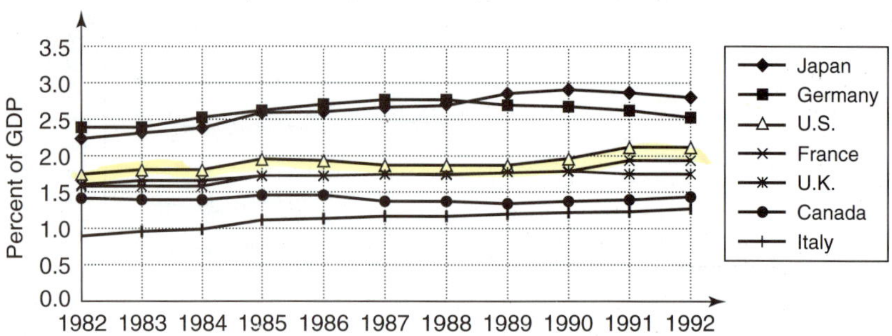


Figures are estimated to show trends

**FIGURE 7-12**
U.S. FEDERAL R&D OBLIGATIONS, 1975–1989
*Source:* Office of Technology Policy 1997.

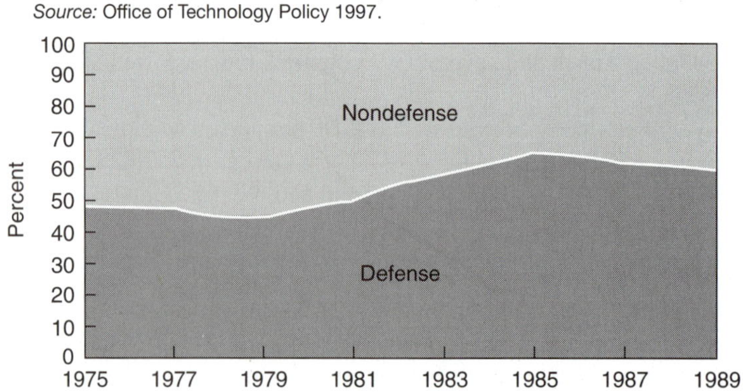


Figures are estimated to show trend

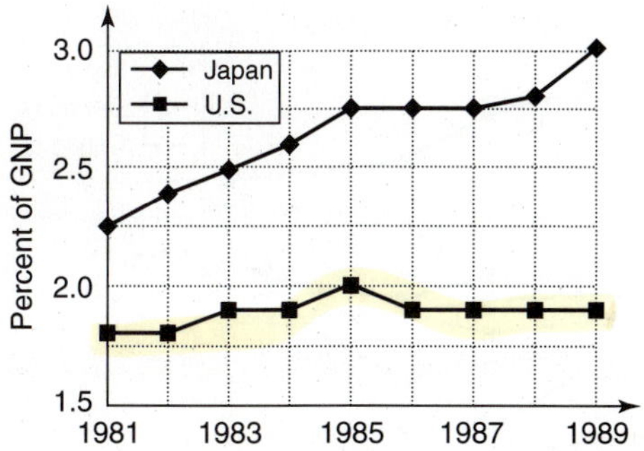


**FIGURE 7-13**
NONDEFENSE R&D EXPENDITURE: JAPAN VS. UNITED STATES, 1981–1989
*Source:* From Chenney and Grimes based on NSF data.

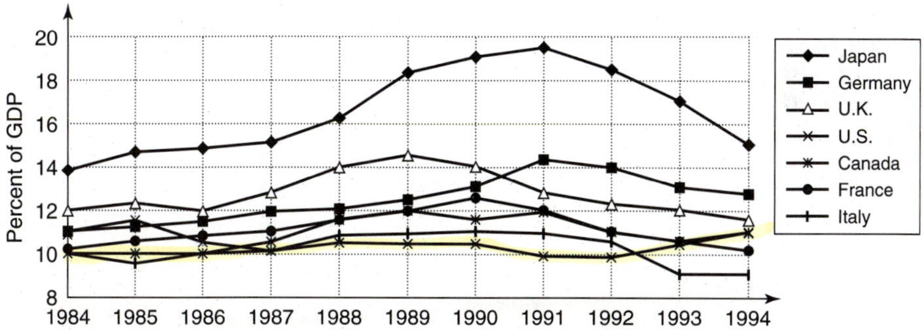


Figures are estimated to show trends

**FIGURE 7-14**
INVESTMENT IN PLANT AND EQUIPMENT, 1984–1994
*Source:* Council on Competitiveness, 1995.

Investment in plant and equipment is an indicator of commitment to improving wealth-creating enterprises. Figure 7-14 shows the trends of investment in P&E for the G-7 countries from 1984 to 1994.

Figure 7-15 shows the savings patterns of the G-7 countries between 1984 and 1995. The United States was significantly weak in this category compared to other countries. This indicates vulnerability in long-term economic growth, an issue for U.S. policy makers to consider.

### Patents Index

Another index of competitiveness is the number of patents granted per year, as patents reflect innovativeness or a country's ability to create technology. In the United States, the share of patents granted to U.S. inventors declined in the 1970s and 1980s but began to increase in the early 1990s (see Figure 7-16). The upswing indicates renewed emphasis on creativity and on the importance of technology in winning the global competition.

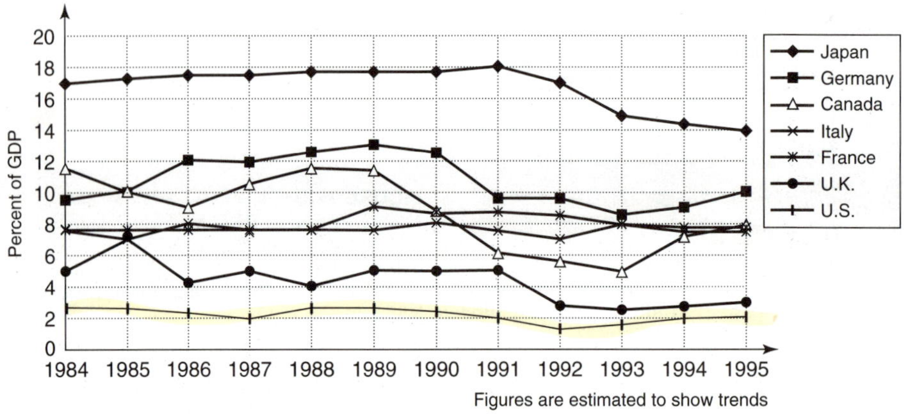


**FIGURE 7-15**
ANNUAL NET NATIONAL SAVINGS RATE, 1984–1995
Net national savings is defined as national disposable income minus national consumption minus depreciation. It is also understood as the sum of private and public sector savings.
*Source:* Council on Competitiveness, 1995.

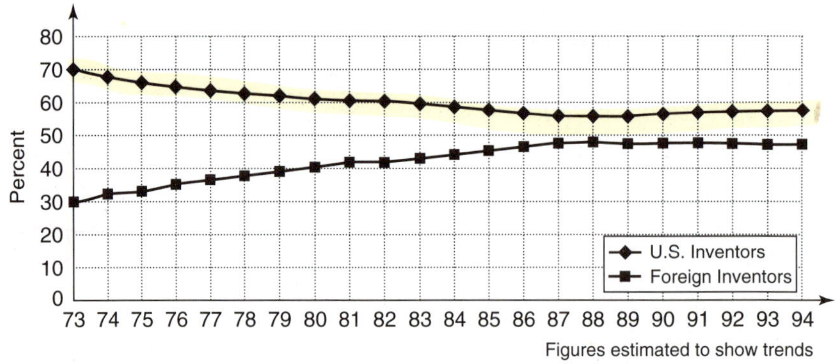


**FIGURE 7-16**
U.S. PATENTS GRANTED TO U.S. AND FOREIGN INVENTORS, 1973–1994
*Source:* Council on Competitiveness, 1995.

## EMERGENCE OF THE TIGERS

Global competition intensified in the mid-1980s with the emergence of a host of newly industrialized countries (NICs) that became known as "the Tigers." Korea, Taiwan, Singapore, and Hong Kong led the way in developing their economies. They were followed by a number of other Asian and Latin American countries that relied on foreign investment and the transfer of production and mature technologies to fuel their countries' economic growth. Figure 7-17 shows the growth of foreign investment in developing countries, and Figure 7-18 shows the growth in the stock market capitalization of developing countries.

The infusion of foreign direct investment (FDI) into newly industrialized and developing countries, combined with those countries' support for better education and their

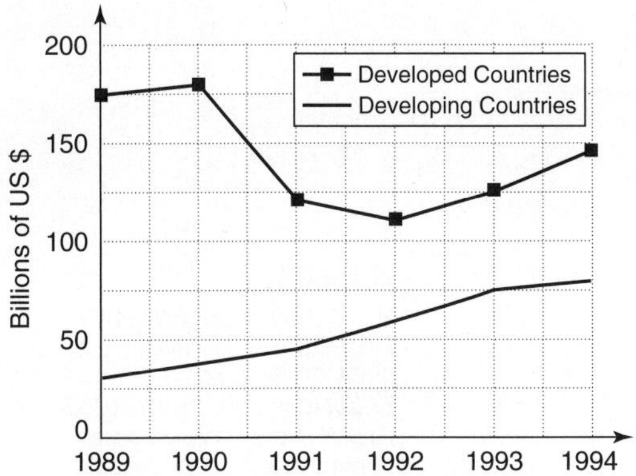


**FIGURE 7-17**
FOREIGN DIRECT INVESTMENT IN DEVELOPED AND DEVELOPING COUNTRIES, 1989–1994
*Source:* Council on Competitiveness, 1995.

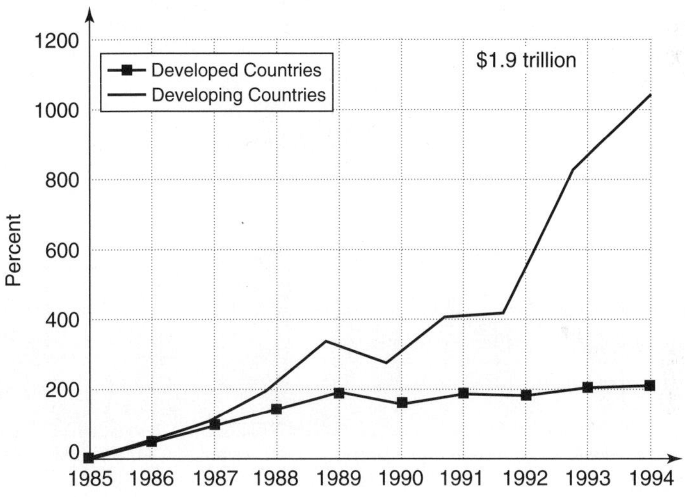


**FIGURE 7-18**
WORLD STOCK MARKET CAPITALIZATION
*Source:* Council on Competitiveness, 1995.

push for technology transfer, resulted in higher rates of real growth in their wealth than the rates in industrialized countries (Figure 7-19). On a regional basis, Asian countries achieved the highest rate of growth in the world, as shown in Figure 7-20. The question now is, Can these countries sustain their economic growth? The answer depends on how well they will manage technology and develop a solid financial and trade support systems to sustain growth.

Countries with developing economies have succeeded in penetrating global markets and have increased their share of wealth. Figure 7-21 shows the real growth of exports in developing economies in the period 1985–1995. The entry of countries such as China and the eastern European nations into the production and export race indicates that global competition will intensify in the future. Figure 7-22 shows GDP growth rates and the 1995 GDP per person for select countries.

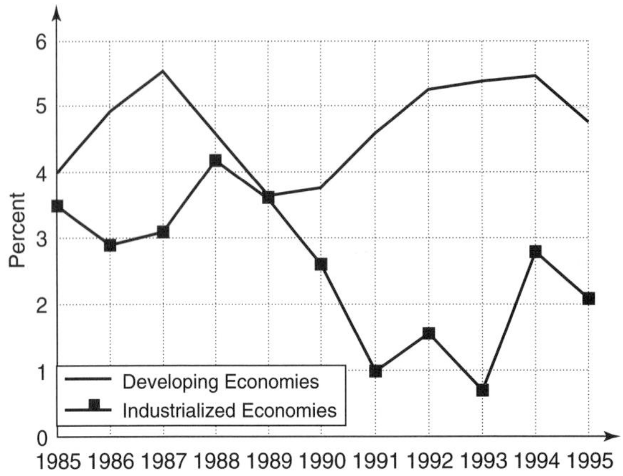


**FIGURE 7-19**
REAL GDP GROWTH IN DEVELOPING AND INDUSTRIALIZED ECONOMIES, 1985–1995
Growth here is based on a constant 1990 U.S. dollar.
*Source:* Council on Competitiveness, 1995.

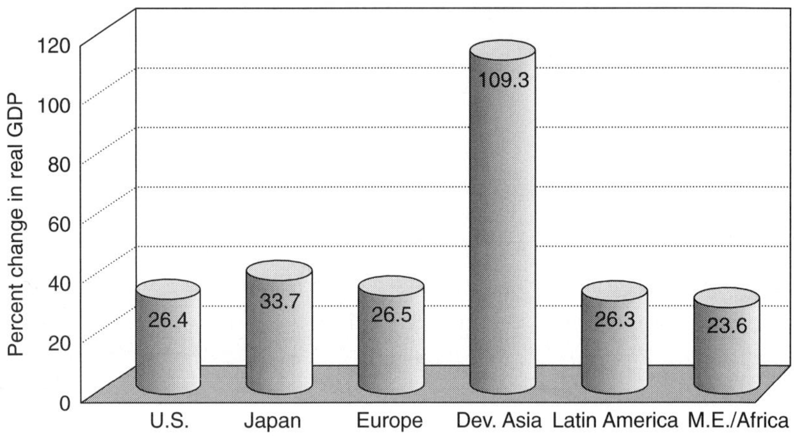


**FIGURE 7-20**
OUTPUT GROWTH RATE BY REGION, 1985–1995
*Source:* Council on Competitiveness, 1995.

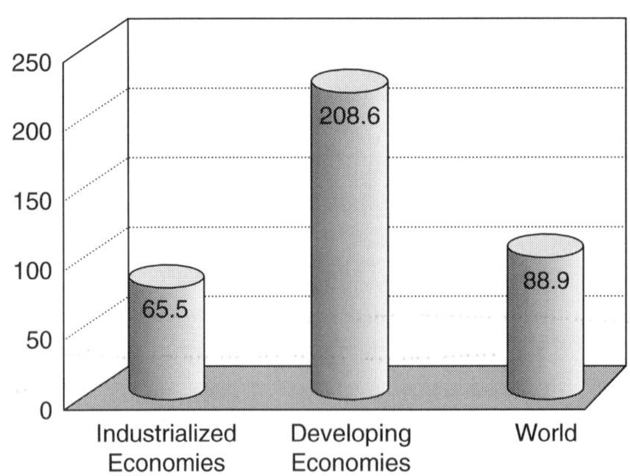


**FIGURE 7-21**
REAL GROWTH OF EXPORTS IN INDUSTRIALIZED AND DEVELOPING ECONOMIES, 1985–1995
*Source:* Council on Competitiveness, 1995.

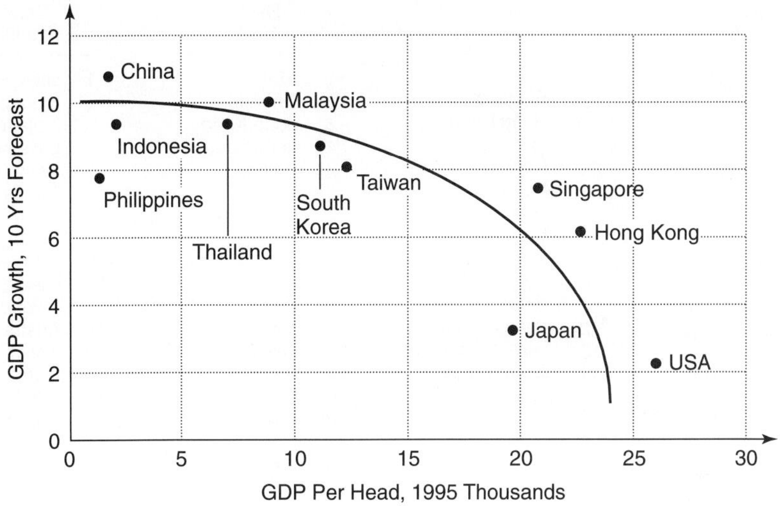


**FIGURE 7-22**
GDP GROWTH RATES AND GDP PER PERSON, SELECT COUNTRIES
*Source:* OTP, 1997.

## COMPETITIVENESS: THE GAME OF NATIONS

Nations of the world are engaged in an economic game. They compete for resources and for the means to harness their resources into production endeavors. The game of nations is like a football game. All nations are invited to participate. Each nation's goal is to win. The trophy is wealth creation and an increased standard of living. Each nation has its team, consisting of its citizens. It has its coaches, the leaders of its government. It can develop its strategies of playing the game. The rules of the game are set through international bodies. They are set based on intensive discussions, negotiations, and agreements. The rules are known to the team members. Each team has the freedom to train its people, practice as a team, and send scouts to watch other teams and analyze their performance. Each team can transfer successful plays of other teams to its playbook. Finally, each team must play well to win a game. They must continue to play well to develop a winning streak and win the Super Bowl or the World Cup at the end of a series of games. The cycle continues for the next round of play. A winner of one game has a chance to lose the next if the team does not continue to innovate and execute. The winner of a cycle, getting a trophy at the end of a year, may lose the next cycle if the team fails to properly manage and execute the plays. The rankings of nations change according to the performance of their institutions, players, and strategy. Public policy is therefore critical to rendering nations more competitive.

Figure 7-23 lists the relative rankings of nations on the basis of an index of competitiveness used by the International Institute for Management Development (IMD) in Switzerland. As can be seen, countries can lose or gain in rank every year. The challenge is to be able to sustain economic growth and win in that game.

| Rank 1996 | | Rank 1997 | | Rank 1998 | | Rank 1999 | |
|---|---|---|---|---|---|---|---|
| USA | 1 | 1 | USA | 1 | USA | 1 | USA |
| Singapore | 2 | 2 | Singapore | 2 | Singapore | 2 | Singapore |
| Hong Kong | 3 | 3 | Hong Kong | 3 | Hong Kong | 3 | Finland |
| Japan | 4 | 4 | Finland | 4 | Netherlands | 4 | Luxembourg |
| Denmark | 5 | 5 | Norway | 5 | Finland | 5 | Netherlands |
| Norway | 6 | 6 | Netherlands | 6 | Norway | 6 | Switzerland |
| Netherlands | 7 | 7 | Switzerland | 7 | Switzerland | 7 | Hong Kong |
| Luxembourg | 8 | 8 | Denmark | 8 | Denmark | 8 | Denmark |
| Switzerland | 9 | 9 | Japan | 9 | Luxembourg | 9 | Germany |
| Germany | 10 | 10 | Canada | 10 | Canada | 10 | Canada |
| New Zealand | 11 | 11 | Britain | 11 | Ireland | 11 | Ireland |
| Canada | 12 | 12 | Luxembourg | 12 | Britain | 12 | Australia |
| | | 13 | New Zealand | 13 | New Zealand | 13 | Norway |
| | | 14 | Germany | 14 | Germany | 14 | Sweden |
| Finland | 15 | 15 | Ireland | 15 | Australia | 15 | Britain |
| | | | | 16 | Taiwan | 16 | Japan |
| | | | | 17 | | 17 | |
| | | | | 18 | Japan | 18 | Taiwan |
| Britain | 19 | | | 19 | | | |
| Ireland | 22 | | | | | | |

**FIGURE 7-23**
RANKING OF COUNTRIES' COMPETITIVENESS
The arrows show the changes of rank from year to year.
*Source:* IMD International, 1999.

Several of the Tigers of Southeast Asia, including Hong Kong, Korea, Thailand, Malaysia, and Indonesia, suffered a strong setback in their economic systems in 1997. Their national debts mounted, stock markets plummeted, and several of their financial institutions went through bankruptcies. Exhibit 7-2 shows the drastic decline in the real GDP growth of several of the Asian countries in 1998 compared to their GDP growth shown in Figure 7-22. A short analysis about the fall of the tigers is given later in this chapter. The tigers seem to have fallen into this crisis because of a combination of managerial problems and a changing technological landscape. Their financial and banking systems did not have adequate controls that permit sustainable economic growth. In the meantime China, a rising star on the competitiveness scale, is outperforming Thailand, Indonesia, and Malaysia. China can produce many products cheaper than its neighbors can. China's wage structure gives them a competitive advantage at the low-technology end. China's cheap labor and cheap currency enables China to undersell its neighbors. China's low-wage structure, however, does not pose the same level of threat to the United States and other countries specializing in high-technology products (Friedman, 1997). The economic crisis in Asia presents an excellent case for the interaction of technology management and public policy in a nation's quest to create wealth.

The following are some MOT guidelines for improving and sustaining a nation's competitive position in the global arena:

- A nation must first have a stable political system that permits economic growth.
- Develop strong institutions to foster proactive involvement in the development, transfer, and implementation of technology.

- Develop strong financial institutions capable of supporting sustained technical progress.
- Strengthen educational and training systems that permit citizens to move up the knowledge ladder.
- Develop technology strategy and support generic critical technologies.
- Support R&D activities.
- Encourage creativity and entrepreneurship.
- Vigorously participate in international debates on technological as well as trade issues to express national points of view and reconcile differences of opinion on issues of importance.
- Predict the social and environmental consequences of technology and develop appropriate public policies to deal with them.
- Develop strategic alliances with compatible countries to enhance technological progress and strengthen trade partnerships.

## COMPETITIVENESS OF FIRMS: THE MICRO LEVEL

National competitiveness is largely dependent on the competitiveness of firms within the nation's boundaries. A company's competitiveness is dependent on its ability to provide goods and services to the marketplace more efficiently than others involved in its arena. This depends on the company's ability to exploit ideas and resources in a timely, cost-effective manner in order to accomplish desired goals and objectives and to create products or services for its customers that meet or exceed their demands and satisfaction. At the firm level, MOT is important not only for improved profits but for survival. Companies that are unable to harness and optimally utilize technology will lag and may not survive in a fiercely competitive environment. To become or to remain competitive firms must be able to:

1. Develop a culture in which the value of technology as a strategic competitive weapon is fully appreciated.
2. Understand the dynamics of the process of technological innovation.
3. Monitor and forecast technological changes.
4. Develop and adopt effective methodologies to measure the impact of new technologies on their business.
5. Facilitate the implementation of new technologies in their operations and build the infrastructure necessary for migrating from one technology to another.
6. Prepare, train, and hire the proper workforce to implement the new technology.
7. Develop an organizational structure that permits effective and efficient implementation of technological changes.
8. Develop an appropriate reward system for employees and managers.

At the firm level, management must develop a strategy for competing. The question frequently asked is, On what basis can a firm compete? The answer follows the most fundamental principle in business: Competitiveness can be achieved by providing value to the customer. This implies setting a strategy by which a firm can achieve a favorable position in the marketplace. Firms can compete in the marketplace using many formulas. Some of these are listed below. One or more can serve as the basis for a strategy that gives a firm an advantage over its competitors.

**1** Offer products or services desired by a customer.
**2** Rely on innovation to introduce new products or services.
**3** Achieve technological superiority in (*a*) products, (*b*) process, (*c*) service, and (*d*) marketing.
**4** Concentrate on quality of product or service.
**5** Reduce cost and/or price.
**6** Be first to market.
**7** Reduce the product development cycle's time from concept to market.
**8** Create and target niche markets for products.
**9** Eliminate waste.
**10** Build in flexibility to change.
**11** Improve efficiency.
**12** Improve customer service.
**13** Promote creativity and entrepreneurial spirit.
**14** Develop and harness employee knowledge and talents.
**15** Follow a progressive culture for the organization.
**16** Encourage teamwork.
**17** Introduce a progressive management style.
**18** Enhance the ability to forecast.
**19** Sharpen the ability to plan.
**20** Focus on increasing market share.

In Chapters 8, 9 and 10 we will discuss methodologies used for developing organizations' strategies, and emphasize the importance of integrating the technology strategy with the business strategy.

Competitive firms can be recognized by a set of characteristics. A successful firm usually has one or more of the following attributes:

**1** Profitable.
**2** Stable.
**3** Capable of leading in innovation and technology.
**4** Has ability to maintain or increase market share.
**5** Capable of developing and introducing innovation in a timely manner.
**6** Is a pacesetter, often setting industry standards.
**7** Has an ability to utilize technology and to capture market share through products, process, information systems, or service innovation.
**8** Has the ability to match its strengths with targeted market needs better than other companies can.
**9** Aggressive in its desire to reach planned goals.
**10** Flexible.
**11** Progressive.

**12** Fair.
**13** Knowledgeable about its core technology.
**14** Knowledgeable about its strengths and weaknesses.
**15** Knowledgeable about its competitors.
**16** Has visionary leaders.
**17** Knows how to fully utilize the capability of its employees.
**18** Motivates and rewards employees appropriately.
**19** Knowledgeable about the technology and business life cycles and knows when to hold and when to fold new projects.
**20** Knowledgeable about its social, political, and legal environment.

Competitive firms can achieve success by setting an appropriate strategy, implementing it, and following up to evaluate results and take corrective actions as needed.

READING 7.1

# International Competitiveness and the Management of Technology

**Tarek Khalil and Javier Garcia-Arreola**

## INTRODUCTION

In recent years, much has been written about the technological competitiveness of nations. Technological competitiveness in the global marketplace helps countries to improve standards of living. International competitiveness is dependent upon a nation's abilities to effectively manage its technical resources and market them globally. Many factors have been associated with improvements in technological competitiveness, such as long-term planning horizons, education, total quality management, engineering research, free trade, etc. A critical problem for many nations, as well as researchers in the field, is the inability to reach a consensus on the relative importance of each factor in improving competitiveness. For the United States, this debate has been raging for years and has been highly ideological and rhetorical.

In 1990, the authors decided to poll the views of experts in the field of MOT on providing a ranked order of problems and priorities. The polling addresses (i) the state of U.S. competitiveness in the global economy, (ii) a ranking of different factors that are said to affect its competitiveness, and (iii) the importance of the different management and policy strategies for improving it.

*Source:* Tarek M. Khalil and Javier Garcia-Arreola. *Proceedings of the 5th International Congress of Industrial Engineering,* Ecole Nationale Supérieure de Génie Industriel et Institut de la Production Industrielle, Grenoble, France, April 1996.

For this purpose, a survey instrument was constructed and conducted on participants at the Second International Conference on Management of Technology held in Miami, Florida in 1990, which, with more than 400 participants, was the largest of its kind. The results were quite revealing in that they showed substantial agreement among respondents regarding what they considered to be pertinent issues of interest regarding U.S. technological competitiveness (Berman and Khalil, 1992).

Because of the dynamic of technology and global marketing strategies, it was decided to repeat the survey in 1994. A similar survey instrument was constructed with some minor but relevant modifications, and the survey was administered to the attendees of the Fourth International Conference on Management of Technology held in Miami, Florida. The results of the new survey were compiled and compared to the 1990 survey. This paper presents the results of the two surveys and provides a comparison of the results, which indicates the changes perceived to have taken place in the four years between the two surveys.

A comparison between the results of the opinion survey of 1994 and actual data provided by the U.S. Council on Competitiveness shows strong similarities. However, the experts' opinion survey reveals many factors that still need to be addressed by American managers and policy makers in order to remain competitive in a global market. These factors are presented and discussed.

## U.S. COMPETITIVENESS IN THE GLOBAL ECONOMY

Respondents were asked to rank U.S. competitiveness relative to the following countries: Japan, Germany, France, Sweden, England, Canada, South Korea, Taiwan, Italy, Singapore, Australia, Spain, China, and Russia. Canada and Russia were not included in the 1990 survey. The criteria in selecting countries were based on their sustained competitiveness or their effort to attain it. The scale was 1 to 12, with 1 = highest and 12 = lowest. Figure 7-24 shows the results for both the 1990 and 1994 surveys. As can be seen from the charts, the 1994 respondents presume that the U.S. has recovered its leadership in global competitiveness, which approximately equals Japan's. Canada entered the 1994 list as number 7. The general groups of countries did not present a significant change from 1990 to 1994: the U.S., Japan, and Germany leading, followed by the group of France, Sweden, England, South Korea, and the newly included Canada. Australia, Spain, China, and Russia were perceived to be less competitive at the time of this survey.

Figure 7-25 shows the perceived change in competitive position of the included countries relative to the U.S. in the five to ten years preceding the survey. A rate of 3 means a very large gain, while −3 means a very large loss. The perceived changes from 1990 to 1994 are significant. Germany has been relegated behind the Asian Tigers—South Korea, Taiwan, and Singapore. China shows a strong appearance and Japan is seen as continuing its competitive advantage. However, it is interesting to note that Japan was perceived to be losing its leadership position compared with its first position in the 1990 survey. On the other hand, the U.S. has maintained its competitive level compared with France, Canada, Spain, and Italy. England and Russia are the only countries that show loss of competitiveness compared to the United States.

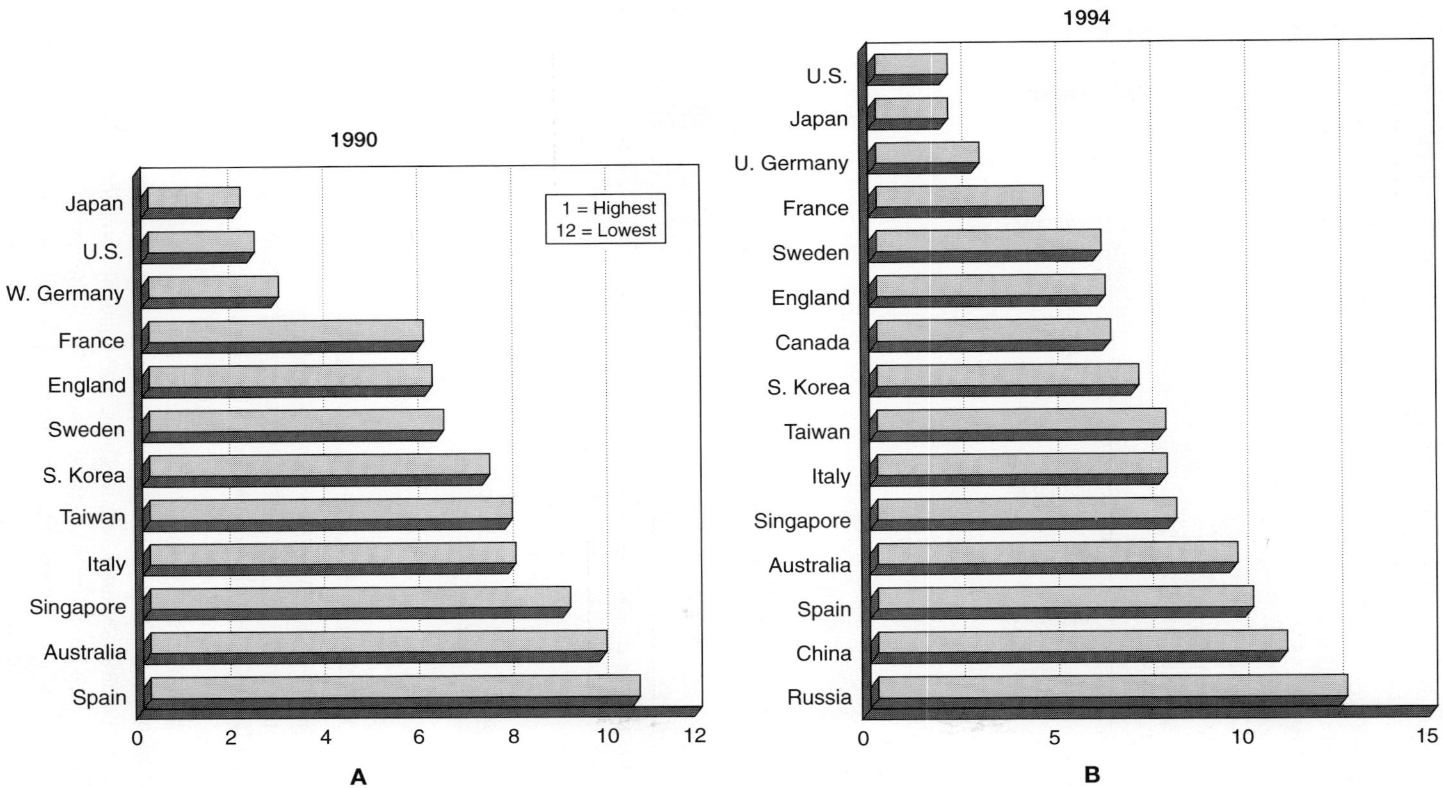


**FIGURE 7-24**
RANKING OF COMPETITIVENESS

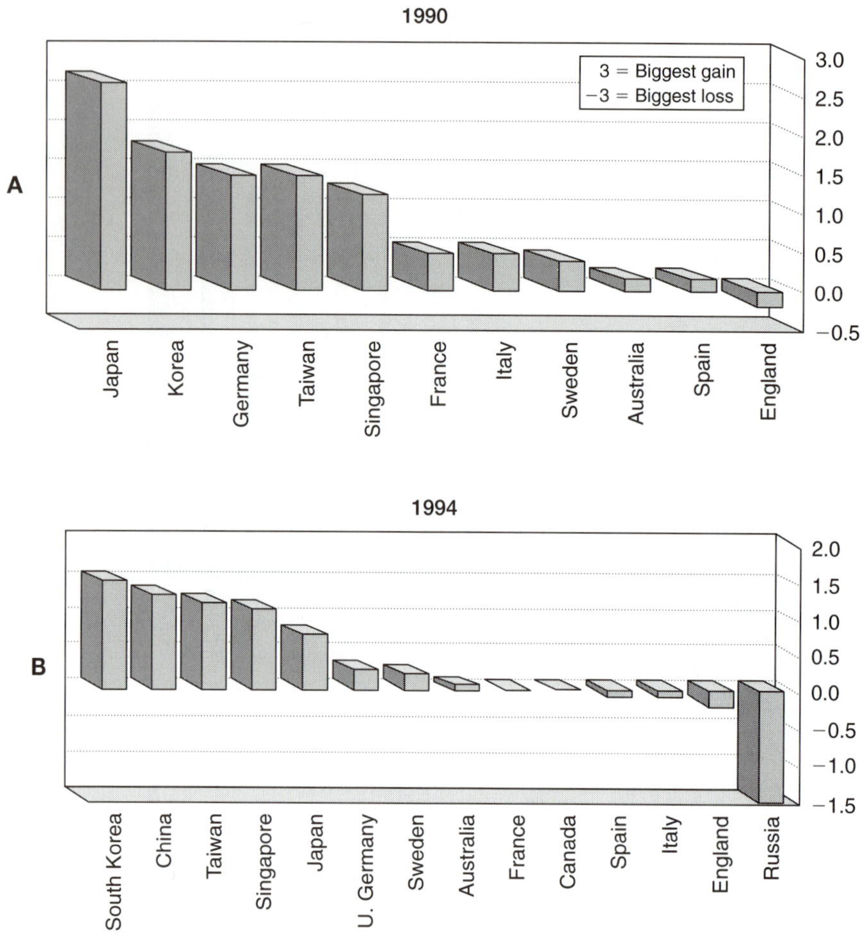


**FIGURE 7-25**
CHANGE IN COMPETITIVE POSITION

## PROBLEMS OF U.S. TECHNOLOGICAL COMPETITIVENESS

What are the problems that interfere with U.S. competitiveness? The experts think that the main problem at the time of the 1994 survey was the short-time horizons used by management when planning the strategies. This was the same number-one-ranked problem in 1990 as well. Figure 7-26 shows the change in the perception of the respondents. The lack of understanding of foreign cultures stands in the second place, one higher than in the first survey. This is a very important issue given the fact that today's global markets demand inter-cultural relationships. Respondents considered the neglect of human resources to be in the third place among the problems for the U.S. Maybe the short-term vision is pushing managers to consider quick profit options and "reengineering" processes that are having an adverse impact on the morale of employees. It can be

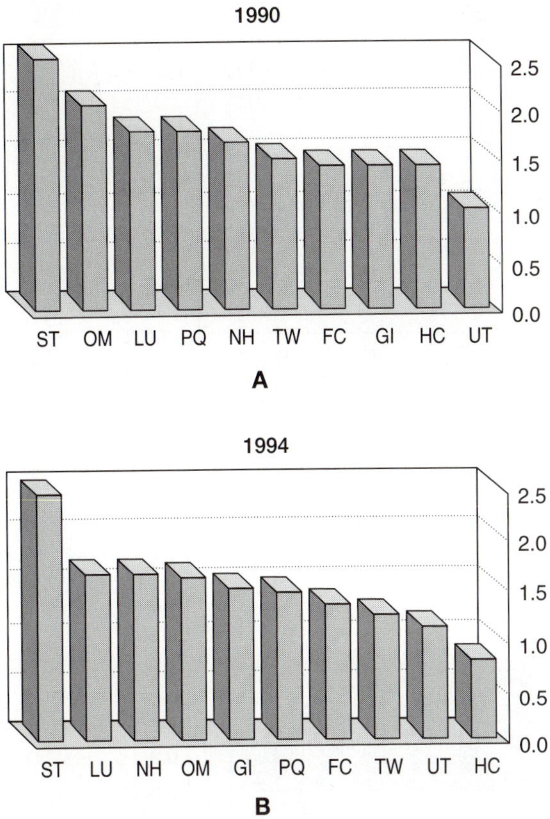


**FIGURE 7-26**
RANKING OF PROBLEMS IN U.S. TECHNOLOGICAL COMPETITIVENESS
Key to abbreviations: ST, short-term horizons; LU, lacking understanding of foreign cultures; NH, neglect of human resources; OM, outdated management strategies; GI, government and industry at odds; PQ, poor quality management; FC, failures of cooperation between companies; TW, technological weaknesses in development and production; UT, unfair trade practices; HC, high cost of capital.

seen that the management style was perceived to have improved from 1990 to 1994: both outdated management strategies and poor quality management went down two positions.

## CHANGES IN COMPETITIVENESS BY INDUSTRY

Figure 7-27 shows the experts' opinion in this regard. Respondents were asked to rate changes in competitiveness for selected U.S. industries. The scale used was from 3 = very large gains to −3 = very large losses in competitiveness. The respondents felt that the U.S. seemed to have restored its competitive position in all industries surveyed, with the computing industry leading the way: it went from number five in 1990 to number one in competitiveness in 1994. Equally important is the movement of communication equipment from a negative perception to a positive one. There is no question that these two industries are essential for the Information Age and represent a very significant aspect of modern industrial and economic growth. It is also important to note the perceived improvement in the motor industry, a sector dominated by foreign automobiles: motor vehicles were the last category in 1990, moving to a more competitive position in 1994.

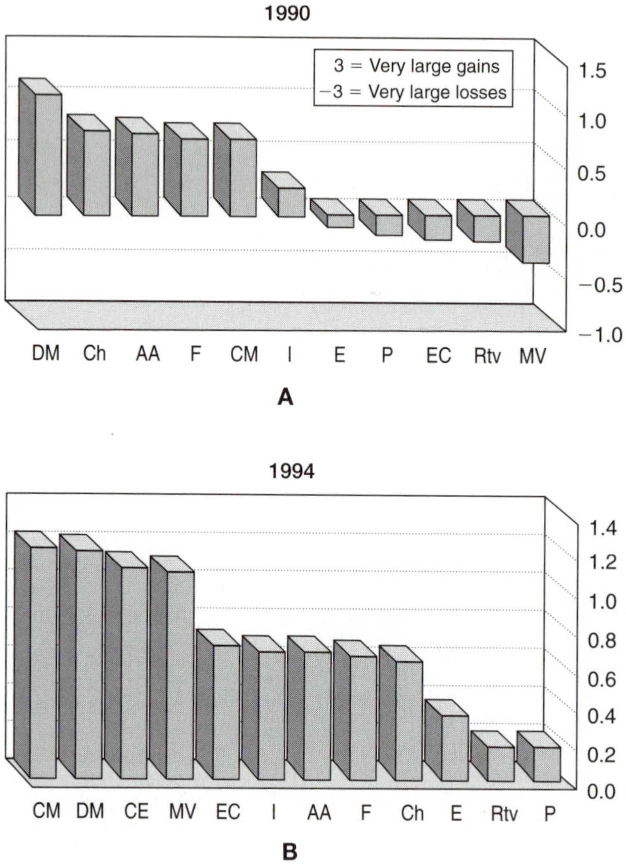


**FIGURE 7-27**
CHANGE IN COMPETITIVENESS OF U.S. INDUSTRIES
Key to abbreviations: DM, drugs and medicine; Ch, chemicals; AA, aviation and aeronautics; F, food; CM, computing machines; I, instruments; E, engines; P, petroleum; EC, electronic components; RTV, radio and TV; CE, communication equipment; MV, motor vehicles.

## MOST IMPORTANT POLICIES FOR THE U.S.

What are the national policies that will help the U.S. to improve its competitiveness? The experts think that the answer is in better management of technology. As Figure 7-28 shows, better technology management curricula, more engineering education, technology transfer, and tax credits for technology occupy the first four places. The following three places (helping industry cooperation, more university research, and increasing venture capital) denote a need for more research. The 1990 survey identified general education and technology transfer as the most important policy that needed to be reinforced.

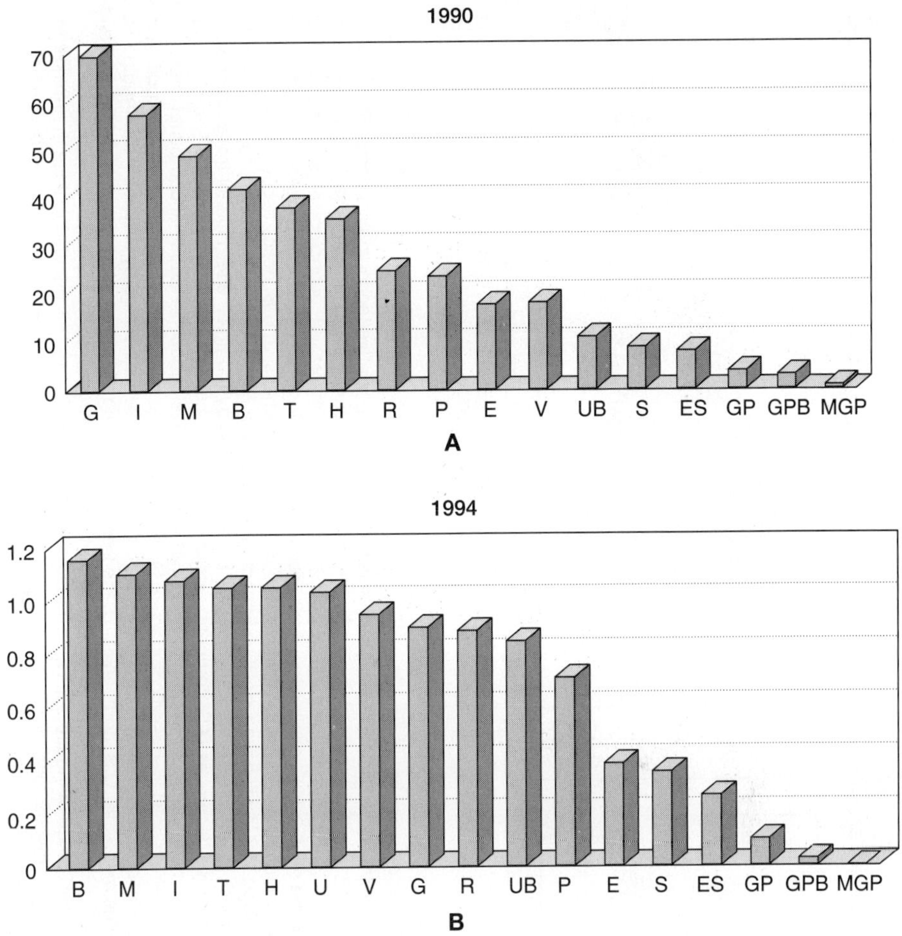


**FIGURE 7-28**
MOST IMPORTANT POLICIES FOR THE U.S.
Key to abbreviations: B, better technology management; M, increased engineering education; I, improved technology transfer; T, tax credits for technology; H, helping industry cooperation; U, more university research; V, increased venture capital; G, improved general education; R, reduced management-labor conflict; UB, increased university-based incubators; P, monetary policies; E, enforcing fair trade policy; S, reduced social regulation; ES, export subsidies and aid in marketing; GP, government procurement performance standards; GPB, government procurement buy-domestic policy; MGP, increased government purchases.

## OTHER PERSPECTIVES

Respondents were asked a variety of other questions. The answers were to be given on a scale from 3 = totally agree to −3 = strongly disagree, shown in Figure 7-29. The statements were slightly changed from the 1990 survey to the 1994 version to reflect new changes in world economics, trade, and industry. The experts point to the short-term-oriented business thinking as the main cause of the loss of competitiveness of the

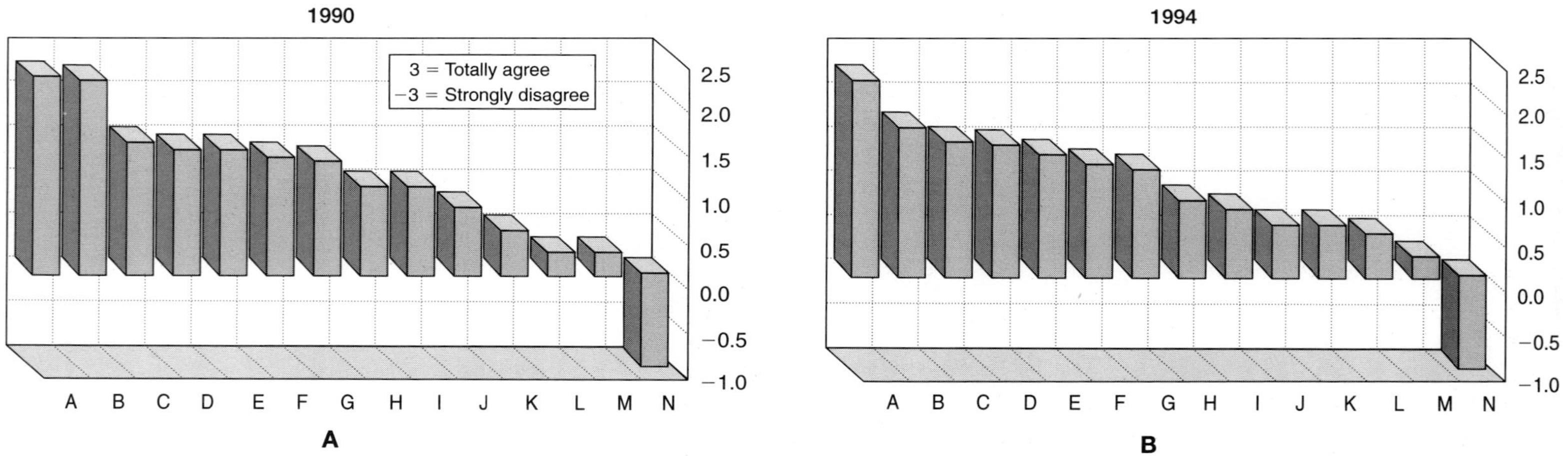


**FIGURE 7-29**
AGREEMENT OR DISAGREEMENT WITH CERTAIN STATEMENT
**A** Key to abbreviations: A, the relative U.S. decline has been increased by short-term business thinking; B, U.S. relative technological competitiveness has declined in the past ten years (in a broad range of industries); C, the relative U.S. decline is primarily the result of poor management; D, the relative U.S. decline has been increased by poor general technical education; E, U.S. policy could learn from industrial policies abroad; F, the increased technological presence by Pacific Rim countries has been an important cause of this decline; G, the relative U.S. decline has been increased by a lack of R&D input in business planning; H, developments in eastern Europe will greatly increase the market for U.S. companies; I, buyouts have harmed U.S. competitiveness; J, U.S. policy assumes that market forces alone will eventually ensure future U.S. competitiveness; K, "Europe 1992" will have a positive impact on U.S. companies; L, U.S. commitment to free-trade policy has hindered its responses to increased competition from abroad; M, the relative U.S. decline has been increased by the unfair international trade practices of other countries; N, collaboration between U.S. and foreign companies decreases U.S. competitiveness. **B** Key to abbreviations: A, the relative U.S. decline has been increased by short-term business thinking; B, the use of total quality management in the U.S. has had a positive impact on industrial competitiveness; C, the North America Free Trade Agreement (NAFTA) will have a positive effect on U.S.-based companies; D, the relative U.S. decline has been increased by poor general technical education; E, U.S. policy could learn from industrial policies abroad; F, the relative U.S. decline has been increased by a lack of R&D input in business planning; G, developments in China will greatly increase the market for U.S. companies; H, the trend of declining U.S. competitiveness has been reversed in the last two years; I, the increased technological presence of Pacific Rim countries has been an important cause of this decline; J, U.S. relative technological competitiveness has declined in the past ten years (in a broad range of industries); K, developments in eastern Europe will greatly increase the market for U.S. companies; L, developments in Russia will greatly increase the market for U.S. companies; M, the relative U.S. decline has been increased by the unfair international trade practices of other countries; N, total quality management is declining in importance in the U.S.

United States. Total quality management (TQM) is considered an important tool that seems to have improved the performance of U.S. corporations. NAFTA was another issue considered significant for U.S.-based companies, with the possibility of a positive impact. The respondents think that the technological presence of countries from the Pacific Rim and developments in eastern Europe will have positive but relatively lower impacts on U.S. competitiveness.

## CONCLUSIONS

There is a significant shift in experts' opinion about U.S. competitiveness from 1990 to 1994. The overall perception is that the U.S. is regaining its competitive advantage in the global marketplace in comparison with other countries and in a variety of industries. China and the Tigers of South Korea, Taiwan, and Singapore are also perceived to be gaining strength. These perceptions are positively supported by a host of quantitative economic indicators published by the U.S. Council on Competitiveness. The experts blame the U.S. relative decline in competitiveness mainly on management's short-term business thinking. They agree that the quality movement has had a positive impact on U.S. industrial competitiveness. They also agree that the most important issues for continuing U.S. improvement in competitiveness will depend on better education and practice in MOT.

### Case Reference

Berman, E. M., & Khalil, T. 1992. "Technological Competitiveness in the Global Economy: A Survey." *International Journal of Technology Management,* vol. 7. Nos. 4/5, pp. 347–358.

READING 7.2

# Can American Manufacturers Compete outside the U.S.?

**Harry A. Hammerly**

Executive Vice President, 3M Company

Why do we ask, "Can American manufacturers compete outside the United States?" Because, for the first time in our history, the field on which we play out our competitive challenges is level—and it is global. The United States has more equal competitors throughout the world today than ever before.

But we should realize that the level playing field on which we now must compete corrects the imbalance in the world that resulted from World War II. From 1945 to the

*Source: Financial Executive,* September–October 1990. Reprinted with permission from *Financial Executive,* Sept./Oct. 1990, copyright 1990 by Financial Executives Institute, 10 Madison Ave., P.O. Box 1938, Morristown, NJ 07962-1938. (973) 898-4600.

late 1960s, the U.S. had an unusual advantage in international business competition. We had a large home market that had not been damaged by the war, and we benefited from the surge in technology that resulted from the war effort.

The country was so strong, in fact, that our foreign policy during the last 40 to 45 years was intended to speed the recovery of Europe and Japan. And our foreign policy succeeded so well that we now have a global market in which we are one of several competitors. We no longer dominate world business.

U.S. business, government, and citizens must face up to the new arena in which we find ourselves. But at the same time we must stop beating ourselves about losing a position that in reality was not a normal one. Looked at objectively, this equality can have a number of advantages for America. The U.S. must treat other nations as equals, but at the same time Japan and Europe must take greater responsibility for maintaining the economic system that has served them so well. If we take this attitude, our approach to negotiating trade issues with other countries will change, and we will be able to pressure those countries into taking their share of the responsibility for maintaining the system. This is a very important point to realize at the outset.

## AMERICA'S STATUS IN THE WORLD

By accepting the premise that parity is replacing dominance, we can examine the current status of the U.S. in international trade and competition from a fresh perspective. To begin with, the U.S. economic engine has not broken down. Annual real per capita income growth in the U.S. is running at about 2 percent, a rate unchanged for nearly the last 100 years. Other economies are growing faster, but the U.S. is still way ahead. If trends of the past six years continue, for example, Japan will need 33 years to overcome the U.S. in annual real per capita income.

Today, America's share of world manufacturing production is 32 percent. That is about the same as it was in 1913 and 1938, just before significant world wars. That is above the 25 percent that was our share in 1900. In fact, the only time our share of world production exceeded our current share was in 1953, when it soared to 44 percent. This supports my earlier statement that the dominance we experienced in the 1950s and 1960s was really abnormal.

Looking at exports and imports we find that our share of global exports is 13 percent, higher than Germany's 12 percent, and Japan's 10 percent. Merchandise exports from the U.S. have tripled in the last 10 years from $180 billion to $360 billion.

## WHAT WENT WRONG?

The problem is, however, that our merchandise imports have increased 122 percent in that same 10-year period. With our imports exceeding our exports by almost 25 percent we should be concerned about our ability to compete in the future.

Compounding the import/export imbalance is the issue of productivity: Our living standard, measured in real GNP per capita, is 40 percent higher than that in Japan and 50 percent higher than that in Europe.

Productivity—output per employed person—is $41,000 in the U.S. compared to $30,000 in Japan and $34,000 in Germany. But productivity growth in the U.S. is lagging behind that of Germany and Japan. Productivity growth is the key to our future prosperity, and the fact that we have fallen behind is indeed cause for significant concern.

Another concern is that during the 1980s we lost technological leadership in some very important areas such as computer chips and machine tools. According to a Ministry of International Trade and Industry (MITI) analysis of the United States' and Japan's prowess in 40 key sectors of commercial technologies, in 1983 Japan was lagging in half of these technologies. But in 1990 Japan trailed in only one—database software (*Fortune,* April 23, 1990). In addition, civilian investment in R&D is 1.8 percent of GNP in the U.S., 2.6 percent in Germany and 2.8 percent in Japan.

Contrary to common belief labor costs in the U.S. are not much higher than those in Japan and other industrialized nations. Our employee health care costs, however, are much higher. These are the issues we need to be concerned with when we discuss our place in global competition.

## MULTINATIONALS BRING HOME THE BACON

A study conducted by Amir Mahini of McKinsey & Co. indicated that in one year, 1987, 2,300 American multinationals accounted for 73 percent of U.S. exports and only 39 percent of our imports. In other words, these 2,300 companies contributed a positive $517 billion to the U.S. trade balance in one year.

To demonstrate how effective U.S.-based multinationals are in the global economy I'd like to refer to the company I know best—3M.

In 1989 3M exported $1 billion worth of merchandise, giving us a three-to-one positive trade balance for our U.S. company. We estimate that one job in six in our American operations depends on our international business.

In 1989 46 percent of the company's $12 billion in sales was generated from business outside the U.S., or OUS. Our target is to raise our OUS business to 50 percent of total sales by 1992. We have operations in 53 countries, we manufacture in 41 countries, and we sell into 135 countries. 3M operates research centers in Europe and Japan staffed with 2,000 technical people. Of our 80,000 employees, 38,000 live and work outside the U.S. Our OUS operating income increased by 74 percent in the last three years and was almost 22 percent of sales in 1989.

3M Our approach to business outside the U.S. has been to start small and grow. We follow a principle we labeled FIDO—"First in defeat others." We build our company on local people though the managing director or the equivalent senior position is usually not a local national. Nor is he or she always an American. For example our managing director in Germany is a Dane; in the Netherlands a Norwegian. An Italian is our executive vice president in Sumitomo-3M in Japan. We have found that when someone 40 or 45 years of age is put in charge of a business in his or her own country it's hard to keep that person motivated and "hungry" for the next 20 or 25 years.

In starting a company overseas we match our product line to the country's needs and objectives, which frequently involves building an infrastructure. We then try to determine which of our products will address those national objectives.

In almost every case we own our overseas companies 100 percent. In 1983 we started a wholly owned company in China where we'll have about $20 million in sales this year. We recently opened a joint venture in India, a wholly owned company in Turkey, and we are starting an operation to assemble road signs with the ministry of highways in the Soviet Union very soon. We also will open a representative office there as part of this agreement.

We have offices in almost every country in Central and Eastern Europe. While these are at present very small we do have plans to expand them as soon as the opportunities justify.

As for 3M's international growth, to generate 50 percent of our total annual sales by OUS business by 1992, we plan to use the following strategies: First, we will expand our product offerings. We are not selling the full 3M line in many countries, so product expansion is the most logical way to increase our business. We also want to continue to invest in local technical resources because our business frequently involves solving customer problems by applying 3M technologies or products to customer applications, which requires local technical support. And when justified we will increase our manufacturing facilities outside the U.S.

Innovation is one of the hallmarks of 3M. It's a building block in our strategy for growth.

## AMERICAN COMPANIES CAN COMPETE

What do American companies need to get their share of the global market?

First, we must as a country make being successful in the global market an important national priority. We must realize that the unique imbalance we all enjoyed for so long will never be seen again by us or any other nation. In fact the world's battles in the future may well be fought in the economic arena.

Second, the manufacturers must commit to quality. The Japanese call it "total quality" but it is really a management system that conveys an attitude ensuring quality far beyond the product. At 3M we call it Q90s. The objective of Q90s is to be the best at everything we do in all aspects of our business using as a guide the seven criteria required for the Baldrige Award.

But Q90s means we have to empower our people to delegate to use all of the skills of all of our people. At the same time our entire organization must buy into the common goals. We can't as management delegate and abdicate our responsibility; we have to delegate with common goals throughout the organization.

A third essential to America's gaining its share of the global market is to have its manufacturing community focus on the customer. At 3M we say, "Think global and act local." This means we have to know the similarities and the differences among the various markets around the world. It means serving the local customer his way, not some way we've generated internally. The balance between thinking global in our business but acting local on the scene is critical.

Focus on customer satisfaction, a key element in our Q90s program and a key element in the effort all U.S. businesses must put forth to win overseas. Customer satisfaction is not a program; it's part of a process, because it never ends.

Fourth, American businesses have to make the required investment in the individual foreign markets. That may be more difficult and expensive today than it was when 3M first did it. For example, when we started in Japan in 1961, it was much easier because that country had not matured economically. For obvious reasons, Japan can be a very expensive market to enter today.

Fifth, innovation is an important element for American manufacturers to emphasize. Innovation changes the rules of the game. It's the way to get advantage and to force the competitor to go back and regroup.

Sixth, American businesses have to get the word out. We must speak to the public and to our representatives in Washington about global competition.

## WHAT THE GOVERNMENT MUST DO

While American business must not expect the government to pave the way for its success, the government can and must support our competitiveness in global markets. It can create a climate conducive to trade and have sensitivity to the importance of manufacturers. Saying the U.S. is going to be a "service" country is a troubling thought. I have nothing against selling hamburgers and computer software, but without a solid base in manufacturing, the U.S. cannot be the kind of economic influence around the world that it wants to be. There is no simple prescription, but the government can take some steps that would set us off in the right direction.

The first is to get our rational financial house in order to reduce the Federal budget deficit. We have to do this by cutting spending devoted to consumption, as opposed to that which is devoted to investment. At 3M, we also favor some type of consumption tax, particularly on gasoline, which is very cheap in the U.S. compared to other countries. In addition, the government can take a hard look at the increases in entitlements. Doing so would ensure that everybody has to pay to put our financial house in order.

Second, we have to ensure access to foreign markets. The current efforts by Ambassador Hills are serving us well in this regard and should be continued.

Third, the government must not penalize trade by taxing foreign profits, which are earned overseas, taxed by foreign governments, and never brought back to the United States. And it must eliminate the double taxation of dividends that favors debt over equity in the financing of American business.

Fourth, the government must use the Export Control Act realistically. We must recognize that we in the U.S. don't own all of the world's strategic technology. And the government must not use trade unnecessarily or ineffectively as a political weapon, such as restricting grain sales to the Soviet Union, enabling the Canadians and Argentines to take our market share. Once a country loses market share these days, it's very difficult to recapture.

Fifth, we must have realistic environmental regulations. We at 3M are strong believers in supporting the environment, and we have a good reputation in that regard. But, in some cases, these regulations are becoming unreasonable. If carried too far, they will restrict our competitiveness around the world.

Sixth, realistic currency values are important, of course. And the government must continue to support a free trade policy and the principles of the GATT.

Seventh, a stronger national technology policy is needed to support our position in manufacturing around the world. More Federal dollars should be allowed to go into technologies and manufacturing processes that have long-term commercial value.

Eighth, the government must do more for industrial research. In 1986, less than 2 percent of the Federal government's $56 billion research budget went into industrial research. We think Federal funds should provide a greater supplement to industry's commitment to research. In addition, instead of supporting a few horrendous-sized projects, the government ought to support many small projects.

Ninth, the government should remove the anti-trust barriers for joint manufacturing efforts. In a global economy, government must realize that all the competitors are not in the U.S. and that other countries do allow their companies to team up to be competitive in world markets.

Finally, stronger support of education is needed. After all, it is a people game. People make things happen. But many comparative studies have shown that our educational results have been slipping behind those of our major competitors.

In short, American manufacturers can compete in global markets. But, to do so, American government and industry must work better together, and American society must commit to being competitive in global markets. We are not in a bad position today, but we must dedicate the nation's business to total quality, to being the best supplier, to being the best at all that we do.

### THE FALL OF THE TIGERS

The economies of countries of Southeast Asia were booming in the eighties and early nineties. The so-called Tigers included a group of countries such as Indonesia, South Korea, Thailand, Malaysia, Taiwan, and Hong Kong. These countries followed an aggressive growth strategy. They were aided by a cheap labor force, good education strategies, government policies favoring rapid development, and liberal investment policies. The Tigers attracted many multinational corporations and large sums of foreign investment capital. Local industry witnessed significant growth, and the GDP of all the Tigers was outpacing that of other countries in the world. The growth rate reached two-digit or high-single-digit growth per year. Stock prices at the Southeast Asian markets hit record highs. Products produced by the Tigers penetrated the world markets and competed very favorably.

These countries competed with products based mostly on mature, low, or medium-technology levels. They excelled in the garment industry, household products and appliances, semiconductor components, personal computer hardware, and certain segments of the automotive industry. These technologies are relatively easy to transfer, and they enabled the Tigers to compete on the basis of price and quality. The Tigers also followed a strong global marketing strategy.

In the second phase in their development effort, large corporations in South Korea, Taiwan, and Hong Kong started to emphasize their R&D efforts and to pursue products in higher technologies such as telecommunications and aerospace components. The Tigers were giving Japan, the United States, and European nations strong competition.

Suddenly in 1997 the crisis began. It started with Thailand and spread like wildfire to all the other Southeast Asian countries. The currencies of Thailand, Indonesia, and South Korea were devalued. The stock markets of Indonesia, South Korea, Malaysia, Thailand, the Philippines, and Hong Kong plunged. The Japanese stock market followed suit. Many companies that overextended their borrowing to support their development efforts were unable to pay their debts. Banks were starting to fail, and the governments were unable to support their currencies.

The crisis of the Asian Tigers is largely a failure of the financial system. But this system is so closely intertwined with technology management fundamentals that it should be carefully examined. Look again at Figure 1-6, which illustrates the dependence of sustained competitiveness upon three systems: (1) the economic and financial system, which places controls on monetary policies, (2) the technological system, which channels knowledge into production, and (3) the trade system. In the countries of Southeast Asia, loose practices in financial dealings were common. Many loans to enterprises were made without adequate rationale for the creation of a strong, competitive productive system. The technology has been shifting, yet companies pursued growth strategies disregarding the fundamental changes in global production and markets. The existing economic and financial system did not have adequate controls to weather the change. Friedman (1997) discussed some of the factors contributing to the crisis in Asia. He attributed the problems of the Tigers to (1) a China crisis, (2) an education crisis, and (3) a political crisis.

The China crisis refers to the emergence of a Chinese economy that has the ability to produce products underselling China's neighbors. China has cheaper wages and cheaper currency than its neighbors. The Chinese are doing to the Tigers what the Tigers did to Japan and what Japan did earlier to the United States. The game is simply to compete favorably on mature low- and medium-technology products and gain a strong market position. For example, Thailand employees involved in the gem-cutting business in Bangkok are being paid $200 a month compared to $50 a month paid to their counterparts in China. Thai business owners are finding it hard to compete under such conditions.

The second crisis is the education crisis. This is due to an education system that does not encourage creativity, independent thoughts, or innovation. To compete with China, the Southeast Asian countries need to move up the technology ladder and produce higher-end products, leaving the low end to China. This requires a strong effort in human resource development in order to create a workforce capable of competing in the global marketplace. Existing educational systems in Thailand, Malaysia, and Indonesia do not seem to be adequately fulfilling this need. Thailand has compulsory education only up to the sixth grade. Students are not equipped adequately to deal with the information and knowledge age.

The political crisis seems to be a source of great danger to the Southeast Asian countries. Their governments have to turn things around quickly before the masses become restless. In Indonesia President Suharto's 30-year dominating government was toppled. There are political crises in other Southeast Asian countries as well. A reform in the economic and financial system is needed, but the reform will be painful and will require sacrifices.

Can organizations restructure to meet the technological challenges that would restore their competitive edge in the global marketplace? This is a question that will be answered in the months and years ahead. According to Friedman (1997), Southeast Asia's people have not yet slammed onto the pavement, but they are expected to hit soon. They don't have to, if they manage their technology system, economic system, and trade system in such a way as to restore their competitive edge.

## CONCLUDING REMARKS

Competitiveness is dependent on how well people manage the system of wealth creation. At the macro level of nations, public policy will determine how well coordinated the economic and financial system is with the technological and production system and the trade practices of a country.

At the micro level of the firm, competitiveness will depend on how well organizations manage their technological resources. Keeping up with changes in product, production, and marketing technology enhances any firm's opportunities of success. Policies and strategies followed by firms at the micro level influence the economic conditions at the national level. At the same time, public policy regarding investment policies, interest rates, tax incentives, education, and trade policies affects industry and businesses. It is therefore essential that policies at both the national and the firm levels be well integrated and harmonized. Governments and businesses should simultaneously focus on creating and sustaining production systems capable of competing in a global environment. Technology is the engine of that system.

## DISCUSSION QUESTIONS

1. Why is it that technology contributes to the competitive level of a country? How did Watt's steam machine contribute to England's leadership during the Industrial Revolution?
2. Is there a relationship between corporate and national competitiveness?
3. What efforts have been taken recently by the U.S. government to ensure competitiveness?
4. Conduct an independent review and write a two-page report on GATT.
5. Conduct an independent review and write a two-page report on NAFTA.
6. Get a copy of the most recent competitive index and observe changes taking place in the various indexes. Discuss reasons for these changes.
7. Analyze the issues in U.S. technological competitiveness, and discuss the experts' rankings of the problems involved.

## ADDITIONAL READINGS

Michael E. Porter. *The Competitive Advantage of Nations.* Free Press, New York, 1990.

In the study presented in this book, Porter presents evidence of how nations achieve leadership in certain industries. Competitiveness is based on synergy between cultural aspects, natural resources, and national abilities.

D. W. Cheney and W. W. Grimes. *Japanese Technology Policy.* Council on Competitiveness, Washington, DC, February 1991.

Council on Competitiveness. Most recent publications.

IMD. *The World Competitiveness Year Book.* Lausanne, Switzerland.
This book, published annually, contains up-to-date statistics about many indexes contributing to competitiveness. Ranking of competitiveness of nations is also provided.

### SUGGESTED CASES

- Praegitzer Industries Inc. Harvard Business School, Case 97A007.
- Bay State Milling Co. Harvard Business School, Case 9-594-080.
- Technology Collaboration in Europe. Harvard Business School, Case 9-389-130.

### REFERENCES

Cheney, D., & Grimes, W. 1991. *Japanese Technology Policy.* Council on Competitiveness, Washington, DC.

Council on Competitiveness. 1994. *Competitiveness Index,* July.

Council on Competitiveness. 1995. *Competitiveness Index,* August.

*Financial Times.* 1999. "US Powers Ahead as Competition Drives Investment," June 25, 1999.

Friedman, Thomas. 1997. "Asia's People Have Not Yet Slammed into Pavement," *New York Times,* Dec. 26.

IMD International. 1999. *The World Competitiveness Year Book.* IMD, Lausanne, Switzerland.

Lee, Thomas H., & Reid, Proctor P. 1991. *National Interests in an Age of Global Technology.* National Academy of Engineering, Washington, DC.

OECD. 1998a. *Economic Outlook,* Dec.

OECD. 1998b. *Quarterly National Accounts,* 1998-4, Dec.

OECD. 1999. *Main Economic Indicators,* Feb.

Office of Technology Policy. 1997. Data provided by special request. U.S. Department of Commerce, Washington, DC.

Sumanth, David. 1984. *Productivity Management.* McGraw-Hill, New York, N.Y.

Wong, Poh-Kam. 1995. "Small, Newly Industrializing Economies Facing Technology Globalization: A Singaporean Perspective." In Lefebvre, L. A., & Lefebvre, E., *Management of Technology and Regional Development in a Global Environment.* Paul Chapman, London, pp. 66–75.

World Bank. 1996. *Plan to Market: World Development Report 1996.* Washington, DC.

World Bank 1999. *World Development Report* 1998/1999. Washington, DC.

# 8

# BUSINESS STRATEGY AND TECHNOLOGY STRATEGY

Events of the past two decades have shown us that business competitiveness is no longer a matter of choice but is a matter of survival in the global marketplace. Achieving organizational goals and competing successfully, in times of fast-changing business environments, require the development of sound strategies. This chapter discusses the fundamentals of strategic management, explains the cornerstones of formulating business and technology strategies, and offers methodologies that are helpful in strategic decision making. The importance of linking technology strategy with business strategy is emphasized both in the text and in Reading 8.1. Chapter 9 provides more details about technology planning and presents methodologies used by leading corporations to guide the planning effort.

## WHAT IS MEANT BY STRATEGY

*Strategy* involves envisioning and planning for the future. It is the means by which long-term objectives will be achieved. In business terms, it represents a broad formula of how an organization intends to succeed. It also details the plan that needs to be followed to compete and win. A strategy entails defining goals, deciding the way to reach these goals, setting action plans to execute specific tasks, and following up on accomplishments to ensure that objectives have been met.

Some people think of strategy as developing a long-range plan, assuming that they will continue to do what they are already doing. There is a fallacy in this assumption, since the economic environment and technological scene are continually changing. A more appropriate way of thinking about strategy is to think what should be done, regardless of what has been done and without saddling the plan with existing practices. Formulating strategy is a continuous challenge, requiring the evaluation of old practices and the search for new ones. Thus, a strategy must first affirm the "core" of the business—what the business knows—and then develop what the business can do.

It is essential that nations, as well as firms, develop competitive strategy and institutionalize strategic planning efforts. This will help them compete more effectively and strengthen their market positions. *Strategic management* is a process consisting of three important and interrelated components:

1 *Strategic planning* includes strategic vision setting and strategy formulation. This component of strategic management concentrates on "strategizing."

2 *Strategic implementations* include detailing actions that need to be followed and designating the functional units responsible for implementing operational actions and strategic projects. This component of strategic management deals with tactics and systematic planning.

3 *Strategic evaluation* involves performance measures, feedback mechanisms, continuous improvement, and the organizational learning process. This component of strategic management permits refinement of the strategy and corrections to the plans.

Strategizing entails envisioning, brainstorming for ideas, thinking about the future, and analyzing existing conditions and trends in society, the industry, and the marketplace. Strategizing should result in a vision, a mission, and a set of objectives that will set the direction of the organization. The planning process charts the path to the objectives and the systems that will follow-up on performance. Many organizations' managers tend to spend more energy on the planning component of strategic planning than on strategizing. When this happens, they tend to lose sight of what they should be doing as opposed to what they have been doing.

Hamel (1996) distinguishes strategizing from planning by the degree of innovation included in the strategy. He points out that planning is about programming, not discovering, and that the world has been more hospitable to industry revolutionaries producing strategic innovation than to industry incumbents. The latter are likely to surrender the future to revolutionary challenges. He describes the characteristics of two companies, one with a revolutionary strategic planning attitude and one that he terms "incumbent," or of the ruling class. The industry revolutionary has a planning process that is inquisitive, expansive, prescient, inventive, inclusive, and demanding. In contrast, Hamel describes the strategic planning process of incumbents as ritualistic, reductionist, extrapolative, positioning, elitist, and easy. A company that performs strategic planning as a routine exercise without periodically questioning its directions or pursuing innovative approaches can become stagnant and may lose its competitive edge.

## FORMULATION OF A STRATEGY

The formulation of a strategy requires a core and operating units to execute the strategy (see Figure 8-1). To use an analogy, the core of a strategy can be thought of as the brain. The brain receives information from the surrounding environment, processes it, sets goals, and coordinates a network of functional units in the human system. These

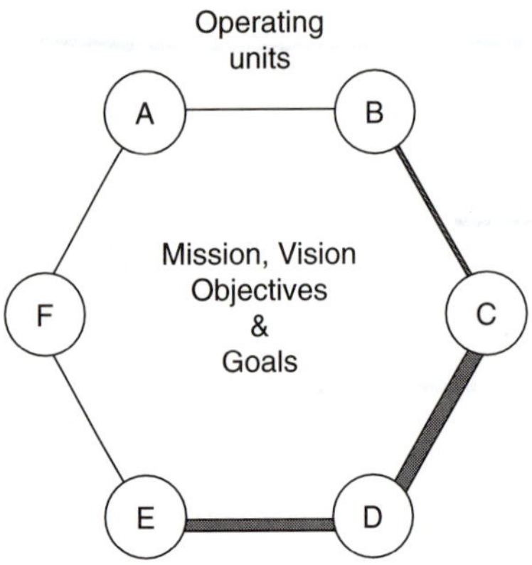


**FIGURE 8-1**
THE CORE AND OPERATING UNITS FOR THE EXECUTION OF STRATEGY

units are the muscles, the heart, the lungs, and other organs. Each of them performs its own function. In a company, central to the formulation of a strategy is management acting as the brain of the business enterprise, receiving information from the surrounding environment, processing it, and setting goals and appropriate policies to be executed by functional units of the organization. The core of the strategy is based on the vision, the mission, and the objectives and goals that need to be realized. The operating units of an organization, just like the organs of the human system, execute the strategy according to established policies. They may be divided by function or by groups of functions, as set up by the organizational structure. The latter should be designed to facilitate optimal attainment of goals and objectives. An organization without a strategy is like a body with a dead brain. It may exist as a structure, but it will go nowhere. An organization with poor strategy is like a sick patient. Strategic management is a critical undertaking that involves formulating, implementing, and evaluating cross-functional decisions that enable an organization to achieve its objectives (David, 1997).

Central to the formulation of a strategy is the existence of a vision and a mission for the business. Peter Drucker, a pioneer of modern management thinking, says, "A business is defined by its mission, not by its name, statutes, or articles of incorporation." Asking the question "What is our business?" is synonymous with asking the question "What is our mission?" (Drucker, 1974). A mission statement specifies the reasons for which the business is established. For example, Florida Power and Light Company's mission is "to supply safe, reliable, reasonable priced electric service to its customers" (Hudiburg, 1991). The mission statement provides a foundation for establishing goals, plans, priorities, and work assignments. A mission statement usually describes an organization's purpose, customers, products or services, market, philosophy and basic technology (David, 1997).

A vision is a paramount component in the core of a strategy. A vision provides a picture of a different reality for the future. Without a vision a business wanders aimlessly. A vision statement of a business answers the question, "What do we want to become?" A vision gives a business direction, a long-term view of what it wants to accomplish, who it wants to serve, and what it would like to be. It also inspires employees of the organization to be committed and to work toward achieving future objectives. Here are examples of short but powerful visions of leading companies:

- *3M:* "Be innovative and satisfy our customers."
- *Microsoft:* "Information at your fingertips."
- *Florida Power and Light Company (FPL):* "We want to be the best-managed electric utility in the USA."
- *GE:* To be "number one or two in every business."
- *Federal Express:* "When it absolutely, positively, must get there overnight, use FedEx."
- *Oracle:* "Enabling the information age."

In these cases, the vision has been converted into a slogan that motivates employees and conveys to customers what the organization is all about. A vision is expressed in the vision statement adopted by the company.

FPL's vision statement, developed in the mid-1980s reads, "During the next decade we want to become the best-managed electric utility in the United States and an excellent company overall and be recognized as such." FPL's chairman and CEO at the time, John Hudiburg, explains that a vision is a dramatic picture of the future that has the power to motivate and inspire. FPL, in its vision statement, specifies a period of time of one decade to realize the vision. This is done so that the vision is more than an open-ended dream. Hudiburg (1991) explains: "An objective without a date is a hope; an objective with a date is a goal."

To accomplish a mission and realize a vision, a company must define objectives and set goals. Doing so will focus the company's effort on executing a set of tasks that help accomplish long- and short-term objectives. The question at this point is, "How do we achieve these goals?"

Porter (1980) used a wheel to illustrate the competitive business strategy of a firm (Figure 8-2). The objectives of the firm are the hub of the wheel, where the firm can state its business goals. These can include achieving a certain level of profitability, growth, market share, or other desired short- or long-term objectives. The operating units, designed to execute the business plans, may be divided according to functions such as manufacturing, distribution, marketing, purchasing, human resources, or other pertinent units. Each one of these units operates according to policies established to guide its actions. A good organization will communicate the central objectives of the firm to all operating units staffed by appropriate levels of managers and employees. Top management's function is to create the vision, state the mission clearly, set the objectives, establish procedures necessary to formulate the strategy, and follow up on the actions of all units. Management should develop a fair system for rewarding employees and reinforcing behavior leading to the achievement of desired goals.

A strategy is not developed in a vacuum. It is formulated in the context of internal and external factors, as depicted in Figure 8-3. Porter (1980) suggested that competitive strategy is formulated on the basis of two sets of internal factors and two sets of external factors. The first set of internal factors pertains to the strengths and weaknesses of the organization. Strengths and weaknesses include resources (financial or human), know-how, market position, and degree of competitiveness. Recognizing

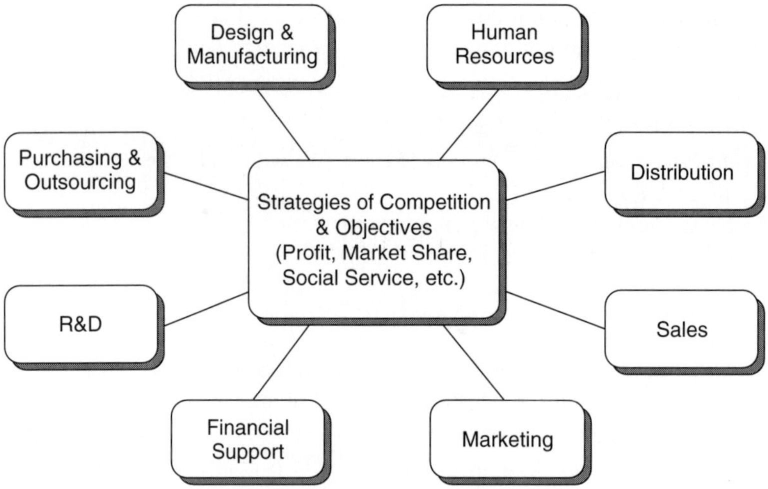


**FIGURE 8-2**
THE WHEEL OF COMPETITIVE STRATEGY
*Source:* Based on Porter, 1980.

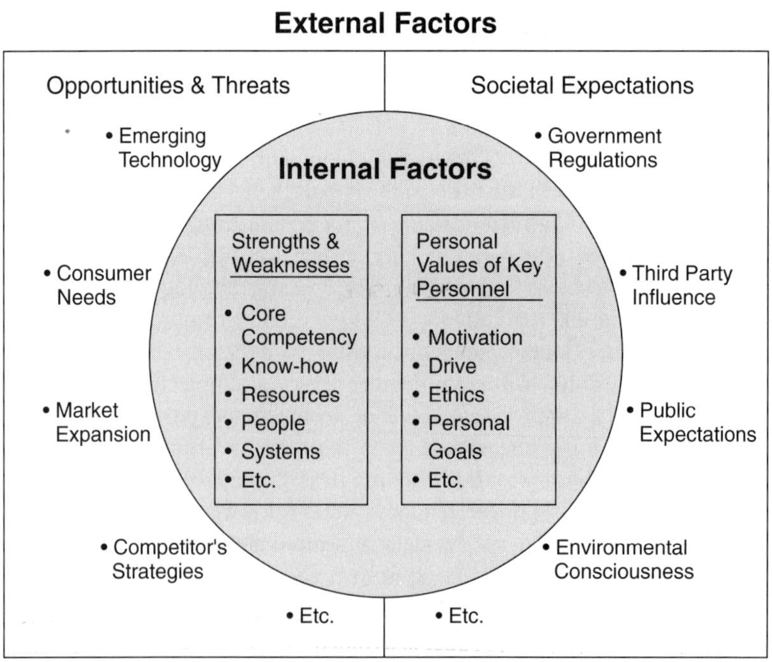


**FIGURE 8-3**
CONTEXT IN WHICH COMPETITIVE STRATEGY IS FORMULATED
*Source:* Based on Porter, 1980.

strengths and weaknesses of an organization requires in-depth analysis and frank, realistic self-evaluation.

A second set of internal factors pertains to the motivational drives, the needs and personal values of executives and personnel implementing the strategy of an organization. Owners and executives of organizations greatly influence the values of their organizations. The internal factors are specific to each organization, and they impact the strategy formulated and the actions taken.

Two sets of external factors need to be considered in the formulation of a competitive strategy. One set includes business opportunities that may exist or develop in the future. The opportunities may be a result of economic trends, emergence of a technology, or social and political change. The second set of external factors is dependent on many issues affecting the business climate, such as societal expectations of the business, the political climate, or the perceived value of the technology. For example, if a society rejects a technology, such as nuclear technology for power generation, because of a perceived danger to people, it might be extremely difficult for a power-generating company to build its strategy entirely on the basis of the exploitation of nuclear technology. External factors are mostly beyond the control of the organization. A strategy must be based on realistic considerations of what is acceptable within a society.

For a company to have a successful strategy, it is important to develop a clear understanding of the business, its strategic objectives, its products, and its targeted customers. It is also essential to understand the market position of the company by conducting a market analysis of its business portfolio. Figure 8-4 provides a comprehensive model for the strategic management process showing its components and the elements needed for the formulation, implementation, and evaluation of a strategy in an organization.

## METHODS USED IN STRATEGIC ANALYSIS AND DECISION MAKING

The available information required for decision making and for selecting appropriate strategies is frequently fuzzy. Strategists are never absolutely sure whether a market will respond in a precise, quantifiable manner to their expectations. In the business world decision making is an imperfect process in a highly dynamic environment. Even so, strategic planners must still devise plans and make decisions, and these should be made on the basis of known facts and the best available information. One approach in this situation is to construct and use decision-matrix techniques to guide decisions according to a selected set of criteria.

The information can be clustered in a matrix cell according to the criteria selected. A decision is associated with each cell. The matrix can be as small as one by two, or as large as *m* by *n*. For example, a decision on the acquisition of a product can be based on one criterion, such as price. Price can be forced into one of two categories, high or low. If the information collected indicates a low price, the decision is to acquire the product. If the price is high, the product is avoided.

If a second criterion, such as quality, is to be considered, it can be classified in two categories, high and low. Thus information about the product can be located in any one of four cells, as shown in Figure 8-5, and a decision is associated with each cell. If a

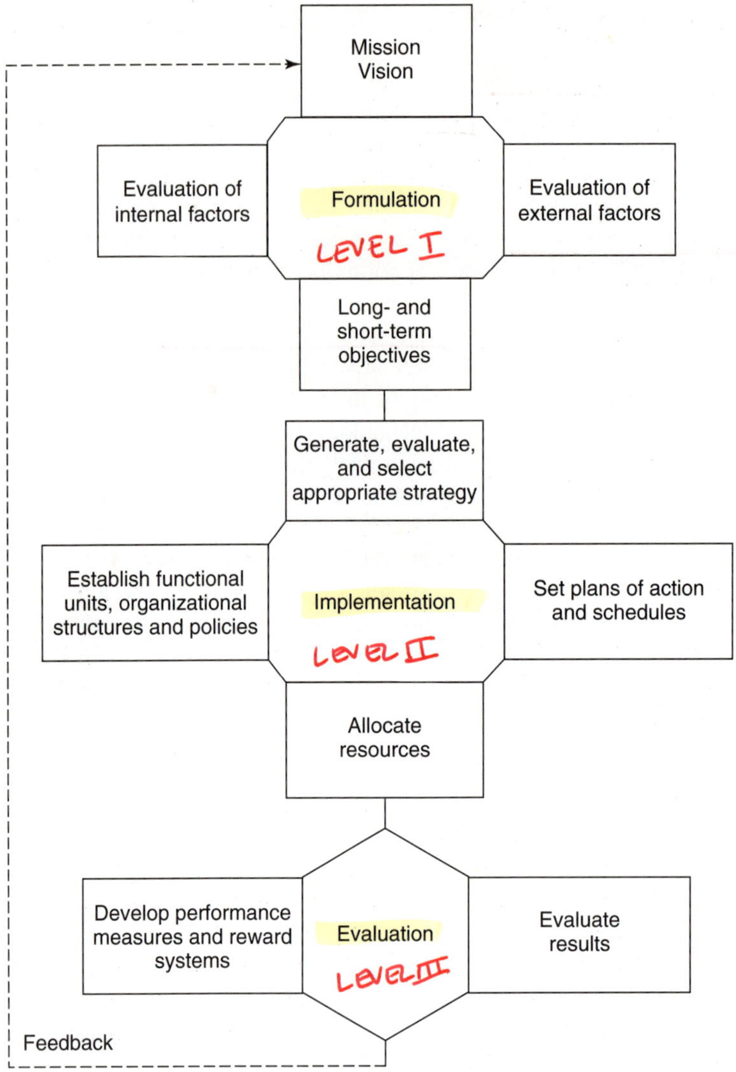


**FIGURE 8-4**
A MODEL FOR STRATEGY DEVELOPMENT

product is located in the upper right cell of an acquisition decision matrix, it is easy to make the decision to acquire it. If it falls in the lower left cell, it is easy to make the decision to avoid it. If it falls in the upper left or the lower right cell, additional information may be required before a decision is made. This means that one has to choose between a good product at high price and a poor product at low price. Such a decision will have to be based on the weight of each criterion, such as the relative importance of quality versus price. If quality is more important, it has more weight and the decision associated with the lower right cell will change to "acquire." If price weighs more, the decision associated with the upper left cell will change to "acquire."

| Price \ Quality | Low | High |
|---|---|---|
| Low | Maybe | Acquire |
| High | Avoid | Maybe |

**FIGURE 8-5**
TWO-BY-TWO MATRIX CONSTRUCTED FOR USE AS AN AID TO ACQUISITION DECISION

| Product new to the world \ Business Competitive Position | No | Yes |
|---|---|---|
| New | 1 Consider acquiring marketing expertise | 3 Most Promising |
| Known | 2 Least Promising: Inventors encouraged to go outside 3M | 4 Should fit one of the existing divisions |

**FIGURE 8-6**
PRODUCT EVALUATION MATRIX USED BY 3M

### Product Evaluation Matrix

Business policy can be set and decisions made on the basis of a two-by-two matrix analysis such as the one shown in Figure 8-6, which is used by 3M Company. New products are placed in one of four cells according to two factors: (1) Is the product new to the world? (2) What is the company's business strength or competitive position for this type of product? A decision rule is attached to each matrix cell. If a product falls in quadrant 1 and it looks like it is a promising product, the company may acquire marketing expertise in this line. If a product falls in quadrant 2, the inventor is encouraged to go outside 3M for exploitation of the product. A product positioned in quadrant 3 is more promising and deserves full support to the extent of its potential, including establishing a new division for it. A product categorized in cell 4 should fit within one of the existing divisions.

The 3M Company is recognized as a leader in innovation. Its philosophy is that any idea is worth pursuing. If the company has strength or can connect the innovation to one of its divisions, the idea is exploited internally. Otherwise, the innovators are encouraged to pursue ideas outside the company.

### Market-Growth–Market-Share Analysis Matrix

One popular type of strategic analysis is based on using the market-growth–market-share matrix popularized by the Boston Consulting Group (BCG). This matrix, shown in Figure 8-7, is also known as the *portfolio matrix* because it can be used to analyze a corporate stockholding portfolio for investment decisions.

Companies that have high market growth and a high market share are positioned in the upper right quadrant of the matrix and are classified as "stars." A company such as Microsoft would be positioned in this quadrant. A company that has a high market share and low market growth is classified as a "cash cow." A pharmaceutical company owning a popular patented drug formula may fall into this category. A company in a high-market-growth sector of the economy with a low market share in that sector is classified as a "problem child." An example is a software company that has a low market share for its products. Apple Computer became a problem child in mid-1997 with a slow sale of its computers in a growing market sector. Finally, a company that operates in a low-growth sector and has a low market share is classified as a "dog." An example here is a company that produces radios, cassette tapes, or typewriters (all are mature, low-growth products) and has a low market share for its products.

Decisions can be associated with each matrix cell. Classifying a company into one of the four categories of the matrix is helpful in deciding on the strategy necessary to improve its performance. The common business wisdom is to continue to nourish the stars, milk the cows, get rid of the dogs, and nurse the problem children to health so that they develop into stars. However, the proper strategy and course of action must be determined in each case on the basis of a much more detailed analysis including the strengths and weaknesses of the company and opportunities or threats in its environment. Charting the proper competitive strategy is the business's insurance for continued survival and success. Management must be able to determine the company's distinctive advantage and exploit it. Doing so requires understanding the company's core technology and its technological capabilities. Understanding the competition and the market needs is also a critical factor.

Some appropriate strategies that can be used by companies (or divisions) located in the four quadrants of the BCG matrix are shown in Exhibit 8-1.

**FIGURE 8-7**
COMPANY PORTFOLIO MATRIX

| | | Market Dominance | |
|---|---|---|---|
| | | **Low** | **High** |
| **Market Growth** | **High** | Problem Children | Stars |
| | **Low** | Dogs | Cash Cows |

**EXHIBIT 8-1**
STRATEGIES TO USE IN CONJUNCTION WITH A PORTFOLIO MATRIX

| Portfolio classification | Strategy |
|---|---|
| Stars | Stars should concentrate on continuing their product development effort, focus on competencies that gave them their competitive advantages, continue market development and market penetration, consider vertical integration, and consider business or product diversification to reduce the risks associated with a potential downturn in one product line or one source of investment. |
| Cash cows | Cash cows are strong companies in slow-growth markets. They should consider diversifying their products or activities in higher-growth areas, pursue joint ventures, utilize market segmentation strategies, and pursue defensive R&D to keep their products competitive. |
| Problem Children | These types of companies should evaluate their marketing strategies and aggressively pursue market development and market penetration. If their cash position is strong, they should pursue some type of integration strategy. If their cash position is weak, they may consider divestiture to acquire funds for development. Liquidation should also be considered to reduce losses. Joint venture with market leaders could also be pursued. |
| Dogs | These are companies that have weak technologies and compete in slow-growth industries. They must pursue aggressive strategies to avoid further drain on their resources and to reverse the direction of their business. Strategies of retrenchment should be pursued. Reengineering of the company to reduce cost and divesting resources into other businesses or different areas are possible alternatives. Liquidation of operation should also be considered. Innovations are needed to save the company from demise. |

*Source:* Based on Christensen et al., 1976.

### *X-Y* Coordinate Positioning Method

In an *x-y* coordinate system, the *x* axis can represent one factor and the *y* axis another factor. The axes will then represent a high-low relationship, as shown in Figure 8-8. A company or a product position with respect to the two factors can be plotted on the *x-y* plane. This positioning methodology assists in recognizing the relative standing of a company or a product with respect to a competitor. For example, if the *x* axis is labeled "quality" and the *y* axis "price," one can plot the relative position of several brands of cars on this plane.

This methodology is helpful in decisions concerning market segmentation and the strategic positioning of new products.

### *M*-by-*N* Matrix

An *m*-by-*n* dimension matrix can be utilized as an aid to decision making when decision criteria have multiple levels, as shown in Figure 8-9. A company can be classified as

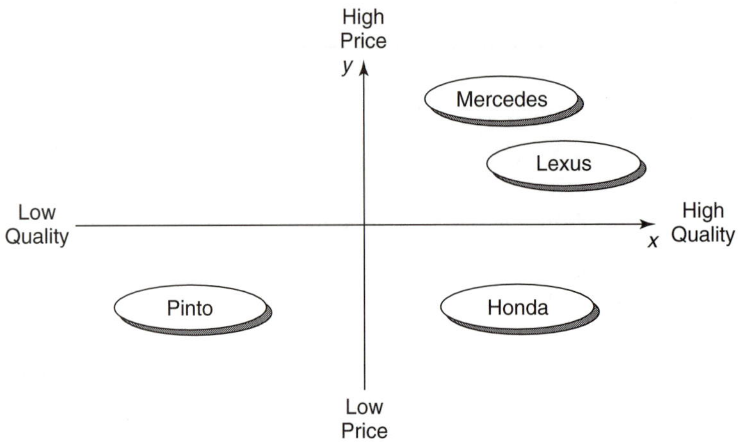


**FIGURE 8-8**
EXAMPLE OF AN *X-Y* PLANE REPRESENTATION

| Technical Competence \ Market Competence | **High** | **Medium** | **Low** |
|---|---|---|---|
| **High Cutting edge** | Star? | Success? | Doubt? |
| **Medium State-of-the-art** | Success? | Doubt? | Failure? |
| **Low Obsolete** | Doubt? | Failure? | Disaster? |

**FIGURE 8-9**
AN M-T MATRIX FOR ANALYZING TECHNICAL AND PRODUCT COMPETENCE
*Source:* Based on Holt, 1992.

having a low, medium or high level of a certain criterion such as market competence. Multiple-level criteria provide a more refined classification and more decision options. Holt (1990, 1992) used a market-technology (M-T) matrix to analyze the technical and marketing competence necessary for strategic decision making in product innovation (Figure 8-9).

An example of using an *m*-by-*n* matrix to evaluate technology and make a decision on adoption is described by McConnell and Khalil (1988). The methodology is based on the four-phase process outlined in Exhibit 8-2. This process can be used to choose an appropriate technology for a particular application.

McConnell and Khalil used this methodology to select an automatic identification technology for use to record time and attendance and for shop-floor-control application. The shop-floor-control application matrix, shown in Figure 8-11, suggests that the bar

**EXHIBIT 8-2**
TECHNOLOGY EVALUATION FOR ADOPTION DECISION

| Phase | Actions |
|---|---|
| *Phase 1:* Identifying the technologies and the attributes (the criteria for selection) | Identify all the possible technologies that are capable of solving a problem. List all the attributes, positive and negative, associated with each technology. |
| *Phase 2:* Rating the technology according to specific attribute | Rate each technology based on the attribute list identified in Phase 1. The following rating scale can be used:<br>5 Excellent technology for the attribute<br>4 Above average<br>3 Average<br>2 Below average<br>1 Poor<br>0 Technology does not possess this attribute<br>An attribute of technology could be its speed, reliability, price, etc. |
| *Phase 3:* Rating the importance of each attribute | Evaluate the importance of each attribute to the application being considered. This is a value assigned by the user of a particular application. The following rating system can be used:<br>5 Must have this attribute<br>4 Extremely important but could adjust without it<br>3 Important<br>2 Not important, but would be beneficial<br>1 Do not need<br>0 Do not want to consider in the decision<br>For example, if the attribute "speed" is extremely important for the application, it should be given a rating of 4. The attribute "price" may be given a 5. The assignment of a value here can be made by a team of engineers, managers, and users, as the case may require. |
| *Phase 4:* Constructing the decision matrix | Construct a decision matrix as shown in Figure 8-10. The technology rating from Phase 2 is placed in the upper left corner of each cell. The attribute rating from Phase 3 is placed next to each attribute. Each technology rating is then multiplied by the attribute rating and the result is recorded in the lower right corner of each cell. Results are summed by column and recorded at the bottom of each technology column. The technologies with the highest ratings are those to be most seriously considered for selection. |

Attribute ranking (where 5 is Most important and 0 is Least important)

Rating of the technology with respect to attribute 1

| Ranking | | Technologies: Device 1 | Device 2 | Device 3 | Device 4 | Device 5 |
|---|---|---|---|---|---|---|
| 2 | Attribute 1 | 5 / 10 | 3 / 6 | 0 / 0 | 4 / 8 | 2 / 4 |
| 0 | Attribute 2 | 2 / 0 | 5 / 0 | 3 / 0 | 4 / 0 | 2 / 0 |
| 1 | Attribute 3 | 4 / 4 | 1 / 1 | 2 / 2 | 1 / 1 | 3 / 3 |
| 5 | Attribute 4 | 1 / 5 | 0 / 0 | 1 / 5 | 5 / 25 | 1 / 5 |
| | TOTAL | 19 | 6 | 7 | 34 | 12 |

**FIGURE 8-10**
A MULTITECHNOLOGY, MULTIATTRIBUTE DECISION MATRIX
*Source:* McConnell and Khalil, 1988; © 1988, Institute of Industrial Engineers.

| | | Voice | Mag slot | Wand | OCR handheld | OCR page | RF active | RF passive | Holographia | Fixed | Laser |
|---|---|---|---|---|---|---|---|---|---|---|---|
| 5 | Price of reader | 2 / 10 | 5 / 25 | 5 / 25 | 4 / 20 | 3 / 15 | 4 / 20 | 4 / 20 | 1 / 5 | 2 / 10 | 3 / 15 |
| 4 | Substitution error rate | 1 / 4 | 4 / 16 | 4 / 16 | 2 / 8 | 2 / 8 | 5 / 20 | 5 / 20 | 4 / 16 | 4 / 16 | 4 / 16 |
| 5 | Price of media | 5 / 25 | 3 / 15 | 4 / 20 | 4 / 20 | 4 / 20 | 1 / 5 | 1 / 5 | 4 / 20 | 4 / 20 | 4 / 20 |
| 3 | Human readability | 5 / 15 | 0 / 0 | 3 / 9 | 5 / 15 | 5 / 15 | 0 / 0 | 0 / 0 | 3 / 15 | 3 / 15 | 3 / 15 |
| 0 | Reading distance flexibility | 2 / 0 | 1 / 0 | 1 / 0 | 1 / 0 | 1 / 0 | 5 / 0 | 4 / 0 | 3 / 0 | 3 / 0 | 3 / 0 |
| 0 | Line of sight requirement | 4 / 0 | 0 / 0 | 0 / 0 | 0 / 0 | 0 / 0 | 5 / 0 | 5 / 0 | 4 / 0 | 0 / 0 | 2 / 0 |
| 2 | Life of media | 5 / 10 | 2 / 4 | 3 / 6 | 3 / 6 | 3 / 6 | 4 / 8 | 5 / 10 | 3 / 6 | 3 / 6 | 3 / 6 |
| 4 | Reader's resistance to harsh | 2 / 8 | 3 / 12 | 4 / 16 | 3 / 12 | 3 / 12 | 4 / 16 | 4 / 16 | 4 / 16 | 4 / 16 | 3 / 12 |
| 3 | Media's resistance to harsh | 2 / 6 | 3 / 9 | 1 / 3 | 1 / 3 | 1 / 3 | 5 / 15 | 5 / 15 | 1 / 3 | 1 / 3 | 1 / 3 |
| 0 | Ability to alter media | 5 / 0 | 5 / 0 | 0 / 0 | 0 / 0 | 0 / 0 | 5 / 0 | 0 / 0 | 0 / 0 | 0 / 0 | 0 / 0 |
| 3 | Density of data | 3 / 9 | 5 / 15 | 3 / 9 | 1 / 3 | 1 / 3 | 5 / 15 | 4 / 12 | 3 / 9 | 3 / 9 | 3 / 9 |
| 0 | Security of media | 0 | 0 | 0 | 0 | 0 | 0 | 0 | 0 | 0 | 0 |
| 0 | Operator requirement | 0 | 0 | 0 | 0 | 0 | 0 | 0 | 0 | 0 | 0 |
| 4 | Speed in reading media | 1 / 4 | 2 / 8 | 2 / 8 | 1 / 4 | 1 / 4 | 5 / 20 | 5 / 20 | 4 / 16 | 4 / 16 | 4 / 16 |
| 4 | Proven technology | 1 / 4 | 5 / 20 | 5 / 20 | 4 / 16 | 4 / 16 | 1 / 4 | 1 / 4 | 4 / 16 | 5 / 20 | 5 / 20 |
| | TOTAL | 95 | 124 | 132 | 107 | 102 | 123 | 122 | 116 | 125 | 126 |

**FIGURE 8-11**
SHOP-FLOOR-CONTROL APPLICATION
*Source:* McConnell and Khalil, 1988; © 1988, Institute of Industrial Engineers.

code wand is the technology receiving the highest score (132 points), and therefore it emerges as the leading candidate for selection.

De Wet (1996) developed an expanded matrix-based methodology linking products to technology through the process of value addition in an enterprise. De Wet's article is recommended as a supplement to this discussion.

### Strengths, Weaknesses, Opportunities, Threats Matrix

Successful strategy formulation depends on creating a match between the resources available to an organization and the opportunities present in its environment. Identification of the internal factors of strengths and weaknesses and the external factors of opportunities and threats is an important step in the strategy formulation process. The development of a strengths, weaknesses, opportunities, threats (SWOT) matrix, shown in Figure 8-12, is a very helpful tool for this purpose. A SWOT matrix includes a list of all the internal and external factors identified as strengths, weaknesses, opportunities, and threats. David (1997) proposes the following approach to the construction of a SWOT matrix:

**1** List the firm's key external opportunities.
**2** List the firm's key external threats.
**3** List the firm's key internal strengths.
**4** List the firm's key internal weaknesses.

**FIGURE 8-12**
SWOT MATRIX
*Source:* Fred David, *Strategic Management,* 6th ed. © 1997. Reprinted by permission of Prentice-Hall Inc., Upper Saddle River, NJ.

| Always leave Blank | STRENGTHS (S)<br>1.<br>2.<br>3.<br>4. List Strengths<br>5.<br>6.<br>7.<br>8. | WEAKNESSES (W)<br>1.<br>2.<br>3.<br>4. List Weaknesses<br>5.<br>6.<br>7.<br>8. |
|---|---|---|
| OPPORTUNITIES (O)<br>1.<br>2.<br>3.<br>4. List Opportunities<br>5.<br>6.<br>7.<br>8. | SO STRATEGIES<br>1.<br>2.<br>3.<br>4. Use strengths to take advantage of opportunities<br>5.<br>6.<br>7.<br>8. | WO STRATEGIES<br>1.<br>2.<br>3.<br>4. Overcome weaknesses by taking advantage of opportunities<br>5.<br>6.<br>7.<br>8. |
| THREATS (T)<br>1.<br>2.<br>3.<br>4. List Threats<br>5.<br>6.<br>7.<br>8. | ST STRATEGIES<br>1.<br>2.<br>3.<br>4. Use strengths to avoid threats<br>5.<br>6.<br>7.<br>8. | WT STRATEGIES<br>1.<br>2.<br>3.<br>4. Minimize weaknesses and avoid threats<br>5.<br>6.<br>7.<br>8. |

**5** Match internal strengths with external opportunities, and record the resultant SO strategies in the appropriate cell.

**6** Match internal weaknesses with external opportunities, and record the resultant WO strategies.

**7** Match internal strengths with external threats, and record the resultant ST strategies.

**8** Match internal weaknesses with external threats, and record the resultant WT strategies.

**FIGURE 8-13**
SWOT MATRIX FOR A FOOD COMPANY
*Source:* Fred David, *Strategic Management,* 6th ed. © 1997. Reprinted by permission of Prentice-Hall Inc., Upper Saddle River, NJ.

| | STRENGTHS (S) | WEAKNESSES (W) |
|---|---|---|
| | 1. Current ratio increased to 2.52<br>2. Profit margin increased to 6.94<br>3. Employee morale is high<br>4. New computer information system<br>5. Market share has increased to 24% | 1. Legal suits have not been resolved.<br>2. Plant capacity has fallen to 74%<br>3. Lack of strategic-management system<br>4. R&D expenses have increased 31%<br>5. Dealer incentives have not been effective |
| **OPPORTUNITIES (O)**<br>1. Western European unification<br>2. Rising health consciousness in selecting foods<br>3. Free market economies arising in Asia<br>4. Demand for soups increasing 10% annually<br>5. U.S./Mexico NAFTA | **SO STRATEGIES**<br>1. Acquire food company in Europe (S1, S5, O1)<br>2. Build a manufacturing plant in Mexico (S2, S5, O5)<br>3. Develop new healthy soups (S3, O2)<br>4. Form a joint venture to distribute soups in Asia (S1, S5, O3) | **WO STRATEGIES**<br>1. Form a joint venture to distribute soup in Europe (W3, O1)<br>2. Develop new Pepperidge Farm products (W1, O1, O4) |
| **THREATS (T)**<br>1. Food revenues increasing only 1% annually<br>2. ConAgra's Banquet TV dinners lead market with 24% share<br>3. Unstable economies in Asia<br>4. Tin cans are not biodegradable<br>5. Low value of the dollar | **ST STRATEGIES**<br>1. Develop new microwave TV dinners (S1, S5, T2)<br>2. Develop new biodegradable soup containers (S1, T4) | **WT STRATEGIES**<br>1. Close unprofitable European operations (W3, T3, T5)<br>2. Diversify into non-soup foods |

Analysts can review the factors listed in the SWOT matrix and develop four types of strategies:

**1** *Strengths-opportunities (SO) strategies,* in which the organization uses its internal strengths to take advantage of external opportunities.

**2** *Weaknesses-opportunities (WO) strategies,* in which the organization attempts to overcome weaknesses by taking advantage of external opportunities.

**3** *Strengths-threats (ST) strategies,* in which the organization uses its internal strengths to fend off external threats.

**4** *Weaknesses-threats (WT) strategies,* in which the organization develops approaches to reducing its internal weaknesses and avoiding external threats.

The matrix cells for SO, WO, ST, and WT contain lists of feasible alternative strategies that the organization may follow to achieve its desired objectives. The letters in parentheses identify the strategy that associates specific internal factors with a specific set of external factors. Therefore, a strategy that takes advantage of strength factors 1 and 2 and opportunity factor 3 is labeled "(S1, S2, O3)." This is helpful in developing strategies to address the identified SWOTs. The most appropriate strategies can then be selected from the list of alternative strategies that was developed. Figure 8-13 shows a SWOT matrix developed for a food company that is attempting to formulate its strategies. On the basis of educated assumptions and intuitive judgment, managers make an evaluation and select the strategic option. Desirable criteria can be established. The degree to which each strategy fits each criterion is determined. This can be done quantitatively by using a weighting scale, such as a Likert scale, and giving scores on that scale. The average scores given to each strategy over the set of selected criteria can be used as a guide for selecting the optimal choice. McConnell and Khalil's methodology, shown in Figures 8-10 and 8-11, illustrates how to quantify alternative choices in order to select the best one available.

## FORMULATION OF A TECHNOLOGY STRATEGY

Technology is at the core of systems designed to satisfy societal or customer needs. Companies are formed to provide a structure and a mechanism that facilitate the spinning out of technology to satisfy those needs. When a company has a vision and develops its mission statement, it is stating the reasons the company exists and the inherent values of the company. When the company develops a strategy and its associated plans of action, it creates the vehicle that moves it toward the fulfillment of its mission and the attainment of its vision. The purpose of business strategy is to gain a sustainable economic advantage. The purpose of technology strategy is to gain a sustainable technological advantage that provides a competitive edge. The two strategies must be closely intertwined and highly integrated. This requires extensive forethought about the firm's distinctive technologies, the products or services it can provide, the potential customers, and where the organization wants to be in the future. The company's technologies must be harnessed and exploited according to a well-designed plan. Effective technology management is based on successfully linking business and technology strategies.

Ford (1988) explains that technology strategy is concerned with exploiting, developing, and maintaining the sum total of the company's knowledge and abilities. Many organizations still seem to underestimate technology's importance. Ford cited a Booz, Allen and Hamilton survey of 800 executives: Two-thirds of them thought that their companies were doing a poor job in harnessing technology to their corporate strategies. In an earlier survey, conducted by the Conference Board in 1982, only 20 percent of CEOs considered their top technology executives to be within their *inner circle* of managers. These findings indicate that in the early 1980s many top executives were caught unaware of the forming clouds that resulted in the technological revolution. Failure to develop and integrate technology strategy and business strategy is a major contributing factor in the decline of a firm's competitiveness. Van Wyk and Sweatt (1994) argued that corporate boards used to concentrate on reviewing financial budgets. Recent indications show that they are becoming more directly involved with the strategic management of technology. Van Wyk and Sweatt suggest that boards should consider ways to improve their technological literacy and capability in order to make informed decisions in an increasingly technological corporate environment.

There are many factors that determine business success; although technology is a very important one, it is not in itself sufficient to ensure business success. Good business is about integrating technological innovation with production, marketing, finance, and personnel to achieve established goals.

Research conducted by Frohman (1982) reveals two commonalities among companies that use technology as a competitive weapon. These are:

**1** Management views technology as a major competitive weapon but does not emphasize it at the expense of other areas.

**2** The criteria used to support any project consist of (*a*) whether the project supports the business goal, (*b*) whether the project protects and/or establishes technological leadership, and (*c*) whether the project solves customer problems.

A basic purpose of strategy in any business is to answer three fundamental questions:

**1** What business should the firm engage in?

**2** How should the firm be positioned in the business?

**3** What technology, production, and marketing will be necessary to attain the desired position?

Technology gives a company a competitive edge. Corporations with inferior technology cannot compete with corporations utilizing superior technology. However, to use technology as a competitive weapon, managers must manage it as part of the business system.

Michael Porter (1985) advocates that technology strategy be formulated within the larger context of business planning. Porter's approach to formulating a competitive strategy is to concentrate on optimizing the efficiency of the value chain. This implies developing and maintaining a competitive advantage by finding the most effective means of carrying out all the activities of the business process so as to offer the customer long-term value. Porter proposes that a technology strategy be formulated using the following steps:

**1** Identify all the distinct technologies and sub-technologies in a value chain.
**2** Identify potentially relevant technologies in other industries or under scientific development.
**3** Determine the likely path of change of key technologies.
**4** Determine which technologies and potential technological changes are most significant for competitive advantage and industry structure.
**5** Assess a firm's relative capabilities in important technologies and the cost of making improvements.
**6** Select a technology strategy, encompassing all-important technologies that reinforce the firm's overall competitive strategy.
**7** Reinforce business-unit technology strategies at the corporate level.

## DIRECTION OF STRATEGY

The vehicle for being competitive and achieving wealth can be visualized as an assembly of the following components: (1) developing a strategy with clear direction, (2) exploiting technological competence, (3) achieving high levels of productivity, and (4) mounting an aggressive and persistent marketing effort. Figure 8-14 illustrates this concept.

A strategy's direction is a vital ingredient in the success of an organization. Setting the direction depends on the changes in technology, customer needs, and environmental factors. A case in point is Microsoft Corporation. This company has long dominated the personal computer software market by offering its disk operating system (DOS) in the 1980s and Windows in the 1990s. Profits have been pouring in because of the booming sale of its products. It would appear that continuing the company's strategy of developing software for the PC industry is the logical thing to do. However, the introduction and

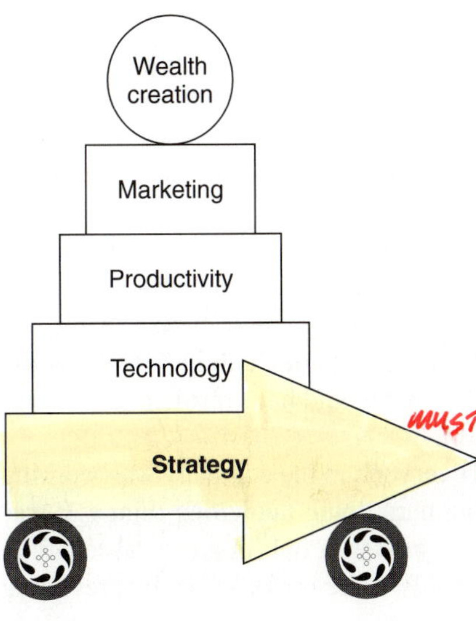


**FIGURE 8-14**
THE VEHICLE FOR CREATING WEALTH

growth of the Internet has created a new medium that is changing the way people use personal computers. Microsoft decided to change the direction of its strategy and has mounted a major effort in developing software suitable for the Internet. This change in strategic direction means that a company worth $70 billion (at the time of making the change in strategic direction), with $6 billion in sales and close to 20,000 workers, has thrown itself into the competition for the Internet business (Ramo, 1996). This change in strategic direction of developing technology for the Internet proved to be very valuable in ensuring the continued success of Microsoft.

Another case is Northwest Airlines, which reversed the direction of its strategy in the late 1980s. The new strategic direction has helped Northwest survive and become competitive in the very tough airline business.

### Northwest Airlines' Changing Strategy

Northwest Airlines underwent a leveraged buyout in the late 1980s. Two Los Angeles investors, Alfred Checchi and Gary Wilson, privatized the company in 1989. They hired three consecutive presidents who could not turn the company into a profitable operation. In 1993, when new president and CEO John E. H. Dasburg took over, he was faced with an accumulated debt of $1 billion that the company was unable to pay: Northwest was close to filing chapter 11 bankruptcy. The company was losing money, but so were many other airline companies. Eastern Airlines had folded its Miami operation, eliminated many point-to-point connections, and vacated its major hub in Atlanta.

**Changing the Strategy** Most airlines at the time were using a strategy based on ubiquity in all markets and the Procter & Gamble "supermarket model" of shelf space. The more shelves you have in a supermarket on which to display your products, the better your chances that the customer will pick up your product, thus increasing your sales. Applied to the airline industry, this philosophy holds that the more markets you serve, the better your position in the industry.

In a speech at the 1996 Institute of Industrial Engineers (IIE) International Meeting, Dasburg commented on a strategy frequently advocated in business teachings: Focusing on maximizing return on equity (ROE) should prove to be more prosperous for companies in the long run. He indicated that this may not be true at all in the real world. Most CEOs are left to their own judgment as to how to improve returns, and their decisions, in turn, are based on the interests of the stakeholder groups. Each business should develop the strategy that guarantees its survival and serves its stakeholders best.

After the demise of Eastern Airlines, Northwest had an opportunity to expand into places where Eastern used to have strength. Replacing Eastern's hub in Atlanta would have followed the ubiquity and shelf-space strategy but would have resulted in a $3 billion loss. This did not make sense to Northwest's CEO.

Dasburg became convinced that even though everyone in the company was working hard and meant well, the personnel were working hard doing the wrong things. It was clear that even though the airline hub-and-spokes model (a hub in a central location, such as Atlanta, with routes connecting to other cities in the region) was the prevailing strategy in the airline industry, it was an expensive one and was leading Northwest to failure.

In a market-driven industry, where cost and customer satisfaction dominate, the old strategy would have led to the collapse of the airline and maybe the entire industry. Rather than following it blindly, Northwest turned to its customers for directions. They were interested in safety, reliability, cleanliness, promptness, reliable luggage delivery, and frequency of operation in their own areas. Dasburg decided to change the strategic directions of the airline. Escaping the failing-strategy malaise requires radical change. Northwest pulled out of flying in many markets and decided that a tighter focus concentrating on core markets and customer satisfaction would reenergize the struggling company. The company closed unprofitable hubs in Milwaukee, Washington, and Seoul, Korea; ended point-to-point service on the East and West Coasts; and reallocated assets to its hubs in Detroit, Minneapolis, and Tokyo (Nuello, 1996). Dasburg's strategy was to discontinue flying in many failing markets and to focus the company's resources, concentrating on reliability and convenience.

In adopting this strategy, Dasburg reversed the direction of the industry strategy. He recognized that less is better than more. As he explains it, the underlying belief sets the vision, but if this underlying belief is incorrect, you may have a fatally flawed vision and a disastrous strategy. This is indeed true in regard to setting strategic directions. If you aim your rifle in the wrong direction and shoot, you accomplish the task of firing but you miss the target. Similarly, if a company's strategy is aimed in the wrong direction, it may create tremendous negative consequences for the company. Dasburg aimed Northwest's strategy toward the right target.

**Results of the New Strategy** Changing the direction of Northwest's strategy created a new formula by which the company could compete. The new strategy helped restore profitable operation. Whereas Northwest lost $1.8 billion from 1990 through 1993 under the old strategy, the new strategy enabled the company to report a net income of $295.5 million in 1994 on revenues of $9.1 billion. Northwest complemented its strategy of concentrating on its core markets by entering into alliances and bilateral agreements with other airlines in order to boost cross-border traffic.

## CORE COMPETENCIES

A fundamental concept in the formulation of a technology strategy is *core competence*. This is the inner strength upon which a strategy should be built. An organization's core competence could be in a technology, a product, a process, or the way it integrates its technological assets. An example of a technical core competence is the creation of a product or service with unique value to customers. An organization may have core competence in marketing with its ability to access and serve markets in a unique way. Another example of core competence is an organization infrastructure that permits managing operations in a uniquely efficient and effective way. Core competence may also be the human knowledge or skill of an organization's employees. Boeing, the giant builder of airplanes, has many successful production and business activities. However, it considers its core

**EXHIBIT 8-3**
CORPORATE CORE COMPETENCIES

| Company | Example |
|---|---|
| Florida Power & Light | Transmission network |
| Sony | Miniaturization |
| Honda | Motors |
| NEC | Telecommunications, semiconductors, and mainframes |
| Motorola | Wireless communications |
| Black and Decker | Fractional-horsepower motors and household appliances |
| Boeing | Large-scale system integration, efficient design and manufacturing, and knowledge of its customers |

competencies to be in large-scale system integration, efficient design and manufacturing, and knowledge of its customers. Honda's core competence is not in car manufacturing as much as it is in motors. Additional examples are provided in Exhibit 8-3.

Core competencies are collective sets of knowledge, skills, and technologies that a company applies to add value for its customers. This is what determines the company's competitiveness. A company can improve its competitive abilities by becoming a learning organization (Machado, 1997). This means continuously learning and building capabilities that (*a*) cannot be easily duplicated by its competitors, (*b*) create new products and services for its customers, and (*c*) generate alliances and relationships with suppliers to provide its customers with cost and value advantage. Prahalad and Hamel (1990) propose that the core competencies of an organization "are the collective learning in the organization, especially how to coordinate diverse production skills and integrate multiple streams of technologies." They use a number of case studies to illustrate that competence is about the harmonization of technology as well as about the organization of work and the delivery of value.

The core competencies of an organization are usually converted to core products, which in turn may be embodied in one or more end products. The end products link the organization to its customers. The perceived value of an end product increases when the organization relates the product to its unique or specific competence. Prahalad and Hamel used a tree analogy to illustrate the idea of core competencies in a diversified corporation: The roots are the competencies of the corporation, the trunk represents core products, the small branches represent business units, and the leaves are the end products (Figure 8-15). Indeed competencies are the roots of competitiveness. The roots of the tree provide nourishment and keep the tree alive.

It is incumbent upon management to identify the organization's core competencies. The following common characteristics of core competencies may help an organization

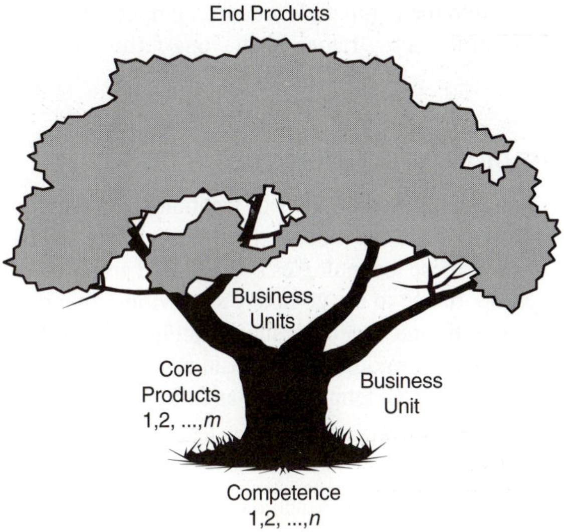


**FIGURE 8-15**
COMPETENCIES:
THE ROOTS OF
COMPETITIVENESS
*Source:* Based on Prahalad and Hamel, 1990.

distinguish areas of competencies from the multitude of its other activities:

1 They provide the distinctive advantage of the organization.
2 They are difficult for competitors to imitate.
3 They make a significant contribution to the end products offered by the organization.
4 They provide access to a wide variety of markets.

To capitalize on strength, a company should strive to exploit its core competencies. Specifically, it must:

1 Clearly identify the following:
   - What it does best.
   - What it can do that no other company can do better.
   - What will permit it to achieve *best-in-the-world* status in regard to what it does.
2 Develop its plans to fully exploit its capabilities.

### Exploitation of Competencies

Management must consider the company's core competencies as its distinct advantage around which to develop technology and business strategy. The following management actions are recommended:

- Develop, cultivate, and enhance the company's core competencies.
- Deploy core competencies as widely as possible throughout the company's products and services.
- Align all other activities in the company around the areas of competencies to create synergy. When synergy exists, the whole is greater that the sum of the parts.
- Develop an optimal plan for technology integration and outsourcing.

• Build barriers to competitors' entry into the company's areas of competencies.
• Overcome temptation for short-term gains rather than long-term strategic positioning.

## TECHNOLOGY AND THE CONCEPT OF CORE COMPETENCE

Products produced by any company either are based on a set of technologies linked to the set of competencies within the company or are dependent on technologies owned by other companies. It is essential that each of these technologies be identified and categorized appropriately as to their relative importance to the company's activities. Technology in a company (or in a product) consists of three layers, as shown in Figure 8-16. The core represents the distinctive technologies; the middle circle, basic technologies; and the outer circle, external technologies. Ford (1988) defined these as follows:

> *Distinctive technologies:* Those technologies in which the company's standing gives it a distinctive competence.
>
> *Basic technologies:* Those survival technologies on which the company's operations depend and without which it would be excluded from its markets. Basic technologies are necessary for a company to stay in business but do not differentiate or distinguish it from competitors.
>
> *External technologies:* Those technologies which are supplied by other companies. These types of technologies are usually available to the market at large.

Distinctive technology is what gives an organization its unique competitive advantage in the marketplace. Organizations must protect it, nourish it, and capitalize on the fact that they have something desirable that others do not have. However, distinctive technology may not be in a form that permits its commercialization. For example, a company holding a patent for a product design that constitutes a distinctive technology has no way of reaching a consumer without the support of basic technologies. These include production technologies, such as manufacturing, or logistic technologies, such as transportation and delivery. Manufacturing in this case will be a survival technology,

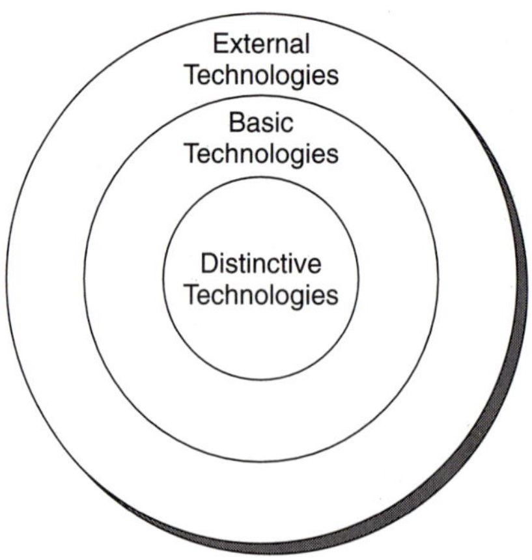


**FIGURE 8-16**
CLASSIFICATION OF TECHNOLOGY AS TO ITS RELATIVE STANDING IN A PRODUCT

without which the company's product will not be produced and reach the market. To complement its technological needs, the company may decide to develop its own manufacturing operation and control its survival technologies too. Alternatively, the company may be able to contract out, engaging another company to manufacture a product based on the distinctive design for which it holds the patent (i.e., outsource its manufacturing operation). Managers can make this decision on the basis of economic criteria and market conditions.

Basic technologies are technologies widely available to many organizations. They are essential for the development of a product but do not give it a distinctive advantage.

External technologies provide a third level of technological need but they are not critical to the company's survival. They have a much lower impact on the company's competitive standing. External technologies usually are more economically supplied by an outside vendor. For example, a company may need standard components for its products, such as bolts and nuts, or need packaging material for shipment of its products. These can be acquired from an outside source. They are important items for the product; however, their technologies do not have to be owned or controlled by the company.

The distinctive, basic, and external technologies of a company can be determined from a technology audit of the company and its products. An audit of Black and Decker, Inc., a company well recognized for its household appliances and hand tools, reveals that its distinctive technology is the manufacturing of fractional-horsepower electric motors. Its basic technology is the assembly process for small hand tools, and the external technology includes plastic parts, which can be brought in from other companies (Ford, 1988).

## INTEGRATION

When a corporation owns or has control over all or most of the technologies that contribute to producing and marketing a product, it is known as a *vertically integrated corporation* (Figure 8-17). This could be the case whether the technology is a product, process, marketing, or integrative type of technology. General Motors is considered a vertically integrated corporation. It owns plants that manufacture the chassis, the transmission train, the engine, and most of the other components of automobiles and trucks. It also exercises strong control over distribution and marketing arms of the business. Vertical integration of a company can be defined at any point on a continuum, with one end designating total ownership of the technology (making the product) and the other

**FIGURE 8-17**
VERTICAL INTEGRATION

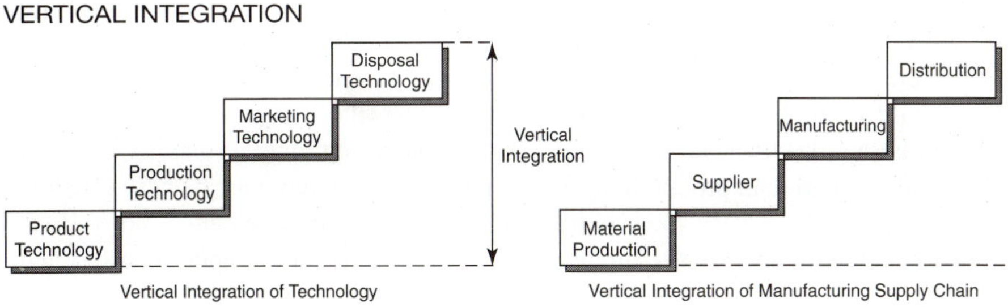

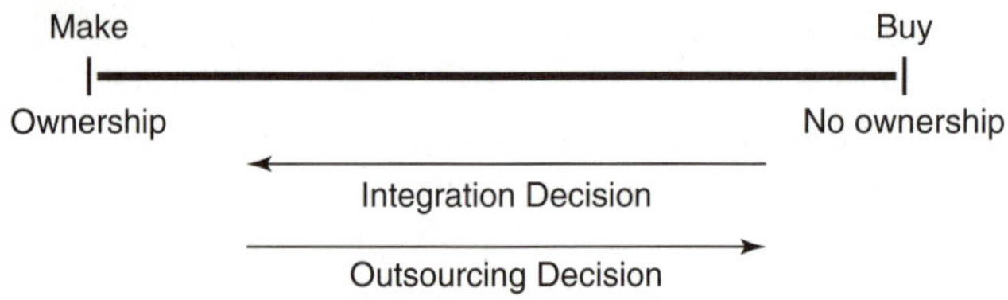


**FIGURE 8-18**
THE BOUNDARIES OF THE INTEGRATION DECISION

end showing no ownership (i.e., having to buy everything, as opposed to owning the technology or making the product within the company) (Figure 8-18).

Decisions as to whether technology should be owned or not, or whether products should be made or bought, must be guided by the company's standing in technology. Therefore, a company must be able to:

**1** Identify its distinctive technologies and choose areas in which to build competence in technology.
**2** Do all it can to acquire or keep itself at the top of these technology areas.
**3** Decide on the level of integration needed for its operation, based on realistic technology and business decision-making criteria.
**4** Be aware of emerging technology that may impact its business.
**5** Modify its business strategy to support its technology strategy.

If one uses the value chain as a framework for analysis, a company's strategy to integrate is made according to the direction of integration desired. *Backward integration* occurs when the company seeks ownership or control of its suppliers. *Horizontal integration* involves increased control over production competitors. *Forward integration* occurs when a company seeks to control distribution, retailing, and postmanufacturing activities. Vertical integration may combine backward, horizontal, and forward integration. It involves ownership or control of activities over the entire value chain. Many companies achieve integration through mergers, acquisitions, and takeovers.

Manufacturing technology requires developing technology-based strategies to deal with the entire value chain. Decisions to integrate an activity should be based on how well the technology fits within the portfolio of the company's core competencies. They also depend on how important it is for the company to control the technology, as well as the cost associated with such a decision. The automotive industry provides many good examples of practices in supply-chain management. While there are only about 20 very large, recognized multinational corporations in the world, there are thousands of small and medium-size enterprises that make up the supply sector of the industry. To what extent should an automotive manufacturer own or control those suppliers? A business decision is based on financial considerations. Outsourcing may prove to be more economically sound than production. The decision to outsource must be based on how critical the technology is to the company. If it is within the company's portfolio of core competencies, either ownership or stronger control over the supplier is warranted.

The same rationale is used when developing strategies for distribution, retailing, and postmanufacturing.

In all cases, management should seek to establish a close, trusting relationship with the company's suppliers and distributors. The Japanese have shown the world that partnership with suppliers helps improve quality, reduce time to market, and improve profitability. Distribution retailing can add a significant cost to a product without providing an added value to customers. Manufacturers that are capable of cutting such costs gain a significant advantage over competitors. The cost of these downstream segments of the supply chain can be streamed with new technologies and innovative practices. For example, it is estimated that marketing, distribution, and retailing represent approximately 20 to 30 percent of the value of a new car, depending on the dealer type and level of sales incentives applied to the product line (Fine et al., 1996). Information technology in retailing and distribution is capable of connecting manufacturers to customers directly. This can lead to a reduction in the finished-goods inventory by dealers and to a reduced cost of sales. Direct sales on the Internet might reduce the nonadded value of intermediate agents and reduce the overall cost of a product.

Figure 8-19 shows a matrix developed for evaluating alternative strategies regarding integration. The two criteria selected for evaluation are level of importance of the integration strategy and difficulty in implementing the strategy in terms of the financial investment required. The matrix used is a two-by-two discrete matrix that forces a classification in one of four cells.

Figure 8-20 shows another matrix developed to evaluate integration implementation strategies according to the two criteria of level of investment required and time required to achieve favorable competitive position.

More complex methodologies of evaluation can be developed by assigning several values for each criterion and developing a larger matrix. Complex approaches require

**FIGURE 8-19**
INTEGRATION EVALUATION MATRIX
This matrix utilizes importance and level of difficulty as criteria.
*Source:* Teece 1987. With permission.

| How important is it to integrate? | Level of Difficulty (ex. investment required): **Minor** | Level of Difficulty (ex. investment required): **Major** |
|---|---|---|
| **Critical** | Integrate (majority ownership) | Move towards integration (if cash constrained, take minority position) |
| **Not Critical** | Discretionary | Do not integrate (contract out) |

| | | Time Required to Achieve Favorable Competitive Position | |
|---|---|---|---|
| | | **Long** | **Short** |
| **Investment Required ($)** | **Minor** | OK if timing is not critical | Full steam ahead |
| | **Major** | Forget it! | OK if cost position is tolerable |

**FIGURE 8-20**
INTEGRATION EVALUATION MATRIX
This matrix utilizes time and level of investment as criteria.
*Source:* Teece 1987. With permission.

that more information be gathered and that there be a certain level of confidence in the accuracy of that information.

## LINKING TECHNOLOGY AND BUSINESS STRATEGIES

Business success depends on the products or services brought to the market. As previously indicated, these have their base in technology. Organizations that know how to link their technology strategy with their business strategy will be more competitive in the global marketplace. Mitchell (1985) argues that the first step toward integrating business and technology strategies is to get the business and technical sides of corporate management to agree on a common set of priorities. Usually the business side perceives technology as a subset of business, while technologists perceive business as a subset of the general technological ascent of human beings, as shown in Figure 8-21. On one side, technology is a subset of a business enterprise. Market demographics influence the success of the business. Here, businesses tend to identify technologies relevant to creating business opportunities that satisfy market demands. On the other side, technology, through its role in the ascent of human beings, is the influencing factor in creating business. Business becomes a subset of technological advances that create significant opportunities for companies. For optimal results both sides must be integrated into one organizational strategy. Metaphorically speaking, integrating technology strategy and business strategy can be thought of as two sides of a coin: Either side is worthless without the other.

Companies that have a one-eye view toward business-oriented functions, such as finance, accounting, marketing, and sales, may face technical obsolescence or miss out on potential growth and profitability. Similarly, companies that focus entirely on technological development without effective strategy for exploiting the technology in a timely manner may not be able to sustain profitability. Management must be able to align its technology and business strategies to focus on achieving its goals and objectives. An interesting illustration of this concept is shown in Figure 8-22.

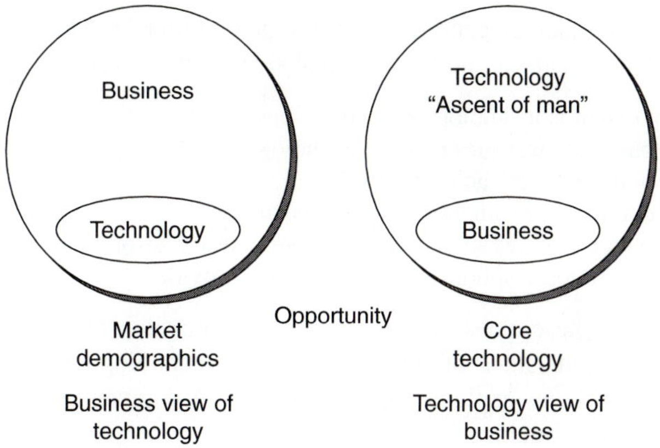


**FIGURE 8-21**
FRAMEWORK FOR FORMULATION OF BUSINESS AND TECHNOLOGY STRATEGIES
*Source:* Mitchell, 1995.

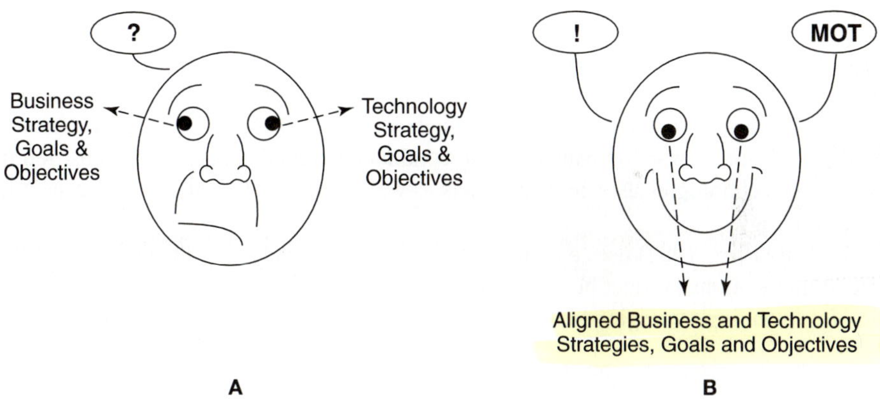


**FIGURE 8-22**
INTEGRATION OF TECHNOLOGY AND BUSINESS STRATEGIES
**A** Organizations that do not integrate technology strategy and business strategy have blurred visions of the future. **B** Well-coordinated and focused organizations are more competitive.
*Source:* Escobar, 1995.

According to Mitchell (1992), the technical community tends to hold the view that technical achievements by peers around the world often provide a more reliable guide to the future than do formally documented business forecasts. By contrast, the business-planning community usually looks at markets and other external trends as a more obvious and direct source of business opportunity. These two perspectives need to be reconciled. Mitchell emphasizes the importance of the linkage between the goals and objectives of the corporation and its technological strategy. Broad consensus and understanding must exist between business and technical managers throughout a company.

Mitchell poses a number of generic questions that should be addressed by strategic planners on both the business and the technical sides of the house:

1 To what extent is technology relevant to business?
2 Which business strategies require technology?
3 Where will we get it [the technology]?
4 What are our core technologies for the business?
5 In which technologies should we focus our research effort?
6 What new strategic options will technologies provide?

In responding to these questions, a company can develop relationships among its high-level strategies, its lines of business, and the technologies that are needed to achieve business goals, as explained in Reading 8.1. A company can then proceed to develop the product-technology-business connection.

## CREATING THE PRODUCT-TECHNOLOGY-BUSINESS CONNECTION

To identify the relationship between products or services and the underlying technology, a company can use any of several methodologies. Fusfeld (1978) proposes the use of a product-technology matrix similar to the one shown in Figure 8-23. In this methodology each product or service is broken down into its constituent technologies. Each matrix cell indicates the company's relative strength in the technology. A company can then determine which technologies it owns, which it would like to acquire, and which it wants to outsource. The company can also conduct a detailed analysis of its position in each of the technologies used in its products as well as of the state of art with respect to that technology.

Mitchell (1988, 1992) uses a similar approach to determine core technologies of the telecommunications product business (see Reading 8.1). This approach was used to aid corporate planning, strategic positioning, and technology investment at GTE Corporation.

**FIGURE 8-23**
PRODUCT-TECHNOLOGY MATRIX
This matrix shows the company's relative strength in each technology.

| Technologies \ Company's Products | **Product A** | **Product B** | . . . | **Product N** |
|---|---|---|---|---|
| **Required Technology 1** | Relative Strength | | | Relative Strength |
| **Required Technology 2** | | Relative Strength | | |
| **Required Technology 3** | | Relative Strength | | Relative Strength |

READING 8.1

# The Changing Agenda for Research Management

**Graham R. Mitchell***

*The research community needs to become better hunters and gatherers of technology at the same time as it plays a more effective and integrated role in the internal commercialization processes of the firm.*

*Overview:* The scope of industrial research in the U.S. and the role of research managers has expanded significantly during the last two decades. With the growth of strategic planning in the early 1970s, many research managers were challenged to work within the mainstream management processes in the corporation to couple the output of their laboratories more effectively to the business needs of operations. Significant progress has been made in many companies by working through the generic questions involved in transforming business objectives and strategies into core technical competence and program priorities for the research laboratory. More recently, global trends and the growth of the information industry have raised new planning challenges for research managers and are leading us to reexamine many of our long-held assumptions about the industrial research process.

Much of today's management practice and operating culture in large industrial research laboratories was firmly established prior to 1970. However, there are increasing signs that some of the intuition and instincts developed in this earlier period are at odds with current realities. Several authors have suggested that it is time to reexamine and augment many of our traditional assumptions on managing the research function. It is being argued, for example, that broader roles and new job descriptions are appropriate to today's technical leaders (1) (2). More generally, it has been suggested that we may be in need of a new paradigm for the industrial research process (3).

What makes the present situation difficult to interpret and disentangle is that we are dealing with two waves of change, both of which require research managers to step out beyond the confines of the laboratory. The first started for many companies in the early 1970s and has intensified during the last decade. It is concerned with the need to couple the research laboratory more directly to the strategy of business operations, and has been part of the broader attention to planning and strategy in United States corporations in general.

The second wave of change, bringing with it the need to further broaden the perspectives and responsibilities of research management, comes about because of the spread of technology itself, and its intensifying influence on global competitiveness. Fast-changing technology on a global scale is altering the rules of operation in all business functions.

*Source: Research · Technology Management,* September–October 1992, pp. 13–21. Reprinted with permission.

*At the time of writing this paper, Dr. Mitchell was Director of Planning for GTE Laboratories, in Waltham, Massachusetts.

| OVERALL BUSINESS OBJECTIVES | → 1970 MANAGING THE RESEARCH FUNCTION | 1970–1990 STRATEGIC BUSINESS MANAGEMENT | 1990 → GLOBAL COMPETITIVE ADVANTAGE |
|---|---|---|---|
| KEY ISSUES | Environment to promote individual creativity, innovation in groups. | Coupling of research laboratories to business operations. | Pervasive external trends: global management and sourcing, rapid commercialization, information technology. |
| FOCUS AND ACCOMPLISHMENTS | Management guidelines for organization, financing, technology transfer, project management, human resources, administration. | Integrated planning process to transform business objectives into technical priorities.<br>Support and extension of business strategy.<br>Increased credibility of research management. | Need for:<br>• New paradigms for industrial research.<br>• New roles for R&D leaders.<br>• Revision of conventional management wisdom. |

**FIGURE 1**
The research management agenda has evolved and broadened from the 1970's concern with managing the research function to today's challenges of global competitiveness.

This evolution and broadening of objectives and roles for research management is illustrated in Figure 1. There is a substantial and well-established body of literature and experience that was developed prior to 1970 on *managing the research function,* and that is routinely passed on to successive generations of aspiring research leaders. It covers many familiar topics, including organization, technology transfer, management and selection of projects, as well as human resource management, financing, administration, and external relations. A principal managerial objective throughout is the establishment of an environment in industrial laboratories that nurtures both individual creativity and innovation by groups. The resulting organizational structures frequently reflect underlying academic disciplines. Significantly, many of these assumptions and shared values were acquired in a period when U.S. corporations dominated many world markets, and at a time when the process and manufacturing industries provided the preeminent models of industrial innovation.

The principal focus of this article deals with some of the generic lessons the industrial research community has learned in integrating the research function with *strategic business management* 1970–1990. With the benefit of considerable hindsight, we can provide some guidance on the tools and approaches that have worked in bringing more effective coupling between research laboratories and business operations. These planning approaches have significantly improved the credibility of research management within U.S. corporations and helped to position the research community to deal with the emerging challenges of *global competitiveness* in the 1990s.

## BUSINESS STRATEGY AND TECHNICAL COMPETENCE

During the 1970s, increasing national and international competition put pressure on corporations in the U.S. to better integrate operations and focus business goals. One major result was that many companies introduced formal strategic planning and management systems to business operations. As a direct consequence, many research directors have been repeatedly challenged to "Get the laboratories coupled to the strategies and goals of the business and work within the mainstream management and planning processes in the corporation."

What appears at first sight to be a relatively straightforward request has in retrospect turned out to be complex. In order to achieve this coupling, research management has not only had to become intimately familiar with the goals and objectives of business operations, but has had to understand the process of strategy development in detail throughout the corporation. Ultimately, many corporate laboratories have had to establish complementary planning activities that are parallel to, and integrated with, those of business operations.

At the outset, an almost universal difficulty encountered in setting up strategic planning systems in technical organizations is that research managers and business planners frequently have differing perceptions as to the fundamental role of technology in business (4). The research community, with its traditional organizational focus on technical and academic disciplines, places great strategic value on the development and maintenance of state-of-the-art internal capability in certain critical or core technical skills. These are often seen as the source of business success in the past, and are presumed to provide the ultimate assurance of new revenues and lower costs in the future.

In keeping with this point of view, researchers typically harbor the belief that technical achievements by peers around the world often provide a more reliable guide to the future than formally documented business forecasts and strategies produced within the company. By contrast, the business planning community, reflecting a more traditional and deductive approach to business strategy, usually looks to markets and other external trends as a more obvious and direct source of business opportunity. This typically produces a shorter-term overall perspective, in which technology and engineering are often considered to be less dominant determinants of strategic direction, and are sometimes more easily thought of as internal functional capabilities, vying for resources with marketing, manufacturing, operations, and other elements of the organization.

A major goal of strategic planning systems for industrial research organizations is, therefore, to *transform* (in a mathematical sense) the objectives and strategies of the business into core technologies and program priorities for the laboratory, so that changes in business direction will be routinely reflected in the laboratory plans.

The transform process must also recognize other complexities. For example, in moving from business needs to technical programs, there are frequently temporal shifts, as the most appropriate technical response may take significant time to implement. There is also feedback. Core technical competence, established in the first place to directly support business strategy, often provides the opportunity to extend and change it. However, whatever the analytical subtleties, it is probably more important to recognize that business leaders throughout the corporation need to be convinced that they and the laboratory management share the same vision for the future of the company, and that the technical programs address the highest business priorities. Thus, the linkage between the goals and objectives of the corporation and its technical strategy must be established with the broad consensus and understanding of business and technical management throughout the company.

While an abundance of planning tools have been created during the last two decades that provide piecemeal *answers* to various aspects of the challenges discussed above, the comprehensive solution must begin with a relatively complete statement of the *problem.* Ultimately, this requires that we work through a logical sequence of generic questions, starting with the discussion of business strategy, proceeding through the identification of technical needs of operations, and progressing to the eventual determination of priorities for the research laboratory. The approach suggested here is to call on the rich legacy of planning tools, charts and matrices to systematically address the following:

- To what extent is technology relevant to the business?
- Which business strategies require technology?
- Where will we get it?
- What are our core technologies for the business?
- In which technologies should we focus our research effort?
- What new strategic options could they provide?

There are two comments on this sequence of questions. The first is that while they may seem obvious enough in retrospect, it has taken GTE a number of planning cycles and several special studies to reduce the issues to this straightforward list. The second is that, given the ultimate purpose of providing a plan for the laboratory, it is perhaps surprising that, in general, research management can supply relatively complete answers to only the last two of these questions. Responses to the first four have, for the most part, to be developed with the planning and technical communities in operations.

## STRATEGIC QUESTIONS

**1.** *To what extent is technology relevant to the business?* Most formal strategic planning processes start with a common series of analyses that deal with the primary factors affecting the business. Typically, studies of the external *environment,* including customer, market and other trends, identify potential opportunities. If not preempted by *competitors,* these may form the basis of future strategies for the business, as suggested by Figure 2. To the extent that technology is a critical element in the environment, a key attribute of major competitors, or a strategic strength or weakness of the business entity being examined, it should emerge naturally from these analyses.

It was not uncommon a decade ago to hear business planners argue that to give technology special attention over other elements or functions of the business would unreasonably distort the outcome of the analysis. However, in recent years, technological breakthroughs and the emergence of nontraditional and fast-moving technologies have transformed the competitive structure of numerous industries in ways that generalized planning approaches have failed to predict. We have become painfully aware that simply including technology in the checklist of topics to be addressed in the strategic

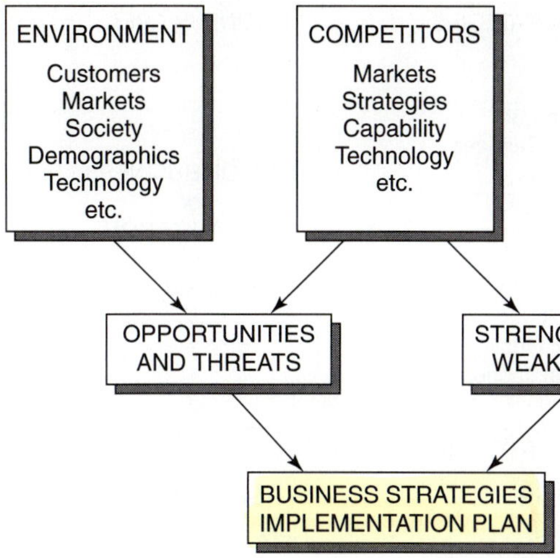


**FIGURE 2**
Strategy development typically begins with studies of the external environment, including customer, market, and other trends.

planning process, in the form suggested by Figure 2, in no way guarantees that U.S. corporations will not be exposed to significant technologically based surprises (5). Recognizing this vulnerability, many corporations, including GTE, have taken the obvious next step and asked directly, "What are our business strategies, and what technology is required for their success?"

**2.** *Which business strategies require technology?* In an ideal world, clearly stated numerical goals are unambiguously and uniquely aligned with strategies and implementation plans, making the identification of needed technology relatively trivial. In practice, overall business objectives and strategies are often written broadly, and typically aspirations for growth, new markets and revenues must be further specified by product line or business area before strategies can be developed and implementation plans formulated. The same is usually true of goals for cost, quality, productivity, and customer support.

Figure 3 illustrates typical relationships between high-level strategies and lines of business, in which the various technologies needed for success may be displayed in the appropriate element of the matrix. In manufacturing businesses, the required technologies that appear in these matrices usually turn out to be expertise of one form or another in materials, product design, and manufacturing or process technology. In service businesses, equivalent entries in Figure 3 usually describe system-level technologies such as network or system design, system operations, or customer applications and services.

Several key points usually emerge from this high-level strategic analysis. First, the need for technology is seldom uniform across elements of the matrix. In real business situations, the demand for technology varies widely between strategy-product combinations. Second, the range of technical disciplines required for success is frequently extensive, particularly when the impact of information technology is included. Third, the

| GOALS AND STRATEGIES | | BUSINESS AREA |
|---|---|---|
| GROWTH:<br>New Markets<br>New Products/Services<br>Extensions<br>COST IMPROVEMENTS:<br>Products/Services<br>Manufacturing/Operations<br>Capital Requirements<br>OTHER:<br>Productivity, Quality<br>Customer Satisfaction | X | LINE OF BUSINESS<br>or<br>ORGANIZATION<br>or<br>MARKET |

| REQUIRED TECHNOLOGY | |
|---|---|
| MANUFACTURING | SERVICES |
| Materials<br>Product Design<br>Manufacturing/Process | Network or System<br>Operations<br>Services and Applications |

**FIGURE 3**
In the typical relationship between high-level strategies and lines of business, the various technologies needed to succeed may be displayed in the appropriate element of the matrix.

scope of technical activity is often broad, ranging from fundamental insights at a research level through design, system application and operating expertise.

Once it is recognized that the successful implementation of existing business strategies requires extensive technical support, this, in short order, raises the obvious question, "Where will we get it?"

**3.** *Where will we get the required technology?* There are, in general, three different sources of technology available to business operations within large corporations: within the business operation; from the corporate or other company laboratories; from outside the corporation.

The immediate benefit of the formal discussion of this question is that it serves as a direct audit on the viability of present technical plans. Any gaps or poorly supported areas stand out and become candidates for increased resources. More important, strategic planning for technology takes a qualitative advance at this stage, when it becomes apparent that the range of technical strategies open to the corporation depends directly on the state of technical knowledge in-house. In the extreme, if the company does not have internal technical capability, the only possible strategy to provide immediate access to technology is to source it externally.

In this context, it becomes clear why technology is not a commodity (or good). The issues associated with sourcing technology are inherently strategic because internal core technical competence can rarely be built instantaneously. Even after the technical specialists are up to speed, additional time is invariably needed to absorb the impact of new technological advances on both internal operations and markets, and to translate these insights to new or modified products, services and competitive advantage.

Recognizing the practical limitations to both the availability and the effectiveness of technology from external sources focuses planning attention on the strategic necessity of maintaining a set of core technical capabilities within the company. The *number* of core areas represent a delicate balance. Choosing too many areas, and in effect overstating the business needs for technology, may reduce the resources per area below the critical mass needed to be effective. Choosing too few areas also leaves the business vulnerable.

**4.** *What are the core technologies needed by the business?* Any discussion of the need for core technical capability must be predicated on a common understanding of how the corporation intends to add value; that is, on precisely which processes must be controlled internally within the business (6). During periods of turmoil, the precise boundary between a company and its suppliers upstream, or customers downstream, may be in motion. At the present time, for example, there are signs that these boundary relationships are changing in the telephone industry as it becomes deregulated. As more manufacturers and equipment choices are provided to the industry upstream, telephone companies have to make more complex and better informed purchasing decisions. This requires not only improved technical insights into the evolution of tomorrow's networks, but a more sophisticated understanding than hitherto by the telephone company of materials, components, equipment design, and rate of obsolescence.

Similarly, at the customer interface, the potential explosion in volume and variety of information and communication services downstream requires improved understanding of market needs and a broadening of technical capability in some areas of information and social sciences. Clearly, the choice of core technologies can be critical in determining precisely how the business will be able to compete.

When asked to *define* technology, scientists and engineers quite naturally tend to talk in terms of technical skills and disciplines, that is, the input to the process. The business community, on the other hand, is more likely to respond in terms closer to products or systems, i.e., the output of the process. Over the years we have learned that both these perspectives need to be combined within the definition of core technologies in industry (7). The skills determine "what it is," and thus who it is we have to have on board, and the applications and markets define "what it does," and lead into the discussion of the strategic importance and appropriate level of resources for the technical area.

The most direct way to identify core technologies is by asking technical staff in *operations* to determine which internal skills and disciplines are essential to the growth and survival of the business. Particular emphasis should be given to those areas where strengthened technical capability leads to improved business performance (sometimes called key and pacing technologies) (8). This assessment may be carried out by individual product line and extended, as appropriate, to cover the complete corporation.

Figure 4 illustrates the result of this kind of analysis for two separate telecommunications businesses: one a products manufacturer, the other a provider of services. Both display a typical pattern in which the need for some technical skills (integrated circuit design, software engineering, network architecture, operations management, in these examples) are critical to several different business lines. This process, illustrated in these charts which couple identified technical skills to their application in products or services, automatically provides the dual perspective on technology and the two-part definition needed for industrial strategy.

| TECHNICAL SKILL | Product Line 1 | Product Line 2 | Product Line 3 |
|---|---|---|---|
| Integrated Circuit Fabrication | x | | |
| Integrated Circuit Design | | x | x |
| System Architecture | x | x | |
| Software Engineering | x | x | x |

| TECHNICAL SKILL | Service Area 1 | Service Area 2 | Service Area 3 | Service Area 4 |
|---|---|---|---|---|
| Network Architecture | x | x | x | |
| Switching Systems | x | | | |
| Operations Management | x | x | x | |
| Human Factors | | | | x |

**FIGURE 4**
The analysis that couples identified technical skills to their application in products or services automatically provides the dual perspective on technology and the two-part definition needed for industrial strategy.

Simply replacing the Xs in Figure 4 with program expenditures gives the analysis significantly more power and allows the overall technical resources throughout the corporation to be simultaneously mapped along two dimensions; that is, vertically by product or business, and horizontally by core technology. While business management is usually familiar with this vertical perspective, the horizontal view often suggests opportunities for technical synergies between separate business operations, draws attention to long-term technical human resource requirements, and clarifies the areas where technical forecasts and scenarios will most benefit the corporation. Often the most important contribution from this horizontal perspective is that it automatically draws direct attention to technical skills that impact multiple businesses, and thus to the technologies where change and evolution will produce the most significant impact on the corporation's operations.

Since these areas have been defined as that combination of skills and applications most critical to the strategy of the business, they are also the most appropriate element against which to benchmark competitive performance. A lagging competitive position in core technologies frequently indicates the demise of a business, long before it is apparent from financial reports (4).

**5.** *Where shall we focus our research effort?* Having identified which technologies the corporation needs, we may rationally approach the determination of strategy for the corporate laboratories by asking, "In which few technical areas shall we focus our research effort?" It is difficult to overestimate the importance of this choice, as it defines the domain where technical staff will make future contributions to the business. These core areas may be chosen to complement, extend, or lead technical activity in operations, but must contain sufficient resources to explore and develop new technology to the point where it can be picked up by business operations for commercial exploitation.

Within GTE, we identify these areas of research focus as "Strategic Technical Areas" (STAs), and formally define them as having two components: (*a*) technical skills and disciplines, plus (*b*) applications, products/services, markets. For example, the following is a definition of an STA that addresses network planning:

a) Major mathematical disciplines include decision, queuing and game theory, simulation, regression analysis, linear programming, as well as algorithmic techniques applicable to the solution of computationally large network problems.

b) These are used to develop network and engineering design tools for public telecommunications networks.

The overall strategy of the corporate laboratories is managed through 10 to 20 STAs, and the level of funding is typically an order of magnitude greater than that needed for individual projects. Dealing with resources at these higher levels of aggregation naturally focuses the discussion on strategy and ultimate benefit of the work to the corporation, rather than on shorter term operational issues which arise in the review of individual projects.

**6.** *What strategic options are provided by technology?* If the strategic area is well chosen, it will not only support the stated strategy of business operations, but will, in addition, result in a continuous stream of benefits to the corporation, from applications only dimly perceived at the outset. It may even provide opportunities to extend or change existing business strategies.

Figure 5 gives two illustrations of how this process has worked in practice. In the first, research into the fundamental properties of fiber-optic materials and devices that initially focused on the development of components for telecommunications transmission *products* has, over the course of several years, produced significant benefits in other businesses that address different market segments. Most notably, this research has been valuable to telephone companies in deploying and demonstrating broadband and video *services* in the local loop.

In the second example, fundamental research into artificial intelligence in the early 1980s has progressed to a variety of applications in switching and outside plant maintenance, and to the integration of heterogeneous databases. In both examples, the decision

**FIGURE 5**
The decision to build core expertise in the strategic technical areas of fiber optics and artificial intelligence was made before most of their commercial applications were well understood.

| TECHNICAL AREA | RESEARCH TOPICS | EVOLVING APPLICATIONS |
|---|---|---|
| FIBER OPTICS | Late 1970's<br><br>Glass Technology<br>Advanced Devices<br>Optoelectronic Devices | 1980's<br><br>Fiber Installation, Know-How<br>Optical Components<br>Broadband Systems<br>Video Distribution Systems |
| ARTIFICIAL INTELLIGENCE | Early 1980's<br><br>Databases<br>Machine Learning<br>Knowledge Representation<br>Distributed Artificial Intelligence<br>Expert Systems | Later 1980's<br><br>Telephone Switch Maintenance<br>Preventative Maintenance on Outside Plant<br>Integration of Multiple Heterogeneous Databases |

to build core expertise in these strategic technical areas was made before most of the eventual applications were well understood.

Positioning the laboratories strategically in appropriate core technologies requires considerable management skill and judgment. Choosing the right areas to establish technical excellence implicitly opens up future business opportunities. Neglecting other technologies may expose the company to significant threats and limit downstream options. While it is relatively straightforward to make the case for strategic positioning when long-range technical trends are relatively well understood, as in some areas of defense, energy and transportation, paradoxically, the need to build core research skills is often more critical in periods of relative turmoil. At such times, when the future refuses to be preordained and business forecasts and scenarios are in frequent need of revision, there is added advantage in being positioned to promptly respond to the impact of changing market conditions, with the capability to follow through immediately.

There are, however, two fundamental problems in making the case for strategic positioning. The first concerns the language used to communicate the purpose of the work to senior management. Considerable care and semantic ingenuity are often required to convey that in choosing core areas for long-range research, we are not so much picking targets that will *not pay off for a long time* (e.g., land a man on the moon in ten years), as we are positioning ourselves in areas in which it will prove *fruitful to work for a long time*. This distinction is often critical to the business case, but is frequently unstated.

The second problem concerns appropriate financial justification. The objective of much of this kind of research is to create technical options; that is, to position the corporation strategically to make investments downstream if and when present technical and business uncertainties are resolved. The obvious difficulty is that many of these are, inherently, multi-stage investment decisions involving significant uncertainty that inevitably fare badly if force-fitted without sufficient care to ROI or similar DCF approaches (9). In many cases, the underlying structure of the decision on research priorities has more in common with valuing a call option on the stock market. Recognizing this parallel with stock options and the limited downside risk involved may well explain why the research community, intuitively and appropriately, often chooses projects having significant uncertainty (10).

## LESSONS FROM TWO DECADES

The experience in developing strategic planning systems for corporate research during the last 20 years has provided research managers with several important lessons:

**1** The research community in total has been able to successfully embrace the larger concerns of the corporation, and to some extent shed its image as single-minded protectors and defenders of the research *function*. This augers well for the challenges ahead.

**2** Experience points to the importance of asking the right questions, and to the need to search for the answers to many of them outside the laboratory; the choice of tools with which to answer them is less critical. Several conceptual transitions are involved in integrating the traditional deductive approach to strategy with that based on technical core competence. The sequence of questions must succeed in exposing the time dependence of technical solutions, as well as articulate the case for strategic positioning and its potential to modify business strategy.

**3** The need for core technologies must be established as a response to the strategies and scope of the corporation's business operations, and the definition of technology should couple specific skills and disciplines to applications and markets. Lists of disembodied disciplines are ambiguous, even to the technical community, and without strategic and business context, provide no path to determine their relative importance.

**4** In most cases, it is necessary to make the case for strategic positioning and to commit resources to core technologies on the basis of fit and alignment with business goals and strategy, that is, *before it is clear from the numbers*. Language needs to be chosen with care to convey that the purpose of the work is often to provide downstream options, and that uncertainty and risk are not necessarily synonymous if funding is managed appropriately.

**5** Credibility with business management will be significantly enhanced by working within the planning systems of the corporation to measure and track research performance against a variety of quantitative and semiquantitative goals. This process, despite reservations about the particular measurement criteria, strengthens the role of research management as corporate team players.

**6** Credibility with business management is further enhanced by making strategic tradeoffs for the laboratory highly visible. The decision to develop new core areas of research invariably involves cutting resources to lower priority areas. Rather than attempting to finesse these choices, laboratory management is invariably better served by using the strategic planning process to highlight the technical directions that were discontinued.

## PLANNING CHALLENGES OF THE '90s

While much of the progress to date has been accomplished by working on processes internal to the company, many of the planning challenges for research managers in the 1990s are driven by trends outside the corporation. The research management agenda is shifting to respond to the impact of global competition, and in the process, researchers are having to reconsider many of the traditional and sustaining notions underlying industrial research. Three trends are particularly important.

**1. Globalization of Markets** While markets for consumer and industrial products are becoming increasingly global, sources of leading-edge technology needed to compete in these markets are often discrete and localized (11). In this environment, the risks associated with "going it alone" and providing "unique" technological solutions are increased, and research managers are learning to actively embrace a wide variety of partnering arrangements, strategic alliances, collaborations, and consortia to gain access to new technology.

A major implication for industrial research strategy in the 1990s is that the role of the laboratory in providing access to sources of external technology may be growing, relative to the traditional function of developing original technical solutions. Planning systems and management practice are evolving to meet these new challenges. For example, competitive benchmarking of technical capability in critical areas is becoming widespread, and tracking of worldwide centers of excellence in the business' core technologies (STAs) is being increasingly formalized and used to focus interaction with universities and external organizations. As the decade continues, planning systems for research can be expected to incorporate guidelines for successful partnering and alliances (12).

The obvious impact of all the above on the existing planning procedures is that it expands the choices in answering one of the six fundamental strategic questions, "Where will we get the required technology?" At a more basic level, however, this expansion in external sourcing underpins the evolution of the strategic role of corporate researchers toward becoming better *hunters and gatherers* of technology.

**2. Rapid Commercialization** Our traditional approach to R&D is also challenged by the trend toward increasingly rapid commercialization cycles. To start with, the process does not fit at all well with the long-established pattern of technical discovery in the laboratory, progressing through various stages of feasibility and scale-up to final market introduction. In this alternative vision, successive new model introductions are arranged like a series of aircraft lined up for takeoff. Giving free reign to the research department to delay those toward the front of the queue is, to say the least, irresponsible. Although the implications may vary significantly by industry, the advent of time-based competition moves the center of gravity for decision making away from the laboratory and into business operations (13).

An inexpensive, flexible, and rapid product introduction process allows the latest technology and market reaction to be continuously included in successive product generations. To the extent that this process provides a way around the need for long-range technical insights, the research community may have some cause to view the adoption of rapid commercialization processes as threatening to its traditional role. Beyond this, the process of project management is undergoing significant change through the extensive application of concurrent engineering and improved management systems. In total, these trends imply that new planning systems for R&D will reflect an even less independent role for the research function in the future, and a significantly greater and structured involvement with all elements of the business at an early stage.

**3. Information and Services** By today's standards, the information industry hardly existed during the period when most of our notions about the proper conduct of industrial research were being established. The explosive growth of the industry has also been accompanied by a related increase in technical sophistication in the service sector (14). Not only is this of enormous importance to the service sector directly, but research directors in formerly traditional manufacturing businesses are also being asked to address services as the complexion of their corporations evolves.

The impact on R&D has several dimensions. For example, many of us automatically expect the focus of industrial research programs to be materials, product design and manufacturing process, as it is intuitively obvious that better materials, designs and processes will lead directly to a better bottom line. These, as pointed out earlier, often reflect the priorities of manufacturing industries. Network or system design, operations (that is, how to run the system), and customer applications may better represent the equivalent generic focus of R&D programs in the more than 70 percent of the economy farther down the value chain, in the service sector (15). Many service industries deal directly with the individual consumer, and a principal aim of the research process is often to determine how information technology can economically address the deep needs and psychology of the general public (16).

These trends in information technology and services suggest that, during the 1990s, planning systems will evolve and enlarge the range of disciplines that are routinely classified under the rubric of industrial research to include human factors and some of the social sciences. More important, they will increasingly recognize that a lot of what looks and feels like industrial R&D is going on in nonmanufacturing organizations.

## IMPLICATIONS FOR LEADERS

A traditional role of research managers as stewards of the industrial research function has been to pass on to the next generation the combination of formal tenets and working folklore that make up the "secrets of the guild." These establish the language and conceptual frameworks that are used to communicate shared experience and best practice across different companies and industries. They have contributed to the building of a strong and independent culture among industrial researchers.

During the 1970s and 1980s, research leaders have had to take on wider responsibilities and broaden their involvement with other business functions to achieve improved coupling with operations and congruence with business strategy throughout the company. By working through strategic planning processes, research leaders have often been able to become more effective advocates for technology, and better accepted members of the corporate management team.

However, this transition has not come without producing some strain within the traditional value system of the research community. For example, it is clear in retrospect that credibility with senior business management is usually enhanced as technical leaders break with some of their earlier instincts and distance themselves from being perceived primarily as protectors and defenders of the laboratory and technical function. Working more closely with operations, working inside mainstream management processes, agreeing to be measured and tracked, being visible in cutting back or closing research areas—all enhance this credibility. However, these actions also have the potential to be seen as limiting the necessary independence of individual researchers, and are thus potential sources of tension or confusion among the research staff.

The evidence seems clear that some of the traditional assumptions of the research community will be further strained during the 1990s. Business and technical management, between them, have to articulate the case for the research community to simultaneously address two seemingly conflicting challenges. These are the need to be better hunters and gatherers of technology with more complete and comprehensive access to global sources, while at the same time playing a more effective and integrated role in the internal commercialization processes of the company. Existing strategic planning systems and management approaches need to be extended to accomplish this, and to encompass the pervasive and rapidly changing impact of information technology.

The adjustment to these new paradigms may be painful, but it appears necessary, as U.S. industry in total is proving resistant to the proposition that the problems of U.S. competitiveness can be fixed by simply providing more R&D dollars. Despite popular support for the notion that technology is increasingly important to the growth and survival of industrial corporations, U.S. industrial R&D expenditures in aggregate have grown annually by less than one percent above inflation over the last five years. To be effective in this new environment, planning approaches in the 1990s should focus less on the narrow issues of R&D and more broadly on the pervasive impact of science and technology on corporate goals and performance. Throughout this process, credibility with the business community within the corporation and beyond will continue to be critical, as success is not going to depend on convincing those who work for the technical function so much as on prevailing with those who do not.

### Acknowledgments

It is a pleasure to acknowledge the support of numerous colleagues from the Industrial Research Institute Research-on-Research Committee, from McKinsey & Company, and from GTE.

### Reading 8.1 References


1 Lewis, William W., and Linden, Lawrence H. "A New Mission for Corporate Technology." *Sloan Management Review,* Summer 1990, pp. 57–67.

2 Uttal, Bro; Kantrow, Alan M.; Linden, Lawrence H.; and Stock, Susan. "Building R&D Leadership and Credibility." *Research Technology Management,* May–June 1992, pp. 15–24.

3 Steele, Lowell W. "Needed: New Paradigms for R&D." *Research Technology Management* July–August, 1991, pp. 13–21.

4 Mitchell, Graham R. "New Approaches for the Strategic Management of Technology." *Technology in Society* (1985), pp. 227–239.

5 Anderson, Philip, and Tushman, Michael L. "Managing Through Cycles of Technological Change." *Research Technology Management,* May–June 1991, pp. 26–31.

6 Steele, Lowell W. *Managing Technology: The Strategic View*. McGraw-Hill Engineering and Technology Management Series, Chapter 3, "Management Conventions—The Ties that Guide and Bind." Michael K. Badawy, Ph.D., Editor-in-Chief, McGraw-Hill Book Company, New York (1989), pp. 69–93.

7 Bitondo, Domenic S. *Technology Planning in Industry—The Classical Approach*. Chapter 4, "Interdisciplinary Planning: A Perspective for the Future." M. J. Dludy, K. Chen, Editors, New Brunswick: Center for Urban Policy Research, Rutgers University (1986).

8 Roussel, Philip A., Saad, Kamal N., and Erickson, Tamara J. *Third Generation R&D*. Harvard Business School Press (1991), Chapter 4, p. 65.

9 Myers, Stewart C. "Finance Theory and Financial Strategy." *Interfaces* 14:1, January–February 1984, pp. 126–137.

10 Hamilton, William F., and Mitchell, Graham R. "R&D in Perspective: What Is R&D Worth?" *The McKinsey Quarterly* (1990), Number 3, pp. 150–160.

11 Perrino, Albert C., and Tipping, James W. "Global Management of Technology." *Research Technology Management,* May–June 1989, pp. 12–19.

**12** Slowinski, Eugene, and Hull, Frank. "Partnering with Technology Entrepreneurs." *Research Technology Management,* November–December 1990, pp. 16–20.
**13** Stalk, Jr., George, and Hout, Thomas M. *Competing Against Time*. The Free Press (1990).
**14** *Managing Innovation: Cases from the Services Industries.* National Academy of Engineering, Bruce R. Guile and James Brian Quinn, Editors, National Academy Press, Washington, DC (1988).
**15** Mitchell, Graham R. "Research and Development for Services." *Research Technology Management,* November–December 1989, pp. 37–44.
**16** Rapaczynski, Wanda. "Developing Technology with a Human Face." *Research Technology Management,* May–June 1992, pp. 34–37.

READING 8.2

# Putting Core Competency Thinking into Practice

**Mark R. Gallon[1], Harold M. Stillman[2] and David Coates[3]**

*Do you know what your core competencies are? Can you recognize potential competencies? Here's a method for finding out, in sufficient detail to be useful.*

*Overview:* Core competency thinking is a powerful and widely promoted approach to focus and mobilize an organization's resources. As a result, R&D and technology executives are increasingly being asked to define the core technical competencies of their companies; unfortunately, they often fail to come up with convincing answers. This is not surprising, as the best methods of employing the thinking in organizations have not been elucidated. What is needed is a method and tools for carrying out a core competency assessment. The method described here is applicable to almost all core competencies, whether they are technical or non-technical in nature and whether they are currently available to the company or will need to be developed in the future. It will

[1]Mark Gallon was a managing consultant in the Technology Management practice of PA Consulting Group, in Cambridge, England when this article was written. A specialist in core competency-based strategy formulation, he played a key role in developing PA's approach while consulting to large manufacturing firms throughout Europe, the U.S. and the Far East. He holds an honours degree in chemistry from the University of York.

[2]Harold Stillman was a director of PA Consulting Group's Technology practice and is now senior vice president, technology and innovation, for ABB Corporate Research, based in Norwalk, Conn. Since he started his consulting career in 1978, he has worked on the technology and new product development issues of over 150 organizations, gaining a perspective on the challenges of developing and commercially exploiting technical competency across most business sectors. He holds a B.E. in mechanical engineering from New York University.

[3]David Coates is a managing consultant in the Technology Consulting and Management practice of PA Consulting Group and is based in Cambridge, England. Over the past two years he has played a key role in developing and applying PA's approach to core competency in a wide variety of European organizations, at both the corporate and business-unit level. He has been a consultant since 1985, and prior to this worked in a number of materials- and chemicals-based companies. He graduated in engineering and has a Ph.D. in materials science.

*Source: Research · Technology Management,* May–June 1995, pp. 20–28. Reprinted with permission.

allow many organizations to put core competency thinking into practice—to great competitive advantage—without unnecessary struggle and disruption to business activities.

In Prahalad and Hamel's landmark paper, a core competency is defined as "an area of specialized expertise that is the result of harmonizing complex streams of technology and work activity" (1). Core competencies have special qualities: They exemplify *excellence* and provide *competitive advantage*. Additional qualifications are that this excellence is translated into *customer-perceived value,* is *difficult to imitate* by competitors and is *extendable to new markets* (provides market mobility).

Putting it simply, core competencies are the things that some companies *know how to do uniquely well* and that have the scope to provide them with a better-than-average degree of success over the long term. A company's current product and service offerings merely represent today's physical embodiment of its competencies. Indeed, the success of a company's products at a point in time does not necessarily relate to its level of competency.

Some major manufacturing corporations have consistently applied core competency thinking and have gained considerable strategic value from the coherency of this approach. Examples of some definitive practitioners and their core competencies include:

- 3M, in developing products that creatively combine flexible substrates and functional coatings.
- Sharp, in manufacturing high-volume, large-area liquid crystal displays.
- Kodak, in developing and applying efficient silver halide imaging materials (2).
- US Surgical, in bringing novel, Class II/III medical devices to market.

These core competencies are the property of the companies *as a whole* (not of individual businesses or functions). They provide a set of unifying principles for the development of all aspects of the organization, and ensure that strategies have continuity, are robust and are flexible to changing circumstances. They are intrinsic to the company's overall vision and are pervasive in all strategies.

The core competency statements capture the essence of the firm's business and technology strategy. However, they are at a high level of abstraction and do not seem particularly useful in determining what should be done to exploit the competency to enter new business areas, to enhance competitiveness in existing product categories, or to focus R&D and technology investments. Executives want to understand the complex streams of technology and work activities that make up core competencies. They want sufficient detail to be able to assess their firm's strength in the competencies and to determine where and how they need to improve, and they want this detail provided in the most efficient and useful way. They would also like to establish a common understanding within their organizations so that appropriate investment and strategy decisions can be made. Here is how we have been able to accomplish these tasks.

## CORE COMPETENCIES AT THE TOP OF A HIERARCHY

We have found it necessary to establish a clear and consistent set of definitions for explaining core competency thinking. It is important that this new language is understood by all who are, or who plan to be, involved in core competency activity.

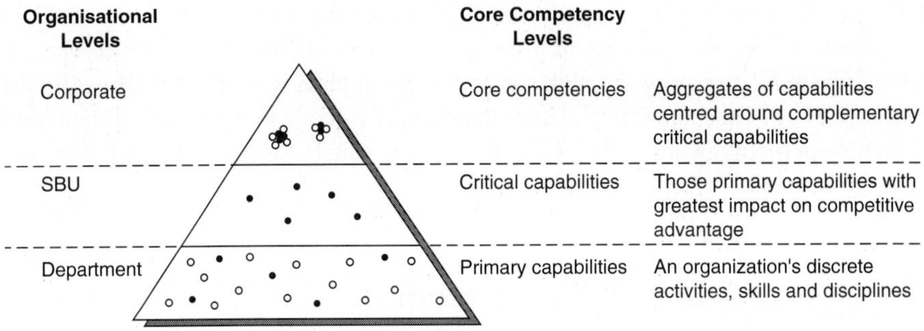


**FIGURE 1**
Core competency thinking involves a hierarchy.

All organizations contain a large and diverse array of discrete activities, skills and disciplines (see Figure 1). These elements—termed primary capabilities—are the building blocks of core competencies. The development and operation of most primary capabilities are the responsibility of individual functions of a company.

Certain capabilities are distinct from other primary capabilities in that they have a direct and significant effect on competitiveness in their own right. These capabilities, termed critical capabilities, can provide reduced cost, improved product or service differentiation, increased speed to market to larger barriers to competition. The development of critical capabilities is often a key element of strategies at the strategic business unit (SBU) level.

A useful way to think of core competencies is as *aggregates of capabilities,* where synergy is created that has sustainable value and broad applicability (see Figure 2).

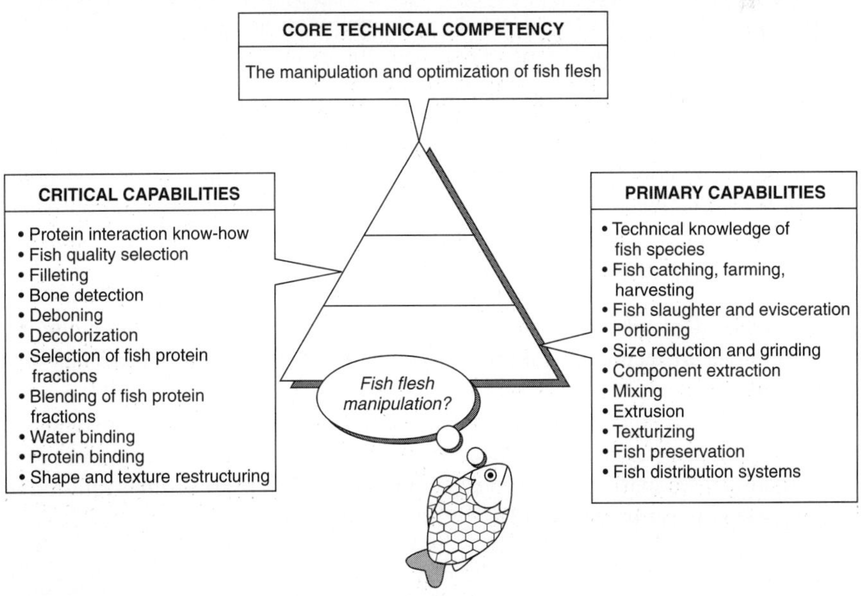


**FIGURE 2**
A core competency is the aggregation of many capabilities.

Within core competency aggregates, there are always at least two complementary critical capabilities; these critical capabilities are the source of the power of the competency and are where a company's strength can really distinguish it from other firms. In 3M's case, several critical capabilities exist at the heart of its core competency—for example, know-how in rheology and capabilities in surface coatings formulations and continuous coating processes.

## DIFFERENT CAPABILITIES/CORE COMPETENCIES

Primary capabilities may be usefully divided into different categories, as follows:

- *Market-interface capabilities*—capabilities that are used in the marketplace or that are clearly visible to it; selling, advertising, consulting, invoicing or customer satisfaction monitoring are generic examples of these capabilities.
- *Infrastructure capabilities*—capabilities that concern the internal operations of the company and that are invisible externally; for example, management information systems or internal training.
- *Technological capabilities*—technical capabilities providing direct support to the product or service portfolio; these may be further subdivided into:

1 *Applied science capabilities*—fundamental know-how derived from basic research; for example, know-how of things like genetics, biocompatibility, demographics, nuclear physics, applied statistics, or ceramic materials.
2 *Design and development capabilities*—disciplines employed in converting a product idea into an operational reality; generic examples are CAD, project management, prototyping, industrial engineering, or software development.
3 *Manufacturing capabilities*—capabilities employed in, or directly supporting, established manufacturing or operations; for example, internal QA systems, environmental control, or final test and inspection.

We have found that the large majority of the capabilities that are critical to organizations are either technological or market interface capabilities. *Most core competencies thus rely on technological and market interface capabilities;* we define two general categories of competency depending upon which group of capabilities predominates within the competency aggregate:

- *Core technical competencies* (CTCs)—where the majority of the underpinning critical capabilities are technological in nature (where technology is the major determinant of uniqueness).
- *Core marketing competencies* (CMCs)—using the term marketing to embrace product management, pricing, communication, sales and distribution (where most of the critical capabilities are market interface capabilities). CMCs are sometimes referred to as non-technical core competencies (3).

Competencies in each category can be equally powerful, but CTCs are especially important because they are more frequently able to cross market boundaries and can provide the basis for significant product superiority. The rest of this discussion focuses on CTCs, as they are also more prone to being overlooked.

## VALID CTCS ARE RARE

In most organizations, only a few areas of technical expertise have the right attributes to be worthy of the term core technical competency (are *valid* CTCs) and even fewer have been developed by these companies to the level of excellence that is necessary to give them broad strategic value. Core competency assessment requires a rigorous and disciplined approach in deciding which candidates are valid and in determining the organization's real competency position.

A candidate CTC needs to be carefully scrutinized and must be deemed to pass all core competency qualification criteria before it is considered valid:

- Does it harmonize streams of critical technological capabilities to provide competitive advantage?
- Does it translate into customer-perceived value?
- Is it difficult to imitate (are there substantial barriers to competitors)?
- Is it extendable to new markets (does it provide market mobility)?

After CTC validity has been verified, one can move to considering the company's status in the competency, and, again, new terminology is required. A company that recognizes and exploits a valid core competency *owns* the core competency. This idea of ownership is useful because it rightly portrays a feeling of value that has been created by investment, is appreciated and requires maintenance. The qualification for ownership of a core competency is excellence in the competency from both external and internal perspectives.

Excellence from an external perspective means that the owner has recognized leadership in the competency; for example, Motorola would certainly be a company that most industry observers would nominate if asked: "Who is the current leader in the development of semiconductor solutions for wireless communications systems?"

Examining excellence from an internal perspective requires an understanding of all the capabilities that make up the core competency aggregate, which capabilities are critical and how well developed each capability is. Excellence from an internal point of view constitutes a set of capabilities where all the critical capabilities are highly refined, with only limited scope for incremental improvement.

## IDENTIFY STRATEGIC CORE COMPETENCIES

When core competency theory is applied as rigorously as we advocate, many companies conclude that they do *not* currently own any CTCs. However, organizations can derive as much value from understanding the core competencies that they do not own as they can from identifying their existing competencies. The major benefit of core competency assessment to the majority of companies comes from the recognition of *potential* core competencies. Potential competencies have all the required attributes of a core competency, except that they have not been developed to a level of excellence by the organization. They are the result of companies' failure to develop fully their key areas of expertise, and are relatively common.

*Desired* core technical competencies are a further category of valid competencies where several of the essential critical technological capabilities are currently missing entirely from the organization. The absent critical capabilities may be well known to the organization (for example, competitors may be applying them already) or they may be much less visible (for example, the capabilities may be emerging in an industry sector that is quite different from the organization's area of activity).

A company's *strategic* core competencies are the complete set of competencies that have been targeted for future development and exploitation; they can comprise a mixture of existing, potential and desired competencies. In total, their selection equates to a technology strategy for the organization as a whole, that integrates and focuses *all* of the company's technical capabilities.

## CTC PROGRAMS NEED CAREFUL PLANNING

Core competency programs have a habit of broadening and becoming unmanageably complex unless they are carefully planned and implemented. The focus of early work—its objectives and scope—and the participants and methods must be determined before work commences.

Planning is facilitated by forming a preliminary Steering Group for the program, to take responsibility for this task. The composition of this group is important; senior representatives of the organization's technical *and* business communities need to be involved, so that findings can be linked to existing strategic planning processes. There will be an ongoing role for the Steering Group during the program—to communicate goals and findings, influence the direction of the program, monitor progress and challenge findings—and it may be appropriate to adjust membership as planning nears completion. Another key early step is the identification of a Program Manager to take responsibility for defining and implementing the work activities.

Central to most CTC programs is a desire to examine whether the organization is leveraging its technical capabilities in an optimum way and investing appropriately in research and technology development; however, initial objectives need to be more specific and short term than this. Participants are unlikely to make a sufficient commitment to a CTC program unless it is used as a vehicle for resolving one or more urgent business challenges.

A decision on the optimum scope to employ during the program (i.e., how many of the company's technical activities to include) also needs to be made and will be driven primarily by the objectives that have been identified. Depending upon circumstances, it may be appropriate to begin work with the organization as a whole or with a subset of its technical activities.

Organizations new to core competency philosophy are well-advised to moderate their short-term ambitions and get some early "runs on the board." Beginning immediately to examine the whole strategic competency set (existing, potential, desired, core marketing competencies, *and* core technical competencies) is rarely successful; phasing of the work is highly recommended.

## A GENERIC METHOD TO INITIATE CTC WORK

Our generic CTC identification process consists of a series of six work modules, and is outlined in Figure 3. The process is systematic and thorough; it requires a large amount and variety of employee participation and demands both creativity and rigorous analytical activity.

**FIGURE 3**
PA's method for identifying core technical competencies consists of six modules.

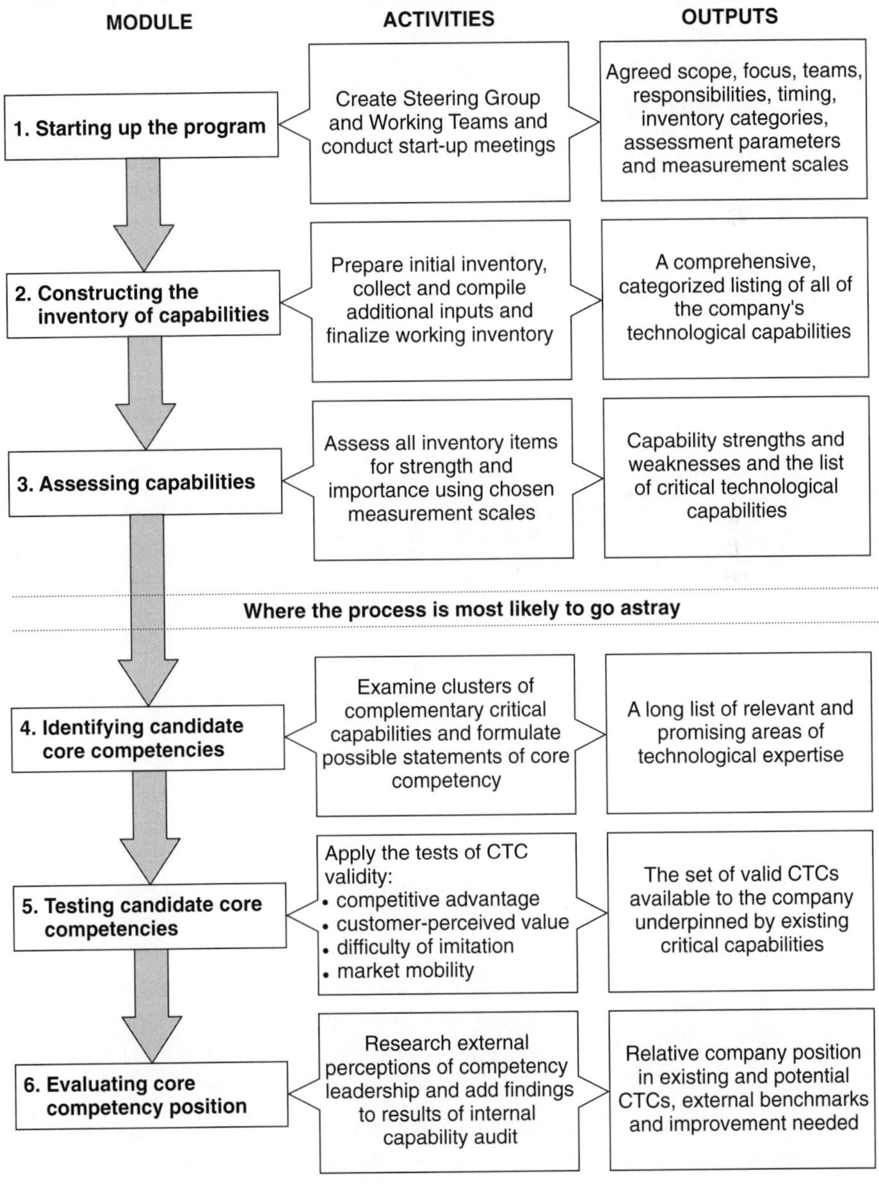

When the process is applied effectively, it provides a highly robust platform for further work, numerous secondary benefits as a result of working through the process and a number of valuable outputs:

- An inventory of all of the company's technological capabilities and indication of those capabilities that are duplicated in different parts of the company.
- An assessment of all of the company's technological strengths and weaknesses.
- A listing of the company's critical technological capabilities.
- An appreciation of the existing and potential core competencies available to the company and external benchmarks of excellence.

### Module 1—Starting Up the Program

Program start-up is the important link between planning and implementing a CTC program; carrying it out effectively goes a long way toward ensuring the success of the program. At the end of this module, all the key participants should have a common set of objectives and a common working language; they should also have an agreed working scope and approach and a set of deliverables and milestones.

It needs to begin with a formal meeting of the Steering Group to confirm the objectives, scope, terminology, definitions, time scales, and structure of the method. The group should then decide how the program is to be introduced inside the organization.

Early and widespread communication of the overall goals of the program by senior management is highly recommended to encourage active and enthusiastic participation. It should also be made clear that the program is the first step in an ongoing process.

The Program Manager should then prepare a structure for the inventory of capabilities; this inventory will define the working space for subsequent analyses and will provide the raw material for identifying CTCs. The Program Manager needs to start by identifying and assembling a Working Team to take responsibility for building and assessing the inventory capabilities. The Working Team must have substantial collective experience and insight into the full range of technologies and company activities included in the program. Team members should be encouraged to treat the program as a work priority and to think beyond their normal areas of responsibility.

In a first meeting of the Working Team, the Program Manager should provide a thorough program briefing and then lead an initial discussion of the structure of the inventory. The normal format of a capability inventory is a large matrix, comprising a categorized list of the company's technological capabilities and additional columns for scoring each capability using various assessment parameters. The team needs to decide which categories to use. They might, for example, be simply (1) applied sciences, (2) design and development, and (3) manufacturing capabilities.

Thought also needs to be given to which assessment parameters to use and the basis for carrying out an assessment. As a minimum, capabilities will need to be assessed for *strength* and *criticality* during Module 3. The team needs to make an initial judgment on how to work with these parameters and to address the following questions:

- How do we define the "strength" of a capability? Should we use a scale based on internal perceptions or assess strength relative to other firms? If the latter, whom should we benchmark?

- How do we define the "criticality" of capabilities?
- What measurement scales should we use for making assessments?
- Should we evaluate the organization's overall capabilities or make individual assessments for every place in which a capability resides?

The information provided in Figure 4 is a good starting point for this debate. Absolute strength scores give an indication of the opportunity for improving an existing capability within the company. Relative strength scores are useful for determining priority areas for investment. Criticality scores are used to establish the most important capabilities now and for the future. Scores of 4 or 5 in any parameter are most likely to be associated with a capability within a core technical competency. Capabilities with criticality scores of 4 or 5 deserve special attention.

Once all of this preparatory work is complete, the initial conclusions of the Working Team are presented to the Steering Group by the Program Manager and discussed. The observations and advice of the Steering Group are then fed back to the Working Team at the beginning of Module 2.

### Module 2—Constructing Inventory of Capabilities

The objectives of this module are to finalize the working structure and contents of the inventory, to confirm the assessment parameters, and to begin to define in detail the approach to be used for making the assessments.

The Working Team simultaneously reviews the categories and identifies some initial capabilities that belong in the inventory. A robust set of mutually exclusive categories is

**FIGURE 4**
Capabilities are assessed using five-point scales.

| | ABSOLUTE STRENGTH | RELATIVE STRENGTH | CRITICALITY |
|---|---|---|---|
| **Score** | **The degree to which the capability has been optimized internally** | **The degree to which the capability constitutes best industry practice** | **The degree to which the capability has a direct impact on competitiveness** |
| 5 | Highly refined, with only limited scope for enhancement | Substantial and undisputable leadership | A major determinant of competitive advantage |
| 4 | Well developed, with moderate scope for incremental improvement | Equivalent to industry best practice but not outright leadership | Has a direct and significant effect on competitiveness |
| 3 | Partially developed, with significant room for improvement | Developed to an average degree for the industry | Important to competitiveness in an indirect or enabling way |
| 2 | At an early stage of development | Substantially inferior to best practice | Rather unimportant to competitiveness but has an indirect effect |
| 1 | In its real infancy or with opportunity for improvement | Significantly under-developed compared to industry norms | Has (almost) no impact on competitiveness |

produced by iteration, with each category characterized by a few initial entries. Individual members of the team are then each assigned categories and given the task of constructing the full lists of capabilities that make them up. A decision should also be made on a common storage and analysis medium to use; employing a spreadsheet or database computer package helps the process.

The capability lists are built up by the team members through face-to-face and telephone interviews with colleagues throughout the company's technical community. The Program Manager coordinates the individual efforts and compiles the results. Once this work is complete, the Working Team reviews the compiled list of capabilities and considers and deals with any overlaps and gaps that might exist.

The team also needs to consider the assessment parameters, choose which to use and decide how assessments are to be made. The objectives, scope and time scales set for the initial work will be influential in resolving these questions. We normally advocate separate processes for assessing strength and criticality. For capability strength, where the true situation is unpredictable, we would recommend that a large number and diversity of opinions be collected. For criticality, where a strategic context is a prerequisite for making judgments, we prefer a process where the Working Team makes initial assessments. In its meeting, the Working Team needs to consider the options, and to formulate recommendations to the Steering Group.

### Module 3—Assessing Capabilities

The output of this module is the final inventory, comprising an exhaustive set of capabilities and assessments and, most importantly, the identification of the critical technological capabilities of the company.

The first step is a meeting of the Steering Group to respond to the Working Team's recommendations and provide guidance. Further communication to the organization may be advisable at this stage, since participation in the process is about to escalate.

In our preferred process, *strength assessment* is the focus of initial activity. It starts with team members conducting individual face-to face interviews with carefully selected respondents to gather initial data and validate the integrity of the inventory. The Working Team meets to review early results and to discuss the alternative approaches for completing the collection of strength assessments. Use of a questionnaire can be both efficient and effective; a draft needs to be produced, tested through a few additional interviews, finalized in a further Working Team meeting, and then circulated to respondents.

The results obtained from all the work streams are tabulated and reviewed by the team for gaps, and, if necessary, additional interviews or questionnaires are used to complete the data set. The team must then make judgments on what scores to assign to each capability. Statistical techniques can be helpful in analyzing the responses but, ultimately, *the team must decide on the assessments* and be prepared to justify them. The team's findings are then reviewed with the Steering Group.

Next, the Working Team should hold a separate workshop to assess inventory items for *criticality*. Working with the definitions and measurement scale previously agreed on, the team needs to examine each technological capability for its impact on competitiveness and make objective and consistent assessments. In preparation for the workshop, team members should read and assimilate the company's current strategic plan; it may also be appropriate to involve additional business or marketing staff in the workshop. Results are rigorously challenged and modified, if necessary, in a session with the Steering Group.

Working in this way will avoid two common potential pitfalls: first, when a long listing of technological capabilities has been assembled and the CTC identification program stalls because there is no established way of selecting the critical capabilities; second, when someone in the organization asserts, "I know we have" a particular core competency but cannot comprehensively and convincingly describe the technologies and work activities that make up that competency.

At the end of Module 3, some people will be tempted to decide on changes based on the information that is already available. Business unit managers have a particular propensity for doing this; they will find from the analysis that some of the critical technological capabilities that reside in their units need to be improved. Their immediate technology investment decisions may not be in the best interests of the company's CTCs, however, and *the program must be allowed to be completed as planned.*

### Module 4—Identifying Candidate Competencies

The challenge for the Working Team in Module 4 is to use the insight and data generated in Modules 2 and 3 to identify the company's candidate CTCs. The result should be a long list of relevant and promising areas of technological expertise that can be tested and evaluated as CTCs.

The most productive approach is to focus on the identified critical capabilities. Different clusters of critical capabilities need to be examined, with the objective of finding those that might have the attributes of a CTC. Mapping critical capabilities according to strength is helpful in guiding analysis. Adopting and applying additional assessment parameters for the critical capabilities (for example, the level of maturity of the capability) can also help. Figure 5 illustrates two useful mapping tools. Note, however, that identifying CTC aggregates always requires some *creativity* and can never be an entirely analytical process; it helps to forget about current products and to think about future business possibilities.

The initial analysis is carried out by the Program Manager, with support as required. It should not be done too aggressively, as the priority at this stage is to identify as many candidate CTCs as possible. Working CTC descriptors need to be formulated for each of the identified clusters of capabilities. Several descriptor variants may be needed for some of the clusters. It is important to take the time to get the wording right.

The Program Manager presents the results of the work at a Working Team workshop. The session is used to review, consolidate and refine the working CTC descriptors. The output is a list of distinct candidate CTCs, expressed in clear and meaningful language, that will be the basis for further work.

### Module 5—Testing Candidate Core Competencies

The objective of this module is to determine which of the identified candidate CTCs meet core competency criteria. The output is the set of valid CTCs that are available to the company based on its current critical capabilities.

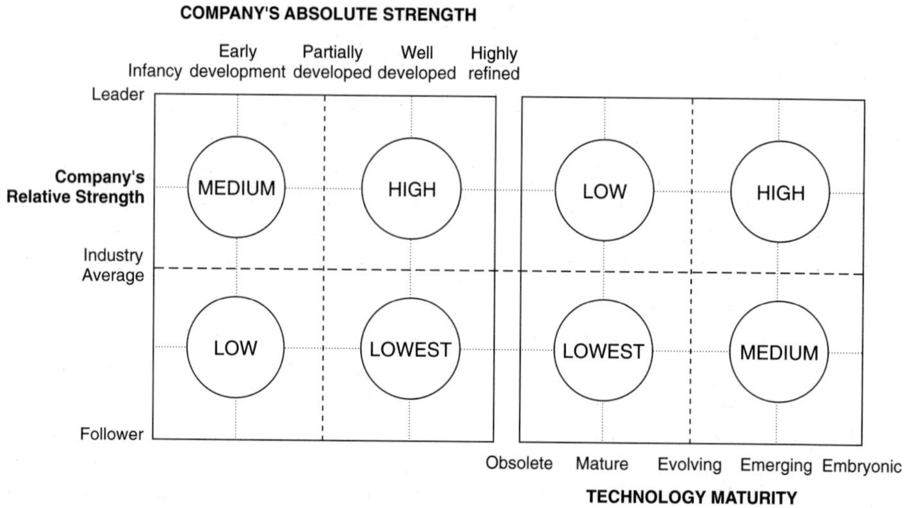


- These two maps are both populated with all of the company's critical technological capabilities
- Clusters of capabilities are examined as prospective CTCs
- Clusters centred in the quadrants marked HIGH have the greatest probability of yielding valid existing CTCs

**FIGURE 5**
Clustering critical capabilities helps to identify core technical competencies.

The work is an extension of the analysis conducted in Module 4 and should be carried out without delay; a Steering Group review of the results of Module 4 is often not necessary. Each candidate CTC is examined by the Working Team for compliance with the remaining tests of core competency; that is, for customer-perceived value, difficulty of imitation and extendability to new markets. Simple yes/no/don't know answers will suffice, but decisions need to be agreed on and documented. If necessary, areas of uncertainty should be assigned to team members to resolve quickly and the team should meet a second time to reach final decisions.

The list of valid CTCs is highly significant as it equates to *all the possible existing and potential CTCs in the company.* It should be thoroughly reviewed with the Steering Group before continuing further. The Steering Group should also guide the Program Manager on which (if any) CTCs should receive preferential treatment or be excluded during Module 6.

### Module 6—Evaluating Core Competency Position

The final module of work determines whether the company has ownership of any of its valid CTCs. It clarifies the company's relative position in each CTC, the best external organizations to consider as competency benchmarks, and the improvement that would be needed to secure ownership of the company's potential CTCs.

To evaluate competency position from an internal perspective, the Working Team reexamines the aggregates of capabilities that make up each valid CTC, refines the set if necessary, and compiles the relevant assessments of capability strength. The result of this activity is a series of easily interpreted competency "balance sheets" that highlight overall status and any capability gaps that may exist.

Dealing with the external perspective is more difficult, but should not be avoided. It requires market research to establish which companies are perceived to be the leaders in each valid CTC, which capability strengths explain leadership positions, and how the organization itself is perceived.

Research involves interviews with selected respondents. Opinions need to be elicited from major customers, suppliers and competitors, and several independent industry analysts and observers should also be included. The work may either be carried out directly by the Working Team or by a well-briefed and competent external firm. Either way, opinions must be well-documented.

The Program Manager needs to bring together the findings from both streams of work. The Working Team then discusses the results and forms a view of which CTCs are owned by the organization, which are owned by other companies and which are currently "ownerless." Conclusions are presented to, and challenged by, the Steering Group.

With knowledge of the organization's existing and potential CTCs, the Steering Group focuses on:

- Beginning to evaluate the strategic value of identified existing CTCs.
- Forming an initial view on which potential CTCs might be strategic.
- Drawing conclusions on capability gaps and deciding on immediate improvement priorities for critical capabilities.
- Deciding where, when, and how to expand the scope of the core competency process.

## IN CONCLUSION

Organizations fortunate enough to have an existing core technical competency without knowing it are missing a big opportunity to create value. Conversely, companies dependent on technology that do not have CTCs should ask themselves *how else they expect to remain competitive over the long term?*

The value of knowing your existing core technical competencies and the areas of technical prowess that could potentially become CTCs is enormous: It can be a sound basis for formulating a technology strategy for the organization as a whole.

### Reading 8.2 References


1 Prahalad, C. K; and Hamel, Gary. "The Core Competence of the Corporation." *Harvard Business Review,* May–June 1990, pp. 79–91.

2 Przybylowicz, Edward, P; and Faulkner, Terrence, W. "Kodak Applies Strategic Intent to the Management of Technology." *Research Technology Management,* January–February 1993, pp. 31–38.

3 Chester, Arthur, N. "Aligning Technology with Business Strategy." *Research Technology Management,* January–February 1994, pp. 25–32.

## DISCUSSION QUESTIONS

1 Analyze the strategies of two successful competitors in an industry. What are the differences? Do they have common elements? Use as many of the tools described in this chapter as you can.
2 Select any company or business whose operation you are familiar with. Develop a strengths, weaknesses, opportunities, threats (SWOT) table for it. Discuss your table in class or with your colleagues.
3 Analyze the strategies of two competitors in an industry, one successful and the other not. What made the difference in their competitiveness levels? Use as many of the tools described in this chapter as you can.

## ADDITIONAL READINGS

Marco Iansiti & Jonathan West. "Technology Integration: Turning Great Research into Great Products." *Harvard Business Review,* May–June 1997.

Reaching synergy among different technologies must be the aim of modern corporations. Since no company today masters every relevant discipline, integration involves both R&D and technology transfer. The authors analyze the case of 30 computer-related companies in Korea, Japan, and the United States.

Michael Porter. "What Is Strategy?" *Harvard Business Review,* November–December 1996.

James C. Collins and Jerry I. Porras. "Building Your Company's Vision." *Harvard Business Review,* September–October 1996.

The first step toward strategy should be a clear vision. Collins and Porras propose a framework consisting of two parts: core ideologies and envisioned future. They link corporate values and purpose with future opportunities.

Karen Patten. "Managing Migration to ISDN: Integrating Strategic, Technical, and Customer Service Perspectives." In T. Khalil & B. Bayraktar (eds.), *Management of Technology III, Proceedings of the Third International Conference on Management of Technology,* vol. I, pp. 243–252. Industrial Engineering and Management Press, Norcross, GA, 1992.

Technology transitions are one of the most difficult tasks managers encounter. Patten (with AT&T Bell Laboratories) describes the use of a new technology implementation model to help achieve integrated management. She analyzes the effect that migration to a new communications technology (ISDN) has on planning, design and installation, ongoing operation, administration, and maintenance.

William B. Werther Jr. "Strategy-Driven Technology in International Competition." In Mueller, Persson, and Lumsden (eds.), *Proceedings of the Sixth International Conference on Management of Technology,* 1997, pp. 13–24.

Technology maturity is an issue that may be disregarded by managers in technology-driven firms, forcing them to become followers or even losers. Werther (professor of management, University of Miami) argues that "strategy formulation for technology-driven firms must embrace both the technological and contextual issues in order to achieve successful technology management." He provides a framework for identifying likely transition points that differentiate technology-driven from strategy-driven approaches to technology management.

C. K. Prahalad & G. Hammel. "The Core Competencies of the Corporation." *Harvard Business Review,* May–June 1990, pp. 79–91.

These two authors have contributed several papers to the HBR. This particular one proposes that a business strategy be built not around products but around competencies. What's a competency? The authors claim competencies provide advantage over competitors, are difficult to imitate, and provide access to a wide variety of markets.

Gideon De Wet. "Corporate Strategy and Technology Management: Creating the Interface." In R. Mason, L. Lefebvre, and T. Khalil (eds.), *Management of Technology V,* pp. 510–518. Elsevier, Oxford, U.K., 1996.

## SUGGESTED CASES

- "New Product Development at Canon: The Contact Sensor Project." Harvard Business School, Case 9-396-247.
- "VeriFone 1997." Harvard Business School, Case 9-398-030.

## REFERENCES


Christensen, Roland, Berg, Norman, & Salter, Malcom. 1976. *Policy Formulation and Administration.* Irwin, Homewood, IL.

David, Fred. 1997. *Strategic Management,* 6th ed. Prentice-Hall, Upper Saddle River, NJ.

De Wet, Gideon. 1996. "Corporate Strategy and Technology Management: Creating the Interface." In Mason, R., Lefebvre, L., & Khalil, T. (eds.), *Management of Technology V.* Elsevier, Oxford, U.K.

Drucker, Peter. 1974. *Management: Tasks, Responsibility, and Practices.* Harper & Row, New York.

Escobar, Camilo. 1997. *MOT and the Alignment of Technology Strategy with Business Strategy.* Internal report, University of Miami.

Fine, C. H., & St. Clair, R. 1996. "Meeting the Challenge: U.S. Industry Faces the 21st Century—The Automobile Manufacturing Industry." U.S. Department of Commerce, Office of Technology Policy, Washington, DC.

Ford, David. 1988. "Develop Your Technology Strategy." *Long-Range Planning,* October, pp. 85–94.

Frohman, Alan L. 1982. "Technology as a Competitive Weapon." *Harvard Business Review,* January–February, pp. 97–104.

Fusfeld, A. 1978. "How to Put Technology into Corporate Planning." *Technology Review,* May.

Hamel, Gary. 1996. "Strategy as Revolution." *Harvard Business Review,* July–August, pp. 69–82.

Holt, K. 1990. "Technology Strategy: Is There a Need for It?" In Khalil, T., & Bayraktar, B. (eds.), *Management of Technology II.* Industrial Engineering and Management Press, Norcross, GA.

Holt, K. 1992. "The M-T Matrix. A New Strategic Tool." In Khalil, T., Bayraktar, B., & Edosomwan, J. (eds.), *Management of Technology III.* Industrial Engineering and Management Press, Norcross, GA.

Hudiburg, J. 1991. *Winning with Quality: The FPL Story.* Quality Resources, A Division of the Krauss Organization, White Plains, NY.

Machado, F. M. 1997. *Technology Management for Leap-Frogging Industrial Development: The Challenge for Developing Countries at the Beginning of the New Millenium.* UNIDO, Vienna.

McConnell, Scott W., & Khalil, Tarek M. 1988. "Evaluation of New Technology: A Methodology and Case Study." In Khalil T., Bayraktar, B., & Edosomwan, J. (eds.), *Technology Management,* Inderscience Enterprises, Geneva, Switzerland, pp. 727–736.

Mitchell, G. R. 1985. "A New Approach for the Strategic Management of Technology." *Technology in Society,* vol. 7, pp. 227–239.

Mitchell, G. R. 1988. "Options for the Strategic Management of Technology." In Khalil, T., Bayraktar, B., & Edosomwan, J. (eds.), *Technology Management I,* Interscience, Geneva.

Mitchell, G. R. 1992. "The Changing Agenda for Research Management." *Research Technology Management,* September–October.

Mitchell, G. R. 1995. "Technology—Business Strategy—Government Policy." Lecture notes, University of Miami, March 3–4.

Porter, Michael E. 1980. *Competitive Strategy: Techniques for Analyzing Industries and Competitors,* Free Press, New York.

Porter, Michael E. 1985. *Competitive Advantage.* Free Press, New York.

Prahalad, C. K., & Hamel, G. 1990. "The Core Competence of the Corporation." *Harvard Business Review,* May–June, pp. 79–91.

Ramo, J. C. 1996. "Winner Take All." *Time,* Sept. 16.

Teece, David J. 1987. "Capturing Value from Technological Innovation: Integration, Strategic Partnering, and Licensing Decisions." In Guile, B., & Brooks, H. (eds.), *Technology and Global Industry: Companies and Nations in the World Economy.* National Academy Press, Washington, DC.

Van Wyk, R., & Sweatt, W. 1994. "The Corporate Board and MOT: From RR to EE."In Khalil, T., & Bayraktar, B. (eds.), *Management of Technology IV.* Industrial Engineering and Management Press, Atlanta/Norcross, GA, pp. 207–213.

# 9

# TECHNOLOGY PLANNING

Technology planning is a central component of corporate business planning. It is needed both at the corporate level and at the strategic business unit (SBU) level (Steele, 1989). Large successful corporations such as GE, GTE, Motorola, and NEC view technology planning as vital to their ability to offer customers superior value based on superior technology. There is a difference between strategizing and planning, which Hamel (1996) considers the difference between discovering and programming. Strategizing should be creative and revolutionary, while planning is systematic and follows established methodologies. Whereas strategy determines the formula by which the firm intends to win, planning charts the procedures and actions to be followed. Planning is essential for successful strategy implementation and evaluation.

The process used in planning is in itself at least as important as the plan developed. The process includes:

- Examining all points of view in the organization.
- Setting clear, realistic objectives.
- Charting a path or paths toward achieving those objectives.
- Obtaining commitment for execution.
- Executing and following up on the plan.

Planning is a central managerial function leading to the other important management functions of organizing, staffing, motivating, and controlling activities of an organization. The strategic planning time horizon may vary according to the organization's objectives. Short-range plans of one to three years, midrange plans of three to five years, or long-range plans of more than five years are common in industry. Several models have been proposed for technology planning. Porter et al. (1991) proposed a technology planning framework based on the work of Madox, Anthony, and Wheatley (1987). This framework, shown in Exhibit 9-1, follows the general process of strategic planning used by many corporations. It entails forecasting the technology and the market to assess opportunities and needs, assessing the organization's strengths and weaknesses, developing and implementing a plan of action to achieve the organization's goals and fulfill its mission.

Figure 9-1 shows a model suggested by Martin (1994), in which technology planning involves top-down, bottom-up, and sideways participation. This approach calls for involving not only corporate or SBU managers, as may be deemed necessary by the structure of the organization, but also R&D, production, and marketing staffs, who should be aware of the state of the art in their respective areas of work. These technology gatekeepers can present valuable contributions from their own perspectives. The arrows

**EXHIBIT 9-1**
TECHNOLOGY PLANNING FRAMEWORK

1 *Forecast the technology:* This is the starting point of technology planning. Project both internally owned technology and that available in the marketplace over the planning period.

2 *Analyze and forecast the environment:* Identify key factors in the organization's environment, potential states of the environment, key uncertainties, major threats (especially competition) and opportunities.

3 *Analyze and forecast the market/user:* The development of a requirements analysis that identifies the current needs of major customers determines the likelihood that these needs will change and specifies explicit demands that these needs make on the organization's products or services. The tools of market research and impact assessment will complement each other. However, analytical tools, no matter their sophistication, will never be adequate. It is imperative that this step includes direct contact with potential customers. Real quality is the fulfillment of customer requirements and desires (Crosby, 1979), and the best way to know those is to get closer to them.

4 *Analyze the organization:* Delineate the major assets and problems; develop a catalog of available human and material resources; and assess recent performance against stated objectives. Understanding the strengths and weaknesses of your organization is critical and cannot be overemphasized. This may be a great time to involve external consultants to avoid the errors that arise when members of an organization assess themselves.

5 *Develop the mission:* Specify critical assumptions; establish overall organizational objectives and specific target objectives for the planning period; and specify criteria by which to measure the attainment of those objectives. This step will provide the central focus of the organization and should include as many participants as possible. Organizations have a much better chance for success when each member understands and feels a sense of ownership in the mission.

6 *Design organizational actions:* Create candidate actions; analyze and debate them; develop a consensus strategy limited to a few key actions, possibly attendant on several key contingencies. This is another excellent time to apply the tools of impact assessment.

7 *Put the plan into operation:* Develop timely sub-objectives, if appropriate; specify action steps, schedule, and budget; develop tracking mechanisms; and specify control mechanisms in case performance falls below established standards. During this step, monitoring can be very useful. Technological marketplaces are dynamic, and each firm must maintain a knowledge base of changes and customer reactions to them.

*Source:* A. Porter et al., *Forecasting and Management of Technology.* © 1991, John Wiley and Sons Inc. Reprinted by permission of John Wiley and Sons Inc.

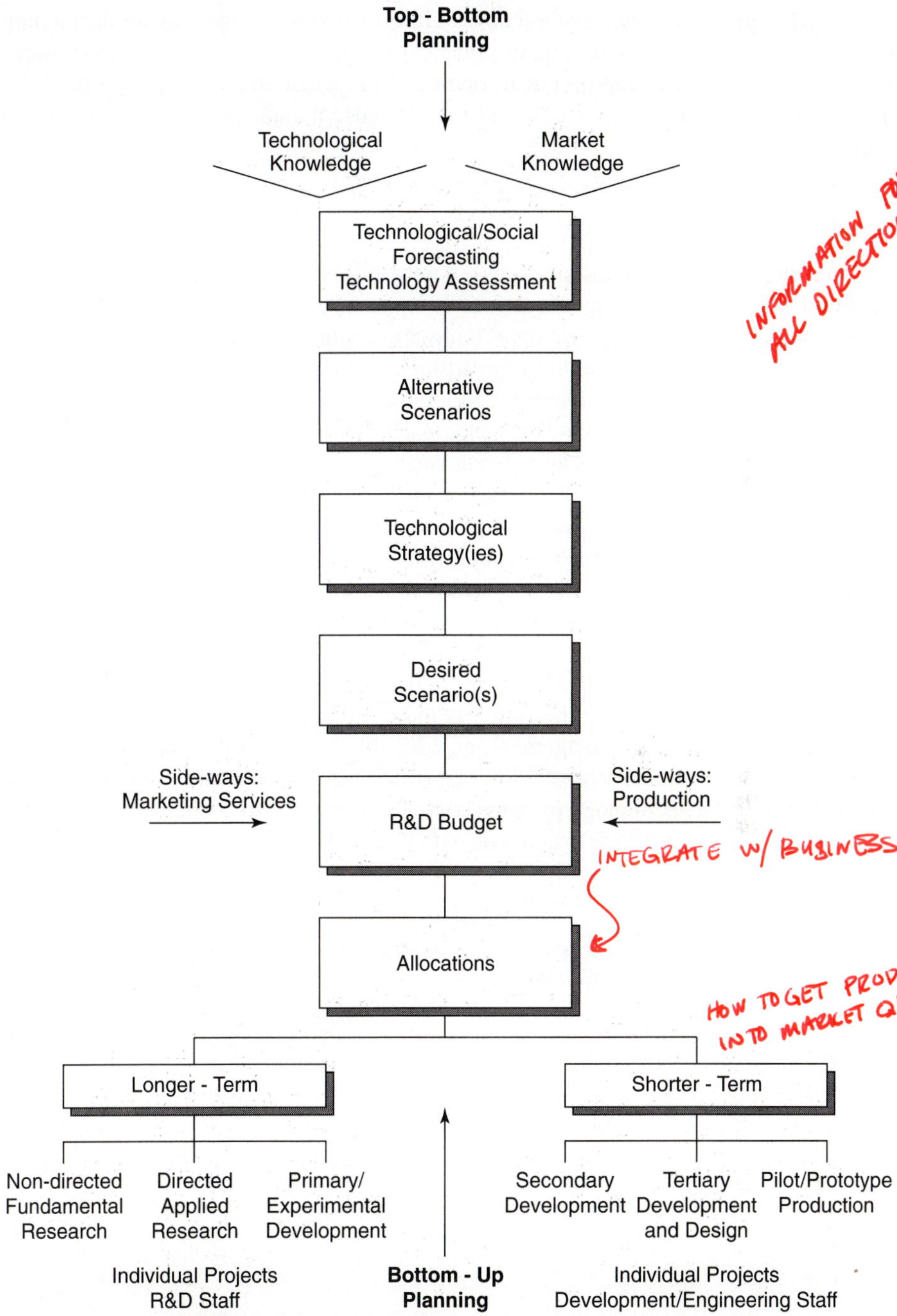


**FIGURE 9-1**
TECHNOLOGY PLANNING

*Source:* M. J. C. Martin, *Managing Innovation and Entrepreneurship in Technology-Based Firms.* © 1994, John Wiley and Sons Inc. Reprinted by permission of John Wiley and Sons Inc.

shown in Figure 9-1 indicate the directions of information flow necessary for developing optimal plans. The outcome is scenarios that identify the technologies needed to achieve business objectives. The appropriate technologies to include in the company's portfolio are selected, after which resources are allocated to ensure implementation of the desired scenarios.

## FORECASTING TECHNOLOGY

The first step in technological planning is forecasting. Forecasting provides visions of the future that can be used to guide actions of the present in anticipation of future states. Those who forecast well can seize opportunities in a timely manner and thus reap the rewards of future changes. *Technology forecasting (TF)* is based on following established methodologies to forecast the character and role of technological advancement.

Traditional forecasting methods depend, to a great extent, on projecting past performance into the future. This has an inherent weakness in that the future may not behave like the past. Figure 9-2 shows three extrapolations of the possible future growth pattern of a technology. A future state depends on the characteristics and physical limits of the technology, the social and environmental factors influencing its development, and market conditions compared to those of its competitors. For example, technology forecasters could have predicted that technology of nuclear power plants would follow the pattern of the top S-curve in Figure 9-2. Environmental concerns and market conditions have forced a change in the growth pattern of that technology.

The problem of predicting the future is more difficult with technology that is experiencing rapid change. Management must be able to predict discontinuities, which occur when one technology threatens to replace another. An example of this is given in Figure 9-3. $S_1$ is the technology progress curve of technology 1. A company committed to that technology may decide to continue with it even though a replacement technology,

**FIGURE 9-2**
GROWTH PATTERN AND POSSIBLE FUTURE STATE OF A TECHNOLOGY

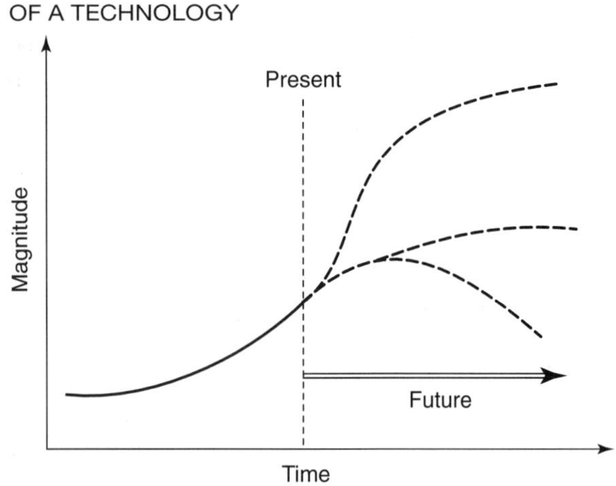

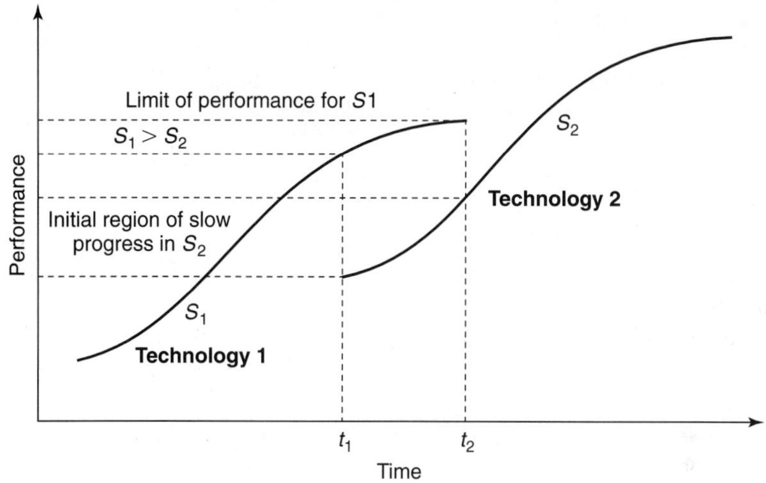


**FIGURE 9-3**
TECHNOLOGY DISCONTINUITY
Leaders became losers if they fail to recognize discontinuity and deal with technology's diminishing returns.

technology 2, designated by the technology progress curve $S_2$, might be looming on the horizon. A competitor using the second technology which has a superior performance parameter, even though starting late, at time $t_1$, will progress on an entirely new path, and its technology will outperform that of the first company. In this case, the first company's strategy to protect technology 1 will be futile in the long run, and management must make the decision to migrate to new technology 2 in a timely manner. There are many historical technological discontinuities—the steamship replacing the sailing ship, the transistor replacing the vacuum tube in electronics, and the personal computer replacing the typewriter.

Figure 9-4 shows S-curves of a series of transportation technologies that contributed to the improvement of transportation speed. Each technology follows its own S-curve to its natural limit of performance. The upper envelope of the set of progress curves of the different modes of transportation technology form a cumulative overall progress-curve pattern for transportation speed. Innovation pushes the envelope of the technology by defining a new natural limit for transportation speed. In the meantime, each emerging technology threatens an older one and may render it obsolete. For example, trains and automobiles replaced the Pony Express.

A technology forecaster predicts the natural performance limit of each mode of transportation technology. Technology managers must use this information to guide decisions about when to start using or abandon a technology before the competition renders it obsolete. They can also use the information to make decisions about investing in a new technology, R&D, or new products or about acquiring a new company that has the technological capability of the emerging mode of technology.

Figure 9-5 is a plot of the technology S-curves of incandescent lamps and fluorescent lamps, showing the rate of change of the performance parameter in lumens per watt over

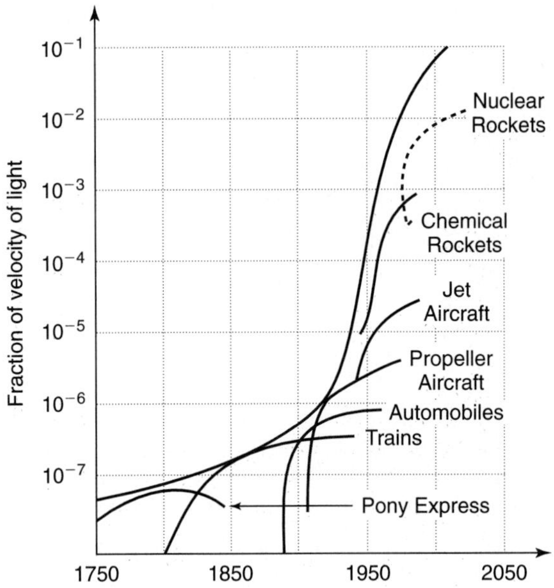


**FIGURE 9-4**
S-CURVES OF TRANSPORTATION SPEED
The cumulative curve is used to forecast maximum transportation speed.

*Source:* S. Millet and E. Honton, 1991. Battelle Memorial Institute Press. Reprinted by permission.

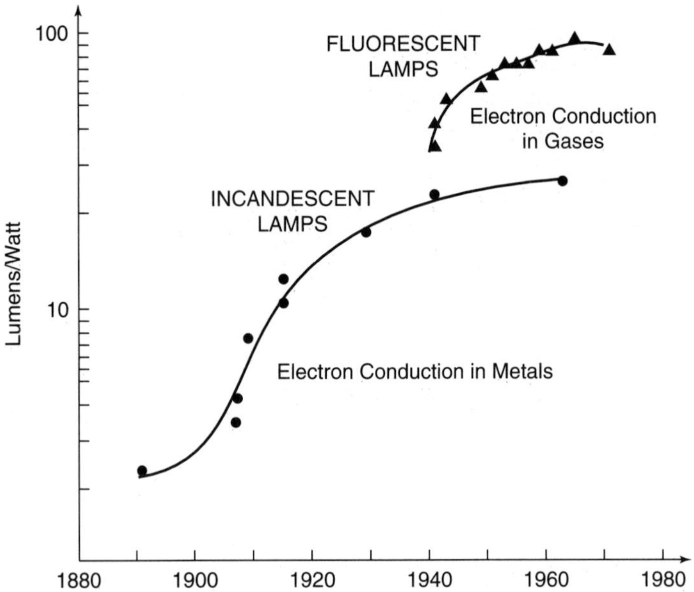


**FIGURE 9-5**
TECHNOLOGY S-CURVES FOR PROGRESS IN LAMP TECHNOLOGY

*Source:* F. Betz, 1996, in G. Gaynor, *Handbook of Technology Management.* Reprinted with permission of the McGraw-Hill Companies.

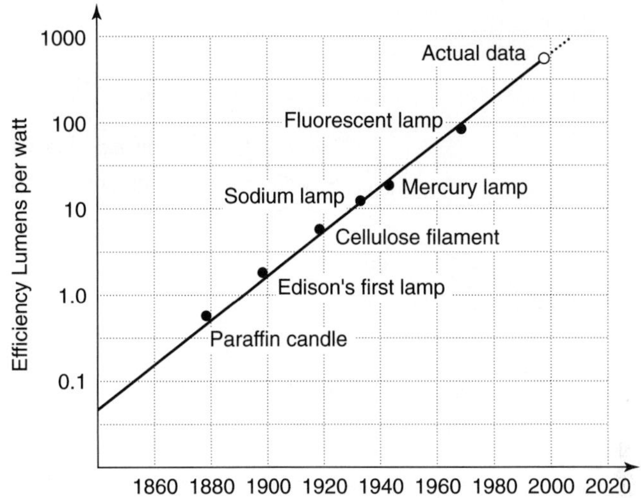


**FIGURE 9-6**
TREND EXTRAPOLATION OF EFFICIENCY OF WHITE LIGHT
*Source:* S. Millet and E. Honton, 1991. Battelle Memorial Institute Press. Reprinted by permission.

time. Figure 9-6 shows historical data for the efficiency of produced white light. When plotted on a logarithmic scale, the efficiency appears to be improving linearly. A forecaster may use this consistent trend to extrapolate the efficiency of white light into the future. While this might be a reasonable assumption to make, there is no guarantee that the trend will indeed continue to hold in the future. The best that one can say in this case is that extrapolating the trend will give added information, permitting an educated guess and better decision making. Management by data is much more effective than management without adequate information. Figure 9-7 shows a technology trend chart forecasting changes in chip density. This chart is helpful in predicting what the characteristics of future products should be. It can provide managers with insight into the future and allow them to focus their research and advanced development (Willyard and McClees, 1987).

In order to develop a good forecast, a technology forecaster must have a good understanding of technology life cycles and the factors that influence technological development and the rate of innovation. It is important for technology managers to understand the inherent strengths and weaknesses of each forecasting technique. A good forecast must have:

1 Credibility and utility.
2 An accurate information base.
3 Clearly described methods and models.
4 Clearly defined and supported assumptions.
5 Quantitative expression whenever possible.
6 A stated level of confidence in the forecasted information.

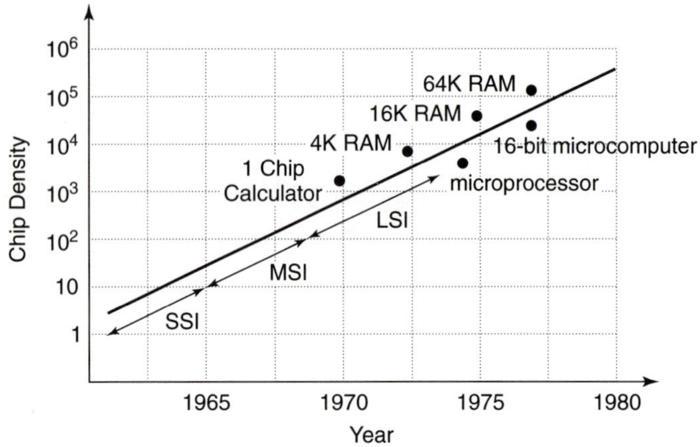


**FIGURE 9-7**
TECHNOLOGY TREND CHART FOR CHIP DENSITY
This type of diagram forecasts changes in the characteristics of future products.
*Source:* Willyard and McClees, 1987. Reprinted with permission.

Porter et al. (1991) indicate that the attributes of technology most often forecast are "(1) growth in functional capability, (2) rate of replacement of an old technology by a newer one, (3) market penetration, (4) diffusion, and (5) likelihood and timing of technological breakthroughs." They describe five methods of technology forecasting: (1) monitoring, (2) expert opinion, (3) trend analysis, (4) modeling, and (5) scenarios. Exhibit 9-2 provides a description of these methods, together with an analysis of their strengths, weaknesses, and uses.

**EXHIBIT 9-2**
FORECASTING METHODS

| | **Monitoring** |
|---|---|
| Description | Monitoring is the process of scanning the environment for information about the subject of a forecast. It is not really a forecasting technique, but rather a method for gathering and organizing information. The sources of information are identified and then information is gathered, filtered, and structured for use in forecasting. |
| Assumptions | The environment contains information useful for a forecast, and the information can somehow be obtained. |
| Strengths | Monitoring can provide large amounts of useful information from a wide range of sources. |
| Weaknesses | Information overload can result without selectivity, filtering, and structure. |
| Uses | To maintain current awareness of an area and the information with which to forecast as needed. To provide information useful for structuring a forecast and for the forecast itself. |

**EXHIBIT 9-2**
FORECASTING METHODS (*Continued*)

| | **Expert opinion** |
|---|---|
| Description | The opinions of experts in a particular area are obtained and analyzed. |
| Assumptions | Some individuals know significantly more about parts of the world than others thus their forecast will be substantially better. If multiple experts are used, group knowledge will be superior to that of an individual expert. |
| Strengths | Expert forecast can tap high-quality models internalized by experts who cannot or will not make them explicit. |
| Weaknesses | It is difficult to identify experts. Their forecasts are often wrong. Questions posed to them are often ambiguous and unclear, and the design of the process often is weak. If interaction among experts is allowed, the forecast may be affected by extraneous social and psychological factors. |
| Uses | To forecast when identifiable experts in an area exist and where data are lacking and modeling is difficult or impossible |
| | **Trend analysis** |
| Description | Trend analysis uses mathematical and statistical techniques to extend time series data into the future. Techniques for trend analysis vary in sophistication from simple curve fitting to Box-Jenkins techniques. |
| Assumptions | Past conditions and trends will continue in the future more or less unchanged. |
| Strengths | It offers a substantial, databased forecast of quantifiable parameters and is especially accurate over short time frames. |
| Weaknesses | It often requires a significant amount of good data to be effective, works only for quantifiable parameters, and is vulnerable to cataclysms and discontinuities. Forecast can be very misleading for long time frames. Trend analysis techniques do not explicitly address causal mechanisms. |
| Uses | To project quantifiable parameters and to analyze adoption and substitution of technologies. |
| | **Modeling** |
| Description | A model is a simplified representation of the structure and dynamics of some part of the *real* world. The dynamics of a model can be used to forecast the behavior of the system being modeled. Models range from flow diagrams, simple equations, and scale models, to sophisticated computer simulations. |
| Assumptions | The basic structure and processes of parts of the world can be captured by simplified representations. |
| Strengths | Models can exhibit the future behavior of complex systems simply by isolating important system aspects from unessential detail. Some models offer frameworks for incorporating human judgement. The model building process can provide excellent insight into complex system behavior for the modeler. |

(*Continued*)

**EXHIBIT 9-2**
FORECASTING METHODS (*Concluded*)

| | |
|---|---|
| Weaknesses | Sophisticated techniques may obscure faulty assumptions and provide a spurious credibility for poor forecasts. Models usually favor quantifiable over non-quantifiable parameters, thereby neglecting potentially important factors. Models that are not heavily databased may be misleading. |
| Uses | To reduce complex systems to manageable representations. |
| | **Scenarios** |
| Description | Scenarios are sets of snapshots of some aspect of the future and/or future histories leading from the present to the future. The scenario set encompasses the plausible range of possibilities for some aspect of the future. |
| Assumptions | The full richness of future possibilities can be reasonably incorporated in a set of imaginative descriptions. Usable forecasts can be constructed from a very narrow database. |
| Strengths | They can present rich, complex portraits of possible futures and incorporate a wide range of quantitative and qualitative information produced by other forecasting techniques. They are an effective way of communicating *forecasts* to a wide variety of users. |
| Weaknesses | They may be more fantasy than forecast, unless a firm basis in reality is maintained by the forecaster. |
| Uses | To integrate quantitative and qualitative information when both are critical, to integrate forecasts from various sources and techniques into a coherent picture, and to provide a forecast when data are too weak to use other techniques. They are most useful in forecasting and in communicating complex, highly uncertain situations to non-technical audiences. |

*Source:* A. Porter et al., *Forecasting and Management of Technology.* © 1991, John Wiley and Sons Inc. Reprinted with permission of John Wiley and Sons Inc.

## CRITICAL TECHNOLOGIES AND TECHNOLOGY MAPS

### National Critical Technologies

Planning for the future requires a deep understanding of the changes in the technological scene. It involves scanning the horizon for emerging critical technologies. This requires plotting technology maps that help planners navigate the landscape. The task of identifying future critical technologies and sifting through the maze of promising existing technologies must be undertaken both at the macro level of nations and at the micro-level of individual firms. The U.S. federal government has established a national critical technologies panel and charged it with identifying technologies that promise to have a major long-term impact on the nation's security and prosperity. In its first biennial report to the president, submitted in March 1991, William Phillips, panel chair, stated: "We most recently have been reminded, by the spectacular performance of U.S. and coalition forces in the Persian Gulf, of the crucial role that technology plays in military competitiveness. It is equally clear that technology plays a similar role in the economic competitiveness among nations." The U.S. National Critical Technologies panel has identified 22 technologies considered essential for the United States.

Several agencies of the federal government also produce lists describing technologies considered essential for their particular sectors. Exhibit 9-3 lists the national critical technologies, as well as the critical technologies developed by the Department of Defense and the emerging technologies developed by the Department of Commerce. A

**EXHIBIT 9-3**

NATIONAL CRITICAL TECHNOLOGIES, DEPT. OF COMMERCE EMERGING TECHNOLOGIES, AND DEPT. OF DEFENSE CRITICAL TECHNOLOGIES

| National Critical Technologies | Dept. of Commerce Emerging Technologies | Dept. of Defense Critical Technologies |
|---|---|---|
| | **Materials** | |
| • Materials synthesis and processing<br>• Electronic and photonic materials | • Advanced materials<br>• Advanced semiconductor devices<br>• Superconductors | • Composite materials<br>• Semiconductor materials and microelectronic circuits<br>• Superconductors |
| • Ceramics<br>• Composites<br>• High-performance metals and alloys | } Advanced materials | } Composite materials |
| | **Manufacturing** | |
| • Flexible computer-integrated manufacturing<br>• Intelligent processing equipment<br>• Micro and nanofabrication<br>• Systems management technologies | • Flexible computer-integrated manufacturing<br>• Artificial intelligence | • Machine intelligence and robotics |
| | **Information and Communications** | |
| • Software<br>• Microelectronics and optoelectronics | • High-performance computing<br>• Advanced semiconductor devices<br>• Optoelectronics | • Software producibility<br>• Semiconductor materials and microelectronics circuits<br>• Photonics |
| • High-performance computing and networking<br>• High-definition imaging and displays<br>• Sensors and signal processing | • High-performance computing<br>• Digital imaging<br>• Sensor technology | • Parallel computer architectures<br>• Data fusion<br>• Signal processing<br>• Passive sensors<br>• Sensitive radars<br>• Machine intelligence and robotics |
| • Data storage and peripherals<br>• Computer simulation and modeling | • High-density data storage<br>• High-performance computing | • Simulation and modeling<br>• Computational fluid dynamics |

(*Continued*)

**EXHIBIT 9-3**

NATIONAL CRITICAL TECHNOLOGIES, DEPT. OF COMMERCE EMERGING TECHNOLOGIES, AND DEPT. OF DEFENSE CRITICAL TECHNOLOGIES *(Continued)*

| National Critical Technologies | Dept. of Commerce Emerging Technologies | Dept. of Defense Critical Technologies |
|---|---|---|
| | **Biotechnology and Life Sciences** | |
| • Applied molecular biology<br>• Medical technology | • Biotechnology<br>• Medical devices and diagnostics | • Biotechnology materials and processes |
| | **Aeronautics and Surface Transportation** | |
| • Aeronautics<br>• Surface transportation technologies | | • Air-breathing propulsion |
| | **Energy and Environment** | |
| • Energy technologies<br>• Pollution minimization, remediation, and waste management | | |
| | | • No national critical technologies counterpart: high-energy density materials, hypervelocity projectiles, pulsed power, signature control, weapon system environment |

*Source:* National Critical Technologies Panel, 1991.

comparison of the lists indicates that there is much agreement among the various agencies on what constitutes emerging and critical technologies.

The technology included in these lists is determined by a set of criteria. For example, in developing the list of national critical technologies, the panel based its decisions on factors such as the nation's vulnerability in a specific technology and the pervasiveness of the technology. The technologies listed were thought to constitute appropriate bases for exploitation to satisfy many of the nation's needs (National Critical Technologies Panel, 1991). They are *generic* technologies, which means that they have the potential to be applied to a wide variety of products and processes extending across many industries. A generic technology usually requires subsequent research and development, generally by the private sector, to result in commercial applications. At the national level, support for generic technology provides a strategic advantage not only on military battlefields but also on the battlefield of the international marketplace.

### Critical Technologies at the Firm Level

At the firm level, technological gatekeepers, forecasters, and R&D managers may develop a map of technologies of potential relevance to their firm's products or services.

Betz (1987) proposed that the concepts of technology push and market pull be used to map the areas and direction of rapid technological change. He developed a technology map of the 1980s showing dominant areas of such change and classified the areas into six categories: (1) components, (2) devices, (3) processes, (4) systems, (5) materials and resources, and (6) services. Exhibit 9-4 shows Betz's technology map in the area of devices.

Technology maps enable planners to identify and focus on technologies that have the highest potential impact on their business. Porter et al. (1991) report on a different format

**EXHIBIT 9-4**
TECHNOLOGY MAP OF THE 1980s: DEVICES

| Device | Technology Push | Market Pull |
|---|---|---|
| Computers | 1. Supercomputer architecture: parallel processing<br>2. Computer peripherals: printing, memory, display<br>3. Computer graphics and three-dimensional display<br>4. Expert systems, software, and user friendliness | 1. Segmented computer markets: mainframes, minicomputers, microcomputers<br>2. Applications market: business and office systems, manufacturing systems, scientific systems, personal computers, home entertainment and information |
| Robots | 1. Manipulation and control<br>2. Sensing: vision, tactile<br>3. Flexible manufacturing: tools, materials, scheduling, handling<br>4. Production and sales system integration | 1. Automobile<br>2. Aerospace and defense<br>3. Electronics |
| Lasers | 1. Lasing techniques and materials: frequency and power<br>2. Laser tools | 1. Laser communications: transmission, fiber optics<br>2. Optical logic devices and circuitry<br>3. Holographic imaging and measurement<br>4. Laser tools<br>5. Laser weapons |
| Scientific Instrumentation | 1. Nuclear magnetic radiation (NMR) measurement and imaging<br>2. Synchrontron radiation<br>3. Milimiter, infrared, and ultraviolet radiation sensing and measurement<br>4. Automated instrumentation<br>5. Remote sensing<br>6. Computerized databanks and shared models<br>7. Automated testing | 1. University research<br>2. Aerospace and defense research<br>3. Chemical and petroleum industries<br>4. Medical and pharmaceutical industries<br>5. Electronic and computer industries |

*Source:* Adapted from Betz, 1987.

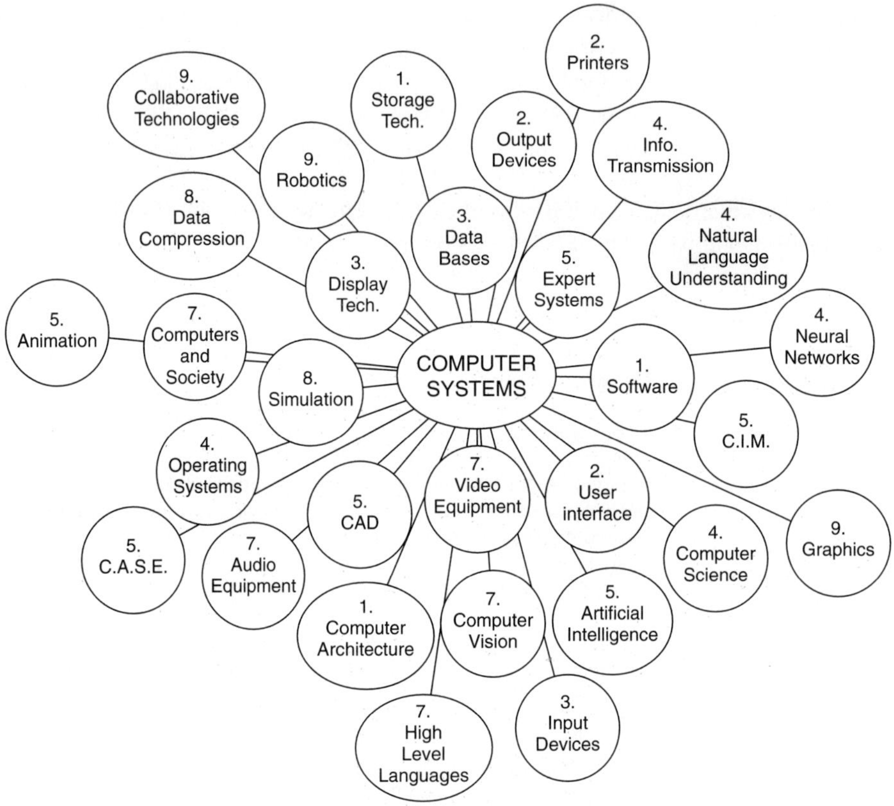


**FIGURE 9-8**
MAP OF TECHNOLOGIES RELATING TO COMPUTER SYSTEMS
*Source:* Prepared by Cunningham. From Porter et al., *Forecasting and Management of Technology.* © 1991 John Wiley and Sons Inc. Reprinted with permission of John Wiley and Sons Inc.

map of technologies related to computer systems, one introduced by Cunningham in 1990 (Figure 9-8). A ranking of importance is given to each one of the related technologies. The creation of this and similar maps can help forecasters and planners identify, monitor and track technologies influencing their own products or services.

## TECHNOLOGY AUDIT

Auditing is a tool used to evaluate the conditions or the existing status of a certain sector in an organization. Accountants often use it to evaluate the financial status of companies. The American Accounting Association defines it as "a systematic process of objectively obtaining and evaluating evidence regarding assertions about economic actions and events to ascertain the degree of correspondences between those assertions and established criteria, and communicating results to interested users." Financial audits are conducted yearly by companies, and the results are reported to the stockholders.

A *technology audit* is an analysis performed to identify the strengths and weaknesses of the technological assets of an organization. Its aim is to assess the firm's position in technology in relation to its competitors and the state of the art. This applies to technologies of the entire value-added functions in the firm, including product technology, production technology, service technology, and marketing technology. The objective is to develop a base upon which technology strategy and associated plans can be formulated. The technology audit is a continuous process of assessment, unlike some accounting audits, which typically are performed on a one-time basis or at a specific date, such as the end of a year.

According to Ford (1988), a technology audit should provide answers to the following questions:

**1** What are the technologies and know-how on which the business depends?

**2** How does the company's technology position compare to its competitors? Is it a leader, a follower, or a laggard?

**3** What is the life-cycle position on which the company depends?

**4** Where is the company's strength? Is it in product or production technologies or a combination of technologies?

**5** Is the company effectively protecting its distinctive core technologies?

**6** What emerging or developing technologies, inside or outside the company, could affect its technological position?

**7** What is the value of the company's technology to its customers? Is there a big technology gap that gives the company an advantage in knowledge as well as in pricing its products?

**8** Does the company have a systematic procedure and a supporting organizational structure that allows optimal exploitation of its technologies internally and externally?

**9** Does the company have technological assets that it can share with other companies? Some of the ideas that need to be explored include selling technology that is no longer of use to the company, creating joint ventures to exploit the company's areas of strength, and transfering technology to another company or country.

**10** What emerging or developing technologies, both inside and outside the company, could influence customers or affect the company's market position?

**11** What social, political, or environmental factors might impede the natural progress of the company's technological plans?

After the audit and assessment, a company can develop a statement of objectives that form the core of its strategy. It should then select an optimal strategy for the acquisition and exploitation of technology. An appropriate organizational structure and clear procedures are needed to enable the company to manage its technology in a way that will achieve a sustainable competitive advantage.

## TECHNOLOGY AUDIT MODEL

Garcia-Arreola (1996) developed a technology audit model (TAM) that includes important areas to be considered in a technology audit. The objectives of TAM are (1) to determine current technological status, (2) to stress areas of opportunity, and (3) to take advantage of the firm's strong capabilities. TAM is a three-level model, with each level going deeper into more specific functions. The upper level is composed of six *categories.* At the second level, 20 *assessment areas* exist. Finally, 43 *assessment elements* constitute the third level. (See Figure 9-9.) TAM assesses the company's position in technology. The model is based on the following six categories:

**1** *Technological environment:* Successful strategies are usually implemented in favorable environments that foster teamwork, creativity, and flexibility. Business environment factors to be examined include leadership, strategies adopted, organizational structure, technology culture, and human resource management.

**2** *Technologies categorization:* Business success is defined in part by the technologies a company utilizes. Critical technologies can be found in either products or processes. However, new technologies emerge continuously, threatening to render or rendering current ones obsolete. It is important to evaluate the company's level of knowledge and appreciation of its own technologies, state-of-the-art technologies, and emerging technologies. This is done across the value chain from upstream R&D to downstream activities of marketing and after-market services.

**3** *Markets and competitors:* A profound understanding of the environment in which the company competes is critical for technology management. Relationships among suppliers, distribution channels, customers, and competitors can change with the creation or adoption of new technologies. Business decisions in this area include pricing, the selection of distribution channels, product positioning, and so on.

**4** *Innovation process:* Transforming ideas into competitive advantage is not a result of luck. Innovation occurs under certain given conditions, which are available for most firms. The ability to bring an innovation to the market in the shortest possible time is as important as the innovation itself. Business decisions in this area include resource allocation, reward systems, product launching time, and so on.

**5** *Value-added functions:* Technology is brought to the market through a value-added chain—activities that add value to the final product, such as R&D, manufacturing, sales, and distribution. Performance evaluation of the functional areas and the entire system is critical. Quality and flexibility are necessary to satisfy current market demands. Evaluations of business decisions in this area include review of capital investments, the policy-making mechanism, the organizational structure, costing, methodologies, and so on.

**6** *Acquisition and exploitation of technology:* The effective adoption of technology requires that knowledge flow from source to receiver. Technology effectiveness depends on how successfully this process is implemented. Business decisions for acquisition and exploitation of technology determine the success of the organization. Important decisions include capital investment, the selection of alliance partners, and so on.

As these six areas indicate, a technology audit can become a very demanding and complex process. A checklist can help guide an auditor through the TAM process. A TAM checklist is shown in Exhibit 9-5.

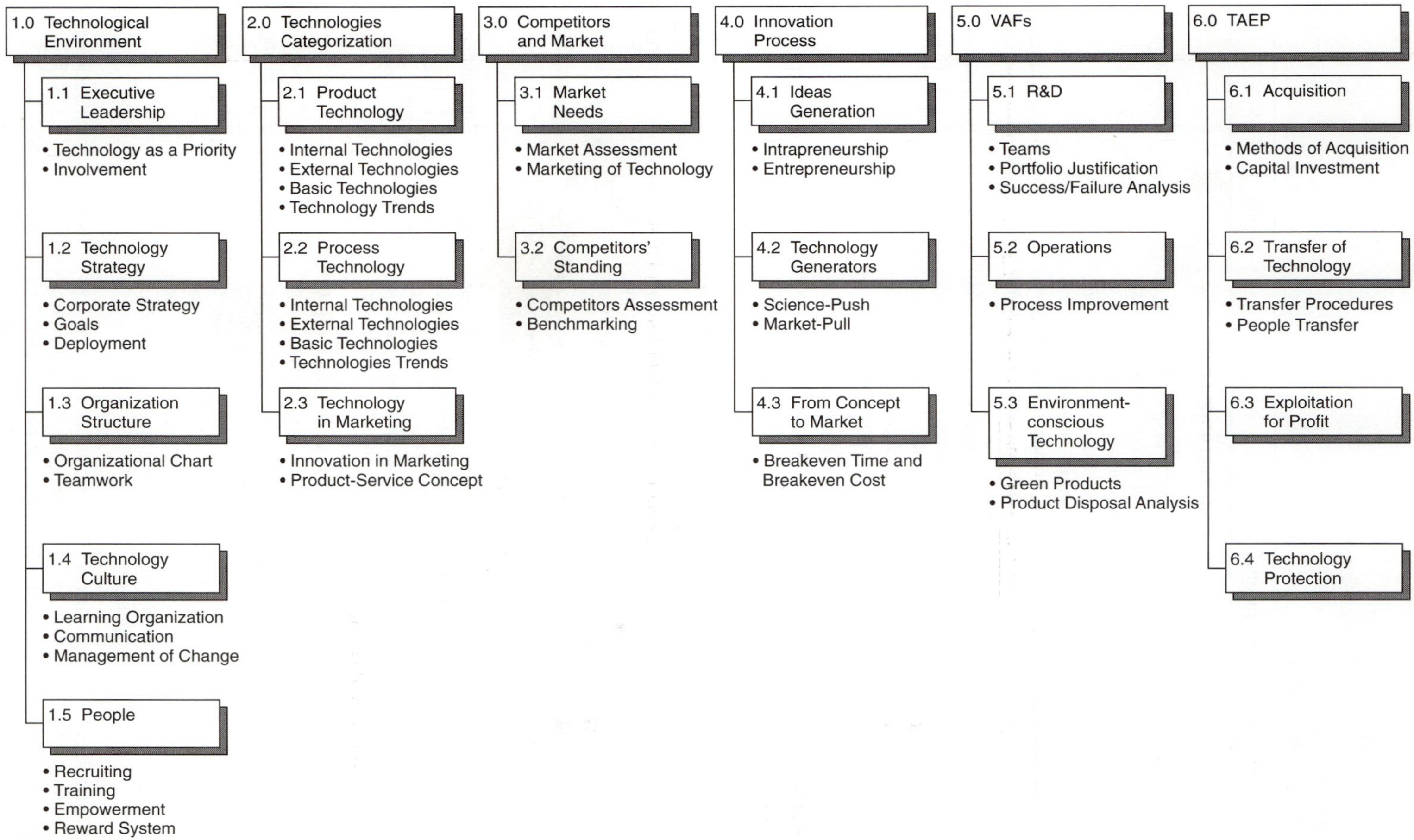


**FIGURE 9-9**
TAM STRUCTURE

**EXHIBIT 9-5**
TAM AUDIT CHECKLIST

| Assessment areas | Elements | Rating |
|---|---|---|
| | **1. Corporate environment** | |
| 1.1 Senior executive leadership and orientation | • *Technology as a top priority:* Technology is appreciated and managed as a key factor in the overall business strategy. There is a chief technology officer, whose judgment has a considerable influence in the decision-making process. The management style is consistent with the maturity of the enterprise. | Poor ☐1 ☐2 ☐3 ☐4 ☐5 Outstanding |
| | • *Involvement and participation:* Managers are active members of the technology culture within the corporation. They have close relationships with the chief technology officer and with technology gatekeepers. | Poor ☐1 ☐2 ☐3 ☐4 ☐5 Outstanding |
| 1.2 Technology strategy | • *Corporate strategy:* There exists a corporate strategy aimed to achieve the corporation's vision. One aspect of this strategy is aimed toward the technologies within the corporation. The technology strategy is a significant contributor to the corporate strategy. | Poor ☐1 ☐2 ☐3 ☐4 ☐5 Outstanding |
| | • *Goals:* There are specific goals directed at establishing technology standards and positioning the company as the industry leader. | Poor ☐1 ☐2 ☐3 ☐4 ☐5 Outstanding |
| | • *Deployment:* The technical strategy is effectively communicated and deployed throughout all levels in the organization. | Poor ☐1 ☐2 ☐3 ☐4 ☐5 Outstanding |
| 1.3 Organization structure | • *Organizational chart:* The organization has a structure that enables agility. It facilitates the decision-making process. Technology is explicitly represented by a chief officer, whose judgment influences the decision-making process. There exists evidence of organizational structure around technologies, not around products. | Poor ☐1 ☐2 ☐3 ☐4 ☐5 Outstanding |
| | • *Teamwork:* The roles and jobs are designed to facilitate teamwork. The teams are self-managed, with only occasional reviews from the manager. The teams can establish their own objectives and measures to support the overall technology strategy. | Poor ☐1 ☐2 ☐3 ☐4 ☐5 Outstanding |
| 1.4 Technology culture advancement | • *Culture:* There are values within the corporation that highlight the importance of technology as a strategic factor. The corporate culture supports and encourages technology. | Poor ☐1 ☐2 ☐3 ☐4 ☐5 Outstanding |

**EXHIBIT 9-5**
TAM AUDIT CHECKLIST (*Continued*)

| Assessment areas | Elements | Rating |
|---|---|---|
| | • *Learning organization:* The organization is skilled at creating, acquiring, and transferring knowledge, and at modifying its behavior to reflect new knowledge and insights. The organization has established methods for systematic problem solving, experimentation with new approaches, learning from its own experiences (both successes and failures) and most successful practices of others, and transferring knowledge quickly and efficiently throughout the organization. Lessons are documented and distributed throughout the organization. | Poor Outstanding ☐ ☐ ☐ ☐ ☐ 1 2 3 4 5 |
| | • *Communication:* There are no organizational barriers threatening the communication top-down, bottom-up, and horizontally. Ideas and concerns can be freely expressed. Information is made available to whoever might need it. The organizational structure is not a barrier when trying to communicate with top management levels. | Poor Outstanding ☐ ☐ ☐ ☐ ☐ 1 2 3 4 5 |
| | • *Management of change:* The organization is effective in dealing with change. People perceive change as an opportunity, rather than a threat. Teams can be easily reorganized to adapt quickly to new corporate needs. | Poor Outstanding ☐ ☐ ☐ ☐ ☐ 1 2 3 4 5 |
| 1.5 People (Employees are considered to be and treated as the company's most important assets. All levels of management uphold a committment to treating people with respect and fairness.) | • *Recruiting policies:* Human resources is in continuous contact with the operative departments to be aware of their needs regarding new employees. Candidates are identified and selected by taking into account their initiative, leadership, and technical skills. | Poor Outstanding ☐ ☐ ☐ ☐ ☐ 1 2 3 4 5 |
| | • *Training:* A process is in place to ensure that the employees are high-skilled, knowledge resources, customer-driven, trainers, and problem solvers. | Poor Outstanding ☐ ☐ ☐ ☐ ☐ 1 2 3 4 5 |
| | • *Empowerment:* Employees are empowered to take direct action when a problem occurs or an opportunity exists. Managers are perceived as facilitators. Data are accessible to the person/team that requires information. | Poor Outstanding ☐ ☐ ☐ ☐ ☐ 1 2 3 4 5 |
| | • *Reward system:* The reward system takes into account the different motivation factors for managers, engineers, scientists, and entrepreneurs, as well as the *flexible nature* of the organization. | Poor Outstanding ☐ ☐ ☐ ☐ ☐ 1 2 3 4 5 |

(*Continued*)

**EXHIBIT 9-5**
TAM AUDIT CHECKLIST (*Continued*)

| Assessment areas | Elements | Rating |
|---|---|---|
| | **2. Technologies categorization** | |
| 2.1 Service/product technologies | • *Internal technologies:* The corporation has clearly identified its core competencies and core services/products. Managers make sure that efforts are focused on strengthening and exploiting them. | Poor ☐ 1 ☐ 2 ☐ 3 ☐ 4 ☐ 5 Outstanding |
| | • *External technologies:* Technology gatekeepers have identified the external technologies included in the products, and make sure that none of them are of strategic importance. The system must be able to identify any important technology and develop it in-house before it becomes a competitiveness factor. There are established systems to forecast future developments. | Poor ☐ 1 ☐ 2 ☐ 3 ☐ 4 ☐ 5 Outstanding |
| | • *Basic technologies:* The basic technologies of the industry are clearly identified and maintained in good competitive position. There are established systems to forecast future developments. | Poor ☐ 1 ☐ 2 ☐ 3 ☐ 4 ☐ 5 Outstanding |
| | • *Technology trends:* Technology gatekeepers know the current standing and trends of the technologies behind the core competencies. There are established systems to forecast the future developments. | Poor ☐ 1 ☐ 2 ☐ 3 ☐ 4 ☐ 5 Outstanding |
| 2.2 Back office/process technologies | • *Internal technologies:* The organization values the development of process technologies as much as the development of product technologies. Managers make sure that efforts are focused on strengthening and exploiting them. | Poor ☐ 1 ☐ 2 ☐ 3 ☐ 4 ☐ 5 Outstanding |
| | • *External technologies:* Technology gatekeepers have identified the external technologies included in the processes. They make sure that the latest developments are included in the processes. There are established systems to forecast future developments. | Poor ☐ 1 ☐ 2 ☐ 3 ☐ 4 ☐ 5 Outstanding |
| | • *Basic technologies assessment:* The basic technologies of the industry are clearly identified and maintained in good competitive position. There are established systems to forecast future developments. | Poor ☐ 1 ☐ 2 ☐ 3 ☐ 4 ☐ 5 Outstanding |
| | • *Technology trends:* Technology gatekeepers know the current standing and trends of the key process technologies that support the manufacturing process of the core products. There are established systems to forecast future developments. | Poor ☐ 1 ☐ 2 ☐ 3 ☐ 4 ☐ 5 Outstanding |

**EXHIBIT 9-5**
TAM AUDIT CHECKLIST (*Continued*)

| Assessment areas | Elements | Rating |
|---|---|---|
| 2.3 Technology in marketing | • *Innovation in marketing:* The company develops sound and aggressive marketing plans to better capitalize on the characteristics of the products, making them more accessible to customers. | Poor ☐ 1 ☐ 2 ☐ 3 ☐ 4 ☐ 5 Outstanding |
| | • *The product-service concept:* The company is able to identify the service customers require from the products and to look for alternative ways to satisfy that need. Products are customized solutions. The boundary between product and service becomes less obvious. | Poor ☐ 1 ☐ 2 ☐ 3 ☐ 4 ☐ 5 Outstanding |
| | **3. Markets and competitors** | |
| 3.1 Market needs | • *Market assessment system:* There are systems which effectively identify the market's needs and its future possible trends. This information is available to R&D leaders, and people within the organization are encouraged to understand it. Market trends are included in the overall corporate strategy. Technology gatekeepers are active participants in this process. | Poor ☐ 1 ☐ 2 ☐ 3 ☐ 4 ☐ 5 Outstanding |
| | • *Marketing of technology:* The marketing department has developed systems to exploit not only products but technologies. Plans must be consistent with exploitation policies and with the overall technology strategy. | Poor ☐ 1 ☐ 2 ☐ 3 ☐ 4 ☐ 5 Outstanding |
| 3.2 Competitors' status | • *Competitor assessment:* Cross-functional teams are in charge of periodically assessing the core competencies, technological status, and possible future capabilities of competitors. | Poor ☐ 1 ☐ 2 ☐ 3 ☐ 4 ☐ 5 Outstanding |
| | • *Benchmarking:* The company periodically looks for the best practices related with its business, wherever they can be found. Internal processes and policies are compared with the benchmarks, and plans are developed to reduce the gaps. | Poor ☐ 1 ☐ 2 ☐ 3 ☐ 4 ☐ 5 Outstanding |
| | **4. Innovation process** | |
| 4.1 Idea generation | • *Intrapreneurship:* Policies exist to permit innovation at all organizational levels. Employees are encouraged to suggest new ideas for products, services, or processes. Reward systems are in place to motivate innovation within the company. Employees know the market needs and build on them | Poor ☐ 1 ☐ 2 ☐ 3 ☐ 4 ☐ 5 Outstanding |

(*Continued*)

**EXHIBIT 9-5**
TAM AUDIT CHECKLIST (*Continued*)

| Assessment areas | Elements | Rating |
|---|---|---|
| | in order to create new products or services. There exists a system that enables intrapreneurs to communicate and develop new ideas. | |
| | • *Entrepreneurship:* Entrepreneurs are motivated to develop their ideas within the organization if the ideas are consistent with the strategy. Otherwise, the system allows the entrepreneur to go elsewhere to develop the idea. | Poor ☐1 ☐2 ☐3 ☐4 ☐5 Outstanding |
| 4.2 Technology generators | • *Science push:* Technology gatekeepers have the resources to be experts within their fields and are empowered to suggest new directions and trends. They are aware of the latest scientific discoveries within their specific fields. | Poor ☐1 ☐2 ☐3 ☐4 ☐5 Outstanding |
| | • *Market pull:* Marketing is able to relate current products to market needs, identifying gaps and opportunities. The information regarding market needs is available to all interested persons/teams. | Poor ☐1 ☐2 ☐3 ☐4 ☐5 Outstanding |
| 4.3 From concept to market | • *Break-even time and break-even cost:* There is evidence of continuous improvement on the time-to-market variable. The teams are able to provide follow-up on their expenses throughout the entire time-to-market period. | Poor ☐1 ☐2 ☐3 ☐4 ☐5 Outstanding |
| | **5. Value-added functions** | |
| 5.1 R&D | • *Cross-functional teams:* Cross-functional and autonomous teams are used to plan, develop, and implement new products, processes, and/or services. Design for manufacturability is achieved through early involvement of all departments in the company. Every new venture has a *champion* leading the effort. | Poor ☐1 ☐2 ☐3 ☐4 ☐5 Outstanding |
| | • *Portfolio justification:* The R&D portfolio is fully consistent with the corporate and technology strategies, with the maturity of the industry, and with the core competencies of the corporation. There is a process to select new projects that will support the overall strategy and its congruency with technology priorities, acquisition, and exploitation. | Poor ☐1 ☐2 ☐3 ☐4 ☐5 Outstanding |

**EXHIBIT 9-5**
TAM AUDIT CHECKLIST (*Continued*)

| Assessment areas | Elements | Rating |
|---|---|---|
| | • *Success/failure analysis:* Projects are analyzed to identify and understand causes of success or of failure; learning is documented and distributed within the company. | Poor ☐1 ☐2 ☐3 ☐4 ☐5 Outstanding |
| 5.2 Operations | • *Improvement:* There are measures related to all the important variables of the processes. There is evidence of continuous improvement in those measures. The organization is able to reach economies of scale and economies of scope to satisfy market needs. | Poor ☐1 ☐2 ☐3 ☐4 ☐5 Outstanding |
| 5.3 Environment-conscious technology | • *Green products and processes:* The company is concerned about designing and producing environment-friendly products. The processes are equipped with filters or appropriate nonpollution devices. | Poor ☐1 ☐2 ☐3 ☐4 ☐5 Outstanding |
| | • *After-life analysis:* The design of the product takes into account the fact that the product will be discharged at the end of its lifetime; its recycling is already considered. | Poor ☐1 ☐2 ☐3 ☐4 ☐5 Outstanding |
| | **6. Acquisition and exploitation of technology** | |
| 6.1 Acquisition of technologies | • *Method of acquisition:* The technology acquisition options (internal R&D, joint ventures, licensed in, or purchase) support the technology strategy. The decisions are based on the life-cycle position of the specific technology. Decisions take into account factors such as the company's standing, urgency of acquisition, investment, life-cycle position, and technology category. | Poor ☐1 ☐2 ☐3 ☐4 ☐5 Outstanding |
| | • *Capital investment:* Capital appropriations are analyzed and approved based not only on financial statements but also on the competitive advantage they may create. | Poor ☐1 ☐2 ☐3 ☐4 ☐5 Outstanding |
| 6.2 Transfer of technology | • *Transfer procedures:* The company has transfer procedures, which allow it to successfully transfer technologies from other institutions, i.e., companies, laboratories, universities. | Poor ☐1 ☐2 ☐3 ☐4 ☐5 Outstanding |
| | • *People transfer:* When a new technology is acquired, people are also transferred to support the transfer process. | Poor ☐1 ☐2 ☐3 ☐4 ☐5 Outstanding |

(*Continued*)

**EXHIBIT 9-5**
TAM AUDIT CHECKLIST (*Concluded*)

| Assessment areas | Elements | Rating |
|---|---|---|
| 6.3 Exploitation for profit | • *Exploitation for profit:* Procedures exist to ensure the optimal exploitation of technologies, whether in product or processes, contracting out manufacturing, joint venture, or licensing out. The decisions are consistent with the overall technology strategy and the technology classification. | Poor Outstanding ☐ ☐ ☐ ☐ ☐ 1 2 3 4 5 |
| 6.4 Protection | • *Protection:* The innovation process is a closed loop requiring that the knowledge be protected either by patenting, secrecy, or other methods. | Poor Outstanding ☐ ☐ ☐ ☐ ☐ 1 2 3 4 5 |

*Source:* Based on Garcia-Arreola, 1996.

A technology auditor should do the following:

**1** Analyze the firm's internal technologies (products and processes) to identify core competencies.
**2** Identify external and basic technologies.
**3** Identify "technology gaps," that is, situations in which new technologies that must be acquired.
**4** Review the technology/science push and the market pull.
**5** Establish whether or not the innovation process takes into account science push and market pull.
**6** Check time to market. Identify constraints in the process.
**7** Review the R&D strategy. Is it consistent with science push and market pull?
**8** Check for consistency between core technologies, R&D, and marketing.
**9** Look for evidence of continuous improvement in manufacturing.
**10** Analyze partnerships and joint ventures. Are they in line with the overall strategy?
**11** Review the technology transfer procedures. How is the company ensuring that knowledge is preserved and transferred?
**12** Analyze the corporate structure. Is it flexible? How is the communication between layers?

Quantitative evaluation for a technology assessment is a challenge. A five-point scale, ranging from outstanding to poor, is proposed as part of TAM. A score of 5 is outstanding, 4 is good, 3 is average, 2 is below average, and 1 is poor. An ideal scenario is set to provide guidance as to what a score of 5 should encompass. This would serve as the standard against which a score is assigned to each element of the model. An overall score can be calculated by adding all the individual scores.

The technology audit should be repeated periodically, at least once per year, to determine progress. TAM provides an overall assessment of the company's leadership approach, methods, strategies, plans, goals, and policies. It is the first step toward technology management excellence. However, although approach is important, results achieved in terms of time to market, profits, market share, and so on, are also essential to evaluating effectiveness. A company must attain results in order to remain in the market. The auditor should not forget that a company must be results-oriented. After a reasonable period of time (e.g., two years), the TAM can be repeated and changes in score can be analyzed. The company's performance results should be the criteria used to evaluate the effectiveness of its technology management capability. If the results are not satisfactory, a change of strategy should be considered.

The TAM process, or any other technology auditing process, can serve as a valuable tool for companies in a number of ways:

- As a diagnostic tool for determining strengths and weaknesses.
- As method of identifying and targeting key opportunities for improvement.
- As a tool for benchmarking with competitors in the same technology or industry sector.
- As a tool for measuring the progress achieved and the effectiveness of implemented programs.
- As a tool for continuous improvement.
- As a self-assessment instrument leading to proper technology planning.

### Motorola's Technology Road Map

Motorola, Inc., is well recognized for its technology-based competitive strategy. The company has developed considerable skill in creating new products and bringing them to the market in a timely manner. Because of the complexity of many of its products, and in order to make sure that no element of technology is neglected in their design or manufacturing, Motorola developed a corporatewide technology planning tool called the *technology road map* (Willyard and McClees, 1987). A road map is developed for emerging technology, and another one is developed for each product. The emerging-technology road map is developed by a small committee of experts from Motorola's technical community. The committee is also responsible for updating the technology road map.

Motorola's experts consider a technology to be a candidate for a road map if it has been demonstrated in a laboratory, whether the laboratory is Motorola's or a university's or a competitor's. According to Willyard and McClees (1987), an emerging-technology road map provides:

- An objective evaluation of Motorola's capability in the technology.
- A comparison of Motorola's capabilities and those of its competitors, today and in the future.
- A forecast of the progress of technology.

The information obtained from a technology road map is quite valuable for technology planners and decision makers. A technical committee at Motorola relies on road maps in deciding whether to recommend that the corporation take a position in the technology or develop a special program to improve its capabilities in the technology.

Motorola's product technology road map documents a product's history, its present status, and its expected future. It tracks the company's progress in product and process development and the response of the marketplace to the product. It analyzes Motorola's technological and business positions compared to its competitors' positions and allows the company to make informed decisions with respect to its products, resource allocation, and market conditions. The product road map is a systematic planning tool that is effective in managing a complex technological environment in each individual business within Motorola.

The product technology road map has eight sections:

**1** *Description of the business:* This section includes the mission, the business strategy, the expected or denied market share, the product's time to market, the product introduction plan, the experience curve that projects the future costs and prices of the product, and a study of the current competition and future position in the technology and the marketplace.

**2** *Technology forecast:* This section focuses on the R&D effort.

**3** *Technology road map matrix:* This is a matrix that combines the technology forecast and the product plans on a time chart. It is an excellent visual tool for depicting future directions and tracking progress. Figure 9-10 shows a technology road map matrix for a broadcast automotive FM receiver.

**FIGURE 9-10**
A TECHNOLOGY ROAD MAP MATRIX
This matrix summarizes the technological requirements for a future product—a broadcast automotive FM receiver.
*Source:* Willyard and McClees, 1987. Reprinted with permission.

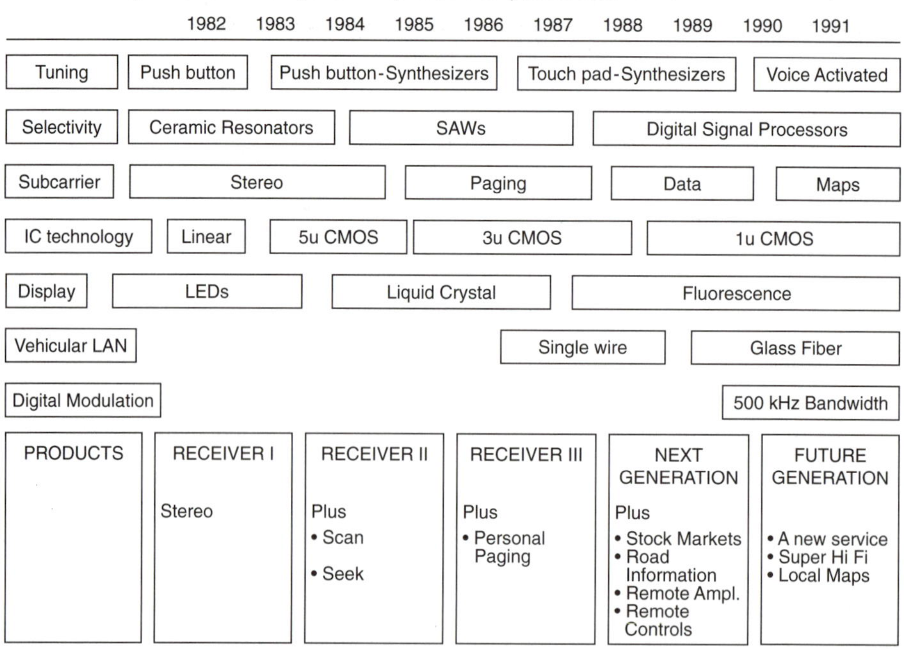

**4** *Quality:* Management can determine the level of quality that is to be built into both products and processes.

**5** *Allocation of resources:* Resources are distributed according to needs in order to achieve desired goals. Resources include people, skills, space, and location of the effort.

**6** *Patent portfolio:* The reward from the effort comes in terms of creating a proprietary position. Information is gathered and communicated to patent attorneys to secure potential patents in a timely manner. Licensing of the patents is a source of income to the company.

**7** *Product descriptions and status reports:* The details of the product's features are documented and the progress of development is tracked.

**8** *Minority report:* A potentially beneficial product, process, or technology that has not been fully considered is brought to the attention of management. This ensures that no technology with a potential benefit can slip by unnoticed.

## PLANNING ACCORDING TO THE TECHNOLOGY LIFE CYCLE

Strategic analysis and planning according to the technology life cycle has been advocated by Arthur D. Little, a well-known consulting firm. In this approach, whether to invest in technology is decided according to the competitive impact of the technology. This impact is dependent upon the position of the technology on the S-curve (Figure 9-11). If the technology is in the embryonic/development phase but has not yet demonstrated potential for changing the basis of competition in the future, it is considered an *emerging technology.* Companies interested in this technology sector should monitor emerging technologies. If the technology is further along the curve of progress and has demonstrated its potential for changing the basis of competition, it is deemed a *pacing technology.* Companies interested in being players in the technology arena should consider investing selectively in pacing technologies (Figure 9-12). Key technologies are those

**FIGURE 9-11**
TECHNOLOGIES AT DIFFERENT STAGES OF THE LIFE CYCLE
*Source:* Based on Little, 1983. With permission.

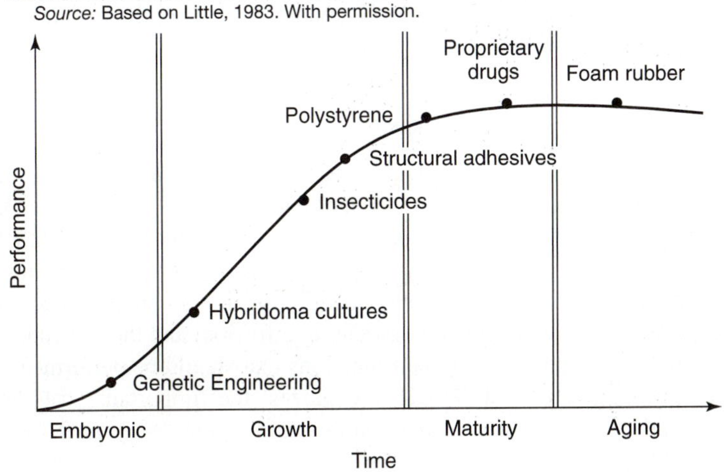

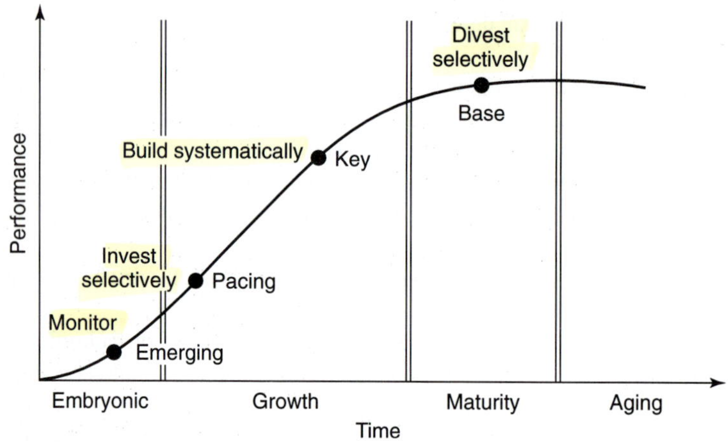


**FIGURE 9-12**
TECHNOLOGICAL INVESTMENT MODE
*Source:* Based on Little, 1983. With permission.

that have a strong impact on the value-added stream of performance, cost, and quality. They allow a company to develop a proprietary position in products or processes. Key technologies are essential to the success of companies. They influence the growth phase of the technology S-curve, and they have a major impact on a company's competitive position. Companies should be prepared to increase their strength systematically in key technologies.

As technologies mature, they become known as *base technologies.* These are necessary for participation in business, but they provide a firm with little or no competitive advantage (Arthur D. Little, 1981, 1983). Base technologies are considered commodities, usually available to all competitors. When a technology reaches this stage of the technology life cycle, companies should start divesting selectively while harvesting the benefits from the mature technology (milking the cash cow). In the aging stage of technology, a company must have already developed its strategic options; otherwise, it will suffer the consequences of going out of business.

## THE B-TECH APPROACH TO PLANNING

Battelle, a leading institute involved in technological innovation and management, developed a comprehensive approach to technology planning. This approach, known as *B-TECH,* was developed by Stacey and described by Bhalla (1987). Battelle views technology planning as a major set of activities much larger than the traditional planning for R&D in companies. R&D is but one part of an overall technology planning effort that encompasses assessment, creation, purchase, diffusion, and the protection of technology. Figure 9-13 illustrates important functions that should be performed in a comprehensive planning effort. B-TECH recognizes the importance of integrating technology strategy with business strategy as shown in Figures 9-13 and 9-14. However, it proposes that each strategy initially follows a separate development path before the

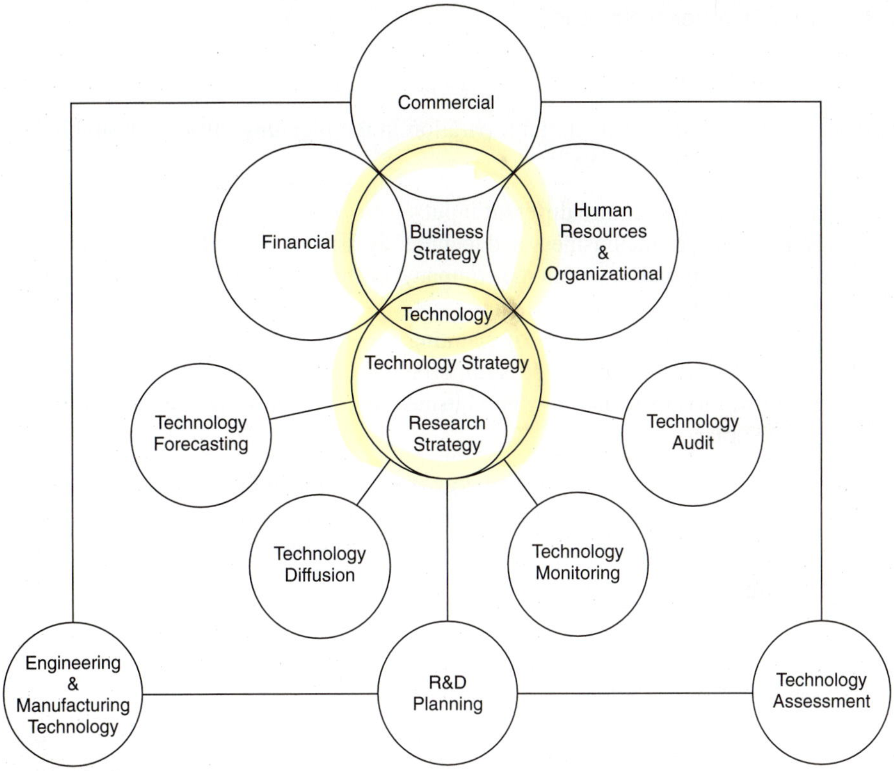


**FIGURE 9-13**
INTERACTION OF BUSINESS AND TECHNOLOGY STRATEGIES
*Source:* S. Bhalla, *The Effective Management of Technology.* Battelle Press, 1987. Reprinted with permission.

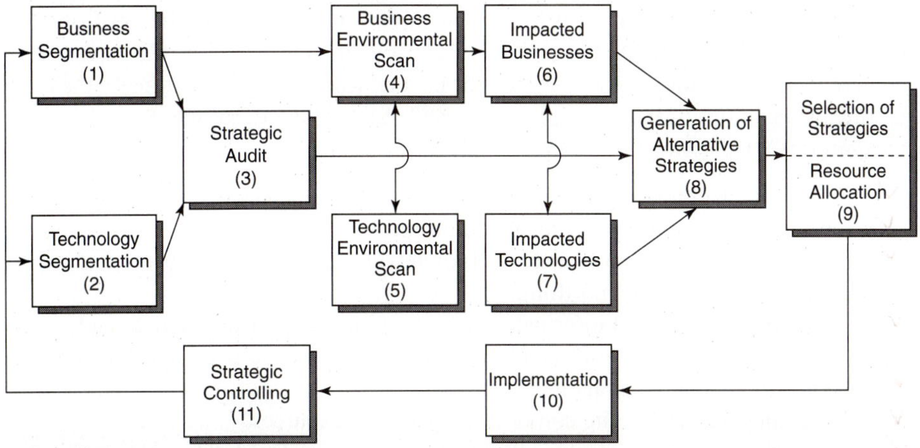


**FIGURE 9-14**
INTEGRATING TECHNOLOGY AND BUSINESS STRATEGIES: THE B-TECH APPROACH
*Source:* S. Bhalla, *The Effective Management of Technology.* Battelle Press, 1987. Reprinted with permission.

two are fully integrated. This initial separation in the planning effort is desirable for a number of reasons (Bhalla, 1987):

**1** The two analyses require different inputs.

**2** The planning for the business and technology aspects of the firm is often in a different state of development within the company, or the two have not been developed in a way that easily permits merging them.

**3** Because the business aspects of planning sometimes submerge the technological ones in the "customary analysis," it is important to create a corporate culture that promotes the integration of technology and business and prevents one element from overwhelming the other.

Several alternative strategies may emerge from this planning exercise. Management selects the optimal strategies, allocates resources, implements the strategies, and follows up to ensure that the strategies are meeting the desired goals.

The B-TECH approach to integrating technology and business strategy has 11 steps, as shown in Figure 9-14.

### The Chief Technology Officer

A relatively new title in corporate America is *chief technology officer (CTO)*. Some companies may use different titles, such as vice president of technology, vice president of R&D, or manager of technology. This position, which could be at the vice presidential level or some other high level of management, reflects recognition of the significant role that technology plays in the company's competitive position. A CTO's major role is to bring a technology perspective to bear on all strategic issues (Lewis and Linden, 1990). Thus the position must be a top management position. The CTO is the orchestrator of the company's technology strategy and is deeply involved in its coordination with the business strategy as well as its implementation. A CTO's role is different from that of a vice president for R&D or a research laboratory director in that the CTO oversees a comprehensive technology plan that extends beyond R&D. A CTO coordinates the forecasting, acquisition, licensing, exploitation, and gatekeeping of all the technologies in the company's technology portfolio. The CTO focuses on:

- Forecasting technology and analyzing prospective acquisition targets.
- Building the company's technical competence.
- Devising an acquisition plan for corporatewide technology resources and maintaining a healthy technology portfolio.
- Developing formal and informal networks and technological alliances and ensuring that corporate cultures, people, and technology mesh well between the allied groups.
- Conducting technology audits.
- Allocating and structuring corporate technology resources.
- Organizing programs of technical education to increase the skill level of employees.
- Ensuring that technologies are transferred and disseminated throughout the company.
- Gatekeeping all technologies in the company's portfolio.

- Protecting the intellectual and technological rights of the company.
- Exploiting other companies' technologies without compromising his or her company's distinctive competitive advantages.

A CTO is much more than a chief information officer (CIO). The latter deals with the company's information technologies. A CTO deals with all technologies that are assembled to enhance the company's competitive posture. Therefore, a CTO must have the confidence and support of the chief executive officer (CEO) and the company's board of directors. He or she should report to them directly and be in the inner circle of decision making.

## CONCLUDING REMARKS

Successful corporations must be able to forecast technological changes. A business enterprise must not be blindsided by a radical shift in technology that may render the business's existing technology obsolete. For example, businesses relying on vacuum-tube technology in the 1950s should have been able to forecast that transistor technology would replace that of the vacuum tube. The transistor had much better performance parameters than the vacuum tube: It was smaller, used less electricity, and was not limited by power or temperature, as was the vacuum tube.

In the 1990s, the forecast of a huge shift in technology, away from personal computer–based applications and toward Internet-based applications, has fueled a surge in the price of Internet stocks traded in the market. Companies such as Microsoft, America On-Line, Amazon.com, and E-bay succeeded in forecasting this shift and were able to adjust their strategies to harness the opportunities. Successful forecasting affords a business the advantage of orderly planning and a smooth transition from one technology to the next.

Technological forecasters extrapolate the rate of improvement in technology from the technology progress curve. Forecasters can track the progress in the technology performance parameter and watch for inflection points. The inflection points may indicate a change in the progress trend of a specific technology.

Technologies exhibiting the traditional S-curve of progress will have two inflection points. The first occurs when the technology performance parameter exhibits growth following an embryonic phase or new invention period: The curve pattern changes from exponential to linear. The second inflection point occurs when the technology moves into the maturity stage: The curve will slow from linear to asymptotic growth. A word of caution here is that some newer technologies, such as software and super-high technologies, may not exhibit a traditional S-curve. For such a technology, the curve of technological progress may shoot quickly into a fast-growth phase and plateau suddenly into maturity. The technology may even move quickly into the obsolescence phase. An example of this is software programs that develop relatively quickly and then reach a plateau defined by their limit of performance.

Forecasters must track historical data on technical performance. They must be able to identify appropriate performance parameters and estimate the natural limits of the technology.

There are several methods of forecasting technological changes, such as monitoring, the use of experts, and Delphi techniques, as well as a variety of mathematical or computer forecasting models. None of them is guaranteed to give an absolutely accurate prediction of what may happen in the future. However, these methods are extremely helpful in making an educated guess, and quite often they are accurate. Many factors may contribute to a change in forecasted information, such as a change in the natural limit of the technology because of expanded knowledge or new research; a change in the markets of the technology; a change in the structure of the industry using or developing the technology; a change of economic conditions; and a change in public perception of the technology.

Other influencing factors are government R&D policies, which can create a technology push and thus accelerate the rate of technological development. Large consumer demands and industrial expansion can also hasten the rate of change, through a market-pull mechanism. Managers of technology must be aware of public policy and government-announced national priorities. Managers should also develop technology maps that list all core technologies and supporting technologies in their products' value chains. Maps can be used to chart likely future directions of these technologies. Managers of technology can also prioritize technologies according to their importance to the core business.

Technology managers should periodically conduct a technology audit to determine the strengths, weaknesses, opportunities, and threats to the company. The technology audit model (TAM) can be a helpful auditing tool for managers. Technological gaps can then be identified, and funding can be allocated to R&D and engineering projects that address the gaps.

Funding priorities can be linked to projects that have the highest potential for creating technological leadership. By closely linking technological development to market needs, a company can attain a faster return on investment. Funding can also be allocated in proportion to the stage of the technology life cycle, as suggested by the Arthur D. Little approach to investment, illustrated in Figure 9-12. This will guarantee a presence in technologies that are of importance to a company.

Forecasting of technology and markets leads the way to strategic planning of the business. Planning for technology requires strong organizational structure that supports the planning effort. The existence of a chief technology officer or someone with a similar title in the company can greatly enhance the corporation's effort in technological planning. In addition, a supporting set of functional units, as suggested by the B-TECH approach to planning, can provide a solid organizational infrastructure and give muscle to the technological planning process within the corporation.

A competitive enterprise must be able to effectively integrate technology planning and business planning. It should develop agile systems responsive to a highly dynamic environment. To do so, leadership must bring critical skills to bear on market opportunities. Developing agile systems requires taking strong strategic management initiatives, setting a clear vision, and plotting the strategies and plans needed to realize that vision. Management should develop the structure needed to support its initiatives and ensure that proper links and interfaces are established among the functional units of the enterprise.

A competitive enterprise should rely on its core competencies to provide real and perceived value to its customers. Key core competencies are those that give the enterprise a substantial competitive advantage that no one else can duplicate. Core competencies are deployed widely in products and services to take full advantage of their value to customers. Meanwhile, a competitive enterprise should be continuously monitoring the environment to identify opportunities and avoid threats. It should benchmark the competition and identify the best practices as well as any gaps in capabilities and competencies. When these are identified, plans should be set in motion immediately to fill the gaps and capture the opportunities. The entire strategic management process must be institutionalized. It must be candid, systematic, and well documented.

**READING 9.1**

# A Structured Approach to Corporate Technology Strategy

**Gary S. Stacey**

Senior Consultant, Battelle Geneva Laboratories

**W. Bradford Ashton**

Senior Program Manager, Battelle Northwest Laboratories

*Abstract:* There is currently a resurgence in recognition of the vital role technology plays in corporate profitability. Unfortunately, interest in technology has been coupled with the difficult problem of the proliferation of technological options, many of which are unknown to the company at the time they become important. This means that much careful attention to acquiring and deploying advanced technology is needed. This paper describes a structured approach called ASSETS, which integrates business and technology actions. The process is based on a set of basic questions that must be addressed in seeking a well-grounded business and technology strategy. Answers to these questions follow a number of conventional planning steps, with both the technological and business issues integrated through the use of a basic technology strategy. The paper describes these steps and shows how they are used to generate and use a strategy based on future impacts. Resource allocation, implementation and control considerations are included.

*Keywords:* technology strategy; innovation; product development; R&D investment planning; technology monitoring; competitive analysis; technology transfer.

*Source:* Stacey, G. S. and Ashton, W. B. 1990. "A Structured Approach to Corporate Technology Strategy," International Journal of Technology Management, vol. 5, no. 4. Reprinted with permission.

## 1 INTRODUCTION

Few companies enjoy long-term success without major commitments to the development and application of new technology. An increasing amount of evidence points to the vital role that advanced technology plays in long-term corporate profitability and indicates that firms are moving actively to exploit this role (1). This is illustrated by more aggressive technology acquisition actions by companies throughout the industrialized world. These moves are aimed at improving the business returns from advanced technology, and, in some industries, at ensuring survival as a viable competitor.

However, the revival of interest in technology has been coupled in recent years with the exasperating proliferation of a vast array of technological options and opportunities. The threat to an organization's success comes from a very wide and disparate set of potential technologies, many of which are unknown to the company at the time they become important. This means that much more careful attention to the commitment of resources aimed at acquiring and using technology is now required.

This paper addresses these needs by recommending more systematic management of technology-related activities. We present a structured approach to corporate decisions regarding acquisition, development and use of advanced technology. Called ASSETS, the framework is designed to integrate technology and business decisions through development and implementation of a corporate technology strategy; the strategy is focused on creating competitive technological assets to generate valuable returns in both economic and noneconomic ways.

## 2 THE NEED FOR STRUCTURED THINKING ABOUT TECHNOLOGY

Companies and organizations acquire, develop and deploy technologies using a variety of approaches, including the following:

- Conducting internal research and development (R&D) activities;
- Directly investing in new equipment or people;
- Divesting or acquiring assets (companies);
- Licensing to or licensing from another company or organization;
- Actively patenting in a technical area to attack or defend for the future;
- Contracting R&D externally with other research organizations or universities;
- Utilizing the results of public sector (i.e., government) R&D;
- Engaging in joint ventures both for the product or production process as well for the necessary R&D for entering a new area;
- Re-training and re-directing existing resources and capabilities.

With literally hundreds of potential technology areas and many alternative management actions, the most important issue for most companies is to focus attention and resources on a few selected areas and approaches. Such focusing necessitates selection from the range of options available. This selection process cannot be accomplished informally. To build a high level of confidence in the results it must be approached with rigor, thoroughness and open communication; all serious candidates must be considered.

In general, a well-known set of actions should be completed to help produce the necessary confidence (2, 3). The objective of these actions is to allocate resources with

**BOX A**

FUNDAMENTAL QUESTIONS FOR EFFECTIVE TECHNOLOGY MANAGEMENT IN BUSINESS

1. How successful is the firm in meeting organizational *goals* and what are the *strengths and weaknesses* that will determine its future prospects?
2. What are the important *market-place needs and opportunities* for technology in products and processes that the firm should target?
3. How should a fundamental technology *game plan* to meet the *future* business and technological environment be prepared and what should it contain?
4. What specific criteria should be used for *technology acquisition or development* investments by the firm?
5. How can favorable technology candidates be *identified* and *evaluated* to implement its fundamental business strategy?
6. How can the background information best be used to establish *priorities* for R&D?
7. How should both financial and non-financial resources be *allocated and committed* in execution of the programs identified in the technology plan?
8. What are the methods for *using the results of technology investments and capturing the returns* in the form of new products, production processes and other applications?
9. How can a firm continually *monitor its environment* for relevant technical and business trends in an efficient and effective way?
10. How does the firm *use information to manage* its business goals and technology approach?

attention to integration of technology and business strategies and plans. However, an integrated strategy is not easy to achieve. In the past, companies have often not experienced satisfactory communication between the technology functions, which involves specialized technical staff, and business and financial strategists. Overcoming this communication gap requires a special effort for most companies.

Effective technical and business communication begins with useful terminology and a structure for evaluating and using concepts and tools. In our approach, a series of essential questions forms the basis for an integrated and practical technology strategy. These questions represent major choices faced by companies, quasi-public and public organizations in producing a strategy to integrate the technological future with the overall organizational and economic future. Addressing these questions can lead to an efficient and effective business and technology planning process.

Our fundamental questions are summarized in Box A. They begin with fundamental business goals and follow conventional planning needs from both the technological and business perspectives. Each question generates information needs which are satisfied through a structured process of analysis, decision and action designed to integrate business and technology concerns.

## 3 THE ASSETS PROCESS

The concept of actively using a structured process for technology strategy and management is an important one. In the past, business success has often come serendipitously, with good ideas being translated into product and production process technology by

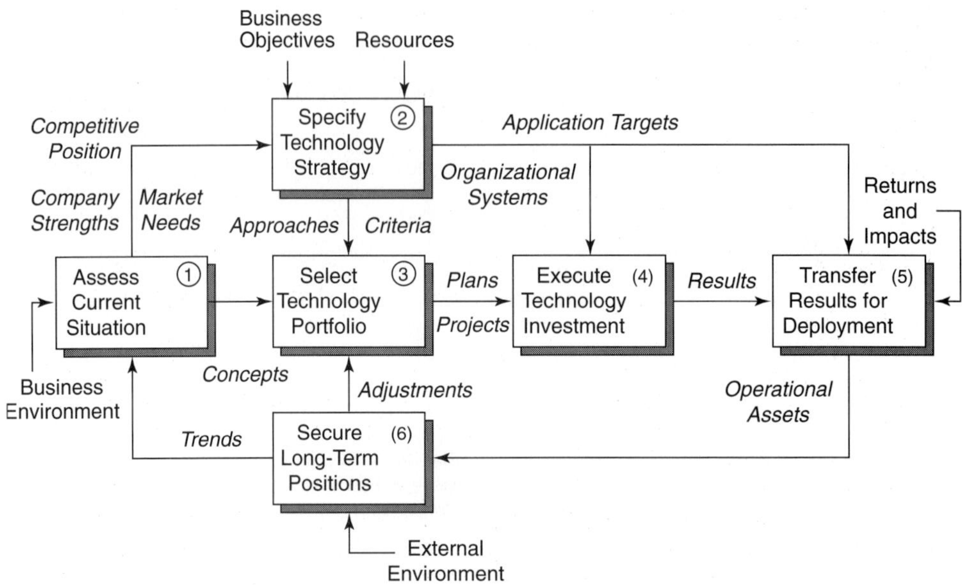


**FIGURE 1**
THE ASSETS PROCESS
The boxes with circled numbers identify steps in strategy development; the boxes with uncircled numbers identify steps in strategy implementation.

energetic, committed people. However, nowadays, the technical complexity of products and processes, the resource requirements and length of time necessary to achieve success are factors that reduce the viability of approaches relying heavily on good luck, unstructured efforts and good ideas from bright people. An organized framework for analysis and planning is required to help ensure that all important aspects of the problem are addressed (4).

A systematic process to develop and implement a technology strategy is shown in Figure 1. The acronym ASSETS was chosen for this process to emphasize the fact that the technological base of an organization is just as much an income-producing asset as any other physical equipment or labor resource. There are six steps in the process; each one is directed towards answering one or more of the questions in Box A. None of the steps is surprising; however, as a group they represent a convenient framework for thinking about technology decisions and acting on the results.

ASSETS is designed to be an on-going process. New ideas or technical areas should be added continuously and out-dated ones dropped as needed. The six steps are integrated in a closed action—feedback loop, with opportunities for learning and correction built into the concept. The first three steps in this process represent the *strategy development* or formation stage; the last three steps represent the *strategy implementation* stage (Figure 1). A variety of supportive tools, both quantitative and non-quantitative, are available to facilitate the conduct of each of the steps. Other tools are under active development. Several examples of these tools and the principal features of each step are summarized below.

### 3.1 Step 1: Assess Current Situation

The background for decisions regarding the future of an organization can conveniently be divided into two main categories. There are both internal and external aspects to technology decisions. The internal background relates to company goals, strategies, capabilities and performance. External background relates to the business and technical environment in which the company operates.

In Step 1, a full understanding of the company's performance and current position is the intended outcome. Since goals are usually expressed in the form of business performance, a review of fundamental company goals is also important to establish the proper basis for all other aspects of the business. In particular, goals that relate to the risk preferences of the organization and to the willingness to work with new technologies should be reviewed.

It is also essential to develop a full understanding of the future business environment: that is, the customers, suppliers, competitors, demand for product, industry financial ratios, and economic trends. At the same time it is also important to develop a similar understanding of the technology environment, including items such as patents, key technologies, centers of research excellence and capability, levels of funding, the potential for breakthroughs, and barriers to technology deployment. The convergence of business and technology planning begins to occur as the environment for each is established and relationships between the two environments are drawn.

An important element of the self-examination process is to review the firm's strategic technology areas (STAs). These are the *areas of functional technology expertise in the firm's operations* (products, processes, support or management) *which give the firm an advantage in the market-place for serving particular customer needs* (5). To illustrate an STA, consider a firm that produces a heavily-used equipment product that is purchased because of its durability; the STAs could be the *design, materials,* or *manufacturing* expertise that prevented premature failure or metal fatigue, thus ensuring high durability. The firm's current and desired STAs are the basic building blocks which form the targets of technology investments.

**Tools** There are a variety of tools that can be used to evaluate performance and goals. These involve various types of measurement or accounting techniques, program reviews or evaluations, and competitor analysis methods (6). Also, it is important to complete informal and external audits of trends and developments which will affect the company's future. The needs of the market-place of interest can be understood by using traditional market analysis tools (7).

An important approach recently used to evaluate the risks from technology replacement or substitution is the S-curve (1). This curve represents technological capabilities on an S-shaped curve which approaches an upper limit asymptotically. A firm's position on the curve is an indication of the availability of advanced competitive technologies and of the possible need to replace existing systems. The curve has an empirical basis in that many of a firm's technologies can be represented in this form as an aid to determining the need to take action, such as re-investing in new technology.

**Results** The results of this step provide answers to Questions 1 and 2 in Box A. Documentation of the company's situation is completed and a description of the future business and technological environment is included. This material is passed on to Step 2, while ideas for specific investments go to Step 3 (Figure 1).

### 3.2 Step 2: Specify Technology Strategy

Specifying a "game plan" for technology development and use is important because it deals with the conversion of the information and concepts developed in Step 1 into realistic plans and actions. In ASSETS the game plan is *technology strategy;* its specifications just address the four main elements which support the firm's basic business strategy: *customers, competitive approach, investments and organizational culture.* The concept of technology strategy is intended to mean "a broad approach to achieve organizational goals through sustained technology advantages in the expected competitive environment." The strategy should provide guidelines (*criteria*) for selecting and implementing specific actions (*tactics*) (8). This involves the simple act of answering questions such as:

- What technologies to develop, license, or buy;
- Whether to seek technology leadership;
- How to protect property rights;
- How to capture economic returns.

Technology strategy must be a consistent part of overall business strategies. Although any classification of strategies by type is necessarily arbitrary, Box B shows several fundamental business strategies that have important technology components [based on concepts in (9) and (10)]. These generic approaches to gaining advantage form the

**BOX B**

GENERIC BUSINESS AND TECHNOLOGY STRATEGIES

**Market Competence: What Value to Offer the Customer?**

1 "Product Differentiation"—distinguish the firm's goods or services from those of its competitors on the basis of either *superior performance* or *unique features.*
2 "Low Cost Producer"—create advantage by being able to consistently underprice the competition in the market-place.

**Market Scope: Which Market Boundaries to Target?**

1 "Largest Market Share"—seek dominant market position through wide breadth of product coverage or entry barriers to competitors.
2 "Specialized Niche Player"—restrict competition in a limited market segment by focusing company products on narrow, specialized customer needs, delivery mechanisms or creating barriers to competition.

**Market Timing: When to Introduce Innovations?**

1 "First Mover"—enter new product-market before other competitors to gain early position and returns; establish early reputation as technology leader and work to maintain lead.
2 "Wait and Improve"—enter a new product after first movers have completed initial penetration and worked out the product and market bugs; strive to gain advantage with attractive follow-on product or process improvements.

basis for identifying the particular elements of technology strategy that will be most beneficial to the firm. For example, the "First Mover" strategy will imply that emphasis be put on developing and quickly introducing "leading edge" technology into a firm's products ahead of competitors. The "Low Cost Producer" strategy, on the other hand, will mean a heavy emphasis on efficient, automated, highly reliable process technology that will reduce total manufacturing cost. Other strategies will have implications appropriate to their characteristics.

Ultimately, strategy must break down to action, and actions must be assignable in such a way that the responsible individual can be held accountable for whether there is success or not. These actions must also be expressed in such a way that they can actually be accomplished. Thus definition of useful guidelines (decision criteria) regarding future customers, product-markets, capabilities and resources are all essential for this step.

The setting of technology strategy is constrained by unique company character and needs. Sometimes the strategy for a company or organization is provided by top management. In such a case, the actions can be "brainstormed" in the context of strategic business goals and strategy. Sometimes there is no good strategic concept. In this situation, a variety of actions can be imagined; relevant possible actions can be collected together and organized into a strategy that provides the needed decision framework. Sometimes a top-down approach is necessary; at other times a bottom-up approach is appropriate. This step is one in which the inescapable requirement is to integrate the variety of perspectives and interests in directions the firm would like to take.

**Tools** There are very few formal tools that are directly applicable to the production of a technology strategy other than general guidelines, conceptual matrices and checklists (7, 8, 9, 11). However, many of the concepts and tools used for conventional business planning are also applicable to technology planning. For instance, forecasts of future implications for candidate strategies can be accomplished using a variety of existing business techniques. However, for subjects that have a long-term future such as advanced technology, it is essential to use a technique that allows a variety of difficult variables and relationships to be included. In our experience, scenario methods that rely on cross-impact matrices for drawing the relationships are especially useful (12). In addition, these techniques are also useful for long-term business forecasts.

A critical factor in developing a strategy is to evaluate the company's position with respect to the competition (6). However, in a technological environment, the evaluation of competitive position is difficult because there is so little data available. To remedy this, work is being done on developing improved methods and software, both in the area of competitor analysis (6) and in more specialized approaches. For instance recent progress has been made in the emerging area of patent trend analysis (13). Using these techniques, data on patent position is evaluated to determine the strength of the competition, the value of current patent positions, and the technology strategy of other organizations operating in the field.

**Results** The results of this step are summarized in the elements of a corporate technology strategy and in guides for its implementation. In practical terms, technology strategy provides approaches to gain product-market advantage, decision criteria for the next ASSETS step of selecting a technology investment portfolio and areas of technical concentration for target applications. Importantly, it also implies use of selected organizational practices, such as "innovation teams," and systems (4, 14), as well as some attention to contingency planning (10). Other outputs include guidance on the composition of the portfolio (e.g., the mix of long-versus-short-term projects, the balance of the types of technical work), the target applications and the approaches to generate returns from the applications (sales, licenses, direct applications, spin-off companies).

It is important to culminate this step with formal approval of a technology strategy and with documentation in technology plans. Documentation should be strategic in nature and does not have to be elaborate or time-consuming to prepare. This strategic technology guidance forms the basis for development of an investment program, including R&D as one of the possible actions.

An example outline for a technology plan to document this step is shown in Box C. This outline is conventional in design but is valuable for developing a technology strategy because it requires consideration of key parameters that might otherwise not be addressed in an implicit planning process. The completion of this step will produce answers to Questions 3 and 4 in Box A.

### 3.3 Step 3: Select Technology Portfolio

Before selecting technical areas for research, it is necessary to identify suitable candidates. The identification process can begin with material from Step 1: a routine scanning of literature; observing the technical environment, monitoring competitors, suppliers and customers; attending conferences, symposia and other meetings on technical subjects. Sometimes it is useful to make assignments to particular technical areas for review and consideration. In addition to the simple process of observation it is necessary to

**BOX C**

TECHNOLOGY PLANNING COMPONENTS

**I** CURRENT STATUS
- **A** Product performance
- **B** Technology and market assessment
- **C** Customers
  - **1** Customer health
  - **2** Future outlook
  - **3** Current and future needs
- **D** Competitive analysis
- **E** Technological strengths and weaknesses
- **F** Existing, key and emerging technologies

**II** FUTURE EXPECTATIONS
- **A** Technology environment
- **B** Product and process trends
- **C** Competition potential
- **D** Capabilities needed
- **E** Problems and opportunities created

**III** BASIC TECHNOLOGY STRATEGIES
- **A** Improvements in the technological base
- **B** New technology acquisition targets
- **C** Technical and deployment risk
- **D** Application timing

**IV** LONG-TERM RESOURCES NEEDED
- **A** Capital and financial investments
- **B** Resource allocation guidelines

**V** MEASURES OF SUCCESS
- **A** Quantitative
- **B** Qualitative
- **C** Milestones and go/no-go decisions

make the connection between what is happening, or may happen, in a technical area and the implications for the company or organization.

The value in this activity comes from the "creative spark" element, not just from the mechanical review of the technical environment. Imagining technologies as they relate to the future business and technological environment is critical to success in establishing the list of candidate subjects for consideration in technology acquisition.

Having identified a series of potential candidates for future development, a process of evaluation (using explicit decision criteria) and selection of a mix of activities supporting company goals and consistent with company business and technological risk preference is necessary. If the screening process is difficult or expensive, there is a tendency to limit the list of candidates. This raises doubt regarding the adequacy of the subjects being considered. The development of an efficient, thorough screening process encourages consideration of all ideas that are discovered. Thus, it is very important for confidence building to have a screening process that is comprehensive and decisive. That is, it should inexpensively eliminate candidates that are not likely to pay off and should permit good candidates to be evaluated quickly and, if appropriate, committed to action.

Another important aspect of this screening process is that the candidates should include existing programs so decisions are made to cease work if the potential pay-off is not adequate. Stopping projects is usually more difficult than initiating work.

**Tools** There is a variety of tools being developed (and some which have been developed) to assist in identification and selection of a technology portfolio (7, 15). In the area of identification, various monitoring programs are available which allow one to use databases and search for key words to identify subjects that are of interest (16). Some companies offer monitoring services for specific technologies.

Establishing priorities can be difficult because of the complexity of the risk involved. For instance, some investments have little technological risk because they are an extension of the known technological base. However, these same investments might be extremely risky from a business perspective because they require the company to enter a completely new business mode. The TECH-RISK array, which is a matrix that can be used to place technology investment activities taking business and technological risk into account, can be used for this purpose (17, 18). This is much like other current "matrices" that are in fashion and it allows the user to realize the riskiness of the portfolio that is held.

Another tool that is useful in analyzing investments is the analytical hierarchy process (AHP), in which the decision-maker considers the importance of a variety of criteria in establishing various investments in technology (15). This process allows any important criterion to be included in the analysis and effectively deals with the problems of multi-attribute decision environments. A hierarchy of criteria is established and the mechanism for evaluating each investment against each criterion is available. Examples of criteria include the following:

- Payback-cost benefit ratio
- Market or spin-off potential

- Business sector priority
- Continuing availability of funding
- Existence of an effective "champion"
- Degree of existing staff capability
- Technological risk
- "Sparkle-factor"— potential for high innovation
- Service to other corporate goals
- Business plan status

**Results** The main result of this step is a prioritized list of attractive technology investment projects. Thus, Questions 5 and 6 from Box A would be answered here. By frequent iteration, this list would be kept up to date to allow selection of additional candidates as financial resources permit.

### 3.4 Step 4: Execute Technology Investments

Resource commitments to technology investments are made in Step 4 and the utilization of the resources to conduct the planned project work is undertaken. The basic approach is embodied in the project management knowledge base and includes the following functions:

- Organizing the work team
- Planning the details of the work
- Staffing the work activities
- Directing and leading the effort
- Controlling the activities to ensure compliance with plans and needs

Generally, considerable effort is required to keep programs moving along rapidly. There is a strong tendency for time-frames to be extended and for decisions to be prolonged. In addition, after the key results have been discovered, there is a tendency for the participants to relax and devote too much time to completion of the innovation process. Therefore, in executing programs, effort is needed to push the program through to completion and on to deployment decisions. In completing this step, the actions that are needed are as follows:

- Plan the program—provide a detailed work plan;
- Evaluate progress—review the detailed elements of the work plan;
- Go or no-go decisions—using progress against milestones, decisions to start and stop projects and programs must be made;
- Slow-down and accelerate—changing the timing is essential so a product or process arrives on the market at a time when it is most likely to be successful;
- Drive to the finish—generally, considerable amounts of resources can be wasted at the end of a program. Pushing to complete early or with reduced budgets can sometimes result in a useful result at considerably reduced cost.

Milestones for success (or decisions regarding continued commitments) are established and periodic review of expectations compared with what actually takes place.

**Tools** There are a number of tools that can be used in the management of projects and programs. These include widely known tools like PERT and GANT charts, work-breakdown structures, budgeting systems and decision-trees. These tools are directed at scheduling, task-definition, budgets and project control. Nowadays many computer software programs are available to facilitate tracking expenditures, status and accomplishments (7, 13).

**Results** The results of effectively completing this task will be to produce well-developed prototypes and technical products that are satisfactory for pushing through the next stage—transferring results to actions. The completion of Step 4 is based on providing answers to Question 7.

### 3.5 Step 5: Transfer Results for Deployment

Ultimately, the efforts of technology acquisition and development must have a pay-off. The utilization of the results in production processes and products is the ultimate goal. This fact is sometimes overlooked by people who are interested in the technology and work on the scientific aspects of a product or process problem. For some researchers, solving the technical problem is the end of the program. But for the company or organization, the solving of the technical problem is only the beginning of solving the problem because more steps are needed to bring the concept into a full state of innovation. The following activities are needed:

- Prototype development
- Scale-up
- Production or purchase
- Distribution or installation

These latter steps are very expensive in comparison with the cost of the technology acquisition. In addition, this is frequently a long and difficult path because of the internal objections and resistance by those who stand to lose as a result of the adoption of a new technology.

The most common way to utilize the results is to employ the new concept in products or production processes. Here the major difficulty faced is the "hand-off." The concept or idea must be effectively moved from the R&D environment to prototype development, pilot plant, scale-up, production and distribution. The hand-off is critical and frequently the source of difficulty because the transmittal of an idea to another person or group is usually accompanied by skepticism at a time in the life of an idea when it needs to be promoted and nurtured (2, 14). In some cases, the management of the intellectual property without the goal of production or new products is appropriate. If successful acquisitions of technology are not easily made consistent with business strategy through processes or products, it may be appropriate to manage the technological assets otherwise by, for example, licensing to earn royalties, divesting, or engaging in joint ventures.

**Tools** There are several tools that are available to complete this step. The most useful tools involve evaluation of commercial potential of products. Also, a number of important operational approaches, organizational in nature, can be used (2). For instance intellectual valuation and traditional market studies to assess new product potential would be applied here (3, 10).

**Results** The results of this step should complete the innovation process and should bring the new technology to use in beneficial applications such as production processes or products. Question 8 will have been addressed in completing the work in this step.

### 3.6 Step 6: Secure Long-Term Position

The final step is a monitoring, review and feedback activity. In managing technology in a rapidly changing environment it is vital to monitor continually the activities of others, and business and technical trends. Such monitoring is needed to provide long-term insurance against unpredictable changes in technology, markets or institutions that can unseat even a well-conceived plan of action.

Moreover, because environments and goals change, business and technology strategies will also change. Thus, Step 6 includes activities to acquire routine data on performance and trends and to communicate to relevant users in the firm, particularly where modifications in investments (Step 3) are required.

This step also provides information inputs back to Step 1 to initiate the ASSETS process in future efforts; maintaining business advantages over the long term requires the ability to update the technology strategy and re-think current technology investments and applications. This means that information systems for storing and massaging data to create useful "early warning" inputs is completed along with data collection. Also, procedures to distribute information and generate discussion are important here. One of the clearest advantages of structured processes like ASSETS is the opportunity for organizational learning (10). To be most useful, the information must have been screened for relevance and validity. The screening process is iterative since needs and environmental trends change so frequently. An important aspect of this step is group interaction, to involve interested stakeholders in the process of developing and distributing information. This makes the results of the step more credible.

**Tools** For these activities, several tools exist and the list is expanding. In the area of scanning the environment and monitoring technology trends, a number of computerized databases exist with interactive software tailored to user-specified information profiles (3, 6).

Communication procedures are also important here. Technical or product newsletters are useful for some rapidly changing areas of technology. Electronic mail systems and bulletin boards are now becoming more common, as is the use of group decision procedures (3). A useful organizational approach to this step is the practice of encouraging "gatekeepers" (10).

**Results** The principal output of Step 5 is information, which is stored for later use and which provides inputs to the current situation assessment in Step 1. The type of information depends on the conditions in a particular firm. In general, the information needs are tied to how the firm monitors its environment and revises its plans and programs. In the framework of Box A, Questions 9 and 10 are addressed in this step.

## 4 IMPLEMENTATION OF AN ASSETS PROCESS

Introducing a process such as ASSETS into an organization is best undertaken as a phased approach. Because success with structured planning approaches in an organization requires careful consideration of cultural and political factors, it usually takes several planning cycles for adoption to be completed. Moreover, even straightforward adoption of an external method almost always involves some modifications to tailor its features to the unique aspects of an organization's needs.

Channeling and focusing technology management efforts through a process that is systematic and which can be followed continuously is essential. There are many advantages to this approach, but of particular importance is the storehouse of knowledge and experience that is built as the process is utilized time after time. The necessity for frequent interactions among those developing the business strategy involves information and experience which should be captured by the organization for future use. Organizational learning is one of the most important benefits of structured processes.

The steps outlined in this paper are designed to be put in place gradually over time. A summary of several suggestions for a phased implementation of an ASSETS-type process is shown in Box D. These steps have been developed from several years of experience of introducing technology strategy concepts in firms in the U.S.A., Western Europe and Japan. All steps are not essential in each organization, and this is not an ex-

**BOX D**

SUGGESTIONS FOR ASSETS IMPLEMENTATION

**What Should Be Done in Phase 1?**

**1** Identify a technology/business strategy working group.
**2** Pick some winner areas to show early success.
**3** Identify the "top ten" technology areas (STAs) for the company.
**4** Write "white papers" on how to get into the top ten and circulate for review.
**5** Require all component business plans to discuss technology strategy.
**6** Develop recommendations on how to proceed with the weak areas.

**What Should Be Done in Phase 2?**

**1** Convene the technology/business strategy group.
**2** Publish a periodic newsletter of technology developments.
**3** Prepare background forecasts and studies for a few key technologies.
**4** Identify additional (long-term) technology candidates.
**5** Show progress on year 1 programs and report on results.
**6** Prepare a written technology plan and coordinate internally.

**What Should Be Done in Phase 3?**

**1** Emphasize development and evaluation of specific programs.
**2** Be prepared to terminate some projects and continue others.
**3** Systematically monitor areas of technology outside existing programs.
**4** Refine an approach to "hand-off" ideas to the deployment groups.
**5** Reinforce an innovation philosophy:
  **a** Use innovation cells
  **b** Allow R&D staff to follow projects to commercialization
  **c** Set up a separate organization for innovations
  **d** Permit rewards to accrue to individuals

haustive list. Rather, the list represents a summary of experience in implementing several formal planning and analysis methods in companies seeking to improve their technology decisions (2, 10).

## 5 FUTURE DIRECTIONS IN TECHNOLOGY STRATEGY

The development of technology strategy and the use of tools such as patent analysis and technological forecasting to support planning is in its infancy compared to the process of setting conventional business strategy. As technological competitiveness is pursued more diligently by companies, much more progress will be made. The key areas where change can be expected include data, methods and organizations.

### 5.1 Data Sources

The databases used in technological decision-making are often limited in scope and have uneven quality. Increases in demand for better information to support decision-making are expected to lead to improved and more specialized databases. At present, access to the literature on technologies is available on-line via commercial databases. Search procedures allow identification of studies that can usefully be analyzed. Future literature analyses will require a referencing mechanism by which the document reviewer identifies how and where the information in the document can be used and applied by the recipient.

To illustrate, an increasingly used fundamental database can now access worldwide patent filings and patent award data files. This data, available for most of today's technologies, provides references, citations, filing and/or award dates, names of inventors, and the companies. Special tools are being developed and applied to permit analysis of the patent data to answer questions regarding technological "position," possible future rates of technology diffusion and markets for technological concepts and areas (19).

### 5.2 Analysis Methods

Future technology strategy and management methods will continue to emphasize tool development for data analysis. A good example is the development of patent analysis tools. Literature analysis in the future will require a mechanism that can search larger data sets for key words (already available in certain text-processing software but not in patent analysis software), and incorporate expert judgement and insight in support of the analysis.

In technology forecasting, improved methods should become available to address typical senior management questions. At present, most forecasting tools require the support of a technical data and methods expert to derive results. In the future, the need for such specialized support should decrease as the tools become more directly usable by the decision-maker.

All forecasting is a process of exercising judgement; new or improved tools and techniques are being designed to facilitate that judgement-making, particularly scenario based techniques now being applied by many companies. These predictive techniques are generally used in conjunction with workshops where those who attend actively participate. Automated methods that take judgements directly from a person and input them into an analytical program are being developed and used now. New software developments are expected to be more time-and-cost-effective in supporting decision-making than traditional analytical models and methods (13, 19).

### 5.3 Organizational Approaches

Companies are finding that to give adequate attention to technology strategy and management, it is essential to assign the person or group in charge designated responsibility, funding and a specific charter. Often a position on the level of Vice-President of Technology oversees the duties. Alternatively, a special task-force may be established. Sometimes this function is placed in the strategic planning area. There is no uniformity in how technology strategy and management actions are organized.

However, certain patterns do emerge. It seems vital that the person appointed be at a level reporting to the CEO (16). This gives the activity the necessary visibility and influence to be successful. The other participants in the process of identification and selection of subjects should come from the business side, including specialized areas like marketing, production or distribution. If the activities are placed in the central R&D organization, technology acquisition recommendations tend to be made in the form of additional R&D, and other methods of acquisition may not be considered.

Supported by a budget that covers the required work and analysis, the staff should help collect, organize, and analyze appropriate data. Generally, a committee is set up to decide on investments in use of technology in production or products. But unless this committee actually produces tangible results—e.g., a series of recommendations and investment successes—or is itself given a budget and staff, it is difficult to convince decision makers of a technology's importance.

Most serious attempts to introduce a broader perspective on technology require commitment that is longer than the typical annual planning cycle for the organization. But here the strategy finds itself in a classic situation—planning based on too short a time horizon. The technology strategy, planning and management functions must have resource commitments that extend beyond the normal budget cycle and are authorized from a level above operating division R&D budgets. This is why the function is sometimes found in corporate planning, assuming a company has a corporate planning organization.

For companies that are fully committed to a systematic technology planning approach the unit responsible for technology strategy and management will probably

- Be attached at a very high level in the company, possibly reporting to the CEO;
- Be relatively small in terms of funding, but have review and sign-off responsibility for technology investment decisions;
- Have a continuing responsibility to develop and maintain strategy and plans and to inform senior management of major shifts and changes in the technological environment as it affects the company's business;
- Have long-term oriented funding in view of the company's business and technological environment, and recommend and promote investments in technologies that will pay-off beyond the normal, expected investment payback period.

A strong case can be made for relying, in part, on sources outside the firm for inputs to technology decisions—to introduce some "disinterested objectivity" into the process and to inject fresh thinking into the firm's traditional culture (20).

## 6 CONCLUSION

A number of issues remain to be addressed in developing technology acquisition and deployment processes that are usable and effective in a broad range of situations. Many companies are struggling with the integration of business and technology in an effective framework for the management of technology. To assist in building the required knowledge, several important factors essential for successful use of a technology strategy have been outlined above. Since the technology aspect of business management is in its infancy compared with activities such as financial management, more tools and experience can be expected in the future. In the meantime, continued attention to purposeful use of technology strategy is essential for most companies to exploit enormous growth and profit potential from effective acquisition and use of advanced technology.

### Reading 9.1 References


**1** Foster, R. N. (1986). *Innovation: The Attacker's Advantage.* Summit Books, New York.

**2** Leonard-Barton, D. and Kraus, W. A. (Nov.–Dec. 1985). "Implementing new technology" *Harvard Business Review* Harvard University, Boston, Massachusetts.

**3** Martin, M. J. C. (1984). *Managing Technological Innovation and Entrepreneurship.* Reston Publishing Co. Inc., Reston, Virginia.

**4** Frohman, A. L. (Jan.–Feb. 1982). "Technology as a competitive weapon." *Harvard Business Review* Harvard University, Boston, Massachusetts.

**5** Mitchell, G. R. (1988). "Options for the strategic management of technology." *International Journal of Technology Management,* Vol. 3, No. 3, pp. 253–262.

**6** Sammon, W. L. et al. (1984). *Business Competitor Intelligence.* John Wiley & Sons, New York.

**7** Bhalla, S. K. (1987). *The Effective Management of Technology.* Battelle Press, Columbus, Ohio.

**8** Kantrow, A. M. (July–Aug. 1980). "The strategy–technology connection." *Harvard Business Review,* Vol. 59, No. 4, pp. 6–8.

**9** Porter, M. E. (1983). "The technological dimension of competitive strategy." in R. S. Rosenbloom, *Research on Technological Innovation, Management and Policy,* Vol. 1. JAI Press, Greenwich, Connecticut, pp. 1–33.

**10** Steele, L. W. (1989). *Managing Technology: The Strategic View.* McGraw-Hill, New York.

**11** Ansoff, H. I. and Stewart, J. M. (Nov.–Dec. 1981). "Strategies for a technology-based business." *Harvard Business Review,* Vol. 45, No. 6, pp. 71–83.

**12** Huss, W. R. and Honton, E. J. (1987). "Scenario planning—what style should you use?" *Long Range Planning,* Vol. 10, No. 4, pp. 21–29.

**13** Ashton, W. B. and Sen. R. K. (Jan.–Feb. 1989). "Using patent information in technology business planning—II." *Research-Technology Management,* Vol. 32, No. 1, pp. 36–42 Industrial Research Institute, New York.

**14** Frohman, A. L. (Nov.–Dec. 1984). "Meshing technology with strategy." *Research Management,* Vol. 27, No. 6, pp. 36–42.

**15** Manahan, M. (1989). "Technology acquisition and research prioritization." *International Journal of Technology Management,* Vol. 4, No. 1, pp. 9–19.

**16** Smith, P. L. (March–April 1988). "Tighten the linkage between research business strategy and marketing." *Research-Technology Management* Vol. 31, No. 2, pp. 6–8.

**17** Stacey, G. S. (1983). "Tech forecasting and the TECH RISK array." *BTIP Review No. 14.* Battelle Memorial Institute, Columbus, Ohio.

**18** Stacey, G. S. and Ashton, W. B. (May 1988). *Integrating Business and Technology Planning in a Global Environment* paper presented at the 1988 International Conference on Strategic R&D Management, Tokyo, Japan.

**19** Tschulena, G. R. et al. (1986). *Databases: Their Development as a Management and Planning Tool.* BTIP Program, Battelle Memorial Institute, Columbus, Ohio.

**20** Drucker, P. (10 Feb. 1988). "Best R&D is business driven." *Wall Street Journal* New York.


## DISCUSSION QUESTIONS

1 Select either Figure 9-13 or Figure 9-14. Write a couple of pages explaining the diagram. Why are the concepts related in the way they are?

2 Play the role of a manager in a growing high-tech firm. How would you go about implementing each phase depicted in Figure 9-14?

## ADDITIONAL READINGS

Marcie J. Tyre & Wanda J. Orlikowski. "Exploiting Opportunities for Technological Improvement in Organizations." *Sloan Management Review,* Fall 1993.

To exploit the advantages of new process technologies, managers must adapt those technologies to fit the organization and its strategy. Adaptation to new technologies is a cycle of successive change and routine. The authors argue that these episodic cycles can be better managed by exploiting the opportunities for change that a new technology brings about, using periods of relative stability in data collection and improvement, and looking for new opportunities of adaptation.

Peg Young. "Technological Growth Curves." *Technological Forecasting and Social Change,* 44, 1993, pp. 375–389.

Nine different growth curve models were each fitted onto various data sets in an attempt to determine which growth curve model achieved the best forecast. The analysis of the results gives rise to a new approach for selecting appropriate growth curve models for a given set of data, prior to fitting the models, based on the characteristics of the data sets.

Vijay Mahajan & Eitan Muller. "Timing, Diffusion, and Substitution of Successive Generations of Technological Innovations: The IBM Mainframe Case." *Technological Forecasting and Social Change,* 51, 1996, pp. 109–132.

Based on the behavioral assumptions of diffusion theory, this article proposes an extension of the Bass diffusion model that simultaneously captures the substitution pattern for each successive generation of a durable technological innovation, and the diffusion pattern of the base technology. Normative guidelines based on the model suggest that a company should either introduce a new generation as soon as it is available or delay its introduction to a much later date at the maturity stage of the preceding generation.

George A. Pogany. "Cautions about Using S-Curves." *Research-Technology Management,* July–August. 1986, pp. 24–25.

The author argues that S-curves are not identical in shape, that not all organizations are suited to inventing and developing revolutionary technologies, and that the S-curve is not as suitable in guiding R&D but is very good in guiding investment policy into hardware.

Philip D. Metz. "Integrating Technology Planning with Business Planning. Five 'Best Practices' Emerge from a Two Year Study of 50 Companies." *Research-Technology Management,* 1996, pp. 19–22.

The author identifies the five "best practices" followed by companies that link technology planning with strategic business planning.

## SUGGESTED ASSIGNMENTS

1 Using the guidelines of the technology audit model (TAM), prepare a checklist and conduct an audit of a local company. Then discuss the result of the audit in class. (Students may work individually or in groups.)

2 Retrieve one of the tools discussed in reading 9.1 from the referenced book or article. Discuss its use in details showing its advantages and limitations.

## REFERENCES


Betz, Frederick. 1987. *Managing Technology: Competing through New Ventures, Innovation, and Corporate Research.* Prentice-Hall, Englewood Cliffs, NJ.

Bhalla, Sushil K. 1987. *The Effective Management of Technology.* Battelle Press, Columbus, OH.

Crosby, P. B. 1979. *Quality Is Free.* Signet/New American Library, New York.

Ford, David. 1988. "Develop Your Technology Strategy." *Long Range Planning,* vol. 21, no. 5, Oct., pp. 85–94.

Garcia-Arreola, Javier. 1996. *Technology Effectiveness Audit Model: A Framework for Technology Auditing.* Master's thesis, University of Miami.

Hamel, Gary. 1996. "Strategy as Revolution." *Harvard Business Review,* July–August, pp. 69–82.

Lewis, William, and Linden, Lawrence. 1990. "A New Mission for Corporate Technology." *Sloan Management Review,* Summer, pp. 57–65.

Little, Arthur D. 1981. "Strategic Management of Technology," *European Management Forum,* Davos.

Little, Arthur D. 1983. *Maturing Chemical Business: An Approach to Renewal.* Arthur D. Little Report No. 831001, Oct., Cambridge, MA.

Madox, N., Anthony, W. P., and Wheatley, W. Jr. 1987. "Creative Strategic Planning Using Imagery." *Long Range Planning,* vol. 20, no. 5, pp. 118–124.

Martin, Michael J. C. 1994. *Managing Innovation and Entrepreneurship in Technology-Based Firms.* Wiley Interscience, New York.

Millet, S., & Honton, E. J. 1991. *A Manager's Guide to Technology Forecasting and Strategic Analysis Methods.* Battelle Press, Columbus, OH.

National Critical Technologies Panel. 1991. "Report to the President," *NCTP 1101.* Wilson Boulevard, Washington, DC.

Porter, A., Roper, A., Mason, T., Rossini, F., & Banks, J. 1991. *Forecasting and Management of Technology.* Wiley, New York.

Steele, Lowell W. 1989. *Managing Technology.* McGraw-Hill, New York.

Willyard, Charles H., & McClees, Cheryl W. 1987. "Motorola's Technology Roadmap Process." *Research Management.* September–October, pp. 13–19.

# 10

# THE ACQUISITION AND EXPLOITATION OF TECHNOLOGY

In the quest to keep up with the pace of technological change, organizations must be able to acquire needed technologies and exploit their own technologies. Decisions in this regard must be included in an organization's strategies and planning effort. This chapter introduces methods of acquiring and exploiting technology and discusses factors affecting decisions in these areas. Special emphasis is placed on research and development as a critical activity in the creation of technology. The role of governments and large corporations in R&D is discussed, and recent global trends are presented. National and corporate strategies for R&D have a profound impact on the competitive posture of industry in any country. These strategies, in turn, influence the level of technological capability and thus can lead to improvements in the products and services offered and to increased wealth.

## ACQUISITION OF TECHNOLOGY

Technology planning encompasses the development of plans for the acquisition of technologies that will impact a firm's competitiveness. Information about these technologies is derived from technology audits that detail all technologies and subtechnologies used in the value chain. The audits also reveal technologies owned by the company and those owned by outside companies. The assessment of strengths and weaknesses in these technologies and the company's flexibility for migrating from one technology to another are important. The forecasting of technological changes is also needed, as well as gaining knowledge of what competitors are doing. The latter can be done by benchmarking the firm's technologies in relation to the technologies of others in the same domain or markets.

Managers must then make a choice as to which technology or set of technologies the firm should immediately acquire or pursue in the future. Deciding when to enter into a new technology or to migrate from an existing one is also important. Questions to be resolved include whether the firm should be a leader or a follower in a particular technology and how this is going to impact competitiveness.

The next issue is how the firm intends to acquire the technologies that it needs. For this decision, a manager should evaluate and examine the pros and cons of each option.

### Methods of Acquiring Technology

There are several recognized methods for the acquisition of technology: (1) using internal R&D, (2) participating in a joint venture, (3) contracting out for R&D, (4) licensing in of technology, and (5) buying the technology from others. These methodologies are described briefly here:

**1** *Using internal R&D:* In this method, the company relies on its own human and technical resources to develop the technology in-house. This requires the presence of a strong technical workforce and strong financial backing for R&D operations. Some large companies such as General Electric (GE), General Motors (GM), American Telephone and Telegraph (AT&T), and Du Pont have their own R&D laboratories to support their efforts to create new technologies.

**2** *Participating in a joint venture:* Two or more firms combine their know-how and technological resources to develop technologies. An example of this is the joint venture between International Business Machines, Motorola, and Apple to develop the Power PC chip, or the joint venture between Motorola and Toshiba, in which Motorola can use its strength in microprocessor technology and Toshiba can use its strength in memory chips.

**3** *Contracting out for R&D:* By contracting out, a company can conduct R&D without having to invest heavily in an in-house R&D effort. Many companies are increasing their use of this approach to cut R&D expenditures. The popularity of this method also increased after the cold war, when many former defense- and military-oriented R&D establishments became underutilized. These laboratories have the necessary human and technical resources and have started making their services available commercially. Many companies take advantage of this opportunity and contract out their R&D projects to them.

**4** *Licensing in of technology:* In this method, a company purchases the right to utilize technologies owned by someone else. In the mid-1950s, the Sony Corporation bought a license for the transistor from AT&T and was able to widely deploy the technology in its products. Observe the number of transistor-based products that Sony produces today. In the service industry acquiring a franchise of a well-known corporation such as Burger King is a common practice. There is a premium to be paid for using the name of that corporation and for offering its products.

**5** *Buying the technology:* In this method, an outright purchase of technology occurs. This is the fastest way to obtain a technology, and does not involve any resource commitment for technology development on the part of the acquirer. However, there is no control over the technology, and no real acquisition is considered to have occurred. Using this method to get access to technology requires building strong bridges with the supplier of the technology to guarantee the continued and timely support of the technology to ensure long life cycle. This method of technology acquisition is suitable for external types of technology.

Ford (1988) developed a very useful matrix that shows the applicability of different acquisition methods under different circumstances. It considers five factors upon which a company can make an acquisition decision. These are (1) the company's relative standing in the technology, (2) the urgency of acquisition, (3) the level of commitment to the acquisition or the level of investment involved, (4) the technology position on the life-cycle curve, and (5) the classification of the technology as distinctive, basic, or external. The matrix is shown in Figure 10-1. Engineers and managers can use such a matrix as a guide during the decision-making process.

The matrix shows the applicability of each method of acquisition according to the criteria listed at the top of the matrix. For example, if a company's relative standing in a technology is high, it makes sense to capitalize on this strength and build new technology internally. This approach allows the company to consolidate its position in a core area of strength and helps in promoting its technology. Conversely, if a company's relative standing in a technology is low, buying the technology is the method preferred. Likewise, if the urgency of acquisition is high, buying or licensing the technology is the preferred method. Relying on R&D in this case is likely to be costly and time-consuming and to lack a guarantee of success. But if the urgency of acquisition is the lowest, the internal R&D method of acquisition is an option to be considered.

**FIGURE 10-1**
FACTORS AFFECTING THE TECHNOLOGY ACQUISITION DECISION

*Source:* David Ford, "Develop Your Technology Strategy." Reprinted from *Long Range Planning,* Vol. 21, No. 5, 1988, p. 91, with permission from Elsevier Science.

| Acquisition Methods | Company's Relative Standing | Urgency of Acquisition | Commitment/ Investment Involved | Technology Life Cycle Position | Categories of Technology |
|---|---|---|---|---|---|
| Internal R&D | High | Lowest | Highest | Earliest | Most Distinctive or Critical |
| Joint Venture | | Lower | | Early | Distinctive or Basic |
| Contracted-out R&D | | Low | | Early | Distinctive or Basic |
| License - in | | High | Lowest | Later | Distinctive or Basic |
| Non-acquisition i.e. Buying final product or part production | Low | High | No Commitment/ Investment | All Stages | External |

## EXPLOITATION OF TECHNOLOGY

Technology can be thought of as an asset or a commodity to be purchased and sold. A company that owns certain technology should include technology exploitation as a component of its technology strategy. The methods of technology exploitation resemble those used for acquisition. Decisions to exploit often contradict those of acquisition.

If a company is strong in a technology, its success in licensing the technology to others is high. Otherwise, it has to prove its technology through internal application. It may have to use it in its own products until the technology proves its worth to the larger market. Licensing out enhances the opportunity for technology diffusion while requiring low commitment for financial investment. Technologies with a wide range of applications are more valuable to license out, given that the company does not have to get into the investment, the support technologies, or the marketing of diversified products.

Technology may need to be exploited as rapidly as possible for a company to have a chance at getting the market's general acceptance of its technology and at defining the industry standard. It may also need to be marketed widely to get good market penetration that increases the technology's market share. This strategy prevents, or at least discourages, competitors from developing a different version of the technology that can change or capture the market. Meanwhile, distinctive technologies must be protected. A company should delay the sharing of its distinctive technology but should not delay too much lest the technology lose its value for exploitation. (See the discussion about diffusion in Chapter 3 and the Apple computer company's case in Chapter 15.)

Ford (1988) developed an exploitation matrix, which includes factors affecting technology exploitation decisions. The matrix is shown in Figure 10-2 and can be used to guide managers in developing their exploitation strategy.

The matrix shows the method of technology exploitation to be used according to the company's position/standing in the technology and six other criteria listed at the top of

**FIGURE 10-2**
FACTORS AFFECTING THE TECHNOLOGY EXPLOITATION DECISIONS

*Source:* David Ford, "Develop Your Technology Strategy." Reprinted from *Long Range Planning,* Vol. 21, No. 5, 1988, p. 92, with permission from Elsevier Science.

| Exploitation Methods | Company Relative Standing | Urgency of Exploitation | Need for Support Technologies | Commitment/ Investment Involved | Technology Life Cycle Position | Categories of Technology | Potential Application |
|---|---|---|---|---|---|---|---|
| Employ in own Production or Products | Lowest | Lowest | Lowest | Highest | Earliest | Most distinctive or critical | Narrowest |
| Contracted-out Manufacture or Marketing | Lower | High | High | | Early | | Narrow |
| Joint Venture | High | Low | High | | Early | | Wide |
| License - out | High | Highest | Low | Lowest | Later | Least distinctive or peripheral technologies | Widest |

the matrix. For example, if the urgency of exploitation (criterion 2) is the dominant concern and the company standing in the technology is high, exploitation by licensing out the technology is the best choice for diffusing the technology.

## TECHNOLOGY CREATION THROUGH R&D

The creation of technology can result from either an individual or a group effort. In the past, individual efforts of inventors and trial-and-error approaches were the dominant modes used to develop technology. In the modern era, these approaches were replaced by more organized efforts. Complex organizations, involving many employees, have been established to undertake R&D activities, and the scientific approach to problem solving has been widely utilized. Contemporary R&D involves coordinating the activities of many disciplines that are collaborating to make a contribution to technological progress.

The rise of organized R&D activities can be traced back to Thomas Edison, who in 1876 established a scientific research laboratory at Menlo Park, Pennsylvania. Edison and his colleagues conducted research and applied their findings to the development of the electric bulb and other revolutionary products. Alexander Graham Bell, the great inventor, needed scientific help before he was able to develop the telephone in the late 1800s. Later, AT&T established its Bell Laboratories for conducting basic research and development to advance technology. Other companies, such as Du Pont, Dow Chemical, and General Motors, created huge R&D laboratories.

The U.S. government also created a number of R&D establishments to support the defense industry and national security effort. Examples are the research laboratories at Los Alamos, New Mexico, and at Wright Patterson Air Force Base in Ohio. Preparing for and participating in World War II created a dramatic rise in the number of scientists and engineers working on R&D. A major effort was mounted to boost innovations in aerospace, electronics, nuclear power, and manufacturing. After the war, the U.S. government was instrumental in establishing a number of R&D agencies to advance both basic and applied research, including the National Science Foundation, the Atomic Energy Commission, and the Armed Services Scientific Offices, as well as the expansion of the National Institutes of Health.

Private industry and academic institutions also expanded their involvement in research, having appreciated the role of science and technology in increasing knowledge and creating opportunities for growth. Nonprofit organizations, such as universities, took advantage of government-funding opportunities to build their laboratories, fund graduate education, and enlarge their research bases.

## STAGES OF TECHNOLOGY DEVELOPMENT

Organized technological development follows a hierarchical progression: (1) basic research, (2) applied research, (3) development, and (4) technology enhancement.

**1** *Basic research:* This is research undertaken to gain new scientific knowledge or understanding; it is not directed toward a specific practical aim or application (Organization of Economic Cooperation and Development, 1970). According to the National Science Foundation (1985), the objective of basic research is to gain a fuller knowledge or understanding of the subject under study, rather than to develop practical applications. Basic research is conducted to advance science, which can be thought of as a process of generating and accumulating knowledge over a long period of time (Allen, 1977). Basic research can be either "pure" or "oriented," depending on whether it was performed at the will of the scientist or was steered by another entity toward a field of particular interest.

While basic research is perceived to be a drain on organizations' resources, and may not lead to an immediate commercial return on investment, it is essential for new discoveries and for the growth of knowledge.

**2** *Applied research:* This is research directed toward a specific practical aim or objective and conducted to develop ideas into operational form (Organization of Economic Cooperation and Development, 1970). According to the National Science Foundation (1985), it is directed toward gaining the knowledge or understanding necessary to meet a recognized and specific need. Applied research is a mix of science and engineering.

**3** *Development:* Development involves the systematic use of the knowledge or understanding gained from research to produce useful materials, devices, systems, or methods, including the design and development of new or improved services. Development work falls more within the realm of engineering than within the realm of science. Development effort is a connecting link between research and the commercial use of ideas.

**4** *Technology enhancement:* This is the continuous effort by scientists and engineers to support and improve existing or developed technologies. It aims to improve the performance parameter of the technology, lengthen the technology life cycle, and foster incremental innovations.

Bhalla (1987) makes some pertinent observations related to science and technology development:

- Science builds on prior science, except for rare random discoveries.
- Technology builds on prior technology.
- Technology development goes through multiple stages, each stage requiring different skills and talents.
- Key technologies require 8 to 15 years to develop. This observation indicates that the time horizon of technology planning is significantly longer than that of business planning. The challenge to managers is to forecast technological change and make necessary provisions for it in their business plans.

## THE TECHNOLOGY PORTFOLIO AND INDUSTRIAL R&D

One of the main concerns of managers is what type of research the firm should undertake and which technologies it should emphasize for development. The answer is not a simple one. It depends on the objectives of the R&D program, the type and sector of the business, its technology base, its customers, its financial and technical resources, and many other pertinent factors. Schmitt (1985) divided corporate research into generic research versus targeted research and market-driven research versus technology-driven research. These are useful classifications for companies because they link research programs to the objectives of the research. Another classification introduced by Merten and Ryu (1982) proposed dividing industrial laboratory research activities into five categories:

**1** Basic research.
**2** Exploratory research.
**3** Development of new commercial activities.
**4** Development of existing commercial activities.
**5** Technical services.

In general, it is believed that a company should engage in R&D to the extent necessary to create a strong technology portfolio to support its activities. A technology portfolio is similar to a business portfolio, in which investment is made in a number of stocks rather than in one stock. This strategy echoes the popular saying: "Don't put all your eggs in one basket." The solution is to diversify investments across a wide spectrum of stocks, bonds, securities, and so on. Similarly, a technology portfolio can be selected to support all aspects of the company's technology, from pure research to development to maintaining and embracing existing business. A generic technology portfolio model is shown in Figure 10-3.

Jain and Triandis (1990) proposed the following R&D needs, which apply to any company technology portfolio:

- *Normative needs:* Here the research is directed toward satisfying the needs of the user, the user being the primary or follow-on beneficiary of the research product.
- *Comparative needs:* Here research is driven by the need to stay in the race with competitors.

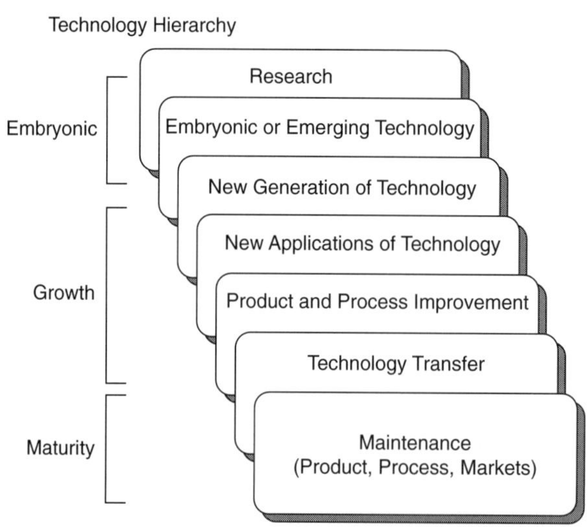


**FIGURE 10-3**
TECHNOLOGY PORTFOLIO CATEGORIZED BY STAGE OF TECHNOLOGY

**EXHIBIT 10-1**
R&D EXPENDITURE, BY INDUSTRY SEGMENT

| | Share of total U.S. Industry R&D* | | |
|---|---|---|---|
| **Industry Segment** | **1981** | **1988** | **1995** |
| Information/electronics R&D | 32% | 42% | 44% |
| Drug/medicines R&D | 7 | 9 | 16 |
| Combined share | 39 | 51 | 60 |

*Approximate.
*Source:* Mitchell, 1997.

- *Forecast needs:* This research is driven by the forecast of future changes in technology, products, consumer behavior, or new regulations.

Corporations use R&D basically to lead and support innovation. Areas where this is relevant are:

- Product innovations.
- Material innovations.
- Process innovations.
- Market innovations for new business development.
- Service innovations.

The R&D effort to support innovation varies from one industry sector to another. Companies working in the sectors of electronics, aircraft, and chemicals invest more in R&D than they do in equipment and plants (Pavitt and Pattel, 1988). As shown in Exhibit 10-1, in the United States, R&D expenditures in information, electronics, drugs, and medicine account for 60 percent of the combined share of total R&D. In some sectors, such as steel, automobiles, and segments of the electronics industry where the U.S. competitive edge declined, Japan's industries spend 30 percent more of their output on R&D than their U.S. counterparts spend.

Corporate-level R&D must devote its energy to highly leveraged opportunities—the ones that create new business or turn entire businesses around (Schmitt, 1983). Leveraging opportunities require up-front R&D investment in a balanced spectrum of programs. These can include:

1. Focused and targeted short-term projects.
2. Focused and targeted long-term projects.
3. Speculative and exploratory work.
4. Supportive research projects for existing products and services.

## JUSTIFICATION OF R&D EXPENDITURES

Conducting R&D requires both human and financial resources. The R&D effort in corporations competes for resources with traditional areas of production, including materials, labor, equipment, facilities, and sales. In corporations with established R&D efforts research directors have to justify the R&D programs undertaken by their laboratories. This is not an easy task since R&D programs by nature are risky endeavors. They require expenditure without a guarantee of profitable return. Top executives may view R&D as a cost without immediate revenues. U.S. managers' tendency to focus on a "short time horizon" (Berman and Khalil, 1992), a practice based largely on a financial rather than technical perspective, compounds this problem for many companies. Management's focus on short-term return for expenditure, which justifies expenditures solely on the basis of their immediate contribution to the bottom-line profit, does not bode well for favorable decisions on R&D projects. It is recognized that longer-term and more risky programs may ultimately bring large benefits to the corporation. Historical and current facts indicate that R&D expenditures are directly linked to innovation, productivity and quality improvement, increased market share, and many of the factors that contribute to organizational competitiveness. In all cases, funding decisions for R&D are dependent on the justification methodology used by R&D managers to get management approval. Existing methods for funding R&D in corporations are usually based on one of the following:

1 *R&D is supported as an overhead (OH) expense.* In this case management considers R&D as a necessary cost of the business. This method of funding indicates that management is committed to R&D. However, this method has practical limitations in terms of determining an appropriate level of funding that will not increase the overhead to a level detrimental to the overall financial performance of the company. Mitchell (1988) indicated that this method of funding is suitable for projects directed toward knowledge building, that is, projects in which exploratory or basic research is undertaken as one end of the continuum in a technology portfolio.

2 *R&D is supported as an investment.* In this case funds are allocated to R&D on the basis of the company's traditional financial criteria justifying capital budgeting. One of the most common criterion used is return on investment (ROI).

The ROI and similar financial justification methodologies are inherently biased against long-term R&D projects, in which the future is uncertain. Returns on revolutionary technological innovations are often underestimated. If the project is considered risky, the endeavor is difficult, if not impossible, to justify under this method. Many promising projects are killed this way. Mitchell (1988) indicates that ROI is clearly appropriate for technical development and engineering programs whose market and financial implications are understood well enough to permit meaningful quantification of the ROI model parameters. Therefore, ROI is suitable for justifying R&D projects at the downstream end, where uncertainty of outcome is reduced or eliminated.

There is an important segment of the technical activity covering applied research, exploratory development, and feasibility that is difficult to justify using either of the two funding models mentioned above. Projects in this area may require large expenditures, which are difficult to accept as a cost of doing business under the OH funding scheme. Meanwhile, the potential impact of the projects is still too uncertain to justify their funding under an ROI investment-funding scheme. These projects are often asso-

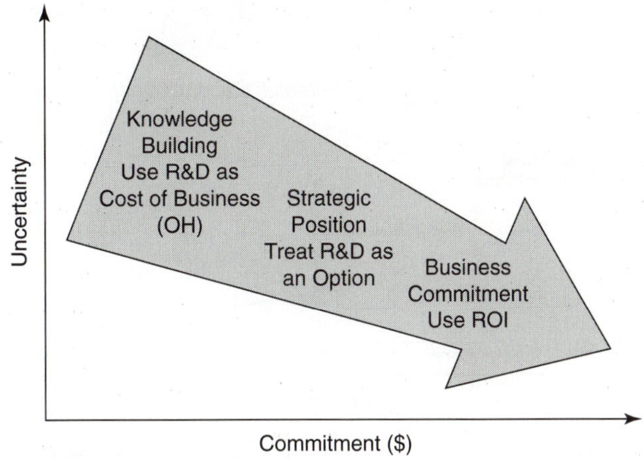


**FIGURE 10-4**
STRATEGIC OBJECTIVES OF R&D AND PREFERRED METHOD OF FUNDING
*Source:* Mitchell & Hamilton, 1988.

ciated with the strategic positioning of the company (see Figure 10-4). Mitchell and Hamilton (1988) recommend that justification of such projects be made according to the American call option of the stock market. The option scheme allows a purchaser to acquire stock at a specified price (exercise price) at any time prior to an agreed-upon expiration date. This method reduces the risk of a major expenditure, thereby permitting the company to delay a decision on investment while maintaining a position in the technology.

The price of the call option is equivalent to the cost of the R&D activities at the time of the call. If the option is exercised in the future before its expiration date, an additional price will have to be paid. This is equivalent to the added cost of R&D that the company will have to invest to consolidate its R&D effort. The value of the stock at that time, which is analogous to the value of R&D, may be higher. Thus the company's return on its initial investment in the price of the call will be rewarded. The risk of losing the cost of the option, should the company decide not to exercise its option, is not overburdening for the company. Management is more likely to accept strategic positioning projects justified financially as an option. Mitchell and Hamilton's paper is recommended reading for this chapter.

Many R&D projects with potential opportunities may exist, but they may require expenditures beyond the range acceptable to an organization. For such projects, government funding, cost sharing with other organizations, or strategic technological and financial alliances with other organizations might be needed to bring the projects into the region of feasibility, as shown in Figure 10-5.

A national technology policy can be formulated to make use of the concepts discussed above. Technology policy can encourage consortia and other cooperative business alliances, provide matching funds to industry, and facilitate wide diffusion of

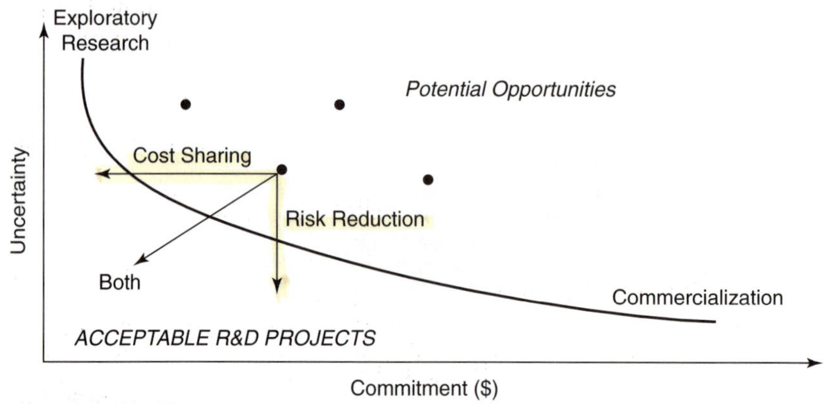


**FIGURE 10-5**
STRATEGIES FOR BRINGING R&D PROJECTS TO ACCEPTABLE LEVEL
These strategies involve changing the cost and risks associated with new technologies.
*Source:* Mitchell, 1995.

federal R&D. By doing so, the policy seeks to drive investment decisions of industry to the range of acceptable R&D projects on the curve shown in Figure 10-5. This places the risks and rewards of particular investments within the bounds of acceptable cost and uncertainty (Brody, 1996).

### General Observations on Industrial R&D

• Large firms undertake the major part of all industrial R&D. In the United States, 300 of the largest companies account for 92 percent of research expenditure. The 40 largest companies account for 70 percent of industrial research expenditure. Large innovative firms usually have strong R&D laboratories.

• The larger the company, the more likely it is to have some research activities.

• Although the great bulk of R&D takes place in large firms, significant inventions have emanated from small firms and from individuals. Examples include xerography and the Apple II personal computer. These innovations occurred because of the entrepreneurial spirit of individuals like Chester Carlson and Joseph Wilson, in the case of Xerox, and Steve Jobs and Steve Wozniak, in the case of Apple.

• Only a small percentage of R&D projects turn out to be a commercial success (a ratio of 1 to 10 in some industries and sometimes as low as 1 to 3,000).

• Small firms are credited with expanding national employment figures; they are effective in matching technology with customer needs. However, they are less likely to hire highly educated scientists and engineers. This is due to either a lack of financial resources or an underestimation of the role of innovation in global competitiveness.

• R&D can be divided into two activities: research and development. Development, in particular, is a costly endeavor, and it is difficult for individuals and, in some cases, small firms to raise the necessary resources to bring an idea to the market. Government, safety, environmental, and legal regulations have increased the cost of the development effort.

- In the early stages of the product life cycle, technology is more crucial than development and marketing activities. Once the technology passes the stage of scientific acceptability and proves its value, effort shifts toward development, production, and marketing.

## THE GOVERNMENT AND MILITARY ROLE IN R&D

Most industrialized countries recognize the importance of R&D and its role in developing technologies to create economic growth and preserve national security. Exhibit 10-2 presents R&D expenditures by country as a percentage of the GNP of each country.

The U.S. government has generously supported research and development activities, particularly in areas connected with the preservation of national security. National R&D expenditure in defense-related endeavors has been greater than or equal to R&D expenditure for nondefense-related endeavors as previously shown in Figure 7-12. In 1993 federal funding for military R&D was $41.42 billion and for civilian R&D was $28.34 billion. Industry funding for the same year was $99.5 billion. The traditional R&D role played by the Defense Department has contributed significantly to the growth of technological knowledge in the United States. Many industries have benefited from defense technology programs over the years, whether through direct involvement in the programs or through technology transfer. Yet there remains a significant amount of defense-related technology that has not found its way into the commercial sector. This technology may contribute to national security but not to the creation of wealth.

In contrast to the United States, Japan devotes a great majority of its R&D expenditure to the commercial sector. This is a significant factor in the emergence of Japanese industry as a fierce competitor in commercial products. It also led to growth in Japan's economy. Japan's investment in R&D has moved Japanese industries from being imitators of technology to being leaders in innovation. U.S. expenditure in the various sectors of R&D has changed over the years as shown in Figure 10-6. The variations in R&D sectors' expenditure depends on many political, social, economic, and environmental factors.

**EXHIBIT 10-2**
NATIONAL R&D EXPENDITURES, AS PERCENTAGE OF GNP

| Year | United States | France | Germany | Japan | United Kingdom | USSR |
|---|---|---|---|---|---|---|
| 1965 | 2.8 | 2.0 | 1.7 | 1.5 | NA | 2.9 |
| 1970 | 2.6 | 1.9 | 2.1 | 1.8 | NA | 3.3 |
| 1975 | 2.2 | 1.8 | 2.2 | 2.0 | 2.2 | 3.8 |
| 1980 | 2.3 | 1.8 | 2.4 | 2.2 | NA | 3.8 |
| 1985 | 2.7 | 2.3 | 2.7 | 2.9 | 2.2 | 3.8 |

*Source:* Extracted from various sources & Department of Commerce data. (Figures are rounded.)

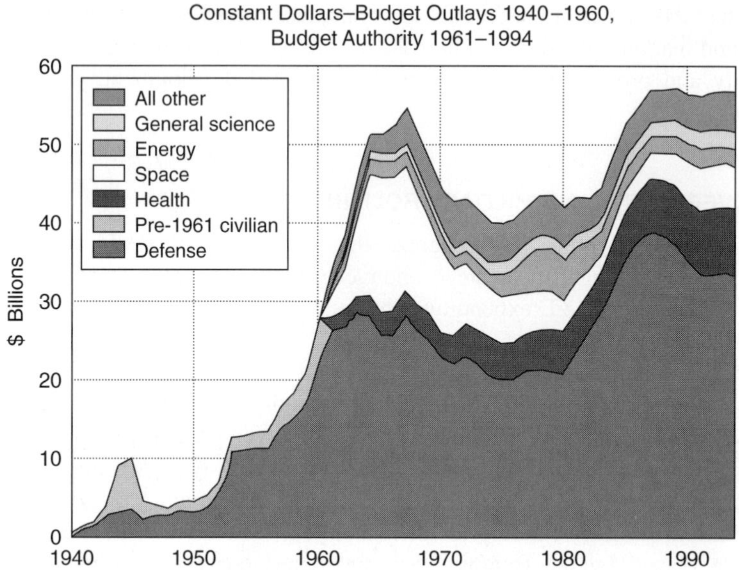


**FIGURE 10-6**
FEDERAL R&D FUNDING FOR DEFENSE AND CIVILIAN FUNCTIONS, 1940–1994
*Source:* Brody 1996.

## GLOBAL MANAGEMENT OF R&D

The complexity of today's technology-based R&D, business dynamics, and markets is compelling companies to reconsider scale, size, and location, as well as the scope and direction of their activities. In a review of corporate trends in R&D, Perrino and Tipping (1989) found the following characteristics:

**1** While markets are global, technology will continue to develop locally in *pockets of innovation.*

**2** Critical mass will be a key factor in successful technology development. Cost has increased because of the need for interdisciplinary teams.

**3** External relationships are becoming more important—survival is no longer a matter of being a *technological island;* rather, it depends on being part of a research consortium, joint venture, or the like.

**4** Companies can get more out of their research by linking it more closely to market needs and customer requirements rather than by increasing spending.

**5** The global network model is the wave of the future. This implies supporting technology core groups in each major market (United States, Japan, and Europe). All groups should be managed in a coordinated way.

Perrino and Tipping found that European, American, and Japanese companies deploy their R&D resources in different patterns:

- The European pattern is based on acquisition of entire companies.

• The Japanese pattern is based on home production and centralized R&D plus "listening posts" for acquiring technology rather than on overseas expansion of R&D.

• The U.S. pattern is based on setting up overseas laboratories staffed by their own company's U.S. or foreign nationals employees.

The six years between 1987 and 1993 witnessed significant growth in the number of foreign-owned businesses in the United States, with a dramatic $8.1 billion increase in their R&D spending (see Figure 10-7 and Exhibit 10-3). U.S. corporations have also

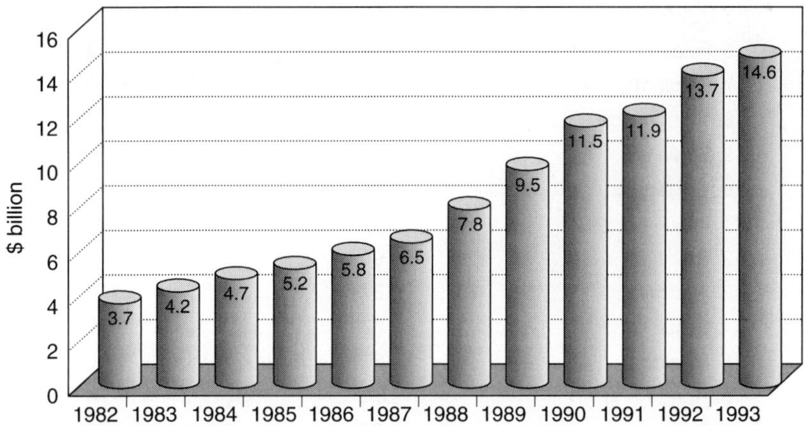


**FIGURE 10-7**
R&D EXPENDITURES BY U.S. AFFILIATES OF FOREIGN COMPANIES
*Source:* Dalton and Serapio (1995).

**EXHIBIT 10-3**
R&D EXPENDITURES AND EMPLOYMENT BY AFFILIATES OF FOREIGN COMPANIES IN THE UNITED STATES

| | Expenditures ($, millions) | | |
|---|---|---|---|
| **Country** | **1987** | **1993** | **R&D Employees (thousands)** |
| All countries | 6,521 | 14,618 | 105.2 |
| Switzerland | 765 | 2,524 | 14.7 |
| Germany | 1,139 | 2,321 | 19.2 |
| United Kingdom | 833 | 2,295 | 20.0 |
| Japan | 307 | 1,781 | 11.8 |
| France | 366 | 1,204 | 9.3 |
| Netherlands | 542 | 691 | 6.3 |
| Canada* | 1,666 | 2,190 | 10.3 |

*Canadian affiliates include a major U.S. chemical company with a minority Canadian investment.
*Source:* Dalton and Serapio, 1995.

**EXHIBIT 10-4**
EXPENDITURES FOR U.S. R&D ABROAD

| Country | 1989 ($, millions) | 1993 ($, millions) |
|---|---|---|
| 1. Germany | 1,496 | 2,568 |
| 2. UK | 1,673 | 1,639 |
| 3. Canada | 914 | 1,030 |
| 4. France | 545 | 942 |
| 5. Japan* | 488 | 862 |
| 6. Ireland | 134 | 669 |
| 7. Belgium | 317 | 460 |
| 8. Netherlands | 360 | 392 |
| 9. Spain | 115 | 321 |
| 10. Singapore | 25 | 312 |
| 11. Italy | 294 | 304 |
| 12. Brazil | 90 | 220 |
| 13. Australia | 181 | 176 |
| 14. Switzerland | 67 | 109 |
| 15. Mexico | 37 | 76† |
| 16. Hong Kong | N/A | 74 |
| 17. Sweden | 33 | 48 |
| Total | 7,048 | 10,954 |

*Of U.S. R&D facilities in Japan, more than half were established or acquired during the past seven years.
†1992 R&D data.
*Source:* Dalton and Serapio, 1995.

made significant commitments to R&D overseas. They almost doubled their R&D expenditure in Japan and Germany between 1989 and 1993 (Exhibit 10-4).

It is evident from Exhibit 10-3 that establishing facilities within a country contributes to the employment of many people and thus to the overall improvement of that country's economy. It is also evident from the continued trend of companies' establishing foreign affiliates in the United States and all over the world that globalization of R&D, production, and marketing is fully under way. Exhibit 10-5 shows the reasons given by foreign executives for locating their R&D activities in the United States. These reasons vary from one industry sector to another. The three major reasons are acquiring technology, being close to the customer, and meeting U.S. environmental regulations.

Most of the countries that are actively seeking to compete in the global arena have increased their levels of support for R&D activities. Figure 10-8 shows the level of national and government expenditures in science and technology in China from 1988 to 1994, where expenditures more than doubled in the six-year period. The rise in the Korean government's science and technology (S&T) investments over the five-year period from 1990 to 1995 is even more impressive (Figure 10-9). The Japanese government

**EXHIBIT 10-5**
REASONS FOR FOREIGN R&D INVESTMENT IN THE UNITED STATES*

| | Electronics | Autos | Biotechnology |
|---|---|---|---|
| Acquire technology | 1 | 2 | 1 |
| Keep abreast of technological developments | 2 | 2 | 1 |
| Assist parent company in meeting U.S. customer needs | 1 | 1 | 3 |
| Employ U.S. scientists and engineers | 2 | 3 | 2 |
| Follow competition | 3 | 3 | 4 |
| Take advantage of favorable research and development | 4 | 4 | 1 |
| Cooperate with other U.S. R&D labs | 2 | 3 | 2 |
| Assist parent company in meeting U.S. environmental regulation | 4 | 1 | 4 |
| Assist parent company's U.S. manufacturing plants in procurement | 4 | 2 | 4 |
| Engage in basic research | 3 | 4 | 2 |

1 = Extremely important
2 = Important
3 = Neutral
4 = Unimportant

* Reasons given by senior R&D/technical executives.
*Source:* Dalton and Serapio, 1995.

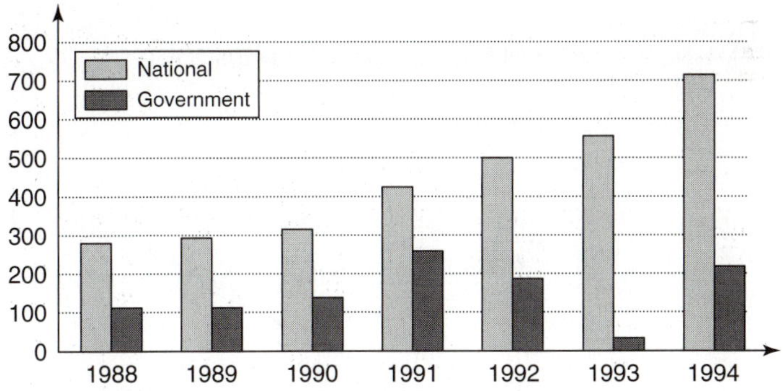


**FIGURE 10-8**
CHINESE NATIONAL AND GOVERNMENT S&T EXPENDITURES
*Note:* Exch. rate: 8.3 Yuan to US $1.00.
*Source:* Office of Technology Policy, 1997.

has been steadily increasing its science and technology budget since the early 1970s, as shown in Figure 10-10. Technology development is a major factor in driving the economic growth of nations. Figure 10-11 shows a number of countries' GDP per capita plotted against their annual expenditure in R&D as a percentage of GDP. There is a strong correlation between high per capita income and high R&D expenditure. Countries around the world have come to realize the importance of R&D in creating and

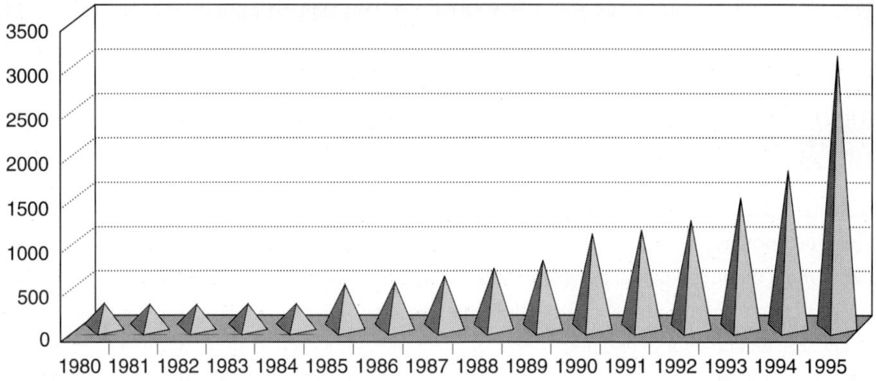


**FIGURE 10-9**
KOREAN GOVERNMENT'S S&T INVESTMENTS
*Source:* Office of Technology Policy, 1997.

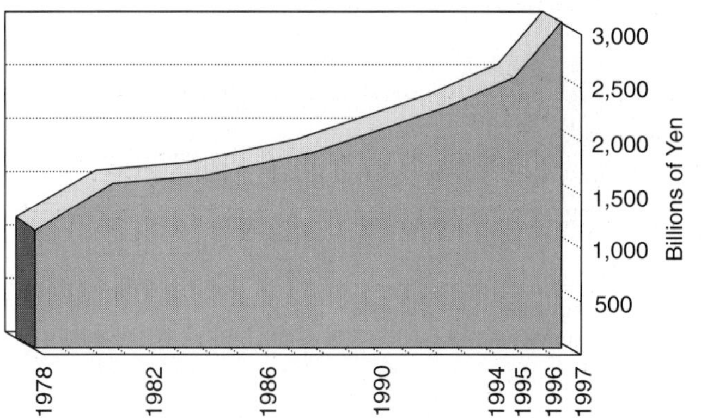


**FIGURE 10-10**
JAPANESE GOVERNMENT S&T-RELATED BUDGET
*Source:* Office of Technology Policy, 1997.

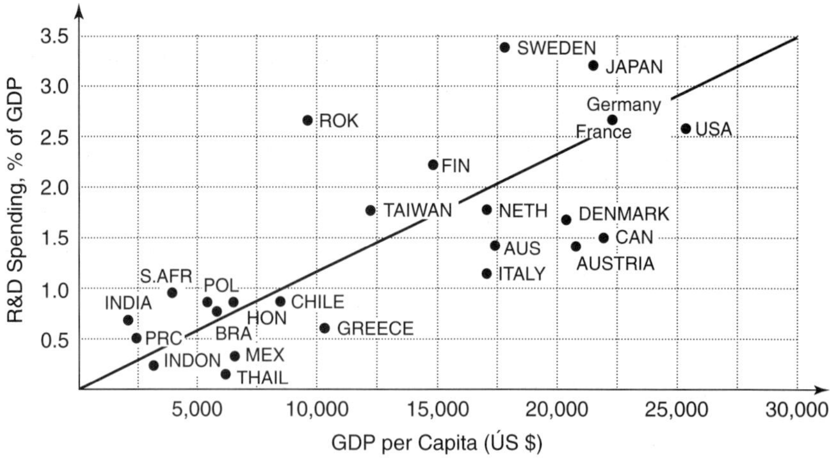


**FIGURE 10-11**
RELATIONSHIP BETWEEN NATIONAL R&D SPENDING AND GDP PER CAPITA, 1994
*Source:* Mitchell, 1997.

developing technologies and in improving the standard of living. Many countries have responded with increased levels of expenditure in R&D.

## CONCLUDING REMARKS

Technology planning requires the development of plans for the acquisition of technologies from outside sources and the exploitation of internal technologies that may have value outside the firm. There are many options available to managers to effect the acquisition and exploitation of strategies. Each one of those options has advantages and disadvantages. Therefore, technology managers must be able to recommend the optimal strategy on the basis of their knowledge of the technologies, the competition, and the markets.

Technology owned by a firm has its value. However, experience has shown that owning the technology often requires strong involvement in R&D. It also requires a special strategy to protect the technology and exploit it for the firm's overall benefit.

R&D has a hierarchical progression, from basic research to applied research to development to technology enhancement. Top management must decide on the company's appropriate level of involvement in R&D and how and where to pursue it. R&D is often perceived to be a risky investment. R&D managers have an important role in clarifying to their business counterparts the benefits of R&D and the risks of not pursuing certain projects. Innovative justification methodologies may be helpful in this regard.

Governments have an important role to play in enhancing the critical technologies needed to support indigenous industries. A government's expenditure on R&D provides essential support for basic research and significant support for the overall environment that permits technological progress in a country. This, in turn, has a positive effect on the nation's GDP and on its citizens' standard of living.

READING 10.1

# Options for the Strategic Management of Technology

**Dr. Graham R. Mitchell***

*Abstract:* Fast-moving technologies are changing the rules of business for many U.S. corporations. New management approaches are needed to better couple business strategy and technology, and to offset the impact of short-term financial perspectives. The paper discusses the development of these approaches in a major U.S. corporation, and their application to the corporate research laboratory.

* At the time of writing this article Dr. Mitchell was Director of Planning, GTE Laboratories Incorporated, 40 Sylvan Road, Waltham, Massachusetts 02254, U.S.A.

*Source:* From T. Khalil, B. Bayraktar, and Edosomwan (eds.), *Technology Management I.* Interscience Enterprises, Geneva 1988.

## THE STRATEGIC IMPACT OF TECHNOLOGY

Fast-moving technologies are changing the rules of business for many U.S. industrial corporations. Advances in electronics are shortening product lifetimes for a wide range of consumer and industrial products. Flexible manufacturing, factory automation, and new control systems are changing the "set points" and historic rules of thumb throughout manufacturing. Advances in software, particularly database systems and artificial intelligence, are increasing productivity and creating many new products and markets in the rapidly growing service sector. Even the boundaries between hitherto distinct market segments are becoming blurred as a result of technological changes. This is occurring between telecommunications and computing. The challenge presented to U.S. general management by this rapid rate of technological change, even when they were originally trained as technologists, is daunting. Set against the context of declining international competitiveness of U.S. industry, many would argue that it is a challenge which is often inadequately addressed.

### Two Issues for Technology Management

Over the last decade, two broad issues have been frequently raised which go to the heart of the strategic management problem.

The *first* is usually described as a concern with the poor coupling between strategic planning and technology, and arises from the failure of strategic management systems to give business adequate warning of fundamental shifts in the competitive balance within an industry and the opportunities for growth brought about by technological innovation [Kantrow (1980)].

The *second* related concern is that even when the potential impact of a new technology is recognized at the conceptual or strategic level, overreliance on short-term measurement and justification within U.S. corporations often biases the implementation process against some of the more strategically important technical programs [Abernathy and Hayes (1980)].

### The Strategic Role of Technology in Industry

These twin problems arise in part because of the largely operational role assigned to technology in most strategic planning and management approaches. To the general management and planning community, technology or engineering is perceived to be a subset of business vying for resources with other functions, such as marketing, manufacturing, operations, etc. From this viewpoint, the appropriate role of the technical community within industry is largely to manage and to carry out those programs necessary to implement business strategy—programs which when implemented will often be justified as part of a business investment, using a capital budgeting framework, possibly ROI, as a basis for choosing between alternatives.

*By contrast, the R&D community within many U.S. corporations may well feel that technical advances by its peers around the world are a much better guide to the long-term direction of the corporation than the formally documented strategies of the planners.*

The strategic importance of technological innovation comes about because it can lead to sustainable competitive advantage, often by extending and improving the corporation's family of products or services, by reducing costs, or by improving the operating systems and the way the corporation does business. However, the process of acquiring the needed technical insight and capability requires not only resources but, more importantly, time. Thus, in order to fully capitalize on developing business opportunities, many corporations must have anticipated the technical needs of the business, and strategically positioned themselves with an understanding of, and access to, the relevant technology.

The choice of the technical areas in which to establish a strong strategic position in anticipation of downstream business investment opportunities is among the most important decisions for U.S. corporations. Often it must be made before the range of commercial applications are well understood. Unfortunately, most formal planning systems provide little guidance on how to make these critical choices, and most accepted financial approaches provide only limited help on how to allocate resources to them.

## KNOWLEDGE BUILDING—STRATEGIC POSITIONING—BUSINESS INVESTMENT

While it is common to classify technical work by activity, such as basic and applied research, development, and engineering, it is often more useful when addressing the management of technology to recognize that these activities are aimed at several different strategic objectives. Most technical work in industry, often development and engineering, is part of a *business investment.* At the other end of the spectrum, many corporations recognize the need to carry out some exploratory research and general awareness activities directed toward *knowledge building.* However, in order to bring this exploratory work to the point where it can be successfully exploited in a business investment, it is often necessary to increase resources and focus research to develop a *strategic position* or capability in the critical technical area.

In industrial R&D organizations, very real limits exist to the rate at which resources or the mix of technical skills can be changed. With relatively fixed resources, the trade-off between knowledge building and strategic positioning in industrial research laboratories is illustrated in Figure 1. The rapid growth of new scientific and technological areas of potential interest to the corporation pressures research management up the curve, to cover more areas at a relatively low level of funding per area. The need for strategic positioning forces decisions the other way, down the curve, toward increased focus and higher resource levels for only the highest priority technical areas. In practice most R&D managements settle for an uneasy balance between the two sets of pressures, recognizing that at any time unanticipated advances such as the recent breakthrough in superconductors may replace previous priorities at either the exploratory or focused end of the spectrum.

The practical problem facing many corporations as they try to strategically manage technology to respond to and anticipate the needs of their businesses is that there is often:

- No generally accepted language for unambiguously identifying and defining the most critical technologies in which the corporation should be strategically positioned for the achievement of its business goals;

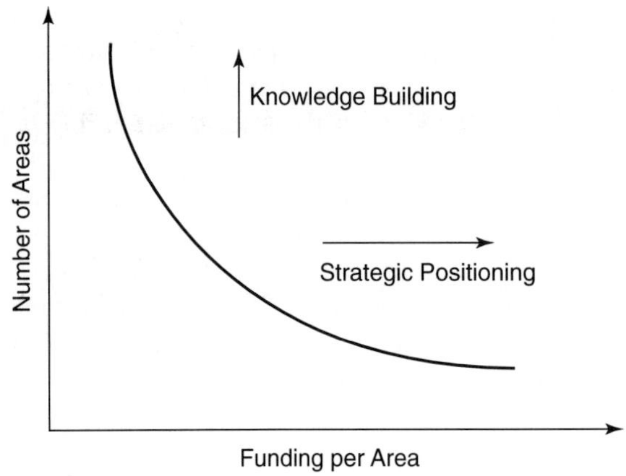


**FIGURE 1**
STRATEGIC PRESSURES ON TECHNICAL RESOURCES

- No way to manage these technologies beyond their implementation in specific business developments;
- No appropriate financial framework for allocating resources to strategic positioning.

## DEFINITION OF TECHNOLOGY

In order to start to address these issues and discuss the strategic impact of technology, it is necessary to define technology in a way in which it can be understood by the technical community within the corporation, and in a way which addresses the priorities as seen by the business management within the operating groups [Mitchell (1985)]. Scientists or engineers tend to define technologies in terms of skills or disciplines, such as circuit design, solid-state physics, or heat transfer. The general business community is more likely to speak in terms of classes of products, such as integrated circuits, central office switches, or even communications systems. The different approaches to the definition are not simply semantic, but reflect alternative viewpoints as to what is most important. The engineer tends to stress the input to the process; management, and for that matter the general public, are more likely to think of the output of the technology.

It is difficult to overestimate the importance of this classificatory language, and the care with which it needs to be developed [Steele (1975)]. In GTE, as in other corporations, we have been able to combine the skills and their application in a common definition. This unit of analysis, the *"strategic technical area"* (STA), is composed of four elements:

1. Skills or discipline → 2. which are applied to a → 3. particular product, service → 4. which addresses a specific market need.

An example of an STA definition that covers *Integrated Circuit Processing* is:

1. The principal skills include lithographic techniques (photo, x-ray, electron beam) for defining fine geometries on semiconductors, as well as high-temperature solid-state chemistry and thin-film processing. 2. These are applied to the fabrication of semiconductor integrated circuits, 3. which are used in a wide range of switching and transmission products, 4. throughout the telecommunications industry.

### Identifying STAs—Building the Network

The simplest way to develop a clear picture of which areas of technical expertise are most important to the corporation as it presently exists is to define the STAs at the operating unit level. For each of the product lines or services produced by the business unit, engineers or scientists are challenged to identify the critical skills necessary to maintain strong or leadership positions and to place particular emphasis on those areas where increased technical capability will produce increased benefits for the business. The number of STAs developed for a product line or service depends on the complexity of the product or service. Usually three or four areas are enough to identify the most critical skills.

As the exercise is repeated for adjacent products or services, it often becomes clear that many of the same underlying strengths or skills are utilized in several different business areas. Thus, definitions can be broadened, and an STA/product line (or service) matrix can be developed, as shown in Figure 2. This illustrates that technical skill, for example, in the design of high-density integrated circuits or software engineering may be recognized as critical in several adjacent product lines. As the process is extended to all business units across the corporation, numerous instances of parallel technical expertise will usually emerge. STAs originally developed for each product or service line may be broadened, and business unit, divisional, group, and eventually corporate-level STAs may be defined. Somewhere in the region of 15 to 20 STAs seems optimum for analysis of technical strategy at each organizational level.

The STAs have been defined to include those technical skills most critical to the survival and growth of the corporation. These are also the target of the corporation's discretionary technical expenditures, the R&D program. It is thus a relatively straightforward process to arrive at the present and planned resource allocation for the entire corporation by STA (Bitondo, 1986).

**FIGURE 2**
STA/PRODUCT LINE MATRIX FOR A TELECOMMUNICATIONS PRODUCTS BUSINESS UNIT

| Strategic Technical Area (STA) | Product Line 1 | Product Line 2 | Product Line 3 | Etc. |
|---|---|---|---|---|
| Integrated Circuit Fabrication | X | | | |
| Integrated Circuit Design | X | X | X | |
| System Architecture | | X | | |
| Software Engineering | X | X | X | |

## MANAGING BY STRATEGIC TECHNICAL AREA

### Organizational Perspective

The aggregation of technical activities into a limited number of STAs automatically focuses attention on strategic issues and longer-term trends for the underlying technology. In addition, the perspective addressed by STAs automatically broadens as the STAs are aggregated to a higher organizational level. At the lowest level of aggregation, e.g., at the individual business unit, the time horizon is shortest, the critical technical issues most immediate, and the coupling to business goals most direct. At the division and higher levels, this view lengthens to the three-to-five-year medium-term perspective, and the discussion of technical strategy often deals with the opportunity to capitalize on the synergies between similar pools of technical expertise revealed by the STAs. In the highest aggregation at the corporate level, appropriate broad strategic issues, such as the provision of long-term technical strength and technical human resource requirements for the corporation for the next five years or longer, as well as the creation of radically new options, emerge directly from the identification of overall corporate technical needs by means of the STA analysis.

### Evaluation of Competitive Technical Strength

One example of the way in which STAs are used to guide resource allocation to technology and influence strategic thinking is to use them to answer the question, "How good are we technically, compared to our major competitors?"

The STAs have been defined as the most important technical areas for the future of the corporation, and are thus the natural unit against which to make this comparison. A simple profile that indicates the number of STAs in which the business unit is behind, equal, or leading these competitors communicates very graphically to general management the technical health of the business. For example, groups with profiles shown in Figure 3a are reasonably healthy insofar as their competitive position is at least equal to the major competitors in over 70 percent of the technical areas most critical to the success of the business. The key issue is future trends; if projections show a deterioration in position, significant additional funding or alternative sources of technology may be required.

**FIGURE 3**
COMPETITIVE TECHNICAL POSITION

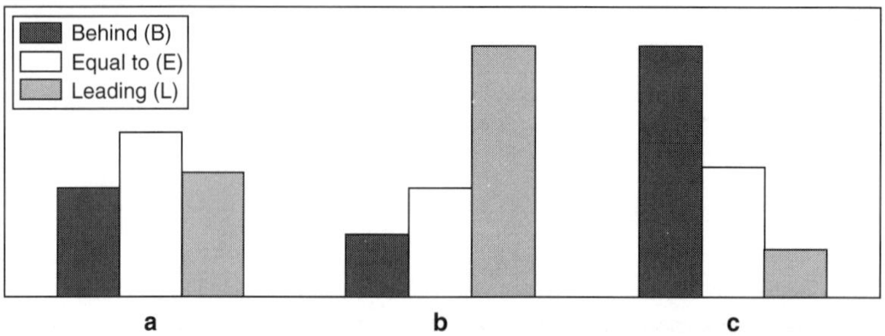

Those businesses which are technically very strong will have a profile such as the one shown in Figure 3b, and here the strategic question most frequently asked is, "Are the businesses effectively using this technical capability to expand position and exploit their full technical potential in the market?"

Businesses with the characteristic shown in Figure 3c clearly face questions of viability over the long haul. This profile may be either symptomatic of a business in poor health or a principal cause of its future decline. In either case, this profile is a good indicator of the need for fairly drastic managerial action.

## DEVELOPMENT OF STRATEGIC POSITIONING TARGETS

A review of the STAs at the corporate level indicates those areas which are most important to the current business profile. In addition, some of the new and emerging areas of science and technology are also critical for future growth and survival and these must be focused to build new strategic capabilities. Within GTE, these combined corporate inputs are used to set the strategic priorities of the central research laboratories. The strategic direction for an R&D program of over 100 individual projects, in telecommunications, software, photonics, electronics, and materials, is developed and focused within approximately 15 STAs.

An illustration of the way work moves from exploratory *knowledge building* through *strategic positioning* to eventual *business investment* is illustrated in Figure 4. Within the general topic of software, artificial intelligence is one of the fast moving areas. There are a number of artificial intelligence topics relevant to telecommunications, such as pattern recognition, speech recognition, speech synthesis, expert systems, self-improving

**FIGURE 4**
EVOLUTION OF TECHNICAL PROGRAMS

| Area of Technology | Knowledge Building (Exploratory Topics) | Strategic Positioning (Research Focus) | Business Investment (Potential Applications) |
|---|---|---|---|
| *Software*<br>Artificial Intelligence | Pattern Recognition<br>Speech Recognition<br>Speech Synthesis<br>Expert Systems<br>Self-Improving Systems<br>Robotics<br>User Modeling | *Expert Systems*<br>Knowledge-Based Systems<br>Local Languages—Lisp/Prolog<br>User-Friendly Interfaces | Operating System Improvements<br>Telecommunications Switching Maintenance<br>Electronic System Design<br>New Services |
| *Photonics*<br>Fiber Optics | Coherent Systems<br>Optoelectronic/Quantum Well Device<br>Advanced Components<br>Low-Loss Glasses<br>Nonlinear Optical Components | *Low-Cost Components*<br>High Performance<br>-- Sources<br>-- Detectors<br>Gallium Arsenide on Silicon | Long-Haul Point-to-Point Fiber-Optic Transmission Systems<br>High-Density Fiber-Optic Transmission Systems<br>Fiber-Optic-Based Services |

systems, robotics, and user modeling, which are addressed at a modest level of exploratory effort. However, significantly higher levels of effort must be devoted to any one topic if it is ever to be developed to the stage where it may have commercial payoff. In the example, expert or knowledge-based systems have been chosen; potential applications include several near-term opportunities in operations, design, and marketing, with the promise of many more, but as yet poorly defined, commercial possibilities. The decision to build a strong strategic position in this area of technology by setting up an STA in Expert Systems must be taken before commercial returns can be demonstrated with certainty. In addition, with limited available resources, the decision to build a strong strategic position by focusing resources in any one area causes significant soul searching, as it necessarily precludes a similar focus on some other promising exploratory research area.

A similar situation is illustrated for photonics, where rapid technical advances in fiber optics are revolutionizing both the design of equipment and networks for telecommunication services. Success in the provision of future networks and in the delivery of new high-bandwidth communications services will be strongly influenced by the rate of technical advance and deployment of this technology. A number of exploratory topics, including the study of coherent systems, low-loss glasses, and nonlinear optical components, are evolving fast and all are potentially very relevant to the future of fiber-optic systems. The decision to focus on (in this example) low-cost, high-performance components arises because a critical element facing the deployment of this new transmission is the performance and cost of the electro-optic equipment at the ends of the fiber. Potential business applications include new transmission products and improved operation of telecommunication systems.

## OVERCOMING SHORT-TERM FINANCIAL BIAS

The practical problem facing most U.S. corporations in the allocation of resources for technology programs is that decisions are made on the basis of one of two funding models:

- R&D as a business investment, using some form of capital budgeting framework, or
- R&D as an overhead.

Most technical work in industry is directed toward implementation and will be funded as a business investment. By contrast, since the determination of returns is often impossible for knowledge building or exploratory research, it is usually treated as an overhead. The problem is that with only two financial approaches, once the level of funding exceeds that with which management is comfortable in allocating as an overhead, the only alternative is to force fit the technical program to ROI or similar criteria [Mitchell and Hamilton (1988)]. This is particularly damaging to strategic positioning programs.

An important first step in dealing with R&D expenditures for strategic positioning is to recognize that they are not so much directed toward an investment as they are toward the creation of an *option.* By this it is meant that the corporation is committing relatively modest R&D expenditures now in the expectation that it will provide the opportunity to make a profitable investment at some later date (Kester, 1984). There are several implications of treating this situation financially as an option, which are counterintuitive from a capital budgeting perspective (Jarrow and Rudd, 1983). For example, because the downside risk is limited to the cost of the R&D program, other factors being equal (e.g., mean value of the expected returns), the option having greatest uncertainty is to be preferred. This is equivalent in the analysis of stock options to the value of a call option increasing with the volatility of the stock. A second implication is that longer term options are preferred over shorter ones. These relationships are very much in alignment with the intuitive position of the U.S. R&D community, which over the years has continually, often unsuccessfully, tried to make the case for the strategic value of longer range programs having high upside potential, even though the potential commercial impact may be uncertain.

Classifying R&D into these three broad strategic objectives helps clarify not only funding approaches, but the difficult issues of technical versus business decision-making responsibility, and the appropriate perspective on markets (Mitchell and Hamilton, 1988). In general, the R&D community assumes the prime responsibility for choosing technical priorities for knowledge-building programs, the business community for business investment programs. Both perspectives need to be combined in defining areas for strategic positioning, and because of the importance of the "bet" to the future direction of the company, the decision should be approved at the highest levels of management within the corporation.

The treatment of markets is also different for the three sets of objectives. In choosing targets in the early stages of most knowledge-building research, the technical opportunity and rate of technical advance are principal determinants; market potential is usually very unclear and consequently of little significance. At the other extreme, the analysis of specific sales or cost reductions is almost certainly needed before committing to technical programs which are incorporated in business investments. The appropriate market perspective for strategic positioning falls between the two extremes and requires a broad view of potential markets, which includes potential market options beyond present commitments for the businesses.

## CONCLUSIONS

There have been two persistent issues which have plagued industrial R&D and business in the United States for over a decade. The first concerns a failure to recognize the strategic implications of technical innovation on business strategy and translate these technical advances into business advantage. The other concerns the adverse effect on the U.S. competitive position which has resulted from the dominance of short-range financial perspectives. It is suggested that they are both linked to a failure to explicitly focus on and strategically manage the underlying technologies necessary for survival and growth of the business. In the present highly competitive environment, many business opportunities in leading industries are open for less time than it takes to fully develop or acquire the technical capability needed to capitalize on them. For success, therefore, it is necessary to be technically positioned with deep insights into some of the critical technical areas *before* the market or other business opportunities fully emerge. This paper discusses the process by which some of these critical technical areas have been identified in a major U.S. corporation and used to guide the strategy of a corporate laboratory.

As the decision to focus resources in a particular area must be taken before the benefits to the corporation can be clearly determined, this paper suggests that the traditional use of capital budgeting and particularly ROI frameworks is inappropriate for selecting strategic positioning targets. The overuse of capital budgeting approaches leads to a conservative short-term overall bias in the selection process. Strategic positioning programs are more accurately described as *options* rather than investments. Using a financial options framework produces selection criteria which more closely fit the intuitive position of the R&D community in that longer-term programs, and programs with high upside potential, receive higher priority than when treated with ROI approaches.

### Reading 10.1 References

Abernathy, W. J. and Hayes, R. H. (July–August 1980). "Managing Our Way to Economic Decline," *Harvard Business Review,* vol. 58, no. 4, pp. 62–77.

Bitondo, D. S. (1986). "Technology Planning in Industry: The Classical Approach," Chapter 4, *Interdisciplinary Planning: A Perspective for the Future,* Dluhy, M. J. and Chen, K., eds., New Brunswick, Center for Urban Policy Research, Rutgers University.

Jarrow, R. A. and Rudd A. (1983). *Option Pricing,* Homewood, Illinois, Dow Jones–Irwin.

Kantrow, A. (July–August 1980). "The Strategy-Technology Connection," *Harvard Business Review,* vol. 58, no. 4, pp. 6–21.

Kester, W. C. (March–April 1984). "Today's Options for Tomorrow's Growth," *Harvard Business Review,* vol. 62, no. 2, pp. 153–160.

Mitchell, G. R. (1985). "New Approaches to the Strategic Management of Technology," *Technology in Society,* vol. 7, no. 2/3, pp. 132–144.

Mitchell, G. R. and Hamilton, W. F. (May–June 1988). "Managing R&D as a Strategic Option," *Research Technology Management,* pp. 15–22.

Steele, L. W. (1975). *Innovation in Big Business,* New York, American Elsevier Publishing Company, Inc.

READING 10.2

# Changing Environment for R&D Leaders: New Challenges, New Responses

**Deb Chatterji**

Managing Director—Technology, The BOC Group

*Abstract:* In recent years, many companies have witnessed dramatic changes in their business environment. Emergence of global markets and competitors coupled with new competitive strategies based on quality, speed and/or alliances have forced business managers—especially in the United States—to adopt new management strategies,

*Source:* This is an updated version of Chatterji, Deb, 1993, "Emerging Challenges for R&D Executives: An American Perspective," *R&D Management,* July. Reprinted with permission from R&D Management 23 (1993), pp. 239–248, copyright by Blackwell Publishers Ltd.

structures and systems. These, in turn, have caused many R&D executives to progress from their traditional agenda of managing R&D activities in domestic laboratories to a new agenda of coordinating and integrating technology development and exploitation on a worldwide basis. This paper discusses this evolution of R&D management agenda in the United States and its implications. It also reviews the approaches being used and the experiences being gained by the industrial R&D management community to address the emerging challenges.

## I. INTRODUCTION

Nothing, it is said, is certain except death and taxes. In recent years, business leaders in general—and R&D executives in particular—have come to face yet another inescapable certainty: change.

While one might argue that change is not a recent phenomenon, the forces and events of the last few years have given new meaning to the word for the R&D executive. Today he is confronting a much broader and stronger spectrum of change and challenge than he has experienced before. The changes are particularly unsettling for many American R&D executives as they may not have looked beyond the United States for expansion or may not have faced intense foreign competitors in the past.

What are the most important changes seen by these R&D executives in recent years and why are they significant? The paper begins on this note and then addresses the related questions: How should an R&D executive respond? What are the constraints he may have to face? And it ends with a brief review of the approaches being used and the insights being gained by the R&D management community in the United States in dealing with the emerging challenges.

## II. RECENT CHANGES AND THEIR SIGNIFICANCE

Globalization. Decentralization. Quality management. Concurrent engineering. Mergers and acquisitions. Leveraged buy-outs. Strategic alliances. Core competencies. Benchmarking. Workforce diversity. Environmental imperative. Information and communication revolution. The list is by no means complete but it gives some idea of the many issues that have entered and altered the management agenda of many business executives.

### Changing Business Agenda

What is the "new" business agenda and how does it differ from the "old" agenda? Recognizing the fact that there is no universal agenda and accepting the risk of generalization, one can make the following observations:

- For most businesses, global market and competition have become harsh realities. The comforts of home market and predictable industry structure and competition have all but disappeared.
- "Think globally—act locally" has become a fundamental management tenet for companies that aspire to succeed as well-integrated worldwide businesses. For most companies, decentralization has become the logical imperative.

• Success now requires strong competitive advantage in the form of innovation, quality, agility and/or productivity (in both labor and knowledge). Cost-leadership alone is not sufficient; it must be combined with other winning ingredients.

• Strategic alliances and partnerships have become popular. Many forms of alliances have emerged ranging from preferred vendor arrangements to technology-based partnerships.

• Information and communication technologies are impacting management systems and practices in numerous ways, creating new threats and opportunities.

• Measurement of performance is being emphasized for every business activity. Organizations are expected to be lean and agile, and individuals are required to add value at every level. "Overhead" is under constant attack.

• Financial success alone can no longer satisfy all the stakeholders, namely, the shareholders, the employees, the customers, the suppliers, the public, and the authorities. For example, good environmental citizenship has become a key issue for many industries and companies.

• Demographic shifts in customer and employee base are directly and indirectly influencing the way managers think and act. For example, companies are learning to adapt to more women and minorities in the professional workforce.

These are some of the most common and significant realities facing today's corporate leaders. Their individual management agenda is, of course, determined by their strategic aspirations—in the context of their own business realities.

### Implications for R&D Executives

To appreciate fully the implications of the new business agenda for the R&D executives, one should first reflect briefly on the evolution of the R&D management value system over the past several decades. Mitchell[1] has suggested that R&D management in the U.S. has evolved through two phases, the first spanning the period 1950–1970 and the second spanning the period 1970–1990. (See Figures 1 and 2.) To quote liberally from Mitchell:[1]

> Much of today's management practice and operating culture in large industrial research laboratories was firmly established prior to 1970. This covers many familiar topics, including organization, management, and selection of projects; technology transfer; as well

**FIGURE 1**
EVOLUTION OF R&D MANAGEMENT, PRE-1970 PHASE
*Source:* From Mitchell.[1]

| OVERALL BUSINESS OBJECTIVES | 1950 → 1970<br>Managing the Research Function |
|---|---|
| KEY ISSUES | Environment to promote individual creativity, innovation in groups |
| FOCUS AND ACCOMPLISHMENTS | Management guidelines for organizing, financing, technology transfer, project management, human resources, administration |

| OVERALL BUSINESS OBJECTIVES | 1970 → 1990<br>Strategic Business Management |
|---|---|
| KEY ISSUES | Coupling of research laboratories to business operations |
| FOCUS AND ACCOMPLISHMENTS | Integrated planning process to transform business objectives into technical properties<br><br>Support and extension of business strategy<br><br>Increased credibility of research management |

**FIGURE 2**
EVOLUTION OF R&D MANAGEMENT, 1970–1990 ERA
*Source:* From Mitchell.[1]

as human resource management, financing, and external relations. A principal managerial objective throughout is (has been) the establishment of an environment in industrial laboratories that nurtures both individual creativity and innovation by groups, and resulting organizational structures frequently reflect underlying academic disciplines. Many of these assumptions and shared values were acquired in a period when U.S. corporations dominated the world markets.

During the 1970s, increasing national and international competition put pressure on corporations in the United States to better integrate operations and focus business goals. One major result was that many companies introduced formal strategic planning and management systems to business operations. As a direct consequence, many research directors have been repeatedly challenged to "Get the laboratories coupled to the strategies and goals of the business and work within the mainstream management and planning processes in the corporations." In order to achieve this coupling, research management has not only had to become intimately familiar with the goals and objectives of business operations, but also had to understand the process of strategy development in detail throughout the corporation. Ultimately, many corporate laboratories have had to establish complementary planning activities which are parallel to, and integrated with, those of business operations.

A direct and important outcome of this change in attitude and approach has been the strong focus on the concept of managing the innovation pipeline[2] (Figure 3). It clearly recognizes that (1) the driver is the business strategy and (2) the cross-organizational and cross-functional interfaces hold the keys to success.

Mitchell[1] concludes that "some of the intuitions and instincts developed in these earlier periods are at odds with current realities, and . . . it is time to re-examine and augment many of our traditional assumptions and managing the research function."

Steele[3] has taken similar views on the subject. He has described and examined the classic paradigm for industrial R&D and concluded that "it is time to go back to our roots, re-examine the fundamental premises that have underlain industrial R&D and consider possible changes in those premises." Specifically, Steele has challenged the following premises that implied that R&D was a rather special activity, beyond the conventional and mundane business justifications and arguments:

- The primary mission of R&D organizations is one of discovering and inventing the big new ideas.

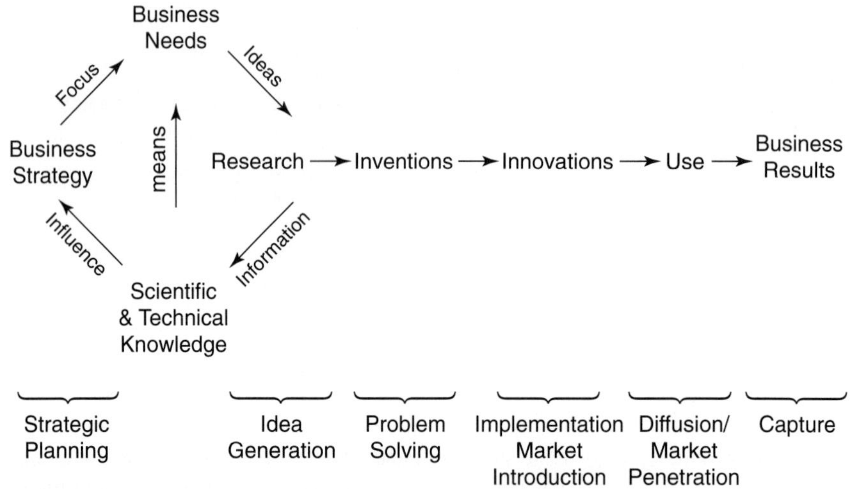


**FIGURE 3**
PIPELINE MODEL OF INNOVATION MANAGEMENT PROCESS
*Source:* From reference 2.

- R&D laboratories should be managed by career R&D professionals for reasons of credibility and creativity.
- R&D workforce must be highly stable, and researchers should develop expertise in their chosen fields through work continuity over many years.
- Funding level for R&D is inadequate. More people and money are needed for R&D to fulfill its mission.
- R&D's contributions are difficult, if not impossible, to measure.
- For R&D organizations to be objective and productive, they should be virtually autonomous within the corporation.

Admittedly, few R&D executives today publicly or even privately embrace these management beliefs. The era of "strategic coupling to business" succeeded in increasing their sensitivity to real business issues[1] and readying them for the challenges to be faced in the 1990s.

The R&D executive must now transform his time-evolved agenda of managing the research function and supporting the business strategies to an agenda of providing meaningful and measurable competitive advantage to a global, decentralized, culturally diverse, but interactive, business enterprise. The Chief Technical Officer must now rise above the role of laboratory director and become a credible and key member of the corporate management team. He must become the spokesman for not just R&D but all the technology/business intersections of interest to the corporate management team: information technology, strategic alliances, technology licensing, manufacturing technology, quality management, and so on. He can no longer just translate the business strategies for his R&D organizations but must participate fully in the development of the strategies. Further discussion on the implications of these changing roles and responsibilities follows.

## III. EXPECTED RESPONSES AND CONSTRAINTS

New realities require new responses. But, is there a basic, structured approach available to the R&D executive to address the long list of issues in a systematic manner, or must he confront the issues as they arise? The author's experience would suggest the following approach for senior R&D executives such as the Chief Technical Officer and the director of the Corporate R&D laboratory:

- Change mind-set and redefine role.
- Treat the world as the new sandbox.
- Develop and implement new processes, systems and tools.
- Add value and insist on value addition.

### Change Mind-Set and Redefine Role

The mind-set that characterized the classic paradigm of industrial R&D laboratories must be abandoned together with the associated intellectual baggage. It must be replaced with a mind-set that places the corporate vision, objectives and values over the "loyalty to science and engineering" culture. The Chief Technical Officer must fully embrace the agenda being pursued by the CEO and his inner sanctum of senior managers and, in time, enrich that agenda through participation and contribution.

One of the first steps the CTO must take is to redefine carefully his role within the CEO's inner sanctum. In most cases, the CTO would be expected to perform the dual role of a generalist and a specialist, i.e., a businessman and a technologist. Frequently he would be expected to be a businessman first, technologist second. He must be comfortable with this duality, and the management team must see some evidence of that. In time, he must emerge as a trusted team member whose assessments of the technological health of the company, convictions about the competitive advantage achievable through technology, and strategies for global management of technology are accepted and valued. Note that the word "technology" was used in this context, not R&D—and that the word "laboratory" was not mentioned at all. Technology is clearly a broader term than R&D and encompasses all the technical capability that an enterprise can and should possess and apply to gain competitive advantage. Technology is not just found in R&D laboratories; it is developed and used at many places in a corporation. Manufacturing plants, distribution facilities, engineering departments, design centers, IT departments are all important to the technological health of the company. The CTO must fully recognize the role of all technology-related functions and organizations and achieve their constructive interplay, not just champion the cause of R&D.

For the CTO to become a key member of the CEO's team, he must successfully deal with several additional dualities: corporate vs. divisional, global vs. local, and internal vs. external.

### Treat the World As the New Sandbox

The dualities of corporate vs. divisional, global vs. local, and internal vs. external require an integrated, world-scale approach to issues, priorities and resources. How should the responsibilities and resources be shared and coordinated between the corporate R&D laboratory and the divisional technology organizations—especially when they are spread around the world? When should the corporation seek external technology alliances, and when should it push ahead with internal development? When should it assign expatriates, and when should it recruit locally?

These are important questions for today's global companies, and they can create confusion and frustration unless resolved with the help of a driving vision of a global, interactive, and organic enterprise. Such a vision in turn requires—and fosters—new systems and tools for technology management.

### Develop and Implement New Processes, Systems and Tools

Many of the management systems and tools developed over the past 30 years by R&D organizations need to be re-tuned, if not totally overhauled, to respond to the new realities. The CTO must be the first to recognize the specific change needs for his R&D organizations and undertake the necessary actions. His greatest change needs are likely to be in the following areas:

- *Establishing a customer management process:* Most R&D organizations in the past did not have a comprehensive and creative process in place to clearly define internal and external customer needs and develop mutually agreed specifications. A well-developed customer management process should not only help project selection and, ultimately, technology transfer but also help the R&D organization in establishing its true criteria for success.
- *Creating a new R&D planning platform:* The new planning platform must give the R&D organization the capability to (a) "transform the objectives and strategies of the business into core technologies and program priorities . . . so that changes in business direction will be automatically reflected in the laboratory plans";[1] (b) integrate the company's worldwide R&D efforts into a coherent game plan; and (c) systematically address collaboration opportunities with external sources of technology.
- *Improving the transparency of R&D decision-making and performance:* The AAAS report "Research and Development: FY 1992" quite correctly points out[4] that "R&D organizations are increasingly required to manage their activities against quantifiable business goals and objectives; against productivity and quality measures common to all company operations." Additionally, R&D executives must accept the fact that their decision-making process must be transparent to both their "customers" and their own employees.[5] Toward that objective, the CTO must energize the management communication system so that everyone knows what is being achieved when and against what goals.
- *Establishing a system to encourage continuous improvement:* While one might argue that not all quality management concepts are directly applicable to R&D organizations, the basic idea of continuous improvement through employee empowerment and teamwork is clearly relevant to all business activities, including R&D. In fact, significant successes are being achieved towards reducing concept-to-market cycle times[6,7] through parallel engineering and other systemic approaches such as "stage-gate" management of projects.

• *Reforming the human resource management system:* Many facets of our HR management system ranging from performance evaluation to internal recruiting of R&D managers should be examined critically. For example, Steele[3] has suggested that "a totally homegrown (R&D Management) team may be too homogeneous to demonstrate the heads-up awareness of external events, system constraints, and resources available outside the enterprise . . . ." A reform of the HR system is very important but difficult to achieve. The system—and all the practices it spawns—is deeply rooted in our culture from the past and is stubbornly resistant to change. Without a reform, however, the HR system can seriously undercut all progressive moves by the senior R&D executives.

The HR needs to change to respond to the changing demographics as well. Entry of women, minority members, and foreign-borns into the R&D workforce requires the management to be sensitive to new issues. It also gives the R&D management the opportunity to be proactive and creative in terms of new initiatives.

### Add Value and Insist on Value Addition

The quest for global competitiveness has inescapably led to the need for organizational agility and frugality. The result has been management de-layering in every function including R&D. Today, R&D executives must constantly ask themselves the questions: What is the value I am personally adding to the management process within the company? What is the value being added by my organization as a whole? How are they separate yet interdependent? How can I and my organization add more value? We must face these questions within the privacy of our mind and soul with a high degree of objectivity. And we must constantly find ways of contributing more through imagination and initiative.

The HR system of the corporation at large—and the R&D organization in particular—should explicitly address the value-addition questions during the annual goal formulation and performance evaluation exercises. The process should not end there, however. The management must constantly raise the issues and search for new ways to replace less-than-essential tasks and steps with meaningful challenges that stretch people in sensible ways.

### Constraints on R&D Executives

Life is not fair, of course. The CTO and his R&D management team must meet the many requirements arising from the rapidly changing business realities, yet accept a number of constraints. The most obvious is the resource constraint: R&D must learn to do more with less. As Steele[3] put it, "There appears to be a virtually universal assumption that doing more with less has become a permanent requirement for survival," and "R&D is not being spared this pressure." Another constraint is our limited skill in speaking foreign languages and experience in dealing with non-Western cultures and value systems. Finally, the governmental regulations place numerous direct and indirect constraints on us, ranging from requirements for chemical hygiene plans for laboratories to limitations on R&D collaborations.

## IV. RESEARCH INITIATIVES AND INSIGHTS

The R&D management community is clearly interested in developing a good understanding of the various challenges arising from the changing business agenda and in gaining insights into successful management principles and practices. Towards these objectives, it has enlisted the help of academics and consultants, fostered peer group exchanges through the Industrial Research Institute (IRI), and supported several industry-university-government initiatives.

It would be clearly impractical for the author to attempt a comprehensive review of the body of knowledge resulting from these research initiatives. Yet it is important to give the reader an idea of the breadth and depth of the emerging knowledge base.

Figure 4 illustrates the basic hierarchical relationships and strategic responsibilities of the CTO. It also provides a logical framework for the author to selectively highlight four major research initiatives by the R&D management community.

### Building R&D Leadership and Credibility

A collaborative project between IRI and McKinsey & Co. has focused on the CTO-CEO relationship. Based on in-depth interviews of 24 CEOs and CTOs (or CTO-equivalents) of major U.S. corporations, the study[5] concluded that the CTOs in these companies played a wide variety of leadership roles. However, not all CTOs had high credibility with their CEOs, resulting in a leadership gap for their organizations. The study found the following three leadership roles to be most common:

- *Functional leadership* consisting of traditional, effective management of an R&D organization (such as a central laboratory) in terms of projects, cost, etc.

**FIGURE 4**
CTO'S RELATIONS AND CONTRACTS WITHIN THE CORPORATION

- *Strategic leadership* focusing on strong understanding of corporate strategies and contributions to corporate success through technology.
- *Supra-functional leadership* encompassing roles and responsibilities above and beyond the R&D function. This relatively rare leadership style "drives to build the sustainable advantage of the corporation by melding the technology function with the business."[5]

The study revealed that each of these types of R&D leadership is based on an explicit or implicit CEO-CTO contract (Figure 5). The CTO should have a clear understanding of his current role, and he must work hard to bridge any credibility gap he may have if he aspires to perform a different (i.e., broader and higher) role.

**FIGURE 5**
SUCCESSFUL CTO-CEO CONTRACTS
*Source:* From reference 5.

| Functional behaviors | |
|---|---|
| CTO | CEO |
| • Consistently meets budgets and schedules | • Rewards performance to commitments |
| • Generates ideas and options for products and processes | • Visits labs routinely to debrief researchers |
| • Interfaces smoothly with other functions | • Encourages close contact among functional managers |
| • Responds to business unit emergencies | • Recognizes R&D for contributions to the business |

| Strategic behaviors | |
|---|---|
| CTO | CEO |
| • Makes substantive contributions to corporate strategy discussions | • Invites CTO to participate in strategy discussions |
| • Aligns R&D strategy with corporate strategy | • Makes R&D planning part of routine corporate planning system |
| • Builds and maintains core technical competencies | • Invests in technical competencies that are shown to be core |

| Suprafunctional behaviors | |
|---|---|
| CTO | CEO |
| • Leads efforts to improve corporate operations through technology | • Grants a broad operational charter |
| • Advises on mergers, acquisitions, and divestitures | • Involves CTO in restructuring activities from the beginning |
| • Drives commercialization efforts across functions | • Creates and empowers a commercialization function |
| • Scouts for technological threats and opportunities | • Acts on scouting reports |

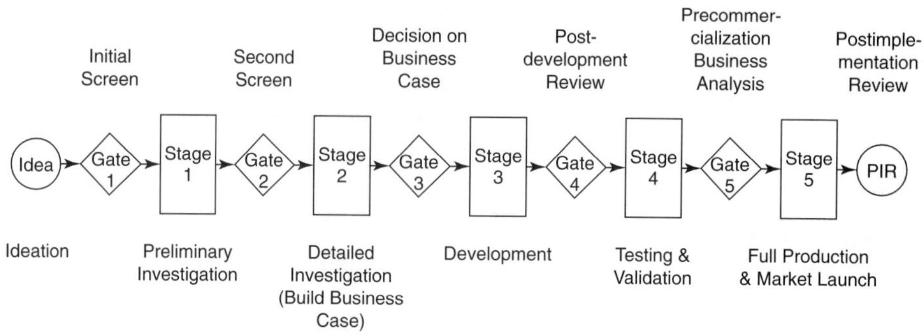


**FIGURE 6**
STAGE-GATE APPROACH TO NEW PRODUCT DEVELOPMENT
*Source:* From Cooper.[8]

### Reducing Cycle Time from Concept to Market

One of the primary responsibilities of the CTO and his R&D management team is to serve the current businesses through development and demonstration of new or improved products and processes. Reducing product development cycle time through cross-functional teamwork has, therefore, become an imperative for R&D management. Several recent publications[6–8] have addressed this subject in some detail. In particular, the "stage-gate" method of project management (Figure 6), introduced by Cooper,[8] has emerged as an effective approach to new product development and has generated a wide following in the industrial R&D community.

### Technology Sourcing though Partnerships and Alliances

The CTO and his R&D management team has an "invisible contract" to develop new business options through technology. That contract requires technology sourcing through alliances and partnerships as an essential complement to in-house R&D. A recent study by Hull and Slowinski[9] of 37 large company/small company partnerships has investigated many aspects of technology-based alliances: benefit expectations, resource skill contributions, structural considerations, communication patterns, etc. This research has provided valuable insights into key success factors—and identified the following common barriers to success in partnering with technology entrepreneurs.

In a paper based on IRI workshops involving over 20 member company representatives, Chatterji[10] has concluded that most companies do not address "technology sourcing" as a properly planned business process, and as a result, run into many difficulties and frustrations. He has proposed a model of the technology sourcing process (Figure 7) consisting of a series/parallel combination of eight discrete steps. Cross-functional, even cross-organizational, teamwork throughout the process is a fundamental ingredient for success. The paper also presents several concepts and tools for finding, evaluating, acquiring and internalizing useful technical knowledge and innovation from external sources.

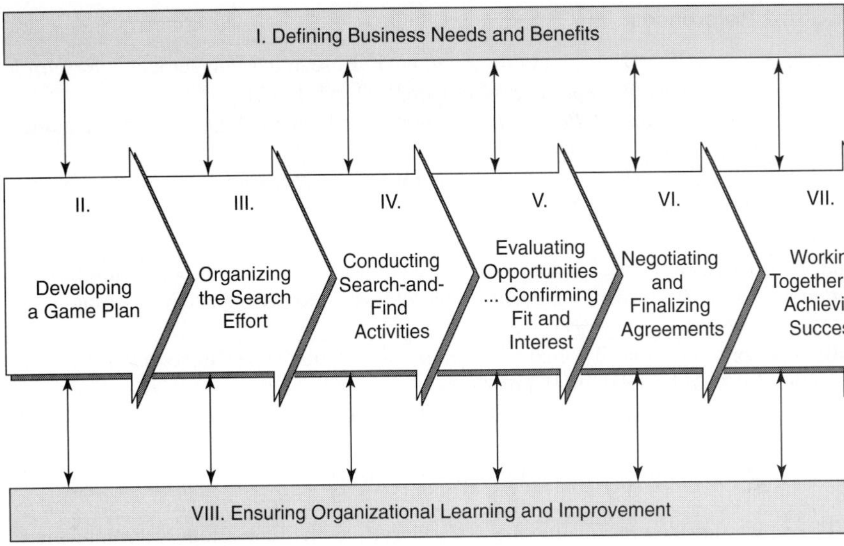


**FIGURE 7**
CONCEPTUAL MODEL OF THE TECHNOLOGY SOURCING PROCESS
*Source:* From Chatterji.[10]

## Managing Workforce Diversity in R&D Organizations

Managing the R&D staff and resources to achieve established program goals is the clearest and strongest contract the CTO has. Historically, that contract has involved a white, male-dominated R&D staff. With rapid changes in male/female ratio and ethnic mix in American R&D laboratories, new management issues are beginning to emerge. DiTomaso and Farris[11–13] have completed the first phase of a major study in this area. This study involving approximately 3,000 industrial scientists and engineers in the U.S. has focused on several key questions: How does this increasing diversity impact on the performance and satisfaction levels of the various subgroups within the R&D laboratory? What are the key cultural, behavioral and environmental factors responsible for the performance differences? What can the R&D management do to maximize the contributions of women and minorities, yet not homogenize the workforce? The results are already beginning to influence our ways of thinking about and managing the R&D workforce.

The above examples of recent research initiatives represent an important but small fraction of the work sponsored by the industrial R&D management community in the U.S. The resulting body of knowledge is already significant in scope and value. The R&D executive can look to the future with some confidence that he will be able to draw from a rich pool of peer experience and insight to address effectively the changing environment.

## Acknowledgment

This paper is based on an earlier publication.[14] The author is grateful to Blackwell Publishers, the owner of the copyright, for permission to reprint significant parts of that paper.

### Reading 10.2 References


1 Mitchell, Graham R. "The Changing Agenda of Research Management." *Research-Technology Management,* September–October 1992, pp. 13–21.

2 *Technology Management: A Research Perspective.* Center for Innovation Management Studies, Lehigh University, 1988.

3 Steele, Lowell W. "Needed: New Paradigms for R&D." *Research-Technology Management,* July–August 1991, pp. 13–21.

4 Research and Development: FY 1992. AAAS Report XVI, 1991.

5 Uttal, Bro; Kantrow, Alan; Linden, Lawrence H.; and Stock, Susan B. "Building R&D Leadership and Credibility." *Research-Technology Management,* May–June 1992, pp. 15–24.

6 Stalk, Jr., George and Hout, Thomas M. "Competing Against Time." *Research-Technology Management,* March–April 1990, pp. 19–24.

7 Smith, Preston and Reinertsen, Donald. *Developing Products in Half the Time.* Van Nostrand Publishing Co., 1991.

8 Cooper, Robert. "Winning at New Products." Addison Wesley Publishing Co., 2nd Ed., 1993.

9 Hull, Frank and Slowinski, Eugene. "Partnering with Technology Entrepreneurs." *Research-Technology Management,* November–December 1990, pp. 16–20.

10 Chatterji, Deb. "Accessing External Sources of Technology." *Research-Technology Management,* March–April 1996, pp. 48–56.

11 Gordon, G. G.; DiTomaso, N.; and Farris, G. F. "Managing Diversity in R&D Groups." *Research-Technology Management,* January–February 1991, pp. 18–23.

12 DiTomaso, Nancy and Farris, George F. "Work and Career Issues for Women Scientists in Industrial Research and Development in the U.S." *Berlin Journal of Sociology,* Vol. 1, 1992, pp. 91–102.

13 DiTomaso, Nancy and Farris, George F. "Diversity and Performance in R&D." *IEEE Spectrum,* June 1992, pp. 21–24.

14 Chatterji, Deb. "Emerging Challenges for R&D Executives: An American Perspective." *R&D Management,* July 1993, pp. 239–247.


## DISCUSSION QUESTIONS

1 In the second half of the 1990s Microsoft has combined both internal research and compulsive buyout. Why is that? Find out about some of the buyouts of another company and comment on their strategic importance.

2 Read about a partnership between companies. What were the benefits each partner obtained?

## ADDITIONAL READINGS

Steven C. Wheelwright & Kim B. Clark. "Creating Project Plans to Focus Product Development." *Harvard Business Review,* March–April 1992.

The authors suggest mapping as a tool to define and manage an effective R&D portfolio. The map classifies a portfolio in five categories: research, breakthrough, platform, derivative, and alliance projects. Following HBR style, the paper presents corporate examples of each type.

John Seely Brown. "Research That Reinvents the Corporation." *Harvard Business Review,* January–February 1991.

Since R&D is a very expensive process, companies are interested in obtaining returns on that investment. The author presents the experience of Xerox's PARC as a contributor to the business. R&D is seen not as a separate entity dedicated to research but as an integral part of Xerox's strategy.

Régis Larue de Tournemine. "Strategic Alliances between Big and Small Firms in the Development Dynamics of Science Based Industries." In T. Khalil & B. Bayraktar (eds.), *Management of Technology III: Proceedings of the Third International Conference on Management of Technology,* Vol. I, pp. 145–154. Industrial Engineering and Management Press, Norcross, GA.

Collaboration among firms is gaining importance as development periods ought to be reduced. The author suggests a dynamic model in which different kinds of small- and large-firm collaborations have to be distinguished depending on the stage of technical evolution.

Graham R. Mitchell & William F. Hamilton. "Managing R&D as a Strategic Option." *Research-Technology Management,* May–June 1988, pp. 15–22.

Brody, Richard J. *Effective Partnering: A Report to Congress on Federal Technology Partnerships.* U.S. Department of Commerce, Office of Technology Policy, Washington, DC, February 1996.

## SUGGESTED CASES

- "Sony Corp.: Car Navigation Systems." Harvard Business School, Case 9-597-032.
- "Du Pont Kevlar Aramid Industrial Fiber." Harvard Business School, Case 9-391-146.
- "Seaman Corp." Harvard Business School, Case 9-396-268.

## REFERENCES

Allen, T. J. 1977. *Managing the Flow of Technology: Technology Transfer and the Dissemination of Technological Information with Research and Development Organizations.* MIT Press, Cambridge, MA.

Berman, E. M., & Khalil, T. M. 1992. "Technological Competitiveness in the Global Economy—A survey." *International Journal of Technology Management,* vol. 7, no. 4/5, pp. 347–358.

Bhalla, Sushil K. 1987. *The Effective Management of Technology.* Battelle Press, Columbus, OH.

Brody, Richard, J. 1996. *Effective Partnering: A Report to Congress on Federal Technology Partnerships.* U.S. Department of Commerce, Office of Technology Policy, Washington, DC, February.

Dalton, D. H., and Serapio, M. G. Jr. 1995. *Globalizing Industrial Research and Development.* U.S. Department of Commerce—OTP, Oct., Washington, D.C.

Department of Commerce. 1985. *Statistical Abstracts of the United States.* Washington, DC.

Ford, David. 1988. "Develop Your Technology Strategy." *Long Range Planning,* vol. 21, no. 5, October, pp. 85–94.

Jain, R. K., & Triandis, H. C. 1990. *Management of R&D Organizations.* Wiley Interscience, New York.

Merten, U., & Ryu, S. M. 1982. "What Does the R&D Function Actually Accomplish?" *Harvard Business Review,* July–August.

Mitchell, G. R. 1988. "Options for the Strategic Management of Technology." In Khalil, T., Bayraktar, B., & Edosomwan (eds.), *Technology Management I.* Interscience Enterprises, Geneva.

Mitchell, G. R., and Hamilton, W. F. 1988. "Managing R&D as a Strategic Option." *Research-Technology Management,* May–June, pp. 15–22.

Mitchell, G. R. 1995. "Technology—Business Strategy—Government Policy." Lecture notes, University of Miami, March 3–4.

Mitchell, G. R. 1997. "The Global Context for U.S. Technology Policy." Office of Technology Policy, U.S. Department of Commerce, Washington, DC.

National Science Foundation. 1985. *Science Indicators.* Washington, DC.

Office of Technology Policy. 1997. "International Plans, Policies, and Investments in Science and Technology," U.S. Department of Commerce, Washington, DC.

Organization of Economic Cooperation and Development. 1970. *The Measurement of Scientific and Technical Activities.* Paris.

Pavitt, K., Pattel, P. 1988. "The International Distribution of Determinants of Technological Activities." *Oxford Review of Economic Policy,* vol. 4, no. 4, pp. 1–21.

Perrino, A. C., & Tipping, J. W. 1989. "Global Management of Technology," *Research-Technology Management,* vol. 32, no. 3, May–June.

Schmitt, R. W. 1983. *When Pendulum Swings toward Applied Research: Research and Development Key Issues for Management.* Report 842, Conference Board, Washington, DC.

Schmitt, R. W. 1985. "Successful Corporate R&D." *Harvard Business Review,* May–June.

# 11

# TECHNOLOGY TRANSFER

A user of a technology does not have to be its creator or inventor. In fact, most inventions are created outside the firms that benefit from them. Innovation may also occur outside a firm's boundaries, and even if it happens within the firm, it may be confined to one department or division.

Transfer of technology is a process essential for the wide application and utilization of technology by one or more users. In this chapter we introduce types and channels of technology transfer. We discuss some cases of international technology transfer, present examples of successful national programs of technology transfer, and examine a model of intrafirm technology transfer.

## DEFINITIONS AND CLASSIFICATIONS

*Technology transfer* is a process that permits the flow of technology from a source to a receiver. The source in this case is the owner or holder of the knowledge, while the recipient is the beneficiary of such knowledge. The source could be an individual, a company, or a country. Jain and Triandis (1990) define technology transfer as a "process by which science and technology are transferred from one individual or group to another that incorporates this new knowledge into its way of doing things." The National Aeronautics and Space Administration (1995) defines it as "the process of providing the technology developed for one organizational purpose to other organizations for other potentially useful purposes."

Technology transfer can be divided into the following categories:

- *International technology transfer,* in which the transfer is across national boundaries. An example of this type is the technology transfer from industrialized countries to developing countries.

• *Regional technology transfer,* in which technology is transferred from one region of the country to another, for example, from Florida to Alaska.

• *Cross-industry or cross-sector technology transfer,* in which technology is transferred from one industrial sector to another. An example is the transfer of technology from the space program to commercial applications.

• *Interfirm technology transfer,* in which technology is transferred from one firm to another. An example is the transfer of computer-aided design (CAD) expertise and computer-aided manufacturing (CAM) machines from a machine tool manufacturing firm to a furniture-producing firm.

• *Intrafirm technology transfer,* in which technology is transferred within a firm from one location to another. An example is the transfer of technology from a company's California division to its Miami location. Intrafirm transfers can also be made from one department to another within the same facility. For instance, if one department uses sophisticated computer technology and another relies on manual work—an imbalance that could hinder the company's operation—technology transfer can balance the system by providing full use of computer technology throughout the firm.

## CHANNELS OF TECHNOLOGY FLOW

Technology is intangible; it flows easily across boundaries of countries, industries, departments, or individuals, provided that the channels of flow are established. There are three types of channels that allow the flow of technology:

**1** *General channels:* The technology transfer is done unintentionally and may proceed without the continued involvement of the source. Information is made available in the public domain with limited or no restrictions on its use. This information is harnessed by users and applied to their purposes. Channels of this type of transfer include education, training, publications, conferences, study missions, and exchange of visits.

**2** *Reverse-engineering channels:* Other channels in which the transfer occurs with no active contribution from the source include reverse engineering and emulation. Here a host, or a traditional receiver of a technology, is capable of breaking the code of a technology and developing the capability to duplicate it in some fashion. This is feasible provided that the host has the knowledge to do this and there is no legal violation of intellectual or property rights. For example, a product that is put on the market by company A can be purchased by company B, reverse-engineered, and introduced to the market as a competitor to company A's product, as illustrated by the case in the accompanying box. This is a powerful method for technology transfer. Its limitation is its inability to transfer the developer's tacit knowledge. Such knowledge is usually gained during the product development process.

**3** *Planned channels:* The technology transfer is done intentionally, according to a planned process and with the consent of the technology owner. There are several types of agreements that are used to effect planned transfers. They permit access to, and use of, technological know-how:

- **a** *Licensing:* The receiver purchases the right to utilize someone else's technology. This may entail an outright purchase or a payment of an initial lump-sum amount plus a percentage of sales.

**REVERSE ENGINEERING AT COMPAQ**

Reverse engineering was used by Compaq Computer Company to develop its first PC clone. Compaq's founders had all the components needed to build a PC except for one piece of technology—a ROM-BIOS chip (read-only memory chip, which stores the basic input/output system computer code). This technology was owned and protected by IBM. The IBM chip was not for sale, so Compaq hired some competent engineers and computer programmers and asked them to reverse-engineer the product (Cringely, 1996). They were successful, and Compaq was able to introduce its IBM-compatible PC at a lower cost than an IBM PC. The product was an instant hit and launched the company's successful entry into the PC market. Compaq selected the niche of portable computers to make its entry into the market.

Compaq's strategy of cloning an IBM computer gave it access to all the software written for the IBM PC. Its strategy for entering the portable market gave it a special niche with little competition. Its lower price gave it an advantage with the customers and with the computer dealers, who could increase their margin of profit. According to Cringely, Compaq set a start-up record, selling 47,000 computers worth $111 million in its first year.

**b** *Franchise:* This is a form of licensing; however, the source usually provides some type of continual support to the receiver, for example, by supplying materials, marketing support, or training. This channel is commonly used in food chains and service organizations, such as McDonald's, Burger King, and Pizza Hut.

**c** *Joint venture:* Two or more entities combine their interests in a business enterprise in which they can share knowledge and resources to develop a technology, produce a product, or use their respective know-how to complement one another. They also share in the rewards of the venture. International joint ventures are frequently used by recipients to acquire technology and by sources of technology to gain access to local markets and distribution skills.

**d** *Turnkey project:* A country buys a complete project from an outside source and the project is designed, implemented, and delivered ready to operate. Special provisions for training or continued operational support may be included in the agreement between the parties. Engaging in a turnkey project is equivalent to buying or selling a machine, but on the scale of an entire plant. Most innovative firms would not sell a plant or license technologies that they intend to exploit themselves.

**e** *Foreign direct investment (FDI):* A corporation, usually a multinational, decides to produce its products or invest some of its resources overseas. This permits the transfer of technology to another country, but the technology remains within the boundaries of the firm (i.e., is still controlled by the firm). This type of investment has advantages for both the investor and the host country. The investor gains access to a labor force, natural resources, technology, or markets. The host country receives technological know-how, employment opportunities for its people, training for the workforce, and investment capital that adds to the development of its infrastructure. The host country will also get tax advantages, since most employees will be contributing to the local economy. The multinational may also gain a tax advantage by locating facilities offshore in a country or territory that gives a tax break. Many U.S. pharmaceutical companies have located facilities in Puerto Rico because of the tax advantage they can get with this arrangement. Some developing countries provide long-term tax relief for foreign companies located on their soil.

**f** *Technical consortium and joint R&D project:* Here, two or more entities collaborate in a large venture because the resources of one are inadequate to affect the direction of technological change. Typically this type of venture takes place between two countries or two large conglomerates. For example, a consortium was formed between France and England to develop a supersonic plane (the Concorde). Both nations needed to combine their technical and financial resources to develop expensive technology and, in the meantime, to compete with their rivals in the United States. Several similar ventures and consortia exist under the auspices of the European Union (EU). European governments have established a number of projects to help national companies compete with American and Japanese firms. Programs the EU supports include "Race," a project to advance communication technology; "Espirit," for information technology; and "Jessi," to bolster semiconductor research. Project "Eureka" is an independent research program involving 24 nations (U.S. Office of Technology Policy, 1997). All these cooperative projects aim to advance research, develop technology, and transfer knowledge to participating member states. The Japanese government, through its Ministry of International Trade and Industry (MITI), fosters alliances between industry and government in projects of national interest and scope. Examples include the VLSI project undertaken to make the Japanese semiconductor industry competitive and the Firth Generation project, which focuses on advancing artificial intelligence and parallel processing (Cheney and Grimes, 1991).

American industry has also changed some of its ways of doing business, moving from a continuous-competition and closed-technology mode of operation to a more flexible, cooperative mode. Examples are the cooperation between IBM, Apple, and Motorola to produce the power chip for the personal computer; between Microsoft and NBC to tap into the future of the multimedia industry; and between Apple and Microsoft to exploit each other's strength in technology.

## INTERNATIONAL TECHNOLOGY TRANSFER

The technological production base previously confined to the industrialized nations of the West and the North has recently spread to a large number of other countries. Developing countries have realized that industrialization is the only means of reaching socioeconomic parity with Europe and the United States. For many Asian and third-world countries, technologies imported from industrialized countries provided the initial base for industrial development. Starting with industries requiring low levels of skills, developing countries are gradually introducing industries requiring both greater levels of skills and more technological capability. Figure 11-1 shows the growth of high-tech goods produced by the Asian Tigers. Today, many of the newly industrialized countries

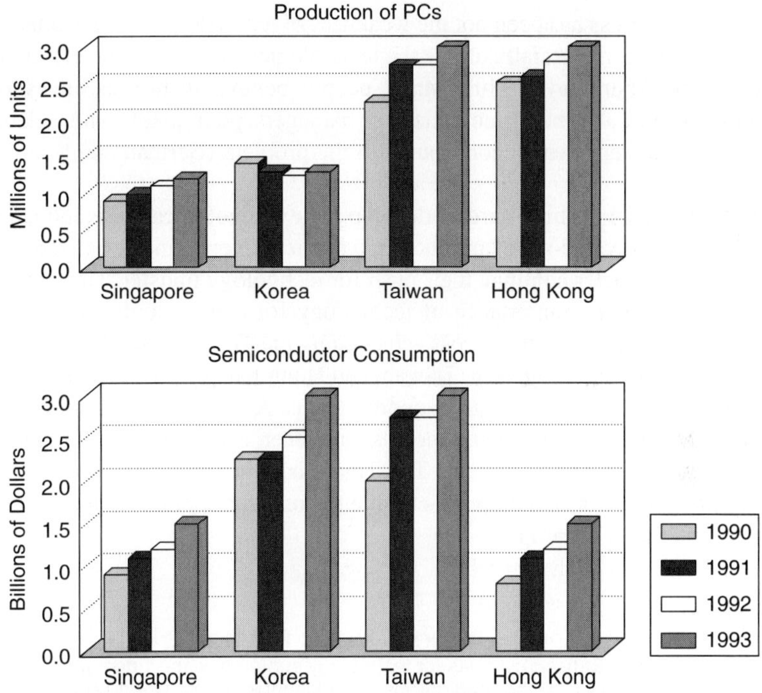


**FIGURE 11-1**
THE ASIAN TIGERS AND HIGH-TECH GOODS
They are making more high-tech goods . . . and consuming more, too.
*Source: Business Week,* Dec. 7, 1992.

(NICs) are equipped with an appropriate industrial and technological base. They have become highly competitive in world markets. In some instances, they enjoy the support of government-instituted fiscal and economic measures to remain competitive in the global arena. They may also possess other advantages, such as lower wages or the availability of natural and human resources. As a result, exports from the Pacific Basin and some developing countries have successfully penetrated the domestic markets of industrialized nations, particularly those of the United States.

In the majority of cases, the migration of technology occurred through international technology transfer, through mutual cooperation agreements, or through direct purchases from the United States, Germany, Japan, and other industrialized countries. In several cases, it occurred through the establishment of overseas manufacturing facilities by multinational corporations. U.S. corporations have frequently chosen to invest in production facilities in other countries to take advantage of lower labor rates and closeness to markets.

The migration of technology and production facilities from the United States to Japan and other Asian countries in the 1970s and 1980s occurred because of a lack of competitiveness in U.S. manufacturing. Migration occurred in a set of mature technologies, such as consumer goods and automobiles. The crucial element resulting in the loss of U.S. competitiveness has been not the weakness of technologies but a managerial attitude that failed to commercially utilize the technologies in a timely and effective manner (Hayes and Abernathy, 1980). Many people believe that weak investment in manufacturing R&D, the short time horizon of managers, poor quality, and lack of focus on technology transfer have all contributed to the problem (Berman and Khalil, 1992; Szakonyi, 1992).

Most countries are pushing hard to develop their technological base and to convert their knowledge into value-added products and services. Newly industrialized countries and developing countries continue their push for technology transfer. They are becoming keenly aware of the importance of technology for economic development. They have seen evidence in the huge success achieved by the Tigers of Southeast Asia—Singapore, Malaysia, Indonesia, Korea, Taiwan, and Hong Kong. In spite of some setbacks these countries have achieved major economic gains. A successful technology transfer effort spurs economic growth. Many models have been used to effect the transfer process with varying degrees of success. The Tigers have targeted niches of technology and turned them into world-class products able to compete in global markets (Engardio and Gross, 1992, see Figure 11-2).

**FIGURE 11-2**
TECHNOLOGY SPECIALIZATION AMONG THE TIGERS
Avoiding head-to-head contests for technology supremacy, the Tigers prefer to target niches where they can quickly turn technologies into world-class products.
*Source: Business Week,* Dec. 7, 1992.

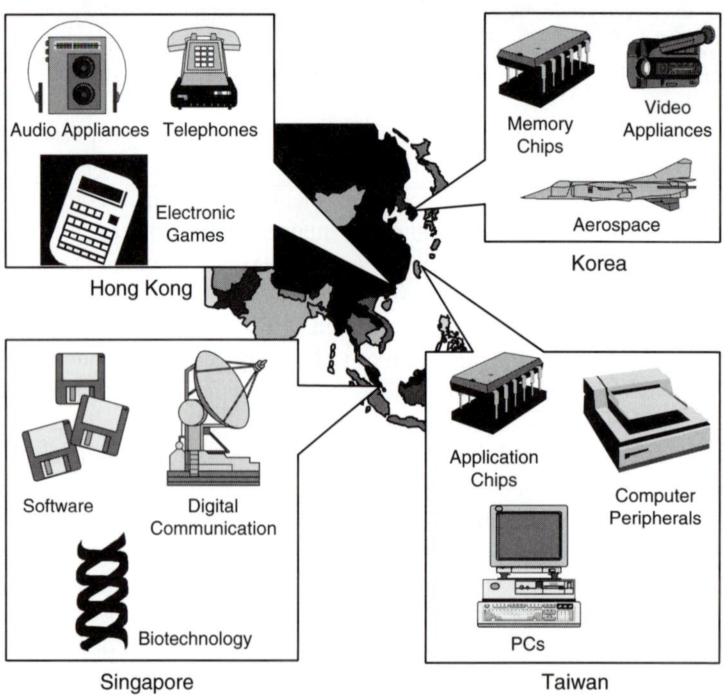

### The Singapore Model

Many lessons can be learned from the success of Singapore's effort for economic development. Factors contributing to its success can be extracted from a speech by Prime Minister Lee Quan Yu during the African Leadership Meeting held in Singapore in November 1993. The prime minister enumerated the essential basis for development. Some of his points are paraphrased below:

**1** Establish/maintain a clean, effective government that is well respected by the people. (Officials must have a philosophy based on understanding and appreciating the development process.) Eliminate corruption and reward officials adequately to protect them from corruption.

**2** Avoid internal squabbles for national unity.

**3** Build on areas of strength (e.g., agriculture or availability of labor force).

**4** Encourage savings to increase investment while avoiding external debt.

**5** Encourage family projects and local industry to create economic opportunities, and keep people from emigrating to large cities.

**6** Do not waste funds on huge projects.

**7** Encourage investment by both small investors and multinationals.

**8** Promote education.

**9** Develop effective strategies for technology transfer.

Singapore built its strategy around becoming a regional business services hub in the Southeast Asia region. It serves as a regional marketing and technical support center, a regional financial and business center, and a regional headquarters for multinational companies (MNCs). It also selected niche industries for specialization, including electronics and computers, ship repair and maintenance, petroleum refining, and aerospace maintenance and repair (Wong, 1995).

## TECHNOLOGY TRANSFER IN TAIWAN

Taiwan's approach to technological development and technology transfer is another success story. In Taiwan, industrial technology R&D is enhanced by a nonprofit corporation known as the Industrial Technology Research Institute (ITRI). This institute conducts technical R&D on targeted projects directed and funded by contracts from the Ministry of Economic Affairs, a central government agency. The research results are then applied to assist or guide the private sectors either through technology transfer or technology expansion (Chen, 1990).

Taiwan has located ITRI next to two of its top universities in science and technology: the National Tsinghua University and the National Chiao-Tung University. A targeted project for technology transfer can draw upon the scientific and technological expertise of the faculty of these two fine institutions. The participation of the private sector through the investment and business planning of industrial facilities completes the team of players needed to spur industrial development.

To facilitate the transfer further, an industrial park is located in close proximity to permit a project incubated in ITRI to be spun off to a privately owned facility in the

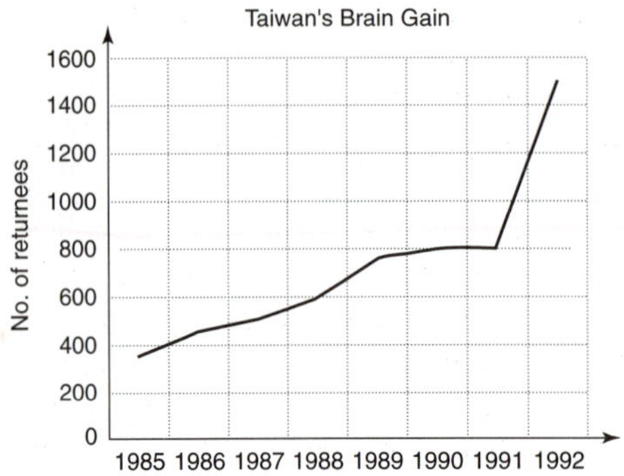


**FIGURE 11-3**
RETURN FLOW OF TECHNICALLY TRAINED TAIWANESE
*Source: Business Week,* Dec. 7, 1992.

industrial park. Such collaborations among government agencies, nonprofit corporations, and private industries have been very successful in transferring industrial technologies within Taiwan, especially in areas of high risk or crucial importance. Taiwan also relies heavily on its well-educated nationals who are trained overseas. Incentives are provided to induce them to return home, thereby bringing technology to Taiwan. Technology transfer through people is a very effective transfer mechanism. Figure 11-3 depicts the return flow of Taiwanese trainees between 1985 and 1992. Examples of technologies successfully transferred are integrated circuit manufacturing, personal computers, and automation technologies (Chen, 1990).

## U.S. NATIONAL TECHNOLOGY TRANSFER

The United States is by far the most intensive technology-producing country in the world. U.S. leadership in science is evident from the nation's large number of Nobel Prize winners in science, more than twice that of each of its closest competitors, the United Kingdom and Germany (Figure 11-4).

The creation of technology in the United States is also superior. The space technology, the defense technology, and the technology available within industry are unmatched by any in the world. How is it, then, that U.S. competitiveness declined in the 1970s and 1980s and that the nation has a huge deficit in its balance of trade? The answer lies within the fundamental principles of MOT. A major factor in competitiveness and wealth creation is how technologies are spun off to commercial and service enterprises. Transferring technology from where it is created to where it can be used effectively is at the heart of this issue. Having realized this fact, the U.S. federal government has made technology transfer a significant activity (National Aeronautics and Space Administration, 1995). The government has also moved to shore up its technology policy through effective partnering and joint investment in technology with the private sector (Brody, 1996).

The National Aeronautics and Space Administration (NASA) has been very active in promoting technology transfer and has created a national network of Technology Transfer Centers. Exhibit 11-1 shows the types and functions of these centers. NASA has also

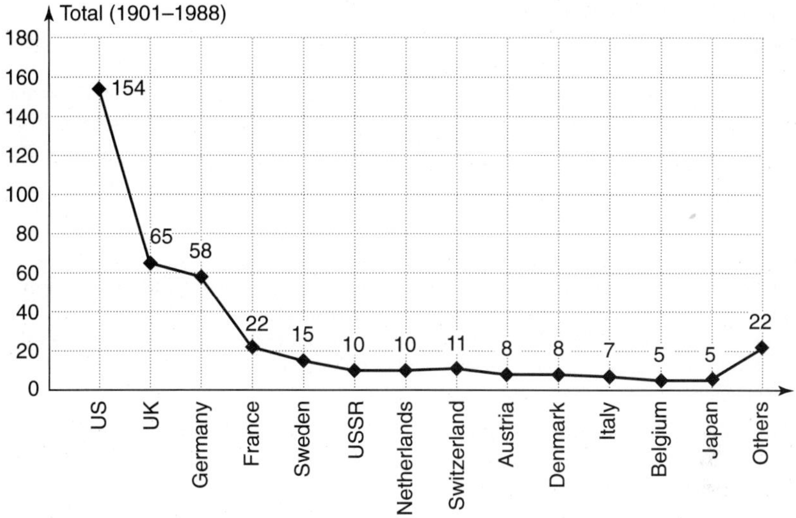


**FIGURE 11-4**
NOBEL PRIZES BY COUNTRY
*Sources:* Glazer, 1992; © 1992 Institute of Industrial Engineers.

established procedures for reporting new technology innovations and publishes the information in its *Techbriefs* magazine. It has developed guidelines for partnership agreements and sponsored several university-based Technology Transfer Centers to effect technology-based economic development locally and regionally. The Southeast Region Technology Transfer Centers located at the University of Alabama, the Georgia Institute of Technology, the University of Florida, and several other academic institutions are examples of technology transfer centers working to link industry users to technology producers.

The post-cold war era sparked a great interest in the transfer of technology from national and defense-related laboratories to the private sector. The federal government operates more than 700 laboratories with an annual budget of over $25 billion (Bloch, 1992). As military budgets decline, many government laboratories must find a new mission or go out of business. One survival strategy is to become more self-supporting through the sale and transfer of technology to the private sector. This was not previously an integral part of the mission of most of these laboratories, and switching to a new form of operation requires a change of organization culture. It also requires a systematic, proactive technology transfer effort. Perrin (1990) noted that the Department of Defense (DOD) spent over $70 billion in R&D and employed 70 percent of the engineers and 38 percent of the scientists working for the federal government. However, utilization of the developed technologies by secondary applications was minimal. Perrin described a proactive approach to the transfer of DOD technologies to solve civil problems. This approach, based on the use of transfer agents rather than passive efforts such as literature dissemination, proved successful upon implementation. Perrin enumerated

**EXHIBIT 11-1**
NASA NATIONAL NETWORK

The national network of Technology Transfer Centers and offices sponsored by NASA is dedicated to the timely transfer of scientific advances and technologies resulting from NASA's aeronautics and space programs and other federal R&D to practical applications throughout the U.S. economy.

*NASA field centers:* Each of NASA's field centers has a Technology Transfer Office to coordinate and manage a full range of technology transfer activities, including new technology reporting, technical assistance, cooperative projects, and industry outreach.

*Regional Technology Transfer Centers:* The NASA Regional Technology Transfer Centers (RTTCs) are staffed by technology transfer experts offering technical consultation services and linkage to other experts in the field. The RTTCs provide services to industry within their designated regions and assist industry clients to locate, assess, and commercialize technologies from NASA and the federal R&D base.

*National Technology Transfer Center:* The National Technology Transfer Center (NTTC) serves as a national clearinghouse/gateway for federal technology transfer and provides services and assistance in training, planning, and outreach.

*Earth Data Analysis Center:* The Earth Data Analysis Center (EDAC) provides technology transfer services in support of the distribution and transfer of remote sensing/geographic information systems data and technology.

*Technology Application Team:* The Technology Application Team (TAT) works with industry to identify and solve critical problems with existing NASA technology and to develop cooperative projects and relationships that address technological needs of national or industrywide significance.

*Computer Software Management and Information Center:* The Computer Software Management and Information Center (COSMIC), operated by the University of Georgia, is NASA's technology transfer program for collecting and documenting computer software technology produced by NASA and distributing it to U.S. private, government, and academic organizations.

*Center for AeroSpace Information:* The Center for AeroSpace Information (CASI) maintains mailing lists and distributes NASA technology transfer publications, including the annual Spinoff report. CASI provides responses and referrals to inquiries about technology transfer. CASI also provides centralized technology transfer documentation support for all NASA centers.

*Source:* National Aeronautics and Space Administration, 1995.

the advantages of the active transfer process as follows:

- Conveys timely information on current and planned developer and user programs.
- Provides real-time feedback and criticism on specific technology or problem items.
- Allows the transmittal of ancillary information and know-how that is not available in formal literature or reports.
- Permits the transfer agents to manage and control in a user-need or applied direction.
- Requires a relatively small expenditure of effort and time relative to the results obtained.

The accompanying box lists a number of U.S. government initiatives undertaken to promote technology transfer.

Examples of success in transferring technology from a federal laboratory to the private sector are many. Wood and EearNisse (1992) described one such transfer—from

---

**U.S. FEDERAL GOVERNMENT INITIATIVES FOR TECHNOLOGY TRANSFER**

The U.S. federal government has addressed the issue of technology transfer from federal laboratories by implementing three legislative acts:

- The Stevenson-Wydler Technology Innovation Act of 1980 (P.L. 96-480)
- The Technology Transfer Act of 1986 (P.L. 92-502)
- The National Competitiveness Technology Transfer Act of 1989 (P.L. 101-189)

These acts resulted in the establishment of the Office of Research and Technology Applications (ORTA) with technology transfer overview responsibility for the Federal Laboratory Consortium (FLC). Mechanisms for cooperative research and development agreements (CRADAs) were established between (*a*) federal laboratories and (*b*) private enterprises, universities, state and local governments, foundations, nonprofit institutions, and consortia of such organizations.

As a supplement to these acts Congress has created the National Technology Transfer Center to augment existing technology transfer mechanisms.

*Source:* Wood and EearNisse, 1992.

---

Sandia National Laboratories, a federal laboratory, to Quartex, Inc., a small, private sector business that commercialized the technology. A shareholder of Quartex learned that a new quartz resonator force sensor had been invented by an employee of Sandia National Laboratories. Quartex invited the inventor to join its staff and be part of a team that would bring his invention to market. Quartex assumed all obligations to file and maintain patents. The waiver process followed a guideline that stipulated a balance between technological commercialization and the preservation of government access to tax-supported inventions.

There are barriers that exist in such a transfer endeavor, including the cultural gap between industry and government laboratories, legal issues, and the need to formulate an innovative business structure to permit a smooth transfer. Wood and EarNisse (1992) mentioned the following factors as contributing to the successful technology transfer project from Sandia to Quartex:

- Technology transfer fit for Quartex's strategic planning.
- Sufficient proprietary rights from Sandia to Quartex to justify the risk of investment.
- Potential for additional proprietary coverage for the technology through improvement patents and new technology.
- A landmark invention.
- Diverse market applications for the technology.
- Incentives for sustained technology transfer and support by a product champion.
- Required resources within Quartex for product commercialization.

## INTRAFIRM TECHNOLOGY TRANSFER

A firm attempting to transfer technology from one site to another or from one division to another must approach the transfer process in a systematic and deliberate manner. For the transfer to be successful, infrastructure, including facilities, equipment, and personnel, must exist or be developed. In addition, a transfer team may be needed to orchestrate the transfer. Arthur Squires calls these teams the "Maestros of Technology" (Bowser, 1987).

In fact, complicated transfer projects may require two teams, one at the source and one at the receiving end of the technology. Each team is led by a "champion" and consists of a number of specialists, depending on the complexity of the technology and the size of the project. All communications regarding the transfer [marketing, quality assurance (QA), production, etc.] are channeled through the transfer-team leaders. Beruvides and Khalil (1990) developed an intrafirm technology transfer model based on experience gained from an actual project involving the relocation of an existing production facility. The model is shown in Figure 11-5.

In this model, the project starts when a company decides to acquire new technology through the purchase of a smaller entrepreneurial company located thousands of miles away. In order to facilitate communication and consolidate operations, the company decides to transfer the entire production facility, including technology and operations, to its headquarters facility. This calls for an organized process of intrafirm technology transfer. The proper infrastructure must be created to permit the transfer, and a competent team must be assembled to execute the transfer process. In this particular case, the team comprises two separate groups: one at the headquarters and one at the acquired-company location. Although these groups are in different locations, they should recognize that they are on the same team with one line of communication that permits clear, open communication. The remaining requirements of the transfer process are indicated on the flowchart in Figure 11-5.

The transfer team is involved in developing schedules and budgets and preparing the new site. Before the transfer, the employees for the new site are selected and given training, at the facility of the acquired company, in the technology being transferred. This culminates with turning over operations to the employees at the new site to ensure a smooth transition.

An appropriate amount of inventory of the product is built up as a reserve against delays or inefficiencies during the transfer. At the optimum time, parallel production facilities are set up, with half the equipment at the old site and the other half relocated to the new site. This ensures continued production throughout the transfer project, an important criterion desired by the company. Once the new site is qualified and the product specifications are achieved at the new site, the remaining production equipment is relocated to the new site. Full production begins at the new site, quality is monitored, and the transfer team is disbanded.

Several useful guidelines for setting up a transfer team are recommended by Beruvides and Khalil (1990):

**1** The smaller the team, the better.

**2** People work best in an atmosphere of trust and healthy competition.

**3** Building teamwork and motivating the team members are critical.

**4** The chain of command and communication channels should be well understood by all.

**5** Success is largely dependent on the quality of the people selected to perform the tasks.

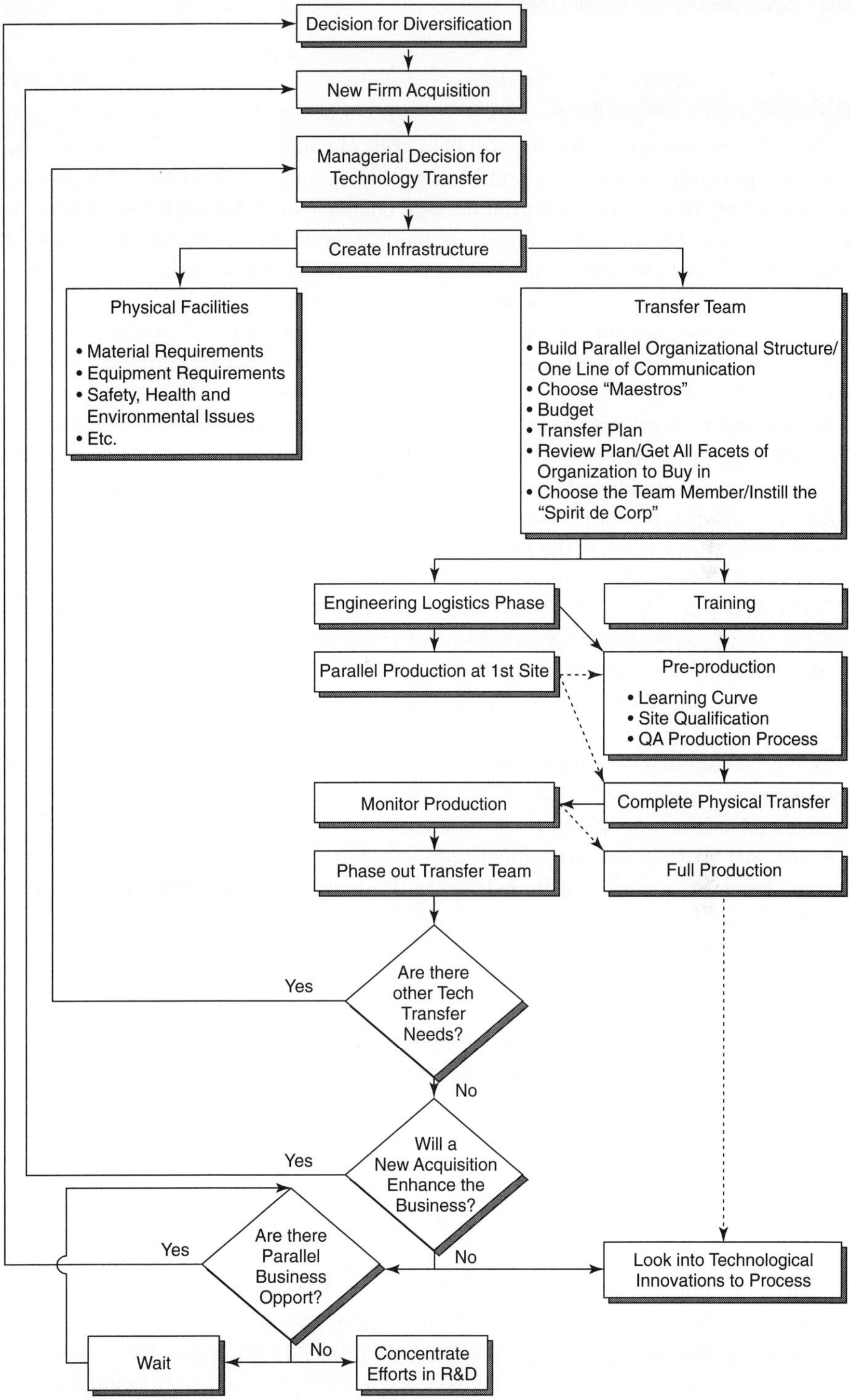


**FIGURE 11-5**
INTRAFIRM TECHNOLOGY TRANSFER MODEL
*Source:* Beruvides and Khalil, 1990; © 1990 Institute of Industrial Engineers.

## CONCLUDING REMARKS

Technology continuously flows across boundaries of countries, regions, companies, and departments within organizations and among individuals. The transfer of technology from one entity to another is effected through channels of technology flow. These may be general channels of contact among individuals and institutions or organized programs designed for the orderly and systematic transfer of technology.

Efficient and effective technology transfer requires the formulation of a strategy and the creation of mechanisms of transfer. These mechanisms can be Technology Transfer Centers, information exchange networks, or organized projects that utilize special teams to effect the transfer.

On the macro level of nations, newly industrialized countries such as Taiwan and Singapore have followed a special niche strategy to become competitive in world markets. They have concentrated on acquiring technologies in which they can gain a comparative advantage over other world competitors. Taiwan relied heavily on its nationals, trained overseas, to transfer the technology. Acquiring knowledge through people can be a very effective means of transferring technology.

The transfer of technology is not, and should not be, a one-time activity. It is a continuous process with follow-up activities. For the technology to take root within the receiving entity of the transfer, it must be nurtured. This requires a program of training, reinforcement, and R&D to keep the technology alive and make it grow in its new grounds. Unsupported technology can fade into obsolescence quickly.

In the United States, a significant number of technologies are created for space and defense-related applications. Efforts are under way to transfer these technologies to the commercial sector, where additional wealth can be created. Technology Transfer Centers are capable of linking government sites and centers of knowledge such as universities and research institutes of knowledge with industry and public sector enterprises.

READING 11.1

# The Profitable Transfer of Technology and Processes from Mature Industrial Nations to Low-Cost Countries

**Jeffrey Herbert**

Chairman and Chief Executive, Charter plc

Transfer of technology and processes from industrial enterprises in Western Europe to countries with low labour costs has been taking place since the 1950s.

The textile industry is a classic example. The relatively uncomplicated processes of cutting out and stitching up have been assigned to garment factories in countries where

*Source: Industry address,* Sixth International Conference on Management of Technology, Gothenberg, Sweden, 1997, Mekanisten, no. 1997:4, pp. 73–82. Reprinted with permission.

labour costs are between 10 and 20 per cent of the corresponding costs in industrialized countries.

This has resulted in advantages both for the country commissioning the work and the country supplying the labour force. The former has been able to obtain goods at lower production costs, while the latter has provided employment for its labour force.

When it comes to products with a more complicated production process and where comprehensive international quality systems must be taken into consideration, matters are not quite as simple. A number of factors must be observed if the company which is contributing its established technology, the recipient company, as well as the country in which the recipient company is situated, are to profit from the transfer.

ESAB AB, the world leader in welding, has drawn up a standard for the transfer of technology and processes, which has been tried and tested by means of a number of company acquisitions in former Eastern-bloc countries. I will briefly describe this standard and illustrate it with ESAB's acquisition of the Hungarian company, Csepel.

What I would like to demonstrate is that lower labour costs alone are not sufficient for the acquisition of an electrode factory, such as that in Hungary, to result in profitability.

But first, a few words about the significance of welding in modern industry, and on the need for top-quality filler metals for welding. The method of joining metals by heating them with an electric arc has been well-known for over a century. But it was through the invention of the coated electrode by the Swedish engineer and industrialist Oscar Kjellberg that welding became a useful method which could be applied to both repairs and new production.

Kjellberg's invention was patented in 1904, and formed the basis of the welding company ESAB. For those who have only come into remote contact with welding, it is certainly difficult to imagine the immense significance of this joining method. Today, the construction of ships, trains and other transport equipment, different types of pressure vessels, bridges and industrial plants would not be possible without the use of welding.

Welding processes have developed in parallel to the basic materials that are to be joined. Modern grades of steel require completely compatible filler metals. The latest welding processes have opened up new applications for welding, which in turn have triggered the development of new filler metals.

In order to guarantee the quality of a weld, it is necessary for the process to be carried out using filler metals with meticulously-tested properties, manufactured in a process controlled by extremely high quality standards.

There are both national and international standards which closely regulate the use of filler metals in certain applications. Even if the costs of the filler metals in a welded construction such as a nuclear power plant only constitute 1 or 2 per cent of the total production costs, the subsequent costs for repairs and rework can be much higher if the use of low-quality filler metals results in the final construction not fulfilling the set quality and safety standards.

It goes without saying that ships, vehicles, bridges and pressure vessels must not break down because of sub-standard welding. No manufacturer is prepared to take that kind of risk. The quality of welding electrodes is of paramount importance. They are either first-rate, that is, they fulfill the applicable quality standards, or they are not of interest to a quality-conscious user. There is no place for "almost-approved" products.

The same applies with regard to the standards for high productivity of the filler metals. This means that the electrodes must be easy to weld and that it is easy to remove the protective slag. In markets where a manufacturer has had a monopoly, these standards have not always been observed.

In a competitive world market, productivity requirements have led to the development of increasingly high-capacity welding electrodes which have benefited the industry and the end-user.

As a result of the exacting standards in the industry, it stands to reason that when ESAB puts its name to a product, that product must fulfill the same high standards, irrespective of where it has been manufactured.

For ESAB to be able to assign the production of welding electrodes to a new production unit, it must therefore make certain that all factors which affect the quality of the end product can be monitored closely. This applies not only to the production process itself, but the entire infrastructure. ESAB is therefore not prepared to assign manufacturing rights if, for example, it does not have control over the management of the manufacturing unit.

When ESAB acquired electrode factories in Hungary and the Czech Republic, great care was taken in educating and training the entire company management in the western concepts of quality. This training could include an apparently elementary matter, such as in-depth education in the English language.

Without an adequate knowledge of English, the management in one country cannot communicate with colleagues and customers in other countries. Good language skills are not normally viewed as an element in the transfer of technology, but on reflection, the lack of a common language creates obstacles in the effective transfer of knowledge and expertise.

Another example of quality control is that raw materials must be able to stand up to the same standards at the factory in Hungary as those applied to a Dutch subsidiary or a Swedish subsidiary. This sometimes meant that important raw materials had to be imported, since those locally available did not comply with ESAB's standards. This resulted in local manufacturers making an effort to attain western industrial standards, and after a few years, acceptance of locally-produced raw materials became possible.

Before I continue with other factors which ESAB is careful to control, let us take a look at what advantages there are for a country such as Hungary in ESAB taking over an electrode factory, then transferring its manufacturing technology there.

During the Communist era, an existing electrode factory to a large extent provided for the Hungarian market. The quality of the products did not correspond to what we in the West consider to be acceptable standards, but amongst other factors, currency problems resulted in Hungarian industry being effectively obliged to use domestic products.

When the Hungarian market opened up to the West, and western companies began manufacturing in Hungary, domestic companies had to compete on equal terms with their new western colleagues. Locally-manufactured electrodes were then, quite simply, not good enough.

Under those conditions, it was consequently of no interest that the Hungarian factory could produce electrodes at a lower cost. The decisive issue was that they could not produce acceptable quality. The only possibility for the Hungarian electrode factory to continue production and be able to provide employment was for someone to supply production technology and instill quality-consciousness, which subsequently resulted in the products being fully comparable to corresponding products manufactured in, for instance, the Netherlands, Great Britain or Sweden.

What could then make it interesting, under these conditions, for a company such as ESAB to buy its way into the Hungarian manufacturing unit and improve it in every respect so that it then becomes fully comparable with its own factories? The answer to this important question is that as a result of its investment, ESAB would be able to secure a significant proportion of the Hungarian market for itself.

The costs of investments in new technology, the introduction of well-tested processes, the training of management and personnel in the use of quality techniques, new manufacturing equipment, and new and often more expensive raw materials must be recouped by sales in the local market.

A prerequisite for the transfer of technology being profitable is that the products resulting from this new technology must be the most attractive alternative both for local manufacturers wishing to compete on the world market, as well as for foreign companies establishing a manufacturing presence.

During production of coated electrodes in an ESAB factory in Western Europe, the proportion of materials used in relation to manufacturing costs is approximately 50 per cent. Before ESAB took over the Hungarian factory, the materials proportion there was 80 per cent, with a significant reason for this being the low cost of labour.

Today, these ratios are more in line with those in the West, but labour costs remain lower. After a while, the total production costs will therefore be lower in Hungary than, for example, in France, when comparing identical products of the same quality.

ESAB has been able to use Hungarian production to its competitive advantage in France, and is currently supplying a number of its standard, Hungarian-produced electrodes to the French market. Please note that this is a secondary effect. In other words, the export of electrodes can never be the primary justification for investment in a country which lacks a significant domestic market.

Which countries are, or could be, of interest for the transfer of technology in the manner which I have just described? Obviously, the former Eastern-bloc countries have been natural prospective candidates, and ESAB has succeeded extremely well in Hungary and the Czech Republic.

The same concept has also been successfully tested in countries such as India and Indonesia. In Russia, ESAB has established a joint venture with a large consumer of filler metals, where in principle the same criteria have been applied. In China, Algeria and Egypt, ESAB has appointed companies to manufacture under license in accordance with the ESAB standards.

As I have mentioned the "ESAB standards" so often, it is perhaps time to describe them in more detail. They can be summed up as follows:

- Concentration of work input
- Participation and influence
- Promptness

What do we mean by concentration of work input? To succeed in achieving profitability with the transfer of technology, a company such as ESAB must constantly be on the lookout for countries, and companies in those countries, that offer potential for acquisition or joint venture.

Just waiting for an opportunity to turn up is not good enough, because although the decision-making processes in these potential countries are considered by many to be extremely slow, a critical moment is always reached, and then you must be well-acquainted with all the facts. That is when it pays off to have done your homework by knowing the market potential, having calculated the acquisition costs and estimated the need for resources at the time of the investment in the company.

I cannot emphasize enough the significance of continually carrying out profitability analyses on markets which could be considered suitable for investment. It is important to have a thorough knowledge of local conditions over a long period of time, and be in contact not only with potential target companies but the national authorities as well.

The takeover of a filler material manufacturing unit and bringing this up to western standards is a question of transferring skills and expertise costing hundreds of millions of kronor. The national authorities will also want to participate in the decision-making process of an agreement. Negotiating with these departments and authorities requires technical, financial and cultural competence.

In accordance with the ESAB standards, sales of welding products on the local market must be adequate to cover the outlay for improvements to the factory and all other measures which I have previously mentioned, and provide ESAB, in its role as owner, with an adequate surplus for it to be able to finance those activities which will not be run locally.

If there is no surplus on domestic sales, there will be no resources for the development of new formulae, new production processes or new quality concepts. The resulting status quo will neither benefit the local company, nor ESAB as its owner.

The transfer of formulae for new products, the introduction of new manufacturing processes, and a likely reorganization of the company mean the creation of a situation filled with potential conflicts and complications. These must not bring matters to a standstill, but must be dealt with by the company's management and ultimately by the company's owner.

In order for it to act effectively, ESAB's intentions are normally to gain at least 51 per cent ownership, with the stated objective of gaining sole ownership after a period of time. Application of the technology and techniques which ESAB transfers to the new company must be according to the ESAB way, and there are many areas where no compromises are possible.

ESAB demands the right to appoint key personnel such as Managing Director, Production Manager or Financial Controller. They are normally recruited from corresponding functions in other ESAB production units, and constitute an important part of the "know-how transfer," which is one of the prerequisites for success. A Managing Director, Production Manager or Controller normally stays in this position for two to three years, and during that time has the opportunity of training his locally-recruited successor.

The second ESAB standard is "Participation and influence." The transfer of technology and processes between two companies involving the type of products which I have chosen as illustrations has a significant impact across all areas of both companies. The research departments and those sections involved in production technology and quality are natural givers and recipients. But we have also found that departments such as marketing, finance and personnel must also take an active part if we are to achieve the desired results.

Our experience has also been that the greatest difficulties are encountered by the side which is to supply the information. A company which wants to devote itself to the transfer of technology must therefore be mindful of the need for a certain amount of "overcapacity" which can be used in such projects.

It is equally important that those who are to transfer new technology and expertise have practical experience of the subjects they are to impart. It is not usually possible to use the method of "Train the trainers," that is, to permit the transfer of technology to take place via a group of instructors specially trained for the purpose. First-hand experience is invaluable and essential for this type of transfer of skills.

It is just as important that it takes place on the right level. Knowledge of the various manual and automatic manufacturing processes is best exchanged between those who deal directly with the particular processes and those who will be dealing with them in the future.

The transfer of technology is just as much the handing-over of well-documented process descriptions as it is the exchange of expertise on the lowest level between persons who are to operate, as an example, an extruder press.

In order to gauge how rapidly the modifications implemented lead to the expected results, ESAB applies widespread benchmarking. This means that for all critical points there are numerical values for productivity and costs, or a quality index. These indicators are continually measured and the results rapidly reported to all in ESAB's production unit management groups.

The Production Manager who finds that his position on the benchmarking scale is low is expected to contact his more successful colleagues to request advice and tips on how they have achieved their results.

At some stage each year, all production managers assemble at one of the factories in order to jointly carry out an on the spot evaluation of that entire company. This always results in a plan of action, based on the overall competence of all the ESAB electrode factories, being drawn up.

Experience has shown us that when it has become evident that ESAB is to take charge of an electrode factory in a new market, the takeover and the introduction of new technology and techniques must take place promptly.

The work must be organized as a high-priority project. This applies especially to that part of the home organization which is to be responsible for the transfer of the technology and techniques. They must be provided with resources in order to be able to make a concrete contribution at the new factory.

It can certainly take a number of years to develop a new production or quality process, but in order to succeed in the introduction of a new, well-tested method or process, much is to be gained by this taking place intensively during a limited period of time. ESAB aims to bring about a replacement of previous methods and processes, not a gradual adjustment. In order to succeed, it is also important to motivate all those who are affected by the changes, (and this is in principle the entire company), so that the changes are accepted in the right way.

To summarize, the transfer of technology and processes for the manufacture of products, where top quality is of paramount importance, by countries in Western Europe to countries with less well-developed economies and industrial structure is something quite different from transferring manufacture to a country with low labor costs.

The company which is to contribute its advanced technology and its proven manufacturing and quality processes, and put its trademark on the products, must have a dominant influence over the manufacturing company. There must also be a local market mature enough to receive the products based on this newly-transferred technology.

Lastly, the local market must be sufficiently large to be able to accommodate such a business venture. Only under these conditions will the transfer of technology be profitable both for the company supplying it and the company receiving it.

When the transfer of new technology and new manufacturing processes leads to the new products bearing a global brand name, the requirements for quality become fundamental. With the current trend towards ever-increasing free trade across national borders, different grades of quality for the same product are simply not acceptable. Putting it another way, an ESAB electrode must always comply with the same quality standards, irrespective of which factory it has been produced in.

What I have described may not be a general rule, but is applicable to companies which manufacture products that form part of comprehensive quality systems. I have mentioned ESAB filler metals as a consistent example, but I could just as easily have used components for airplanes or nuclear power stations as my examples. A distinguishing feature of these products is that they constitute an important infrastructure for the manufacture of finished products.

Clarity of vision and an understanding of modern industrial conditions are also needed, both from the side of the company management and from the authorities of the country in question, if the transfer of technology is to succeed. ESAB's takeover and re-equipment of the electrode manufacturing units in Hungary and the Czech Republic are shining examples, for both the company and the countries involved, of how the transfer of technology and processes can produce factories that successfully compete in the international market.

READING 11.2

# Technology Transfer in the Americas

**Sergio R. Lopes, Jr.**

National Technology Transfer Center, Wheeling, West Virginia

*Abstract:* This paper focuses on a particular technology transfer area (the Americas—Anglo-Saxon and Latin) and presents possible answers to two questions: (1) What are the technologies most needed in the developing nations of Latin America today? (2) How can the transfer of these technologies from the United States and Canada be improved

*Source:* From T. Khalil and B. Bayraktar (eds.), *Management of Technology IV,* pp. 396–405. Industrial Engineering and Management Press, Norcross, GA. 

profitably? In addressing these questions, this paper identifies key technologies driving economic development throughout the world today and their particular importance for Latin American countries. To better analyze the transfer process, the paper proposes a general model designated as the relocation/absorption paradigm. This model evidences the necessity for an active approach to technology transfer in order for key technologies to arrive in Latin America. This active approach identifies a conceptual knowledge exchange as the essential catalyst in the absorption process of the identified technologies.

## INTRODUCTION

The intention of this paper is to explore issues related to the transfer of very recent technologies from the United States and Canada to the developing countries of Latin America. By leaving the political discussion aside, this paper approaches the matter from the perspective of the management of technology, aiming to articulate useful suggestions on why and how to increase the technology flow to "south of the border."

The background that surrounds this topic has been the focus of much debate. In fact, there are still discussions about the characteristics of technology, channels for its transfer, ways to measure this transfer, and its impact on economic development, especially of developing nations. In spite of this discussion, two facts cannot be ignored. First, global manufacturing is undergoing dramatic changes, bringing us to what has been called a Third Industrial Revolution. These changes are a result of major advances in informatics, and new-process technologies in general. These technologies may be the only source of sustainable competitive advantage for countries in the twenty-first century (Thurow, 1992). Secondly, whatever technology and technology transfer may be, they are never desirable for their own sake. They are only valued in the expectation that they will drive economic growth, a phenomenon equated with economic development (Stolp, 1993).

While the relationship between the two ideas may be obvious at first, the significance of this relationship increases as it is analyzed with respect to the economic integration of the Americas. In order for the American continents to continue to cooperate economically, Latin American countries need access to the vital technologies that will enable them to participate in today's technology-based global economy. If this does not occur, there is a danger that the technological gap between developed and developing countries of the Americas will increase rapidly, which will in turn cause the developing countries to function at a different and much lower level of technological skills and capability (Marton and Singh, 1991). In this case, the economic cooperation would also shift to a much lower level. To avoid this technological devolution, both sides require an active approach to technology transfer. This approach considers the nature of the vital technologies for Latin America in order to enhance the efficiency and effectiveness of the transfer process. Furthermore, it highlights the benefits proper technology transfer can bring to donor organizations.

## TODAY'S TECHNOLOGICAL REALITY

The revolutionary technological developments of the 1980s have brought the world a new order. Today, technologies in the areas of informatics, factory automation, environmental technologies, advanced materials, and biotechnology are primary drivers of development in all industrializing and industrialized nations (Marton and Singh, 1991; Pool, 1991; and Stolp, 1993). These technologies, however, are still generated and utilized primarily in developed countries.

The technologies in the area of informatics are the ones that most directly affect the competitive advantage of a country in the short run. They involve a close interrelationship between computers, telecommunications, and system applications (including information technologies). Informatics technologies are enabling key industries in the United States and Canada not only to accomplish rapid transition from design to manufacturing, but also to utilize the resources of their affiliates throughout the world for product development and manufacturing. The Texas Instruments branches worldwide are an example of how informatics technology has affected product development. Detailed designs are sent instantly over a computer network enabling the branches to work simultaneously on separate parts of a project. The advanced telecommunications, information and computer support available have cut down design and development time significantly. In addition, Texas Instruments is now able to tap into all the resources offered by each branch in different parts of the world, further enhancing the quality of its products (Magnet, 1992).[1]

New-process technologies are also vital in today's global economy. They are closely related to informatics (heavily computer- and microelectronic-based) and are continuously being incorporated in all industries. Some examples are: the strong absorption of computer systems into manufacturing; the integration of highly advanced process control devices into the continuous-process industries, such as chemical and steel; and the increasing automation of data processing in the service industry. These technological advances have contributed to the production of faster, cheaper, and better goods and services.

In the specific case of manufacturing, computer aided design (CAD), computer assisted manufacturing (CAM) and computer integrated manufacturing (CIM) are distinctively penetrating factory floors to contribute to vitally flexible, high quality production strategies. These concepts require high-tech support from industrial robots, computer controlled machine tools, advanced instrumentation and sensors, as well as other microelectronic devices. Such high-tech integrated production systems are spreading into almost every industry due to their capabilities to meet the highly competitive requirements of today's market. In fact, Thurow (1992) predicts that in the twenty-first century there will be high-tech and low-tech final products, but almost every product in every industry—from fast food to textiles—will be produced with high-tech processes.

Altogether, these process technologies possess two very important characteristics: (1) they are continuously displacing the emphasis on cheap labor as production of goods and services becomes more capital intensive; and (2) their development is closely interconnected with new approaches to production strategies and organizational structures. The first characteristic is a crucial one. The utilization of advanced, heavy-automated

[1]One interesting aspect to notice is that East Asian countries are the group of developing countries most heavily participating in such global activities. For several reasons, primarily the favorable political structure, these countries are ahead in the implementation of the necessary informatics infrastructure, giving them a significant competitive edge when compared to most developing countries.

production technologies not only replaces the cheap labor competitive advantage, but the cheap, unskilled labor itself. In this dynamic production environment, firms need to constantly retrain workers for the skills that are necessary to be continuously absorbing new technologies. The latter is also a vital component of the utilization of the technological basis. If a company has a high-tech facility but does not have the right organizational structure that best fits the technology, this company is under-using its resources and wasting money. The competitive advantage generated from the adoption of technological tools is closely related to the management changes that come along with it.

The technologies related to the environment also directly affect the competitive advantage of countries. Because of the research and development efforts they require, these technologies may decisively impact the pace at which businesses of a certain country implement technical change. These technologies include replacement for substances hazardous to the environment, waste removal and remediation, and waste clean-up technologies. One example is the Montreal Protocol signed in 1987 by 31 nations of the developed and developing world. The Protocol calls for the phasing out of use of chlorofluorocarbons (CFCs) no later than 2000, in order to save the Earth's ozone layer. In the United States and Canada, the Protocol has been forcing industries to research and develop new alternatives to refrigerants, degreasing agents and insulating foams. The development of such alternatives does not raise productivity levels, nor contribute to more efficient production of goods and services in the short run. It impacts, however, the long-term development of a country's industries.

Biotechnology is also rapidly advancing with several applications in medicine, agriculture and environmental remediation. These applications may impact important areas such as vaccine development and medical diagnosis (Chakrabarty, Kamely and Kornguth, 1991). In the long run, the commercialization of such applications will give developed countries a significant competitive edge. Biotechnological developments are also closely related to developments in the area of advanced materials, especially in production of complex composites. Advanced materials are vital for providing industry with technologically advanced support for the production of innovative products.

## THE REALITY IN THE DEVELOPING COUNTRIES OF LATIN AMERICA

For Latin American developing countries (LADCs), such as Argentina, Brazil, Chile and Mexico, the time is crucial to start interacting closely with companies that possess the above technologies. In fact, there is an increasing commitment from Latin American leaders to make these technologies more available for import through less protectionist legislation.

LADCs are ideally suited to receive leading edge technologies. During the early 1970s, countries such as Brazil and Mexico were primary receivers of mature technologies in various industrial sectors, mostly via large multinational corporations' subsidiaries and foreign holdings of up to 49 per cent (Marton and Singh, 1991). These countries were relatively successful in absorbing foreign technologies and today have a fairly solid technological base. The vital technologies they need are the most advanced ones that will update their current infrastructure.

Information processing and telecommunications form the first group of vital technologies. They are the foundation for LADCs to participate in the global business arena. Furthermore, they are essential to give companies access to domestic and worldwide databases of information on available and alternative sources of technologies that will increase the frequency and quality of technology transfer decisions.

The operational technologies for factory automation are also pivotal. They encompass instrumentation, hardware (robots) and software (computer systems), as well as sensors and micro-electronic control devices necessary to promote quality, efficiency, and flexibility in manufacturing. These are key characteristics that need to be incorporated into the manufacturing frame of Latin American countries for their competitive survival in the global trade market (Stolp, 1993).

Finally, environmental technologies should also require significant attention. As more Latin American countries participate in regulatory agreements, access to technologies to comply with these regulations will be indispensable. An example is the stipulations of the Montreal Protocol for phasing out the use of CFCs. Although manufacture of CFCs is relatively low-tech, and several developing countries have their own manufacturing plants, the manufacture of its replacement will require high-tech investments. The challenge, then, is to transfer technology for CFC replacement from the developed countries, where it is being devised, to developing nations. Interestingly, Brazil and Mexico have already pointed to the need for an informational database to guide manufacturers of CFC in their search for alternative technology sources (Pool, 1991). As stated above, this service can be greatly enhanced with the use of modern telecommunication and information technologies that provide companies with worldwide access to databases containing alternative technologies.

The agricultural and medicine applications of biotechnology may also be of extreme importance for LADCs in the long run (Marton and Singh, 1991). These applications will be essential to assist in the more efficient production of food and improved design of vaccines.

## THE COMMON NATURE OF THE VITAL TECHNOLOGIES

Before considering the transfer process, it is worthwhile to examine the characteristics shared by the vital technologies (viz., telecommunications, factory automation, environmental and biotechnologies).[2] Their foremost common aspect is their great degree of interdependency. This linkage should not be broken when they are transferred to LADCs. In other words, these technologies need to be transferred simultaneously. Secondly, they are highly complex technologies because they so decisively affect organizational and human conditions around them. This makes their transfer to LADCs both crucial and critical. Lastly, leading edge technologies are constantly being updated. This increases the number of available options a firm has to solve a particular problem. To

[2]Dahlman (1989) takes a similar approach in analyzing the technologies that make an impact on the economic development of industrializing and industrialized nations. He explores industrial trends resulting from these technologies, examining the issues they pose for government economic policy.

counterbalance this ever-changing nature, commitment to make the right technological decision is important from all parties involved in the transfer process.

## THE TRANSFER PROCESS

In order to offer suggestions for improving the transfer of the vital technologies from the United States and Canada to LADCs, it is fruitful to employ a model which highlights the most important aspects of that transfer. The model is a relocation/absorption paradigm because the transfer process is differentiated into two general stages: technology relocation and technology absorption. The main point the use of this model reveals in regard to the vital technologies is that the flow of conceptual knowledge must parallel the flow of the "hard" technology in the transfer process. The conceptual knowledge must address human resource issues and is an imperative catalyst in the absorption process for the vital technologies.

The relocation phase encompasses the identification of a specific technology and a donor organization, the negotiations between the parties involved, the legal agreements to accomplish the transfer, and the actual relocation of the "hard" technology to a foreign site. This first phase is rather easy to measure. Possible indicators of the level of technology relocated by a country, either imported or exported, include fees and royalty payments and figures for direct foreign investment. In contrast, the absorption stage is often lengthier, has no specific breakdowns, and is usually subjectively measured. The process begins after a technology arrives in the foreign country and is completed when the receiver masters that technology. This technological mastery is, perhaps, the most important aspect of the entire transfer process.

The relocation/absorption model is applicable to whatever channels are used to deliver technologies across national boundaries. The most common channels are: (1) direct investment in wholly owned subsidiaries or joint enterprises; (2) licensing contracts; (3) turnkey projects; and (4) the installation and servicing of purchased equipment. Some alternative forms of growing importance in several developing countries include non-affiliate technology licensing and contractual arrangements for supply of technology (Marton and Singh, 1991). The sources of these technologies can be either multinational corporations or smaller, medium-size firms. In all these forms, however, a technology is always relocated and should always be absorbed. The following are elements that can be improved in both stages.

The first element is the choice of technology, followed by the choice of the donor organization. The quality of the choice of a technology best suited to solve a particular problem is directly proportional to the access a recipient organization has to information on that and other similar technologies. Differences regarding the possibilities for keeping a technology updated also present options within a technological field, which requires thorough research about the technology prior to a final decision. A company should always be able to analyze different options before deciding on a technology, in order to select the one that best suits its technological needs. Next, finding the right partner organization to supply the technology is also heavily dependent on access to information about potential sources. Information about available technologies increases the bargaining power of the recipient organization because, if the market for that technology is a competitive one, the price of the transaction may be negotiated with many potential suppliers (Greer, 1981).

Most importantly at this stage, however, are the person to person interactions. The basis for any technology transfer transaction is commitment between people. Several interactions between the parties involved in the negotiations are necessary, and total commitment to the project cannot be over-emphasized. These interactions will check similarities and compatibility of both organizations as well as the possible strength of the relationship to be developed.[3] Most often the affinity between organizations early on in the process will shape the contractual agreements benefiting both organizations. In the case of smaller Latin American companies, this is an opportunity to create innovative methods to overcome the possible lack of initial investment capital. For example, a California-based instrumentation company licensed the production of their industrial computers to a Brazilian electronics manufacturer. Instead of the usual fee payment, the American company collects a percentage of the revenue from the sales of the new technology (Hock, 1991).

It is sometimes argued that medium-size rather than top ranked, large firms seem to be better partners in technology transfer operations to developing nations. Smaller donor organizations have the most to gain and the least to lose from engaging in innovative transactions with developing nations. Furthermore, their organizational structure is often more flexible thus better adapting to dynamic markets of developing countries.[4]

Again, it is important to observe that the choice of technology and of the donor organization requires a high-tech telecommunication and information infra-structure. The relevance of this capability goes beyond the investigations that companies should conduct on potential technological sources. It also permits the exchange of technical information electronically (technical drawings for example) and provides companies with worldwide access to databases (Glaser et al., 1983; and Trevino, 1989).

During the negotiation stage both organizations should keep one goal in mind: the achievement of a partnership. This is a paradigm shift for both organizations. Although the goals of each organization may be different at first, an ultimate harmonization is essential. In today's global business arena, the organizations need to share as many resources as possible. This is a significant opportunity for LADCs' technical personnel to engage in the development of technologies with partners from the U.S. and Canada. For the Americans and Canadians this is the opportunity to tap into the intellectual resources of Latin America. The pre-contractual negotiations are the foundation for the building of this alliance. Often it is important that this collaboration is spelled out in the legal contracts to make it an integral part of the transfer transaction.

Finally, for proper absorption of a particular technology the success of two elements already discussed, choice of technology and the negotiations stage, becomes very important. This is primarily so because proper absorption is indirectly dependent on the

[3]Walter (1981) provides a comprehensive commentary on interactions between organizations for the achievement of goal harmonization.

[4]Niosi and Rivard (1990) and Greer (1981) discuss in detail technology transfer to developing countries via small and medium-size enterprises.

degree to which organizational goals and cultures are similar, or become similar during the transfer process. At some point, the absorption stage needs to integrate both organizations' production and marketing strategies, quality philosophy, degree of employee involvement, and degree of commitment to customers' needs. This integration is particularly relevant to the vital technologies (specifically new-process and factory automation technologies) due to their substantial impact on the work force, quality of the final product, and production strategies in general. In other words, technology needs to be exported in a package which includes the "hard" technology itself and the conceptual knowledge that surrounds it. The latter goes far beyond the operational skills usually associated with traditional training programs.

Because today's vital technologies for Latin American countries require comprehensive readjustments of the human resources needed to fully utilize them, a knowledge channel should be open to continuously deliver the basic concepts that make it possible to absorb a new technology. The use of the word "channel" implies the existence of a two-way exchange rather than a one-way flow. Through the empowerment and involvement of employees from both organizations, the donor also absorbs knowledge from the receiver. This knowledge is particularly useful to devise enhancements for current products, assist with new product development and strengthen innovation capabilities. Another way to describe this exchange is parallel transfer. This parallel transfer, then, should carry the following objectives: (1) train personnel who will utilize the new technology in order to bring the recipient organization to an elevated level of operation and understanding of possible alternative applications of the technology; (2) educate technologists (those who, by definition, are able to extend their knowledge and skills to utilize technologies to solve different problems); (3) exchange ideas on production methods and management techniques to best utilize particular technologies (these include issues related to quality, timeliness, flexibility and customer service); and (4) empower and involve employees of both organizations in new product and technology development as well as innovation. This process requires high levels of personal contact and dedication and its results are longitudinal in nature. The two organizations should work closely to constantly address these issues.

Usually, large multinational corporations are the ones most heavily engaged in education and training activities to carry on this parallel exchange. Some of these efforts occur through internal multinational management programs, and also through information consulting companies (Kim, 1990). Smaller multinational firms, however, should utilize the same resources in their quest to exchange knowledge with a partner organization.

A study by Braga and Willmore (1991) on the impact of imported technologies on the technological basis of Brazil indicates that imports of technology positively effect indigenous technological efforts as a whole. Dahlman and Westphal (1983) confirm this idea and emphasize the importance played by proper absorption in the process. They state that effective assimilation has a positive effect on a firm's innovation capability. This effective assimilation requires an increase in local capabilities (some are explored in the next section) which encourages participation of indigenous industry in other transfer projects. The accumulation of such experiences can also lead to the creation of specialized firms to assist in the transfer process, which, in turn, permits greater local participation in future transfers. This increased capability contributes to an economy's capacity to undertake independent technological efforts, including adaptation of foreign technologies as well as creation of new technologies.

Finally, the building of strategic alliances for complete transfer of technologies along with the creation of an intense parallel exchange of conceptual knowledge is justified by the very nature of today's global economy. Operating globally means operating with partners—local government, local business and industry and local citizens—which in turn means the extension of one's market through the further spread of technology (Ohmahae, 1989; and Kim 1990). In other words, proper absorption may be the best generator of more business for both the donor and the recipient organization.

## ADDITIONAL RESOURCES

Additional support by government, consultants and independent institutions is important to complement the efforts of organizations engaged in technology transfer, especially those of the receiver. Altogether, this support should focus on enhancing the efficiency and effectiveness of the transfer process. For example, medium- and small-size companies usually have limited access to information services and legal counseling on, technology transfer. These activities, then, should prove particularly useful in enhancing their initial decision on the source of technology and donor organization. Furthermore, access to extensive information on the alternatives offered by different suppliers of technology can increase the bargaining power of the receiving organization during the negotiation stage, creating a better match between organizations.

Since in most LADCs there is a lack of solid links between the research community and the productive sectors, information services can also provide extended help in linking these two entities (Dahlman, 1989; and Trevino, 1989). This gap needs to be bridged so that new foreign technology can blend with domestic research, generating new technological capabilities for LADCs and their developed partners. Technical information services can provide help to strengthen the links and interactions between research institutes, universities and firms (Dahlman, 1989). Other services that can be provided by the government include prescription and maintenance of standards and metrology support.[5]

## CONCLUSION

Among developing countries, present trends show an increasing polarization between those who can successfully adjust to technical change and heighten international competition and those that cannot (Dahlman, 1989). In order for Latin American countries to successfully bridge the technological gap that separates them from international competition and more intense trade with their developed American partners, an active approach to technology transfer is necessary. This approach should focus on the proper absorption of technologies via the building of new alliances. Alliances require future

[5]For more information on government organizations to support technology transfer activities, see Marton and Singh (1991). For a study of the Mexican case regarding this subject, see Trevino (1989).

partners to be willing to adapt, be open to new approaches, willing to spend time in person to person interactions, and willing to build collaboration, not competition.

Altogether, these factors are main contributors to the generation of economic development and indigenous research and development capabilities in the developing nations of Latin America. Moreover, the contributions of proper absorption and alliance building positively impact donor organizations. It provides them with access to new markets and intellectual resources that may decisively contribute to faster product development and innovation.

### Reading 11.2 References

Braga, H. and Willmore, L. (1991). "Technological Imports and Technological Effort: An Analysis of Their Determinants in Brazilian Firms," *Journal of Industrial Economics,* vol. 39, no. 4, pp. 421–432.

Chakrabarty, A. M., Kamely, D., and Kornguth, S. E. (1991). *Biotechnology: Bridging Research and Applications,* Norwell, Kluwer Academic Publishers.

Dahlman, C. J. (1989). "Technological Change in Industry in Developing Countries," *Finance and Development,* June 1989, vol. 26, no. 2, pp. 13–16.

Dahlman, C. J. and Westphal, L. (1983). "The Transfer of Technology," *Finance and Development,* Dec 1983, vol. 20, pp. 6–9.

Glaser, E., Abelson, H. H., and Garrison, K. N. (1983). *Putting Knowledge to Use,* San Francisco, Jossey-Bass, 1983.

Greer, D. F. (1981). "Control of Terms and Conditions for International Transfers of Technology to Developing Countries," *Competition in International Business,* Schachtel, O., and Hellawell, R. (eds.), (New York, Columbia University Press), pp. 41–83.

Hock, S. (1991). "Local High Tech Seals Deal with Brazilian Firm." *San Diego Business Journal,* vol. 12, no. 28, p. 1.

Kim, E. Y. (1990). "Multinationals: Preparation for International Technology Transfer," *Technology Transfer: A Communication Perspective,* Williams, F. and Gibson, D. (eds.), (Newbury Park, Sage Publications), pp. 259–273.

Magnet, M. (1992). "Who's Winning the Information Revolution," *Fortune,* vol. 126, no. 12, pp. 110–117.

Marton, K. and Singh, R. K. (1991). "Technology Crisis for Third World Countries," *World Economy,* June 1991, vol. 14, no. 2, pp. 199–213.

Niosi, J. and Rivard, J. (1990). "Canadian Technology Transfer to Developing Countries Through Small and Medium-Size Enterprises," *World Development,* vol. 6, no. 3, pp. 47–62.

Ohmahae, K. (1989). "The Global Logic of Strategic Alliance," *Harvard Business Review,* vol. 67, no. 2, pp. 143–154.

Pool, R. (1991). "A Global Experiment in Technology Transfer," *Nature,* May 1991, vol. 351, no. 6321, pp. 6–7.

Stolp, C. (1993). "Technology, Development, and Hemispheric Free Trade," *The Annals of the American Academy of Political and Social Science,* Mar 1993, vol. 526, pp. 151–163.

Thurow, L. C. (1992). "The New Economics of High-Technology," *Harper's Magazine,* vol. 284, no. 1702, pp. 15–17.

Trevino, M. (1989). "Regulation of Technology Transfer: The Mexican Experience," *Technology Transfer,* Winter 1989, pp. 46–51.

Walter, I. (1981). Commentary on "Control of Terms and Conditions for International Transfers of Technology to Developing Countries," *Competition in International Business,* Schachtel, O., and Hellawell, R. (eds.), (New York, Columbia University Press), pp. 143–159.

READING 11.3

# In from the Cold: Prospects for Conversion of the Defense Industrial Base

**Maryellen R. Kelley and Todd A. Watkins**

At the end of the Cold War, the manufacturing operations involved in making military equipment and commercial goods are commonly believed to intersect hardly at all. Our analyses of 1991 survey data from a large sample of establishments in the machining-intensive durable goods sector show that there are few technical and competitive conditions separating the defense and commercial industrial spheres. Commercial-military integration of production is now the normal practice among the majority of defense contractors in this sector. Moreover, we find little difference between defense and commercial producers in the competitive conditions they face or in the diversity of their customers. However, defense contractors have an advantage over their strictly commercial counterparts because of their greater use of productivity-enhancing technologies.

During the Carter-Reagan buildup (1979 to 1987), the U.S. Department of Defense (DOD) became an increasingly important customer for domestic manufacturers, particularly in durable goods industries (1). By the end of 1993, however, reductions in orders for weapons already in production and the elimination of entire programs reduced DOD's real (inflation-adjusted) procurement budget by 58% from 1985 levels (2). In this transition to a post-Cold War economy, policy discussions about the competitiveness of U.S. manufacturing and the restructuring of the defense industrial base are intertwined.

At the peak of the recent defense buildup in 1987, defense purchases were responsible for nearly 12% of the total sales of durable goods manufactured in the United States (3). Much of the concern about the economic consequences of a continued drawdown from these high levels of defense spending stems from questions about the capabilities (and willingness) of defense contractors to successfully function in the commercial economy. The defense industrial base is widely believed to have become isolated and disconnected from the commercial manufacturing base. A host of studies and reports argue that defense contractors have little experience with commercial customers and are unfit for the rigors of competitive markets (3–8). As a consequence,

*Source:* Reprinted *In from the Cold: Prospects for Conversion of the Defense Industrial Base,* with permission from *Science,* Volume 268, Apr. 28, 1995, pp. 525–532. American Association for the Advancement of Science.

conversion of defense manufacturing facilities to commercial uses is expected to be costly and have little chance for success. Moreover, some go so far as to warn against further reductions in defense spending on the assumption that if conversion is successful, defense-specific technical capabilities in the manufacturing supplier base will be irretrievably lost (4).

For the most part, previous research on defense manufacturing has been limited to case studies of a few leading companies and top-down analyses of government contracting practices, particularly as they affect corporate accounting and purchasing procedures. No analysis of a large sample of defense manufacturers has been conducted. The last systematic comparison of the practices of defense contractors and their counterparts operating strictly in commercial markets was conducted by Peck and Scherer 30 years ago (9). At the end of the Cold War, widely held suppositions about the singularity of defense production and its isolation from commercial practices have not been subject to rigorous empirical tests. With data from our 1991 survey of U.S. manufacturing plants from 21 durable goods industries, we demonstrate that structural and behavioral barriers thought to divide defense contracting from commercial manufacturing are actually quite rare. The defense industrial base is far-reaching and substantially "dual-use"; that is, meeting commercial customers' requirements and military specifications in the same facilities—indeed, using the same equipment and work force.

## THE CONVENTIONAL WISDOM: DEFENSE MANUFACTURING AS AN ISOLATED AND DISTORTED SYSTEM

Since President Eisenhower first employed the term in 1961, the "military-industrial complex" has conjured up an image of defense manufacturing as taking place in a specialized set of firms separated from the rest of the economy. Melman, one of the most widely cited critics of Pentagon spending during the 1970s, characterized the manufacturing of defense products as a "permanent war economy" where "whole industries and regions that specialize in military economy are placed in a parasitic economic relationship to the civilian economy" (7). Although there has been considerable academic debate over the question of whether defense expenditures have had a positive or negative impact on economic growth, there has been remarkably little discussion focused on the issue of how specialized and isolated defense manufacturing is from the rest of industry. At the end of the Cold War, even the most knowledgeable defense analysts assert that there is little overlap between defense and commercial manufacturing activities. Instead, much of the current concern focuses on procurement reform and the identification of those government contracting regulations or military technical requirements that are believed to be responsible for the divide that is assumed to separate the two industrial spheres. According to Alic et al., for example, special technology requirements, unique products, and intrusive government oversight have led firms to "conduct military business in divisions that are managed separately from commercial operations, often with separate work forces, production and research facilities, accounting practices, engineering design philosophies, and corporate culture" (3). Although their explanations differ from those of Alic et al., Markusen and Yudken also believe that there is an unbridgeable divide between commercial and military manufacturing, which they describe as "a wall of separation—a business culture on the military side that is ill-suited to engage in commercial production, and vice versa" (6).

The practice of isolating defense operations is not assumed to be limited to the large multidivisional corporations that are the recipients of major prime contract awards. Without reference to any empirical evidence, Markusen and Yudken claim that "subcontractors have become more, rather than less, specialized in military projects, as the 'wall of separation' reaches down into their ranks" (6). Similarly, former Deputy Assistant Secretary of Defense Gansler speculates that the high costs of weapons systems can be at least partly attributable to the dedication of lower tier suppliers to serving defense needs to the exclusion of commercial customers. At the end of the Cold War, he believes that "only a few suppliers remain in the lower tiers of the defense industry and they are highly specialized. The specialization of these firms in defense subcontracting means that DOD loses the economies of scale that could be realized in combining defense and non-defense production in the lower tiers of the industry" (5). In sum, most of the dollars spent by DOD on its weapons systems are commonly assumed to go to plants in which the entire organization—its technology, workers, and management systems—is dedicated exclusively to serving that military customer.

Government contracting practices are believed to be largely responsible for the differences separating the defense and commercial industrial spheres. Certainly, the contracting relationship between the government as "buyer" and defense contractors as "sellers" of weapon systems departs in significant ways from the conditions associated with a market system of exchange (9, 10). The stylized market system of exchange presumed to operate for commercial transactions is characterized by many buyers and sellers. Key features of this system simply do not apply to government purchases of military weapons manufactured by private companies. Rather than many buyers and sellers, there is only one buyer (the government) for military weapons. The buyer also has the political power to restrict the sale or use of products to other potential customers. A company that makes a new high-tech weapon for DOD cannot sell that weapon to another customer (such as another government) without DOD's permission. DOD even forbids commercial use or sale of some of the components of these systems.

For a substantial share of contracts for weapons systems, the government makes payments to defense contractors on the basis of costs rather than on competitively set market prices. The main reason for cost-based contracts is the uniqueness of the products that defense contractors make. Moreover, the government (as buyer) exercises considerable control over sellers' internal operations through its direct involvement in the development of new weapons systems and its auditing of suppliers' costs. These peculiarities of the defense contracting relation have led some analysts to conclude that there must be little potential for overlap between a production system that satisfies military needs and one designed for commercial transactions, causing companies to "spin away" their defense operations from their commercial activities (11). Instead of being organized to satisfy the diverse demands of many customers, defense contractors are believed to be "captive" suppliers to the government, oriented solely toward compliance with its regulations.

The burden of regulatory compliance is also thought to induce behavioral distortions. Special accounting rules and unique or esoteric technical requirements are blamed for a wall of separation dividing production for the military from commercial manufacturing. This division between commercial and defense activities is thought to extend from headquarters to the shop floor, serving to insulate a defense contractor's commercial activities from the rules affecting its defense operations. In the presumably rare instances when companies make commercial products alongside their military products, cost-based pricing rules are expected to provide perverse incentives with respect to subcontracting and investment decisions. As a result, defense contractors are thought to subcontract out less, employing more direct labor than do enterprises that make products only for commercial customers (9, 12). Moreover, because there is assumed to be little or no competitive pressure to reduce costs, defense contractors are also thought to underinvest in productivity-enhancing technologies (13).

Although there may be any number of other differences in the management styles and routines that distinguish companies with close ties to the Pentagon from other enterprises, our focus is on the underlying market structure and behaviors that are so frequently assumed to separate defense production from the commercial industrial world. Our study is the first to make systematic comparisons of commercial enterprises with defense contractors from the same set of industries and the same production processes for the period after the Carter-Reagan buildup. With data from our 1991 survey, we investigated four propositions concerning structural and behavioral characteristics thought to distinguish defense contractors:

**1** Defense contractors tend to operate facilities that are largely dedicated to military contract work.

**2** Compared with commercial enterprises, defense contractors and their managers and workers face less competition and are more highly dependent on a few customers (DOD and a few large prime contractors).

**3** Defense contractors do less subcontracting of production operations than do commercial enterprises.

**4** Defense contractors tend to invest far less than commercial enterprises in productivity-enhancing technologies that are relevant to nonmilitary production.

## DATA DESCRIPTION

Our analysis of the differences separating defense from commercial manufacturing is based on data collected in a 1991 survey of a randomly selected, size-stratified sample of manufacturing establishments. Eighty-four percent of the production managers we contacted completed the survey, yielding a final sample of 973 plants. The questionnaire focused on the competitive conditions, technology, and other practices affecting products manufactured at least partially through the machining process at the plant. The sample was selected from the sector we define as machining-intensive durable goods (MDG), which includes 21 industries at the three-digit level of the standard industrial classification (SIC) system of the Department of Commerce (14). Collectively, these industries account for virtually the entire capital goods sector (excluding computers) and certain consumer goods. The manufacture of high-tech military hardware in the form of aircraft, ordnance, navigational equipment, satellites, and missiles is concentrated in this sector. Overall, durable goods industries accounted for 82.5% of defense purchases of manufactured goods in 1990 and more than half (51.3%) of all defense purchases of durable goods in that year came from the MDG sector (15).

## THE EXTENT OF DEFENSE MANUFACTURING

DOD is the final customer (through prime contracts or subcontracts) for an enormous number of production facilities in the United States. For the MDG sector alone, we found that 48.8% of all plants had defense contracts in 1991. We estimate that nearly 40,000 manufacturing plants in this sector throughout the United States were engaged in defense contracting at that time. This estimate of the extent of the defense industrial base in the MDG sector in 1991 corresponds closely to results obtained from the Bureau of the Census's 1988 survey of 10,000 manufacturing plants employing at least 20 workers (16). Using this government data source, we computed the percent of plants with defense contracts in 1988 for the same set of industries. Nearly half (49.7%) of all establishments with 20 or more employees in the MDG sector reported to the Census that they had defense prime contracts (selling directly to one of the federal defense agencies) or subcontracts to defense prime contractors. Despite declines in defense spending in real terms between 1988 and 1991, there is no statistical evidence of a decline in the share of the overall manufacturing base in the MDG sector serving DOD during this period.

In U.S. manufacturing, there is a vast, hidden defense industrial base consisting of a large number of subcontractors that have no direct dealings with the Pentagon. As Table 1 shows for the MDG sector, most of the plants (64.1%) with any defense-related sales in that year did not sell directly to DOD but rather served only as subcontractors or suppliers to defense prime contractors.

**TABLE 1**
1990 SHIPMENTS FROM DEFENSE CONTRACTORS IN THE MDG SECTOR BY TYPE OF CONTRACTOR
Our calculation of the share of total defense shipments originating in the sector coming from subcontracts should be considered a low estimate, because the reports of prime contract shipments do not exclude the value of subcontracts let by the prime contractor.

| Type of defense contractor | Share of all defense plants (%) | Share of defense subcontract shipments (%) | Share of total defense shipments (%) | Share of total commercial shipments (%) |
|---|---|---|---|---|
| Only prime contracts | 9.63 | 0.0 | 0.84 | 9.94 |
| Prime and subcontracts | 26.23 | 46.0 | 77.04 | 51.26 |
| Only subcontracts | 64.14 | 54.0 | 22.12 | 38.80 |
| Contribution of column total to commercial and military shipments from defense contractors (%) | | 14.80 | 36.09 | 63.91 |

There is substantial pass-through of defense spending from major prime contractors to lower tier suppliers. Subcontracts alone accounted for 41% of all defense-related sales and shipments in the MDG sector during 1990. From one year to the next, the distinction between first (or prime) and lower (sub) tier contracting status will vary, because defense contractors often span tiers, making some products as a prime contract and others as a subcontract to another defense prime or subcontractor. Nevertheless, more than half (54%) of the value of shipments from subcontractors to prime contractors comes from lower tier suppliers; that is, those that had no prime contracts with a federal defense agency in 1990. Lower tier subcontractors contribute over one-fifth (22.12%) of all defense-related sales and nearly two-fifths (38.8%) of all sales from defense contractors to commercial customers in the MDG sector.

Our sample estimates of the extent of the pass-through from DOD prime contractors to subcontractors are well within the range of reports from government sources and from prime contractors about the extent of dependence on subcontracting. Using data on subcontracts to small enterprises that were provided by major prime contractors to the Pentagon, the U.S. Congressional Office of Technology Assessment estimates that 35 to 37% of all defense purchases in the 1980s went to enterprises that met one or another criterion as "small" (17). Of course, some subcontracts go to large companies as well. Our interviews with manufacturing managers at several major prime contractors (such as General Electric, Pratt & Whitney, Lockheed, and McDonnell Douglas) indicate that subcontracts account for 60 to 75% of major prime contractors' costs, depending on the product.

Only a few of the largest defense contractors are really very dependent on defense sales. Over the 5-year period ending in 1988, among the 100 largest defense prime contractors, the 67 that are publicly traded derived only 9% of their total sales from defense prime contracts, on average (3). Moreover, only 9 of those 67 firms had 50% or more of their sales coming from defense contracts during the peak years of the buildup. Yet, because some of these companies have set up a division for their defense business, indicating a formal separation between the reporting chains of command in their other product markets, previous studies have often assumed there to be little connection between the defense and commercial sides from the top to the bottom of the enterprise. However, in matrix organizational structures, the same work groups and organizational units may report to more than one product or market division and a functional department as well, such as manufacturing or engineering, that cuts across product market lines. Only establishment-level data can inform us about the extent to which activities undertaken to manufacture products for the military occur alongside those for commercial customers in the same organizational unit.

Drawing on our 1991 survey data for manufacturing establishments, we measured the extent to which defense procurement is dependent on a manufacturing base that is substantially isolated from commercial activities. Our indicator of the degree of defense segregation was the percent of total 1990 shipments from the plant that was sent directly to a federal defense agency (including any branch of the U.S. Armed Forces, the Defense Logistics Agency, depots of the services, and the Department of Energy) or to a prime contractor of one of those agencies.

Contrary to the conventional wisdom, we found that in 1990, the typical defense contractor was not especially dependent on the Pentagon. The median defense share in 1990 was only 15% for plants with any defense contracts in the MDG sector. The vast majority (80.4%) of establishments integrated commercial and military production in the same facility, selling more than half of their 1990 output to commercial customers. As Fig. 1 shows, only 21.4% of plants with prime contracts had more than 50% of their sales going to DOD in 1990. For the lower tier subcontractors, only 18.5% shipped more than 50% of their 1990 output to defense prime contractors. Moreover, as Fig. 2 shows, less than one-third (32.7%) of the total shipments of military goods from the MDG sector in 1990 came from plants that were highly dedicated to defense production (with more than 80% of their output going to a defense agency or a prime contractor).

The defense industrial base in the MDG sector includes both large and small companies. Multi-plant companies have the option to place all of their defense orders in one facility and their commercial work in another. If multi-plant corporations adopt such a segregation strategy, we should find a higher incidence of dedicated facilities among branch plants doing defense work than among single-plant enterprises. However, as shown in Fig. 3, there is no difference between these two types of companies and the proportions of facilities that are high-specialized in making defense products. We did several statistical tests (at $P = 0.05$) to examine the relation between the size of a plant or firm and defense dependence, as measured by the percent of total shipments from the plant in 1990 that went directly to a defense agency or a prime contractor. We found no significant correlation between size, as measured by sales or employment, and the

**FIGURE 1**
PERCENTAGES OF DEFENSE CONTRACTORS BY DEGREE OF DEPENDENCY ON SALES TO U.S. DEFENSE AGENCIES OR PRIME CONTRACTORS
The degree of defense dependency is measured by the percent of the value of all shipments from an MDG establishment in 1990 that went to defense agencies or prime contractors to defense agencies. "Prime contractors" are plants that shipped at least some of that year's output directly to a federal defense agency. "Only subcontractors" are defense contractors that did not ship any output directly to a federal defense agency. For more than 75% of both types of defense contractors in this sector, shipments of products for the military account for less than 50% of the value of all shipments from the plant.

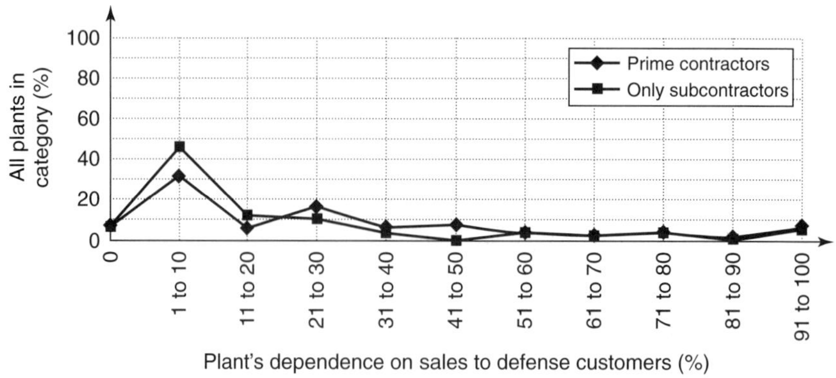

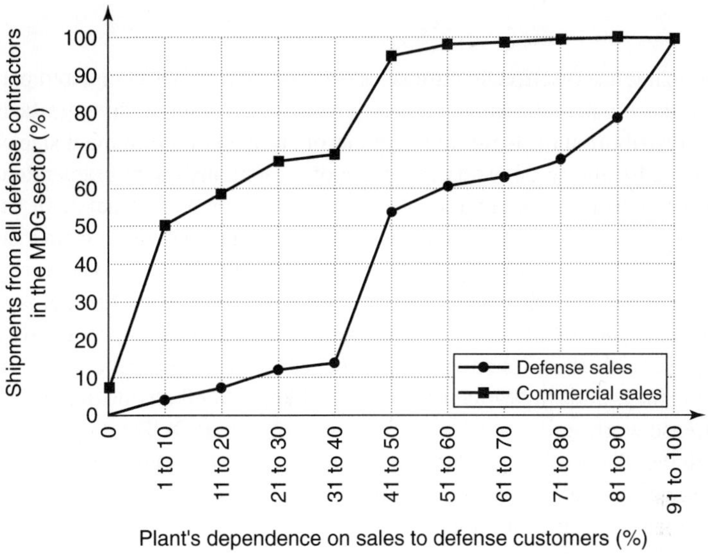


**FIGURE 2**

CUMULATIVE DISTRIBUTIONS OF 1990 SHIPMENTS BY DEFENSE CONTRACTORS TO DEFENSE AND COMMERCIAL CUSTOMERS

The distribution labeled "Defense sales" shows the estimated cumulative percent of the total value of shipments to defense agencies or prime contractors to defense agencies from defense contractors, ordered by the degree of the plant's dependency on sales to defense customers. The distribution labeled "Commercial sales" displays the estimated cumulative percent of the total value of shipments to nondefense commercial customers from defense contractors. Plants that depend on defense contracts for more than 80% of their 1990 shipments contribute only 32.7% of the total defense shipments from the MDG sector.

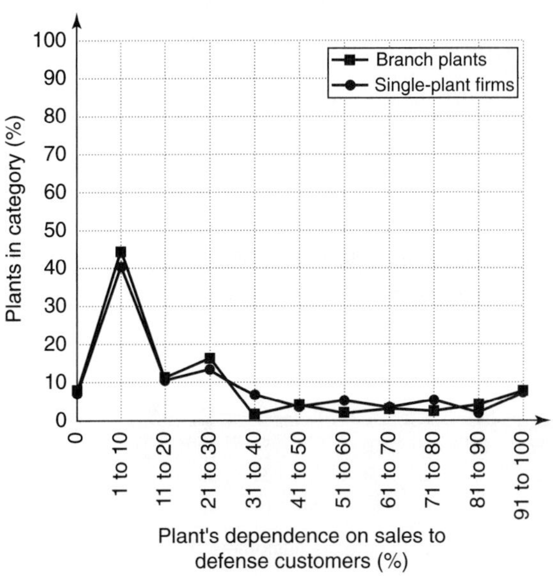


**FIGURE 3**

PERCENTAGES OF BRANCH PLANTS AND SINGLE-PLANT FIRMS WITH DEFENSE CONTRACTS BY DEGREE OF DEPENDENCY ON SALES TO DEFENSE CUSTOMERS IN 1990

Branch plants are establishments belonging to companies with multiple plant locations. Single-plant firms operate only one establishment. For both types of plants, there are no statistically significant differences between the distributions of the percentages of plants by degree of dependency on defense sales. Fewer than one in five plants in the MDG sector sell more than 50% of their output to federal defense agencies or defense prime contractors.

degree of dependence on defense purchases. We considered both plant and parent company size in these calculations. Moreover, $\chi^2$ tests fail to show any significant differences in the distribution of plants among plant or company employment size categories (1 to 49, 50 to 249, and $\geq 250$) and the extent of the establishment's dependence on defense sales when grouped by categories, reflecting 10% intervals (that is, 0, 1 to 9%, 10 to 19%, 20 to 29%, and so on). We also tested the difference between group means, comparing the mean size of single-plant enterprises as a group to that of branch plants of multiunit companies. We found no statistical differences between the practices of large firms (represented by branch plants) and those of small firms (represented by single-plant enterprises) in the sample plants' dependence on defense purchases. For the plants of multi-plant firms and single-plant enterprises alike, fewer than one in five of the plants that did defense work sold more than 50% of their output to DOD or a prime contractor.

Although the plants of large firms are not more defense dependent, on average, than those belonging to small firms, we did find that facilities dedicated to defense production were somewhat more common among those branch plants of larger companies that received prime contracts. As we show in Fig. 4, which looks only at branch plants of multi-plant firms, defense plants that have any prime contracts are significantly more dependent on sales to DOD, on average, than are branch plants that only have subcontracting ties to DOD. For example, a larger fraction of prime contractors (22.3%) than of subcontractors (12.1%) depend on DOD (or other prime contractors) for 50% or

**FIGURE 4**

PERCENTAGES OF BRANCH PLANTS BY CONTRACT STATUS AND DEGREE OF DEPENDENCY ON SALES TO DEFENSE CUSTOMERS IN 1990

Branch plants are establishments belonging to companies with multiple plant locations. "Branch plant primes" shipped at least some of their output in 1990 directly to federal defense agencies. "Branch plant subs." did not ship any of their 1990 output directly to federal defense agencies. On average, branch plants with prime contracts are more dependent on defense sales than are branch plants that have only defense subcontracts. Among both groups of branch plants in the MDG sector, however, the majority depend on defense sales for less than 50% of their total output in 1990.

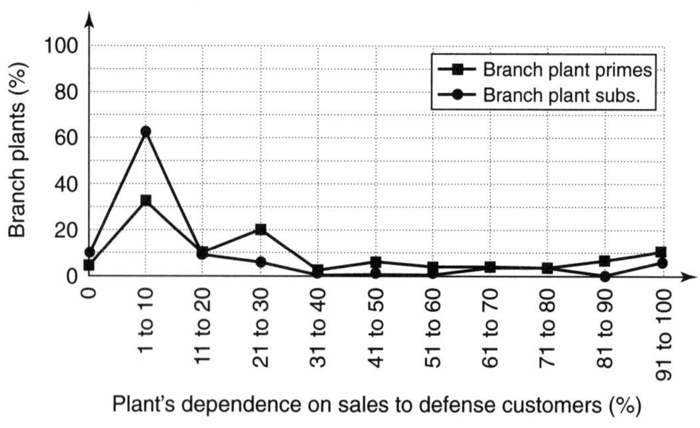

more of their sales. These differences are statistically significant ($P = 0.05$) by several tests. Yet facilities that serve both commercial and military customers are still the norm for defense plants that are part of multiunit companies.

In short, at the level of the plant, we find considerable integration between the commercial and military industrial spheres in the MDG sector. Large multi-plant firms that do defense prime contracting tend to be slightly more dependent on average than are subcontractors. But overall, we find that defense production in the MDG sector (whether directly for DOD or indirectly through subcontracts) usually takes place in facilities in which the majority of shipments go to commercial customers.

## CUSTOMER DIVERSITY AND COMPETITIVE PRESSURES

That defense contractors also serve some segment of the commercial market may not imply a broad capability to address a diverse set of customer demands. For example, defense contractors may conceivably be occupying specialized niches in commercial markets that are substantially different from those commonly filled by companies without the shelter of defense contracts. In this section, we address several questions about customers and competitive conditions in the MDG sector. First, we ask how many different customers defense contractors ordinarily serve and how that diversity of customers compares to that of plants operating in strictly commercial markets. Second, we investigate whether defense contractors are more dependent on sales to a small number of leading customers than are establishments with no defense contracts. Third, we consider whether defense contractors serve only a specialized niche in competitive environments that are more benign; that is, characterized by fewer rivals and less aggressive actions by competitors than those experienced by enterprises that are exclusively engaged in commercial transactions.

In our 1991 survey, we asked the plant managers to tell us how many different customers purchased products made by their plants in the previous year (1990). As Table 2 indicates, plants in this sector serve over 300 customers, on average, and there is no statistical difference in the number of customers reported by defense contractors as compared to enterprises serving strictly commercial markets. However, a substantial number of plants in both groups are niche producers, serving only a small number of customers. Fifty percent of defense contractors and their commercial counterparts have 30 or fewer customers. Moreover, establishments in the MDG sector depend on a small number of key customers, selling 60% of their total output, on average, to their largest three customers in 1990. The point is that on the whole, defense contractors have as diverse a customer base and are as dependent on a few key customers as non-defense establishments are.

Turning to the specific features of the product markets for machining output from these plants, we learn that custom-built products are the norm for this sector. The typical plant produces nearly half (46.9%) of its machining output in small lots of only one to nine items. Moreover, we find no evidence that defense contractors are more specialized in making highly customized machining products than are establishments making products solely for commercial customers. In fact, we find the opposite: Strictly commercial

**TABLE 2**
SELECTED CHARACTERISTICS OF CUSTOMERS AND PRODUCT MARKETS OF PLANTS IN THE MDG SECTOR
Group means are shown for defense contractors, for plants with solely commercial customers, and for the overall sample.

| Customer and market characteristics | Plants with defense contracts | Plants with solely commercial customers | All plants |
|---|---|---|---|
| Number of customers for machining products in 1991 | | | |
| Mean | 281.7 | 346.2 | 314.5 |
| Standard deviation | 1476.5 | 8200.8 | 5944.3 |
| Median | 30 | 30 | 30 |
| Number of plants | | | 920 |
| 1990 sales revenue coming from the plant's top three customers | | | |
| Mean | 60.7% | 59.5% | 60.1% |
| Standard deviation | 25.4 | 28.5 | 26.8 |
| Number of plants | | | 889 |
| Machining output in small lots (one to nine items) | | | |
| Mean* | 40.0% | 53.6% | 46.9% |
| Standard deviation | 36.7 | 39.2 | 38.6 |
| Number of plants | | | 959 |
| Plants with 50% or more of machining output in small lots | | | |
| Mean (% = yes)* | 43.1% | 56.2% | 49.7% |
| Number of plants | | | 959 |

*$P = 0.0001$.

plants produce significantly more of their output in small lots ($P = 0.0001$); and compared with defense contractors (43.1%), a greater share of plants with no defense contracts (56.2%) specialize in customized products, making 50% or more of their total machining output in batch sizes of fewer than 10 items.

In assessing the competitive environment, we considered several indicators, including the number of competitors and the extent to which rivals are particularly aggressive in competing for the same customers in terms of price, quality, or service. On average, we find that defense contractors report having a significantly larger number of competitors than do enterprises that have no defense contracts ($P = 0.0008$). But, as is shown in Table 3, a substantial share of both types of plants operate mainly in markets with few competitors. Fifty percent of defense contractors report six or fewer competitors; the median for non-defense enterprises in this sector is five or fewer competitors. In the MDG sector, the competitive environment for half of the enterprises in strictly commercial product markets consists of only a few rivals rather than the many sellers assumed to prevail in commercial markets.

**TABLE 3**
CHARACTERISTICS OF COMPETITIVE ENVIRONMENTS FOR MACHINING PRODUCTS OF PLANTS IN THE MDG SECTOR
Group means are shown for defense contractors, for plants with solely commercial customers, and for the overall sample.

| Characteristics of competitive environment | Plants with defense contracts | Plants with solely commercial customers | All plants |
|---|---|---|---|
| No. of competitors for machining products | | | |
| Mean* | 65.8 | 18.1 | 42.4 |
| Standard deviation | 267.5 | 96.5 | 203.9 |
| Median | 6 | 5 | 5 |
| Number of plants | | | 745 |
| In 1989 or 1990, did your competitors ever: | | | |
| Undercut your price with an important customer? | | | |
| Mean (% = yes)** | 68.0% | 50.6% | 59.3% |
| Number of plants | | | 657 |
| Introduce services or assistance you do not offer? | | | |
| Mean (% = yes) | 37.5% | 41.2% | 39.3% |
| Number of plants | | | 662 |
| Try to limit your business | | | |
| Try discouraging your customers or distributors? | | | |
| Mean (% = yes)*** | 30.3% | 22.4% | 26.4% |
| Number of plants | | | 650 |
| Introduce a similar product or service but with higher quality or performance? | | | |
| Mean (% = yes) | 16.0% | 12.5% | 14.3% |
| Number of plants | | | 656 |

* *P* in group means at 0.0008.
** *P* in group proportions of 0.0001.
****P* in group proportions at 0.02.

In sum, many of the features thought to be peculiar to the defense contracting relation also apply to a substantial share of the strictly commercial producers in this sector: a high dependence on a small number of customers, an evident willingness to custom-build products, and very few competitors.

In the 1991 survey, we asked about four different actions of competitors over the preceding 2 years. The most common competitive pressure came from price reductions offered by rivals to important customers. Nearly three-fifths (59.3%) of plant managers in the MDG sector reported that competitors had undercut their prices sometime during the previous 2 years. Offering new services or assistance to customers is another common way in which companies attempt to win business away from rivals in this sector. Less common are reports of predatory actions by rivals to discourage distributors or customers. And even though product quality has been touted in the business press as an important competitive pressure, few plant managers reported that their rivals were outcompeting them in quality.

Overall, we find no indication from these data that defense contractors are especially insulated or sheltered from competitive pressures experienced by companies operating in strictly commercial product markets. Indeed, in terms of two of the four indicators measuring the severity of competitive pressures, defense contractors experienced a significantly higher incidence of aggressive actions from competitors than did non-defense enterprises. Price undercutting behavior ($P = 0.0001$) and targeted attacks by competitors to undermine their ties to customers and distributors ($P = 0.02$) were more frequently experienced by defense contractors than by other manufacturers. Heightened rivalry among contractors for declining Pentagon orders may be part of the explanation for these differences, as might procurement reforms undertaken after the 1984 Competitiveness in Contracting Act that were designed to deliberately introduce greater price competition in defense contracting.

## SUBCONTRACTING

All of the establishments surveyed in the MDG sector make products with precision machine-tool technologies. Although we do not have information on all types of subcontracting practices at these plants, our survey did ask about subcontracting of operations from the machining production process at the plant. Our maintained hypotheses were that cost-based pricing rules in defense contracting should contribute to hoarding of direct production labor, and that defense contractors should be less likely to engage in production subcontracting and to spend less on subcontracts when they did contract out, as compared with the strictly commercial enterprises.

Table 4 compares machining subcontracting practices in 1989–90 between defense contractors and plants with no contract ties to DOD. We find that, on average, defense contractors are actually significantly more likely than non-defense enterprises to rely on machining subcontractors ($P = 0.0001$). For this key production process, 66% of defense contractors subcontract out at least some part of that work to other firms, as compared with only 51% of plants that do no defense contracting.

Among those that do contract out, we find no statistical difference between defense contractors and their strictly commercial counterparts in the MDG sector in the amount of subcontracting they do, as indicated by the amount of purchases from machining subcontractors in 1990 as a share of the total value of shipments from the plant. Similarly, we find no difference between defense contractors and non-defense producers in this sector in the average number of subcontractors they employ.

With respect to the machining process, at least, we find no support for the presumption that defense contractors are reluctant to engage in subcontracting as compared with their strictly commercial counterparts. It is therefore unlikely that government accounting and pricing procedures deter defense contractors from subcontracting.

**TABLE 4**
COMPARISONS OF MACHINING SUBCONTRACTING PRACTICES OF DEFENSE CONTRACTORS AND PLANTS WITH SOLELY COMMERCIAL CUSTOMERS

| Features of subcontracting | Plants with defense contracts | Plants with solely commercial customers | All plants |
|---|---|---|---|
| Do you usually contract out machining work to other firms? | | | |
| Mean (% = yes)* | 66.1% | 51.3% | 58.5% |
| Number of plants | | | 940 |
| Total of 1990 sales revenue spent on machining subcontracts | | | |
| Mean | 6.9% | 5.9% | 6.5% |
| Standard deviation | 8.3 | 8.3 | 8.3 |
| Number of plants with any spending on subcontracts | | | 520 |
| How many machining subcontractors did your plant use in 1990? | | | |
| Mean | 7.5 | 7.2 | 7.4 |
| Standard deviation | 22.5 | 30.4 | 26.3 |
| Number of plants with any subcontractors | | | 618 |

*$P$ in group proportions at 0.0001.

## TECHNOLOGY INVESTMENT PRACTICES

Hoarding of direct labor and the failure to make investments to improve productivity have long been identified as a possible source of high costs among defense contractors. Indeed, as early as 1976, a major Pentagon review of procurement practices concluded that defense contractors used only 42% as much capital equipment and facilities per dollar of sales as did durable goods manufacturers overall (18). In 1980, the House Armed Services Committee drew similar conclusions about the lack of investment in new manufacturing technologies by defense contractors (19).

During the 1980s, information technology applications in which computer software and microelectronic control devices are used to direct and monitor such ordinary production operations as machining, welding, testing, and inspecting were first introduced in the United States and elsewhere. These technologies have been heralded as providing cost, performance, and flexibility advantages for a wide range of uses (20). Cross-national comparisons of the adoption and use of certain applications, particularly for the machining process in the form of numerically controlled (NC) and computerized numerically controlled (CNC) machine tools flexible manufacturing systems (FMS), have come to be taken as indicators of the relative strengths of the manufacturing sectors of industrial economies (21).

Our survey results confirm a statistically significant difference ($P = 0.0001$) in the adoption rates of these types of advanced manufacturing technology related to defense contracting. But the differences we find, as shown in Fig. 5, are not what we would expect if defense contracting practices were a deterrent to investment in productivity-enhancing technologies. Sixty-six percent of plants with defense contracts have programmable machine tools (CNC, NC, or FMS), compared with 50% of plants that have no contract ties to the DOD or any of its prime contractors. Moreover, defense contractors that adopt this technology employ a much higher fraction of programmable machines in their total machine tool stock than do establishments engaged in the same manufacturing process but having no defense contracts.

For each of the five common uses of computers in manufacturing shown in Fig. 5, defense contractors have higher rates of use. In addition to programmable machine tools, these applications include computer-aided design (CAD), computer-aided manufacturing process control systems (CAM—used to plan and monitor inventory, work-in-process, and materials flow), computer-aided materials planning, and the use of programmable automation in other production processes. For every one of these technologies, we find significantly higher adoption rates ($P = 0.0001$) among defense contractors than among plants serving exclusively commercial markets.

Although it is difficult to single out a particular cause for these differences, we believe that government policy initiatives and programs directed at the defense industrial base are at least partly responsible for the large technological gap we find between

**FIGURE 5**
RATES OF ADOPTION OF SELECTED ADVANCED MANUFACTURING TECHNOLOGIES FOR PLANTS WITH DEFENSE CONTRACTS (SOLID BARS) AND FOR PLANTS WITH NO DEFENSE CONTRACTS (OPEN BARS)
The selected technologies are: programmable automated machine tools in the form of computer numerically controlled (CNC), numerically controlled (NC), or flexible manufacturing systems (FMS); computer-aided design (CAD) systems; computer-aided manufacturing process (CAM) control systems; computer-assisted materials planning systems; and programmable automation used in other production processes at the plant. In the MDG sector, for each of these technologies, plants with defense contracts have a significantly higher rate of adoption than do plants that operate strictly in commercial product markets (that is, they have no defense contracts).

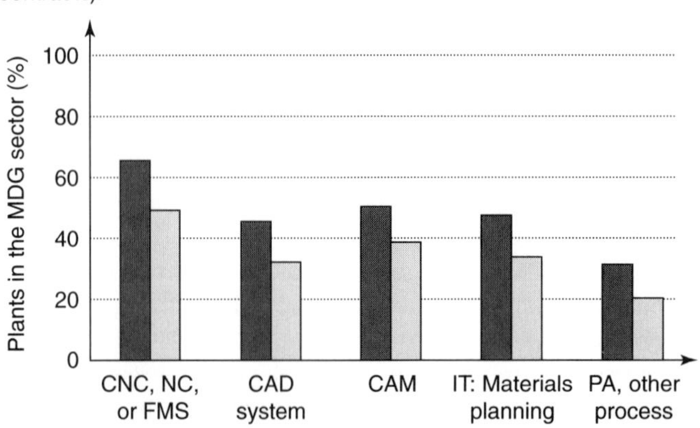

defense contractors and other U.S. manufacturing establishments in the MDG sector. From 1982 to 1992, the Industrial Modernization and Incentives Program of DOD provided technical assistance to contractors in assessing the applicability of advanced manufacturing technologies to defense contractors' operations. Through its manufacturing technologies (ManTech) program, DOD has also supported the development of advanced technologies and improvements in process technologies among defense suppliers. The Pentagon spent between $150 million and $200 million annually throughout the 1980s on these programs, which exceeded spending by all state governments on technical assistance programs aimed at manufacturing firms during the same period (22). Hundreds of defense contractors were directly assisted by these programs. DOD also sponsored annual conferences and workshops on manufacturing practices to highlight the lessons learned from the experiences of the early adopters of these advanced manufacturing technologies, providing an opportunity for representatives from the larger defense industrial community to become acquainted with the difficulties in implementing technical changes and the strategies employed by lead users to solve them. We believe that such forums promoted the dissemination of information about the implementation process that was not as readily available to manufacturing firms outside the defense contracting system. Our research also indicates that major prime contractors provided technical assistance and support to their suppliers that were less commonly available to companies with no contractual relation to DOD or its prime contractors.

Access to the technical assistance and supplier development activities of prime contractors and DOD can be construed as providing a competitive advantage to defense contractors that is not widely available in other supplier production chains. Research on supplier relations in the auto industry, for example, suggests that customer-supplier relations are not characterized by the type of information-sharing and technical assistance that we find to be so common among defense contractors (23). Other research also indicates that institutional mechanisms that foster information sharing and interorganizational learning can accelerate the diffusion of new technologies (24). Thus, the higher rates of adoption of advanced manufacturing technologies we find among defense contractors are at least partly attributable to the greater opportunities for interorganizational learning fostered by such government-sponsored activities.

## CONCLUSIONS

Defense spending reaches a broad segment of manufacturing in the MDG sector, affecting nearly one-half of all establishments. Contrary to conventional wisdom, commercial-military integration is not only feasible but is largely the normal practice at the end of the Cold War. The vast majority of defense contractors in the MDG sector manufacture military products in the same plants with the same workers and equipment employed in producing items for commercial customers. In fact, commercial customers dominate the sales of most defense contractors in this sector. Moreover, defense plants, on average, face as much competitive pressure as do those that produce only for commercial markets. Also, defense contractors use more modern and flexible manufacturing technologies at a higher rate than their strictly commercial counterparts do.

We conclude that the legacy of the 1980s defense buildup has been the generation of an industrial complex poised to exploit certain quite common kinds of commercial markets—those involving customized durable goods—in a post-Cold War era of flexible manufacturing. In the MDG sector, DOD has provided a more supportive environment for long-term investments and the transfer of technology than occurs for firms engaged in strictly commercial customer-supplier relations. Moreover, we find little evidence to support the widely held contention that government contracting procedures have forced a divide in the organization of military and commercial production for the vast majority of contractors. The policy challenge will be to find new ways to promote such supportive inter-firm exchanges outside the defense contracting network.

The integration of defense and commercial manufacturing activities may not be viewed as uniformly beneficial to society or even to the economy as a whole. For instance, the degree of integration we find at the end of the Cold War may reflect as much on the weaknesses of producers in commercial markets as on the capabilities of defense contractors or the influence of the Pentagon as an important buyer for this sector during the 1980s. We have focused here on the narrower questions involved in identifying the extent to which integrated dual-use capabilities exist among defense contractors and the degree of overlap between the competitive and technical environments of the defense and commercial industrial spheres.

Further research is needed to inform debates concerning the need for post-Cold War industrial technology policies. Policy discussions about the feasibility of the integration of military and commercial production and the barriers to defense conversion and diversification would benefit from more realistic assessments of the nature of the competitive environment that commercial enterprises face and the kinds of interdependencies among firms that are important to industry performance. Our study is the first to do so for a large cross-section of U.S. industry in a key sector. We think that other studies should be pursued, particularly in such processes as microelectronics and telecommunications. In our view, too much attention has been given to a few high-profile cases and too little attention to analyses of the broader industrial base. If our findings for the MDG sector hold true for manufacturing as a whole, we see few technical or organizational barriers to converting most defense plants to further serve commercial markets.

## Reading 11.3 References and Notes


1 R. Blank and E. Rothschild, *Int. Labour Rev.* **124,** 677 (1985); D. K. Henry and R. P. Oliver, *Mon. Labor Rev.* **1987,** 3 (August 1987).

2 R. A. Bitzinger, *Adjusting to the Drawdown: The Transition in the Defense Industry* (Defense Budget Project, Washington, DC, 1993).

3 J. A. Alic, L. M. Branscomb, H. Brooks, A. B. Carter, G. L. Epstein, *Beyond Spinoff: Military and Commercial Technologies in a Changing World* (Harvard Business School Press, Boston, MA, 1992).

4 D. Blair. *Ann. Am. Acad. Polit. Soc. Sci.* **517,** 146 (1991).

5 J. Gansler, *Affording Defense* (MIT Press, Cambridge, MA, 1989).

6 A. Markusen and J. Yudken, *Dismantling the Cold War Economy* (Basic Books, New York, 1992).

**7** S. Melman. *The Permanent War Economy* (Simon and Schuster, New York, 1974).

**8** *Deterrence in Decay: The Future of the U.S. Defense Industrial Base* (Defense Industrial Base Project, Center for Strategic and International Studies, Washington, DC, 1989); *Integrating Commercial and Military Technologies for National Strength: An Agenda for Change* (Center for Strategic and International Studies, Washington, DC, 1991); T. Lundquist, *Harv. Bus. Rev.* **70,** 74 (1992).

**9** M. Peck and F. M. Scherer, *The Weapons Acquisition Process: An Economic Analysis* (Harvard Univ. Graduate School of Business Administration, Boston, MA, 1962).

**10** J. S. Demski and R. P. Magee, *Account. Rev.* **67,** 732 (1992).

**11** R. J. Samuels, *"Rich Nation Strong Army": National Security and Ideology in the Technological Transformation of Japan* (Cornell Univ. Press, Ithaca, NY, 1994).

**12** W. P. Rogerson, *Account. Rev.* **67,** 671 (1992).

**13** In particular, the increased use of fixed-price contracts during the 1980s is thought to have discouraged long-term investments in productivity-enhancing technologies by defense contractors [K. W. Tyson, J. R. Nelson, N. I. Om, P. R. Palmer, *Acquiring Major Systems: Cost and Schedule Trends and Acquisition Initiative Effectiveness* (Institute for Defense Analysis, Alexandria, VA, 1989)].

**14** Machining involves the use of precision tools to cut and shape metal and includes grinding, drilling, milling, planing, boring, and turning operations. It is a process found in many manufacturing industries. Based on the industry-occupational matrix for 1985 constructed by the U.S. Bureau of Labor Statistics, we identified 21 industries specializing in this production process. Each industry accounted for at least 1% of all employment in machining occupations in manufacturing, and employment in machining occupations constituted at least 10% of all production employment in the industry. A size-stratified random sample of plants from these industries was selected and surveyed in 1987 and 1991. The industries are: nonferrous foundries (SIC 336); cutlery, hand tools, and hardware (SIC 342); heating equipment and plumbing fixtures (SIC 343); screw machine products (SIC 345); metal forgings and stampings (SIC 346); ordnance and accessories not elsewhere classified (SIC 348); miscellaneous fabricated metal products (SIC 349); engines and turbines (SIC 351); farm and garden machinery and equipment (SIC 352); construction and related machinery (SIC 353); metalworking machinery and equipment (SIC 354); special industrial machinery, excluding metalworking (SIC 355); general industrial machinery and equipment (SIC 356); miscellaneous machinery, excluding electrical (SIC 359); electrical industrial apparatus (SIC 362); motor vehicles and equipment (SIC 371); aircraft and parts (SIC 372); guided missiles and space vehicles (SIC 376); engineering and scientific instruments (SIC 381); measuring and controlling instruments (SIC 382); and jewelry, silverware, and plateware (SIC 391).

**15** These figures are based on the estimates of direct and indirect effects of defense spending in 1990 as reported in *Industrial Output Effects of Planned Defense Spending, 1990–1994* (Office of Policy Analysis, Economics and Statistics Administration, U.S. Department of Commerce, Washington, DC, February 1991).

**16** *Current Industrial Reports: Manufacturing Technology 1988,* SMT(88)-1 (Bureau of the Census, U.S. Department of Commerce, Washington, DC, 1989).

**17** This estimate applies to all DOD purchases, including those for services as well as manufactured goods. For certain defense contracts, an enterprise is considered to be "small" if it employs fewer than 500 workers. In other contracts, the definition of "small" includes companies that employ up to 1000 people [*After the Cold War: Living with Lower Defense Spending,* OTA-ITE-524 (Office of Technology Assessment, U.S. Congress, Washington, DC, 1992)].

**18** *Profit '76: Summary Report* (Profit Study Group, Office of the Assistant Secretary of Defense Installations and Logistics, U.S. Department of Defense. 1976).

**19** *The Ailing Defense Industrial Base: Unready for Crisis,* Report of the Defense Industrial Base Panel, H.R. Doc. No. 29 (Committee on Armed Services, House of Representatives, U.S. Congress, 1980).

**20** R. U. Ayres and S. M. Miller, *Robotics: Applications and Social Implications* (Ballinger, Cambridge, MA 1983); L. J. Hirschhorn, *Beyond Mechanization: Work and Technology In a Postindustrial Age* (MIT Press, Cambridge, MA, 1984); R. Kaplinsky, *Automation: The Technology and Society* (Longman, Harlow, UK, 1984); M. J. Piore and C. F. Sabel, *The Second Industrial Divide: Possibilities for Prosperity* (Basic Books, New York, 1984).

**21** C. Edquist and S. Jacobsson, *Flexible Automation: The Global Diffusion of New Technology in the Engineering Industry* (Basil Blackwell, Oxford, UK, 1988); M. R. Kelley and H. Brooks, "Diffusion of NC and CNC Machine Tool Technologies in Large and Small Firms," in *Computer-Integrated-Manufacturing, Volume III: Models, Case Studies, and Forecasts of Diffusion,* R. U. Ayres, W. Haywood, I. Tchijov, Eds. (Chapman and Hall, London, 1992); *Making Things Better: Competing in Manufacturing,* Report No. OTA-ITE-443 (Office of Technology Assessment, U.S. Congress, Washington, DC, 1990).

**22** P. Shapira, *Modernizing Manufacturing: New Policies to Build Industrial Extension Services* (Economic Policy Institute, Washington, DC, 1990).

**23** S. Helper, *Sloan Manage. Rev.* **32,** 15 (1991).

**24** M. R. Kelley and A. Arora, *Service Provider or Institution Builder? An Assessment of the Role of Industrial Modernization Programs in U.S. Technology Policy,* MIT IPC Working Paper 95-004WP, Industrial Performance Center, Massachusetts Institute of Technology, Cambridge, MA; N. R. Rosenberg, *Explor. Econ. Hist.* **3,** 3 (1972); E. von Hippel, *The Sources of Innovation* (Oxford Univ. Press, New York, 1988); T. A. Watlws, *Res. Policy* **20,** 87 (1991).

**25** We thank H. Brooks, E. Gholz, B. Harrison, R. Lester, R. J. Samuels, F. M. Scherer, and E. Skolnikoff for their advice and comments. Supported by grants from NSF (grants SES-8911141 and SES-9122155) and the Office of Technology Assessment of the U.S. Congress.

## DISCUSSION QUESTIONS

1 The ISO 9000 standard demands that every procedure within a company be recorded and included in a quality manual. Based on Figure 11-5, develop general guidelines for a technology transfer manual.

2 Discuss the technology transfer effort in an organization (industry or university). How structured is that effort?

## ADDITIONAL READINGS

Perrin, J. R. "Active Technology Transfer—History, Methodology, Results." In T. Khalil & B. Bayraktar (eds.), *Management of Technology II.* Industrial Engineering and Management Press, Atlanta/Norcross, GA, 1990.

Wood, O. L., & EearNisse, "Technology Transfer to the Private Sector from a Federal Laboratory." In T. Khalil & B. Bayraktar (eds.), *Management of Technology III.* Industrial Engineering and Management Press, Atlanta/Norcross, GA, 1992.

## REFERENCES

Berman, E. M., & Khalil, T. 1992. "Technological Competitiveness in the Global Economy: A Survey." *International Journal of Technology Management,* vol. 7, no. 445, pp. 347–358.

Beruvides, M. G., & Khalil, T. M. 1990. "Intra-Firm Technology Transfer: A Model and Case Study." In Khalil, T., & Bayraktar, B. (eds.), *Management of Technology II.* Industrial Engineering and Management Press, Atlanta/Norcross, GA.

Bloch, Erich. 1992. *Competition: Challenge for the 1990s.* Keynote speech at the Third International Conference on Management of Technology, Miami, FL.

Bowser, H. 1987. "Maestros of Technology: An Interview with Arthur M. Squires." *American Heritage of Invention and Technology,* vol. 1, pp. 24–30.

Brody, R. J. 1996. *Effective Partnering: A Report to Congress on Technology Partnership.* U.S. Department of Commerce, Office of Technology Policy, Washington, DC.

Chen, Y. Y. 1990. "The Technology Derivation Mode for Enhancing the Transfer of Industrial Technology Research and Development in Taiwan." In Khalil, T., & Bayraktar, B. (eds.), *Management of Technology II.* Industrial Engineering and Management Press, Atlanta/Norcross, GA.

Cheney, D. W., & Grimes, W. W. 1991. "Japanese Technology Policy: What's the Secret?" Council on Competitiveness, Washinghton, DC.

Cringely, Robert. 1996. *Accidental Empires.* 2nd edition Harper Collins, New York.

Engardio, P., & Gross, N. 1992. "Asia's High-Tech Quest: Can the Tigers Compete Worldwide?" *Business Week,* Dec. 7.

Glazer, H. 1992. "An International Comparison of Japanese Corporation R&D." In Khalil, T., & Bayraktar, B. (eds.), *Management of Technology III.* Industrial Engineering and Management Press, Atlanta/Norcross, GA.

Hayes, Robert H., & Abernathy, William J. 1980. "Managing Our Way to Economic Decline." *Harvard Business Review,* July–August, pp. 67–77.

Jain, R. K., & Triandis, H. C. 1990. *Management of R&D Organizations.* Wiley Interscience, New York.

National Aeronautics and Space Administration. 1995. *Technology Transfer and You.* NASA Center for Aerospace Information, Linthnicum Heights, MD.

Perrin, J. R. 1990. "Active Technology Transfer—History, Methodology, Results." In Khalil, T., & Bayraktar, B. (eds.), *Management of Technology II.* Industrial Engineering and Management Press, Atlanta/Norcross, GA.

Szakonyi, Robert. 1992. "Ten Blind Spots in Most American Companies' Management of Technology." SRI International, Menlo Park, CA.

U.S. Office of Technology Policy. 1997. *International Plans, Policies, and Investments in Science and Technology.* U.S. Department of Commerce, Washington, DC.

Wong, Poh-Kam. 1995. "Small, Newly Industrializing Economies Facing Technology Globalization: A Singaporean Perspective." In Lefebvre, L., & Lefebvre, E. (eds.), *Management of Technology and Rural Development.* Paul Chapman Publishing, London.

Wood, O. L., & EearNisse, E. P. 1992. "Technology Transfer to the Private Sector from a Federal Laboratory." In Khalil, T., & Bayraktar, B. (eds.), *Management of Technology III.* Industrial Engineering and Management Press, Atlanta/Norcross, GA.

# 12

# THE MANUFACTURING AND SERVICE INDUSTRIES

## WORLD-CLASS MANUFACTURING

Manufacturing organizations are vehicles for the creation of goods. To compete in the global environment, a manufacturing firm must be a world-class organization. In order to become world-class, it is essential that the firm develop an effective manufacturing strategy.

In the early phases of technology development, the competition is focused on product technologies, but it will eventually shift to process technologies. As the technology matures, the competition shifts drastically toward process innovations and the implementation of sound, well-integrated technology and marketing plans. Competition based on price and quality intensifies.

The competitive position enjoyed by Japan and some newly industrialized countries, such as Korea, largely depends on their ability to implement effective production technologies. German industries traditionally have had strength in manufacturing. Germany has continued to build on this strength to attain leading positions in world markets.

Many observers have attributed the decreased competitiveness of U.S. industries in the 1970s and 1980s to neglect of manufacturing technologies. This may be partially true, because in this arena the technology is available in most industrialized countries. If the technology does not exist in one country, it can flow from another country through recognized methods of technology transfer. Companies move their production facilities to locations where they can get the most for their investment. The competitive advantage of any country is based on its ability to attract investments and to effectively manage its technological resources. Japan is a small country with few natural resources, yet it is one of the world's industrial leaders. Through good strategic planning and the application of innovative production technologies, Japan's economy now occupies a leading position in the world. Japan excels in the area of quality. Meticulous attention to quality gave Japanese industry a strong presence in the market. The Japanese attitude of "get it right the first time," combined with a philosophy of continuous improvement, helps Japanese firms maintain efficiency and ensure customer satisfaction.

Another area in which Japanese industry excels is time to market. In the 1970s, and 1980s, it took Japanese automakers significantly less time than U.S. or European manufacturers to develop a new car model from concept to market. Time-based competition requires well-designed and well-run organizations. Japan's industries have also been known for their ability to respond to market trends. They strike while consumer desire is high. This practice is helpful in capturing large market shares.

Exhibit 12-1 shows the engineering hours and project lead time required to produce a car in Japan, the United States, Europe, and Korea. It indicates the effectiveness of Japanese manufacturers in reducing the cycle time compared to their counterparts around the world in the 1980s. American manufacturers were able to reduce the gap in the 1990s.

U.S. manufacturers have recognized the need to respond to global competition in manufacturing. They are paying more attention to developing world-class manufacturing practices. A surge of interest in quality and reengineering occurred in the United States in the 1980s and 1990s. U.S. industry has become more focused on taking advantage of technological progress and lean manufacturing methods (Noori, 1990). The U.S. National Research Council's National Academy of Engineering undertook a series of studies, starting in the 1980s and continuing into the 1990s, to examine issues and challenges in U.S. manufacturing. Drawing on findings from these studies, Heim and Compton (1992) reported that "most efforts to develop a discipline of manufacturing have concentrated on understanding the performance of unit operations and activities, in the belief that maximizing the effectiveness of separate parts would result in an optimized system." This approach to systems optimization neglects the interactions and dependencies of the components and processes in complex manufacturing systems.

Viewing the manufacturing enterprise as a set of interacting and overlapping activities, Heim and Compton have compiled 10 operating principles that constitute the foundations of world-class manufacturing industries.

**EXHIBIT 12-1**
PRODUCT DEVELOPMENT PERFORMANCE BY REGION

| | United States | Japan | Europe | Korea |
|---|---|---|---|---|
| Engineering hours (millions): | | | | |
| 1980s | 3.366 | 1.703 | 2.915 | NA* |
| 1990s | 2.297 | 2.093 | 2.777 | 2.127 |
| Project lead time (months): | | | | |
| 1980s | 60.9 | 44.6 | 59.2 | NA* |
| 1990s | 51.6 | 54.5 | 56.1 | 54.5 |

*NA = not available.
*Source:* Dept. of Commerce. Office of Technology Policy (Fine and St. Clair, 1996).

READING 12.1

# Operating Principles of World-Class Manufacturing Organizations

**Joseph A. Heim**

National Academy of Engineering, Washington, DC

**W. Dale Compton**

Purdue University, West Lafayette, Indiana

*Abstract:* The discipline of manufacturing has largely focused on the functional elements of the manufacturing system. However, an understanding of the separate components, irrespective of how complete, is not sufficient; one must understand the totality of these functions and the interrelationships among them. Driven by the absence of such an understanding, operational paradigms have often evolved from beliefs or rules of thumb derived from personal experience and individual interpretation of empirical data. This so-called know-how varies widely in extent and validity from company to company and from industry to industry and is almost always impossible to generalize or apply to new situations. This paper presents the results of a study to identify some of the generic principles that improve the effectiveness of the operation of manufacturing systems and that we suggest must be recognized, understood, and adopted by manufacturing organizations that aspire to be world class.

## INTRODUCTION

Manufacturing is a complex activity drawing upon many disciplines and technologies, reflecting management attitudes and philosophies, organizational structures, and influenced by the customers for manufactured products and the suppliers of many of the components used to produce those products. Most efforts to develop a discipline of manufacturing have concentrated on understanding and improving the performance of unit operations and activities, in the belief that maximizing the effectiveness of the separate parts would result in an optimized system. Although our endeavors have provided a greater understanding of the fundamental phenomena underlying the individual components and an increased awareness of the details needed to direct and control them, we are beginning to realize that the complexity of the myriad relationships, interactions, and dependencies of the components and processes precludes such an approach to system optimization. It is now clear that ignoring the many interactions prevents good predictions of system performance and improved controls.

In the absence of an understanding of the totality of the manufacturing system, operational paradigms have often evolved from beliefs or rules of thumb that derive

*Source:* From Tarek M. Khalil and Bulent A. Bayraktar (eds.), *Management of Technology III.* Industrial Engineering and Management Press, Atlanta/Norcross, GA. 

from personal experience or individual interpretation of empirical data gathered from day-to-day operations in uncontrolled environments. This so-called know-how varies widely in extent and validity from company to company and from industry to industry. It is frequently situation dependent and, therefore, often impossible to generalize or to apply to new situations.

But this lack of emphasis on system issues is not the result of a lack of appreciation for the importance of the problem. Rather, anyone attempting to address these system issues is immediately confronted by the intricacy of the problem. The tools that are available for resolving the complicated combinations of physical systems, human workers, and overwhelming volumes of data generated by these systems are limited. Information on the performance of manufacturing systems is often fragmentary and incomplete, and even where the data are excellent, competitive pressures prevent the data from being disseminated and made available for research. Moreover, metrics used to evaluate the performance of the manufacturing enterprise seldom address system performance, and the predictability of the system is rarely established from formal definition of the relationships among the many variables involved.

This study was undertaken to identify a body of operating principles that are being used by world-class manufacturers. In examining the actions and procedures that the most successful manufacturers have taken as they have evolved to world-class status, we found that many of them have adopted common approaches. This suggests that manufacturers who aspire to world-class capabilities should understand and follow, to the extent possible, the successful approaches of others and learn to make their own improvements. In the following sections we define some key terms, consider how such a collection of foundations might be used by manufacturers, and briefly discuss a set of principles—foundations of world-class manufacturing systems—identified in this study.

## THE MANUFACTURING SYSTEM

The manufacturing system can mean many things, depending on the viewpoint taken. Figure 1 presents one possible view. Operations have been placed at the center of the enterprise, overlapping and interacting with administration and management, the product and process engineering activities, the applied sciences, and the marketing, sales, and service activities. Overlapping and interacting with all of these functions are the customers for the products or services, the vendors and suppliers that provide materials, components and services to the enterprise, the community in which the enterprise exists, and the government that establishes regulations, rules, and opportunities for the enterprise.

Figure 1 is intended to illustrate the interrelationships that exist in the manufacturing system. Although particular technologies are not identified in this description of the manufacturing system, it must be understood that they establish or enhance many of the capabilities of the various functions contained in the system. As noted, the applied sciences provide the technical base for many of the areas. The material transformation processes, sometimes referred to as unit processes, are the means used to convert materials into components and subsystems. Computer-based systems provide the tools to

**FIGURE 1**
THE INTEGRATED MANUFACTURING ENTERPRISE
The diagram illustrates overlapping functions, disciplines, and activities.

enhance the capability and performance of the design, planning, scheduling, control, and sales of the products. The product and process engineering, the unit manufacturing operations, the marketing, sales, and services, the vendors and suppliers, and the management and administration each benefit from these systems and their capabilities to describe system performance. Describing the total enterprise in this way draws attention to the fact that no single unit operation or function can exist in isolation from all other components of the system. It is the realization of the interdependencies among these many components of the system that has created the impetus for "simultaneous engineering" or "concurrent engineering." A successful product realization process recognizes the interdependencies and the overlapping of interests of these many unit operations. Figure 1 also suggests how different viewpoints, values, and objectives for the system can develop, depending on the discipline or functional group in which individuals work. People working in applied science, finance, marketing, service, or engineering may have very different views about their role and where it fits in the system. They may also have very different perspectives about the system than do the production people on the factory floor.

An important conclusion to be drawn from this diagram is that if a manufacturing enterprise is to succeed, there can be no basic difference in viewpoints, values, and goals among its constituent groups. It is clear that the areas of responsibility are not neatly separated from one another but overlap to an important degree; financial and accounting systems, for example, have a major impact on operations and engineering. Perceived or artificially created boundaries between organizational units, such as those between marketing and engineering, production and purchasing, production control and marketing, or employees and management, both restrict and complicate communication and cooperation. The performance of the system suffers. The challenge to management is to find ways to take advantage of the strengths of the various unit operations and functional groups while discouraging any tendencies to work at cross purposes or toward conflicting goals. Focusing on the integrated system, as opposed to the individual functional parts that make up the system, is critical to understanding the key relationships and interactions in the overall performance of the enterprise. Achieving true involvement among the various activities requires, of course, more than simply reducing the barriers between groups.

Although the subsequent discussions are addressed to all readers concerned with the competitive position of U.S. manufacturing, it should be noted that no attempt has been made to provide immediate operational solutions to today's manufacturing problems. Such actions would, of necessity, be specific to an industry, a firm, or a plant. This does not suggest, however, that every plant or firm must treat its problems as if they were unique or unprecedented. Nor does it imply that principles of operation are too general to be useful in the daily operation of the manufacturing system. The complex operational problems and the numerous interdependencies among the functions that make up the manufacturing system present a challenge to all who aspire to improve its performance. It is of prime importance that manufacturers constantly remind themselves that a new "system view" of the manufacturing enterprise is critical to accomplishing the desired objectives—the development of globally competitive manufacturing organizations depends upon our enhanced understanding of manufacturing systems and a willingness to persist in a continuous examination of the conventional wisdom for managing and controlling them.

## FOUNDATIONS OF MANUFACTURING

The foundations for a field of knowledge provide the basic principles, or theories, for that field. Foundations consist of fundamental truths, rules, laws, doctrines, or motivating forces on which other, less general principles can be based. While the foundations need not always be quantitative, they must provide guidance in decision making and in operations. They must be action oriented, and their application should be expected to lead to improved performance. In our view, the "foundations of manufacturing" should be universal to manufacturing industries—at least to companies in the same industry—and they should be culture free, although certain cultures may more effectively enjoy certain of the foundations.

Examples of foundations can be found in many fields of engineering. The laws of thermodynamics are used to determine the theoretical limits of efficiency of various heat cycles. Maxwell's equations and quantum mechanics provide the electronic designer with the structure within which to understand and predict the operation of solid-state electronic components and systems. In the design of chemical reactors, the various laws describing fluid flow and mixing have led to the development of certain "scaling laws" that assist the designer in moving from a laboratory scale model to commercial-size systems. Linear and nonlinear mechanics form the basis for understanding the behavior of materials under load. Viscosity, boundary layer phenomena, and molecular surface phenomena are important elements of the foundations of lubrication.

The foundations of manufacturing differ in important respects from those just described. For the manufacturing system, one is dealing with a complex combination of disciplines and technology, management attitudes and philosophies, organizational issues, and the influences of an environment that includes the customers for the product that is being produced. In dealing with this complexity, we have constructed a framework that allows the foundations to be grouped under three topics:

- Management philosophy, practice, and organizational relationship issues.
- Rigorous description of systems and prediction of their performance.
- Developing organizational learning capabilities and adopting appropriate technologies to improve the performance of the manufacturing system.

In the following sections we present 10 foundations that should be viewed as operational guidelines—a set of principles that can be applied in a wide variety of circumstances by those who wish to be a part of an enterprise whose goal is to be a world-class manufacturer. They represent criteria by which actions can be judged, goals and objectives established, and progress measured.

### Management Philosophy and Practice

The foundations related to management explicitly recognize that actions, decisions, and policies advocated and implemented by all levels of management are critical determinants of the success of an enterprise. Included in this grouping is the critical operational philosophy that emphasizes the importance of continuous improvement of all operations in the enterprise and the importance of employee involvement in achieving this form of improvement. There is the role of employee empowerment in achieving the timely solution to problems. There are the interactions that the manufacturing enterprise must have with other activities in the company, with their suppliers, and with the customer. There is the importance or organizational structure, communications, and goal setting. While these elements of the foundations of manufacturing are not quantitative in the usual sense, it is abundantly clear that world-class manufacturers have generally recognized and are applying these management practices and that these practices have contributed critically to their success.

**Goals and Objectives** What should be a manufacturer's goal if it is to compete successfully in the global marketplace? This goal is often referred to as being a "world-class" manufacturer, a term used to convey the sense of excelling. The Japanese describe it as striving to be the best-of-the-best.

> *Foundation:* World-class manufacturers have established as an operating goal that they will be world-class. They assess their performance by benchmarking themselves against their competition and against other world-class operational functions, even in other industries. They use this information to establish organizational goals and objectives, which they communicate to all members of the enterprise, and they continuously measure and assess the performance of the system against these objectives and regularly assess the appropriateness of the objectives to attaining world-class status.

**The Customer** A manufacturing organization serves a variety of customers. In addition to the customers who expect to purchase high-quality products and services, the owners or stockholders may also be thought of as customers in that they expect a reasonable return on the investment that they have made in the company. The employees are customers in that they expect an employer to recognize their contribution to the success of the company and to provide them with a reasonable reward for their efforts. These are the stakeholders in the organization in that each has made a personal commitment to its success. The stakeholders have special expectations and needs that must be met.

*Foundation:* World-class manufacturers instill and constantly reinforce within the organization the principle that the system and everyone in it must know their customers and must seek to satisfy the needs and wants of those customers and other stakeholders.

**The Organization** The complexity of the manufacturing system arises from many directions: the interdependence of the elements of the system, the influence of external forces on it, the impact that it can have on its environment, and the lack of predictability in the consequences of actions.

*Foundation:* A world-class manufacturer integrates all elements of the manufacturing system to satisfy the needs and wants of its customers in a timely and effective manner. It eliminates organizational barriers to permit improved communication and to provide high-quality products and services.

**The Employee** Creating a world-class manufacturing organization begins with recognition that the most important asset of the organization is its employees. When properly challenged, informed, integrated, and empowered, the employees can be a powerful force in achieving the goals and objectives of the organization.

*Foundation:* Employee involvement and empowerment are recognized by world-class manufacturers as critical to achieving continuous improvement in all elements of the manufacturing system. Management's opportunity to ensure the continuity of organizational development and renewal comes primarily through the involvement of the employee.

**The Supplier or Vendor** It is essential that the barriers that have existed between supplier and purchaser be attacked as actively as are the barriers between the elements in a manufacturing organization. The sharing of goals, the exchange of information, the interchange of people, and the making of long-term commitments are some of the ways in which these barriers are being overcome.

*Foundation:* A world-class manufacturer encourages and motivates its suppliers and vendors to become coequals with the other elements of the manufacturing system. This demands a commitment and an expenditure of effort by all elements of the system to ensure their proper integration.

**The Management Task** Imaginative, creative leadership at every level of an organization is critical to building on the foundations of world-class manufacturing systems.

Management creates the culture within which the organization functions. Management must exhibit the concern for the health and well-being of the organization's human resources. Management must insist that the organization look beyond its borders to interact with its customers, its suppliers, and the educational systems that are training its present and future employees. It is a challenge to the organization to find the proper management for the circumstances in which it finds itself.

*Foundation:* Management is responsible for a manufacturing organization's becoming world-class and for creating a corporate culture committed to the customer, to employee involvement and empowerment, and to the objective of achieving continuous improvement. A personal commitment and involvement by management is critical to success.

### Measuring, Describing, and Predicting Performance

It is difficult to conceive of improving the current status of the system without first having a clear description of its status and character. This requires identifying the interrelationship and theoretical limits of the operational variables. It demands that important system parameters be identified and measured. Identifying cause-and-effect relationships that help predict the consequences of actions provides a basis for developing general tools and procedures that will allow the practitioner to extrapolate beyond current operating experience and to anticipate more accurately how a future system may respond or perform. The extent to which modeling, simulation, and analysis can be developed to provide these capabilities is an important element of the foundations of manufacturing. Although some of these quantities could be explored through experiments in the laboratory, we also recognize that some may need to be validated by techniques similar to those employed in microeconomics, social science, and cultural anthropology.

**Metrics** Performance evaluation is a process applied throughout the manufacturing organization to measure the effectiveness in achieving its goals. Because of the variety, complexity, and interdependencies found in the collection of unit processes and subsystems that define the manufacturing system, appropriate means are needed to describe and quantify rigorously the performance of each activity.

*Foundation:* World-class manufacturers recognize the importance of metrics in helping to define the goals and performance expectations for the organization. They adopt or develop appropriate metrics to interpret and describe quantitatively the criteria used to measure the effectiveness of the manufacturing system and its many interrelated components.

**Models** It is difficult to conceive of improving the current status of the system without first having a clear description of its status and character. This requires identifying the interrelationship and theoretical limits of the operational variables. It demands that important system parameters be identified and measured.

*Foundation:* World-class manufacturers seek to describe and understand the interdependency of the many elements of the manufacturing system, to discover new relationships, to explore the consequences of alternative decisions, and to communicate unambiguously within the manufacturing organization and with its customers and suppliers. Models are an important tool to accomplish this goal.

### Improving Performance

The objective of maintaining and achieving enhanced system performance requires an environment in which an organization can learn and benefit from its past experiences. As operating practice becomes more efficient through the application of the foundations of manufacturing, technology will become a more critical element in maintaining the status of the world-class competitor. The arrangements for acquiring, developing, and introducing new technology will become increasingly important as U.S. manufacturers continue to develop their abilities to compete in the world marketplace.

**Experimentation** Organizational learning is a broad-based strategy for capturing and making available to members of the organization information and knowledge that enable them to benefit from experimentation and the experience of others. Too often in manufacturing, sources of information become scattered and isolated and individual learning experiences are not automatically converted to organizational memory and made available for all members to draw and build upon. The rate at which an organization improves its performance as a result of learning is perhaps one of the principal determinants of whether it can become best-of-the-best.

*Foundation:* World-class manufacturers recognize that stimulating and accommodating continuous change forces them to experiment and assess outcomes. They translate the knowledge acquired in this way into a framework, such as a model, that leads to improved operational decision making while incorporating the learning process into their fundamental operating philosophy.

**Technology** U.S.-based manufacturers have often adopted the view that technological prowess is a viable means of compensating for other shortcomings. It is our strong conviction that a manufacturer can make the best use of technology only after it has embraced and is practicing the foundations described above. Only then can technology become a powerful force in achieving a competitive advantage.

For management, selection of the proper technologies from among technological opportunities is becoming a complex challenge that may be different for each manufacturer and for individual facilities. Each manufacturer must develop a strategy that encourages the search for the best and most important technologies, develops a procedure for effectively analyzing technological opportunities, creates or acquires the expertise needed to implement those technologies, and commits the necessary financial and human resources to introduce the new developments when they become available.

*Foundation:* World-class manufacturers view technology as a strategic tool for achieving world-class competitiveness by all elements of the manufacturing organization. High priority is placed on the discovery, development, and timely implementation of the most relevant technology and the identification and support of people who can communicate and implement the results of research.

## BENEFITS AND OPPORTUNITIES

What are the potential benefits we can expect to gain from identification and adoption of these fundamental principles of manufacturing? The following advantages appear to be realizable by applying the foundations.

First, a foundation provides a body of knowledge—a basis for understanding—that industrial and manufacturing executives could use to improve their ability to predict the outcome of specific product, process, and operating decisions. An immediate benefit should be the development of better generic tools for analyzing, designing, and controlling systems. One might hope, for example, that it would become common practice to explore thoroughly the impact of product complexity on the efficiency of the manufacturing operation instead of focusing only on the impact that additional product offerings will have on the marketing and sales activities.

Second, an understanding of the elements of a foundation should indicate some of the opportunities for more meaningful interdisciplinary interactions, for example, among scientists, engineers, production managers, and those who are associated with sales and marketing. Research programs and applications could share a common vocabulary, report on empirical measurements or experiments that test or validate new principles, and identify future research issues. The successful implementation of simultaneous engineering critically depends on a common understanding of many of these interdisciplinary issues.

Third, a foundation can help guide the experimentation and learning process that is important to achieving future improvement. In addition, it can help focus the exploration and use of technology to improve a company's competitive position in the world marketplace.

A foundation may be in a primitive state, such as a collection of empirical observations, that relates variables or outcomes and assists the manufacturing leader with actions and the manufacturing researcher with a context for discovery. However, it must be recognized that an enterprise derives no great advantage from the identification and understanding of a foundation of manufacturing unless it recognizes the strategic importance of manufacturing. Effective use of the foundations demands an organizational environment that encourages inclusion of manufacturing as a necessary strategic tool in becoming a world-class competitive force.

## SUMMARY

This study argues that the modern manufacturing organization cannot be competitive if it continues to operate as a loosely coalesced group of independent elements whose identity depends on a discipline or a detailed job description. The study explores principles that have been demonstrated as a generic to improving the effectiveness of manufacturing systems and draws heavily on the experience of U.S. manufacturers and a rapidly growing body of scholarly work linked closely to changing industrial practices. Particularly relevant research and publishing includes such topics as concurrent product and process engineering, total quality management, just-in-time manufacturing and distribution processes, quality function deployment, lean production, and incorporation and management of innovation.

In this paper we presented a group of operating principles that must be recognized, understood, and adopted by manufacturing organizations that aspire to be "world-class." Because of the universality of these principles, we have designated them as "foundations" of manufacturing. These foundations are generic in that they are not specific to a particular industry or company; they are universal in that they can be applied in a wide variety of circumstances; they are operational in that they lead to specific actions and show directions that should be taken.

Although success in implementing the foundations depends on many things, we believe that they represent a system of actions that cannot be embraced piecemeal. The foundations are as interrelated and as overlapping as are the elements of the manufacturing system they are intended to improve. The foundations must be viewed as a system of action-oriented principles whose collective application can produce important improvements in the manufacturing enterprise.

The elements of the framework presented are purposefully ordered, reflecting a belief that world-class competitiveness can be achieved only by properly applying all of these foundations, starting with management and progressing through the foundations related to metrics and technology. Unless the foundations of management have been put in place, the remaining foundations are not likely to be of lasting benefit to the enterprise. While rules and laws—combined with the continual measurement of important operating parameters—provide the capabilities to set goals and measure progress, these are unlikely to have the desired effect unless the management issues have been addressed. Enhancing an organization's ability to learn from experience is critical to its success, but the value of this learning will depend on how well the enterprise is managed and how thoroughly it understands its current operations. While technology may well become the ultimate tool for achieving a competitive advantage, the success that an enterprise has in using may depend on how well it adopts and integrates the other foundations.

**Acknowledgments**

This paper summarizes findings by the National Academy of Engineering Committee on Foundations of Manufacturing and represents the views of the paper's authors. The purpose of the study was to explore the disciplinary nature of manufacturing. Funding for the study was provided by the Intel Foundation and the National Academy of Engineering Technology Agenda Program. Many manufacturing practitioners, executives, academicians, and senior members of the policy community contributed to the study through participation in committee workshops and meetings, and writing background papers. The committee's full report titled *Manufacturing Systems: Foundations of World-Class Practice* is available from the National Academy Press, Washington, D.C.

### Reading 12.1 References

Camp, R. C. 1989. *Benchmarking: The Search for Industry Best Practices that Lead to Superior Performance.* Milwaukee, Wisc.: Quality Press.

Clark, K. B., and T. Fujimoto. 1991. *Product Development Performance.* Cambridge, Mass.: Harvard Business School Press.

Cohen, S. S., and J. Zysman. 1988. Manufacturing Innovation and American Industrial Competitiveness. *Science* 239:1110–15.

Compton, W. D., ed. 1988. *Design and Analysis of Integrated Manufacturing Systems.* Washington D.C.: National Academy Press.

Cooper, R., and R. S. Kaplan. 1991. Profit Priorities from Activity-Based Costing. *Harvard Business Review* May–June.

Dertouzos, M. L., R. K. Lester, R. M. Solow, and the MIT Commission on Industrial Productivity. 1989. *Made in America: Regaining the Productive Edge.* Cambridge, Mass.: The MIT Press.

Drucker, P. F. 1990. The Emerging Theory of Manufacturing. *Harvard Business Review* May–June.

Eccles, R. G. 1991. The Performance Measurement Manifesto. *Harvard Business Review* January–February.

Edmondson, H. E., and S. C. Wheelwright. 1989. Outstanding Manufacturing in the Coming Decade. *California Management Review* Vol. 31, Number 4.

Garvin, D. A. 1984. What Does "Product Quality" Really Mean? *Sloan Management Review* Fall: 25–43.

Harrington, J. 1984. *Understanding the Manufacturing Process.* New York: Marcel Dekker, Inc.

Hatvany, J. 1983. The Efficient Use of Deficient Knowledge. *Annals of CIRP* 32(1):423–425.

Hauser, J. R., and D. Clausing. 1988. The House of Quality. *Harvard Business Review* May–June: 66–73.

Hayes, R. H., and R. Jaikumar. 1988. Manufacturing's Crisis: New Technologies, Obsolete Organizations. *Harvard Business Review* September–October: 77–84.

Hayes, R. H., S. C. Wheelwright, and K. B. Clark. 1988. *Dynamic Manufacturing: Creating the Learning Organization.* New York: The Free Press.

House, C. H., and R. L. Price. 1991. The Return Map: Tracking Product Teams. *Harvard Business Review* January–February: 92–100.

Imai, Masaaki. 1986. *Kaizen (Ky'zen): The Key to Japan's Competitive Success.* New York: McGraw-Hill Publishing Company.

Jaikumar, R. 1986. Postindustrial Manufacturing. *Harvard Business Review* November-–December.

Johnson, H. T., and R. S. Kaplan. 1987. *Relevance Lost: The Rise and Fall of Management Accounting.* Boston, Mass.: Harvard Business School Press.

Kaplan, R. S., ed. 1990. *Measures for Manufacturing Excellence.* Boston, Mass.: Harvard Business School Press.

Koska, D. K., and J. D. Romano. 1988. *Countdown to the Future: The Manufacturing Engineer in the 21st Century.* Society of Manufacturing Engineers.

*Malcolm Baldrige National Quality Award.* 1991. U.S. Department of Commerce, National Institute of Standards and Technology.

Merchant. M. E. 1988. The Precepts and Sciences of Manufacturing. *Robotics & Computer-Integrated Manufacturing* 4(1/2):1–6.

Merchant, M. E. 1961. The Manufacturing-System Concept in Production Engineering Research. *Annals of CIRP* 2(1):77–83.

Mize, J. H., and T. G. Beaumariage. 1988. A Nation at Risk: Our Eroding Skill Base in Manufacturing Systems. pp. 42–51 in *The Challenge to Manufacturing: A Proposal for a National Forum.* Washington, D.C.: National Academy of Engineering.

National Academy of Engineering. 1991. *Manufacturing Systems: Foundations of World-Class Practice.* Washington, D.C.: National Academy Press.

National Academy of Engineering. 1991. *National Interests in an Age of Global Technology.* Washington, D.C.: National Academy Press.
National Academy of Engineering. 1985. *Education for the Manufacturing World of the Future.* Washington, D.C.: National Academy Press.
National Center for Manufacturing Sciences. 1990. *Competing in World-Class Manufacturing: America's 21st Century Challenge.* Business One Irwin.
National Research Council. 1991. *The Competitive Edge: Research Priorities for U.S. Manufacturing.* Committee on Analysis of Research Directions and Needs in U.S. Manufacturing. Washington, D.C.: National Academy Press.
National Research Council. 1990. *The Internalization of U.S. Manufacturing: Causes and Consequences.* Washington, D.C.: National Academy Press.
National Research Council. 1986. *Toward a New Era in U.S. Manufacturing: The Need for a National Vision.* Washington, D.C.: National Academy Press.
Nevins, J. L., and D. E. Whitney. 1989. *Concurrent Design of Products and Processes.* New York: McGraw-Hill.
Nonaka, I. 1989. Creating Organizational Order Out of Chaos: Self-Renewal in Japanese Firms. *California Management Review.*
Peters, T. 1987. *Thriving on Chaos: Handbook for a Management Revolution.* New York: Alfred A. Knopf.
Pirsig, R. M. 1974. *Zen and the Art of Motorcycle Maintenance: An Inquiry into Values.* Toronto: Bantam Books.
Schonberger, R. J. 1986. *World Class Manufacturing: The Lessons of Simplicity Applied.* New York: The Free Press.
Senge, P. M. 1990. *The Fifth Discipline.* New York: Doubleday/Currency.
Shingo, S. 1989. *A Study of the Toyota Production System from an Industrial Engineering Viewpoint.* Cambridge, Mass.: Productivity Press.
Stalk, G., Jr., and T. M. Hout. 1990. *Competing Against Time.* New York: The Free Press.
Stata, R. 1989. Organizational Learning—The Key to Management Innovation. *Sloan Management Review* Spring.
Stewart, T. A. 1991. The New American Century: Where We stand. *Fortune* Magazine, Special Issue, Spring/ Summer.
Striving for Manufacturing Excellence. 1990. *AT&T Technical Journal* 69(4) July/August.
Taguchi, G., E. A. Elsayed, and T. C. Hsiang. 1989. *Quality Engineering in Production Systems.* New York: McGraw-Hill Book Company.
Womack, J. P., D. T. Jones, and D. Roos. 1990. *The Machine that Changed the World.* New York: Rawson Associates.

### Comments on the Operating Principles

Heim and Compton draw attention to several important arguments pertinent to the management of a technological system. To optimize the performance of an organization, it must be looked at as an integrated system of people, machines, processes, methods, information, energy, and management. This has been the premise of the industrial engineering field since its inception. However, it is the execution of these principles in industry that counts, not just the knowledge of them. Japanese industry has taken these principles and executed them well over the past few decades to gain competitive advantage. The Deming and Juran principles of quality have been known to American manufacturers for many years; however, it was the Japanese who took advantage of this knowledge. They implemented quality teachings into their organizations even before American corporations were jolted awake by a declining market share in industries once thought invincible, such as the automobile industry.

As indicated in an earlier chapter, winning in the game of manufacturing can be viewed metaphorically as winning in the game of football. It depends on the assembling of proper resources—technical and human—and on the perfect execution of a plan to win. In professional football, talent can be bought and skills can be developed. On any Sunday afternoon, teams of the National Football League square against each other. Most teams have acquired talented players and have good resources of equipment and facilities. Winning depends on the game plan and its execution. Any team can win on any Sunday. The state of manufacturing technology today is such that technology can be bought and transferred from one place to another. Educational systems have produced well-educated employees as well as managers to coach the system. On any day one organization can outperform another if it knows how to integrate all its resources and use them properly. Synergy between system components renders the totality of the system greater than the sum of its parts.

In the declining decades of the 1970s and 1980s, the pursuit of unit-operation optimization took priority over the strategic intent developed by our predecessors when they introduced the principles of industrial engineering that called for the design of integrated systems. The industrial superiority and lack of competitive presence in the period following World War II created among management a sense of invincibility and a relaxed attitude toward competition. Several sectors of the U.S. economy, such as the automobile and consumer products industries, were late in embracing change in order to sustain economic growth in the age of the technological revolution. As a consequence they suffered declines in their competitive advantages. The challenges of the twenty-first century require the adoption of new paradigms by management. Likewise, education and training programs need to be revamped in order to prepare workers for organizations of the future.

## THE SERVICE INDUSTRY

The engines of the U.S. economy can be divided into several clusters, as shown in Figure 12-1. The service sector is the largest engine in the economy. Although this sector has been growing steadily since the turn of the twentieth century, its rate of growth has been particularly dramatic over the last 30 years. The service sector now accounts for more than 70 percent of U.S. GNP (Mitchell, 1990). Exhibit 12-2 lists the amount of U.S. GNP contributed by each major industry in 1993. It illustrates the relative importance of the different sectors to the U.S. economy. Note that the economic growth in the United States in the second half of the 1990s has propelled its GNP to new heights, exceeding \$8 trillion, and the growth is continuing.

Figure 12-2 shows the change in employment trends in the United States as a percentage of the total labor force. Employment in the U.S. service sector reached approximately 80 percent in 1995, while employment in the agricultural sector declined to below 3 percent. This represents a dramatic change from the employment figures of the

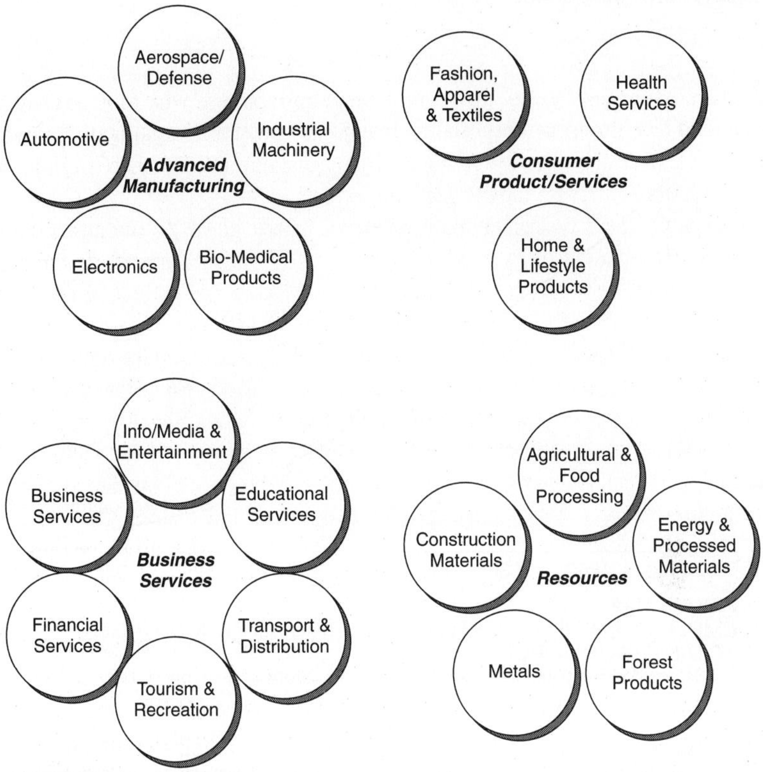


**FIGURE 12-1**
ENGINES OF THE U.S. ECONOMY
*Source:* U.S. Office of Technology Policy 1997.

**EXHIBIT 12-2**
U.S. GNP BY INDUSTRY, 1993

| | Amount ($, billions) | Share of Total (%) |
|---|---|---|
| **Total economy** | **6,650.2** | |
| Agriculture, forestry, fisheries | 105.3 | |
| Mining and construction | 332.5 | |
| Manufacturing | 1,116.3 | |
| **Total goods sector** | **1,554.1** | **24** |
| Finance, insurance, real estate | 1,214.0 | |
| Retail trade | 571.1 | |
| Wholesale trade | 423.1 | |
| Transportation and public utilities | 392.8 | |
| Communications | 173.4 | |
| Other services | 1,266.1 | |
| Total private services | 4,040.5 | |
| Government and government enterprises | 900.2 | |
| **Total services sector** | **4,940.7** | **76** |

*Source:* Office of Technology Policy 1997.

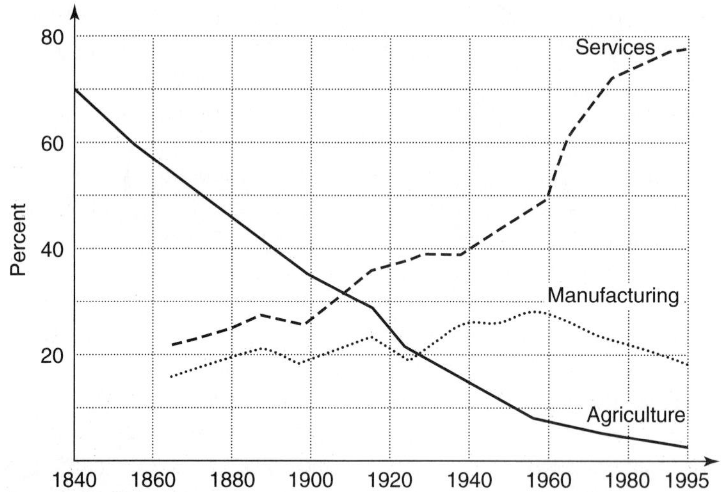


**FIGURE 12-2**
EMPLOYMENT IN INDUSTRIAL SECTORS AS PERCENTAGE OF TOTAL LABOR FORCE
*Source:* Modified from Quinn, 1987.

early twentieth century. The service sector has also contributed significantly to the U.S. trade balance, with the service surplus rising from under $7 billion in 1985 to more than $80 billion in 1995 (Figure 12-3).

While there may not be a consensus on a single definition for the service sector, it can be thought of as a heterogeneous grouping of many diverse industries. The common features among them are that (1) the primary output is not a product or construction, (2) the output cannot be inventories, and (3) the value added for customers does not come from

**FIGURE 12-3**
BALANCE OF TRADE IN SERVICES, 1985–1995
*Source:* Council on Competitiveness 1995.

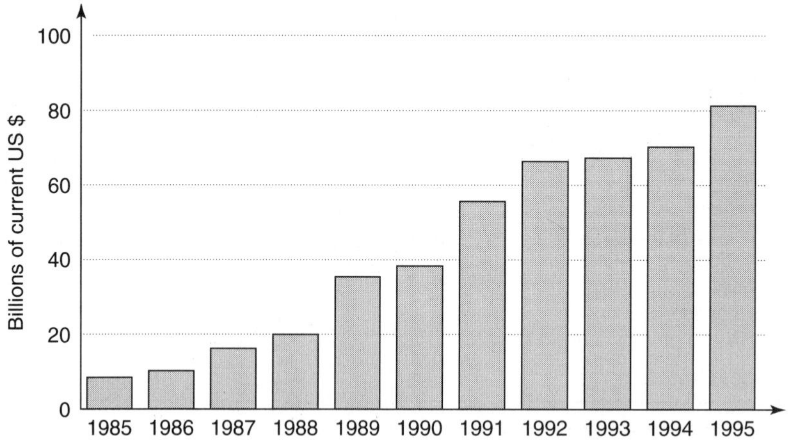

physical objects such as consumer products. A service adds value by enhancing products or quality of life, through forms such as amusement, comfort, convenience, and satisfaction. The service sector includes diverse industries: education, health care, utilities, retail, transportation logistics, packaging, entertainment, tourism, information, consulting, legal services, insurance, banking, and many more. The service sector is as important to the creation of wealth as manufacturing is. A product cannot reach a customer unless it is transported—a service activity. Workers cannot perform if they are sick, and productivity will not increase without education and training. Money flow needs banking, and automated teller machines need software designers.

Some service industries, such as utilities, transportation, and communication, are capital-intensive; others, such as hospitals, consulting, and entertainment, are labor-intensive; and still others are technology-intensive, such as public utilities, airline travel, health care, and communication. However, one thing is consistent: All service activities can use technology to add value for customers. In particular, the infusion of information technology into the U.S. service sector has resulted in improvements in a wide range of services, including more efficient airline reservation systems, reduced delivery times by Federal Express, and faster emergency response times by police officers. All these have resulted in a significant rise in service productivity and a dramatic growth in the trade surplus in this sector.

### Wal-Mart—A Focus on People and Technology

The legendary success of Wal-Mart Stores, Inc., provides an excellent example of a service sector enterprise that has been able to integrate technology and business strategies to become America's number-one retail corporation. The company's management kept one eye on marketing and the other on technology and has been able to sharply focus both eyes on creating value and offering friendly service to customers. Success for Wal-Mart has resulted from its down-to-earth concepts, entrepreneurial spirit, and the personal and professional idiosyncrasies of its founder, Sam Walton.

Having returned from service in World War II, Walton invested $45,000 to purchase a Ben Franklin variety store in Newport, Arkansas, a town of 7,000. He spent a lot of time in the subsequent years visiting competitors and learning their ways. He had no problem copying winning sales tactics. In 1962, Walton opened the first Wal-Mart store in the rural town of Rogers, Arkansas. This was approximately the same time that chief competitor K-mart opened its first store in Garden City, Michigan.

Walton was a people's person. His philosophy was to do whatever it takes to satisfy the customers. He thought of his employees as associates and enabled them to feel important in the organization. Suppliers were partners, and managers were servants of the customers. Walton visited his stores periodically and talked casually to his employees and customers.

He adopted corporate ethics based on maintaining the highest standards of honesty, morality, and business ethics in dealing with the public. He was successful in creating a corporate culture that permitted employees to embrace a corporate mission focusing on "customers' needs and delivering high quality merchandise at low everyday prices, while constantly striving to improve human relations among customers, employees, neighbors, and the communities in which the company operates." Walton's attitude and energy infected his employees, who consistently and characteristically displayed an unmistakable fervor for their jobs.

Historically, Wal-Mart's strategy is to build its stores in rural areas with a population under 5,000. By saturating demands in these markets with everyday low prices, the company dominates the market and precludes rivals. More recently, the company has abandoned the strategy of locating in small towns and has started competing in larger communities as well. The focus continues to be on everyday low prices and friendly service. Wal-Mart has maintained an exceptional growth rate, as shown in Figure 12-4.

Wal-Mart relies heavily on technology to support its expansion scheme. The company has linked all stores to one of the most sophisticated marketing information systems in the world. The company relies heavily on computers, making it the leading purchaser of large computer systems in the United States with the exception of the federal government. Wal-Mart relies on its computer power to analyze the buying patterns of customers and collect real-time data on sales and inventory. Thousands of point-of-sale terminals are directly connected to the company's headquarters to continuously update information. Wal-Mart has also wired 1,800 suppliers into an integrated system for matching demand in the stores with in-transit and in-warehouse merchandise.

Another area of major strength is Wal-Mart's control over the manufacture, movement, and deployment of products to its outlets in the right quantities and at the right time. Its logistics system is second to none in the industry. While the average cost of distribution in a comparable organization is 4 to 5 percent of sales, Wal-Mart's cost is less than 3 percent. Moreover, with the number of stores the company owns and the size of its transactions, it is one of the most powerful bulk buyers. This certainly is helpful to its goal of passing some savings to its customers.

Wal-Mart has been able to continuously embrace change. In 1983, it opened Sam's Wholesale Club. In 1990, it embarked on an aggressive corporate strategy featuring upgrading, refurbishing, and remodeling of stores. It opened many new stores and several

**FIGURE 12-4**
WAL-MART STORES, INC.: GROWTH STATISTICS

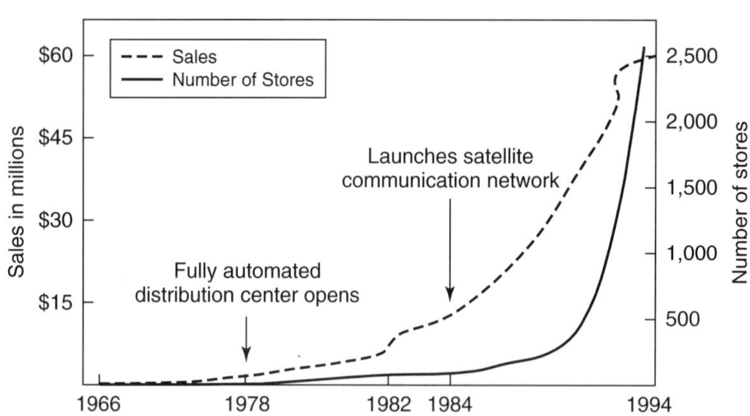

distribution centers. It also entered into a number of joint ventures and expanded its stores and discount clubs globally. Now Wal-Marts exist in Mexico, Canada, South America, the Republic of China, and Indonesia (Rasmussen, 1997).

### Lessons Learned from Wal-Mart

A review of Wal-Mart's case reveals many important lessons in MOT that deserve careful attention:

**1** The use of technology to gain competitive advantage is as effective in the service sector as it is in the manufacturing sector.

**2** Technology can reframe the competitive challenge. Wal-Mart is the world's largest retailer, but the battle for competition takes place in transportation and logistics. Wal-Mart developed a system superior to its competitors. It is winning.

**3** A two-pronged approach focusing on customer satisfaction and on mastery of technology is a winning combination.

**4** Treating employees and suppliers as partners improves efficiency and responsiveness, and leads to mutual commitment to the overall goals of the organization.

**5** The leader of an organization has a strong influence on its culture.

**6** In Wal-Mart, better logistic technology lowered the cost of merchandise and contributed to greater customer demand. Meanwhile, better service resulted in satisfied customers, more customer traffic, and more demand for products. This, in turn, resulted in improved bulk buying, which again resulted in lower prices and continued customer satisfaction. Wal-Mart has a sales figure per retail square foot of $389, compared to $185 for K-mart and $282 for Target, two of its main competitors.

## CONCLUDING REMARKS

At the macro level, there are many sectors that form the engines of the economy in any country. Natural resources, manufacturing, and service are all important sectors in creating national wealth. Technological progress, however, has led to a major shift in employment and production, toward the service sector. In the United States the service sector now accounts for more than 70 percent of GNP and employs about 80 percent of the workforce. This does not mean that output from agriculture, mining, or manufacturing has declined; on the contrary, output may actually have increased in most sectors. This is largely due to enhancement of productivity, propelled by improvements in technology.

At the firm level, globalization and continuous change in the business environment are dictating new mandates for corporate success. Developing world-class organizations capable of competing domestically and worldwide is assuming greater importance in corporate management. Being competitive requires that firms develop effective strategies integrating technology and business planning and, at the same time, pay strong attention to employees and customers. The foundations for developing world-class organizations in manufacturing and service have been recognized. The onus is on management to follow through faithfully. The success of Wal-Mart Stores, Inc., indicates that focusing on both technology and people can be a winning approach. Wal-Mart has shown that even in a traditional service industry, the use of technology can give a company a strong competitive advantage.

## DISCUSSION QUESTION

**1** Read the *Business Week* article "The Best Performers" (March 29, 1999). Identify whether the companies discussed are manufacturing or service sector enterprises. Discuss the important factors that may have contributed to the success of these corporations, and identify any common factors. (This exercise can be done with any list of top performers.)

## ADDITIONAL READING

Noori, Hamid. *Managing the Dynamics of New Technology.* Prentice-Hall, Englewood Cliffs, NJ, 1990.

This is an excellent textbook for managing technology in the manufacturing sector.

## SUGGESTED CASES

- "IG Laboratories, Inc (B): The Paradox of Growth—1994–94." Harvard Business School, Case 9-796-114.
- "America Online: Using Information Technology to Better Serve the Customer." Harvard Business School, Case 9-396-290.

## REFERENCES

Council on Competitiveness, 1995 Competitiveness Index, Council on Competitiveness, Washington, DC.

Fine, C. H. & St. Clair, R. 1996. "Meeeting the Challenge: U.S. Industry Faces the 21st Century—The U.S.A Automobile Manufacturing Industry." Office of Technology Policy, U.S. Department of Commerce, Washington, DC.

Heim, Joseph A., & Compton, W. Dale. 1992. "Operating Principles of World-Class Manufacturing Organizations." In Khalil, T., & Bayraktar, B. (eds.), *Management of Technology III,* pp. 765–776. Engineering and Management Press, Atlanta/Norcross, GA.

Mitchell, G. R. 1990. "R&D Strategy for the Service Sector." In Khalil, T., & Bayraktar, B. (eds.), *Management of Technology II,* Engineering and Management Press, Atlanta/Norcross, GA.

Noori, Hamid. 1990. *Managing the Dynamics of New Technology.* Prentice-Hall, Englewood Cliffs, NJ.

Office of Technology Policy 1997. Data provided by special request. Department of Commerce, Washington, DC.

Quinn, James B. 1987. "The Impact of Technology in the Service Sector." *Technology and Global Industry,* pp. 119–159. National Academy of Engineering, Washington, DC.

# 13

# THE DESIGN OF ORGANIZATIONS

The proper management of technology requires an organizational structure capable of fostering innovation and ensuring the effective use of technological assets. In the current environment of increased dependence on technology, organizations must be able to take advantage of technological progress for their competitive advantage. The speed and rate of change in technology necessitate a paradigm shift in the structure and function of modern and future organizations. The traditional structure of multilayered vertical organizations, with units arranged by functional areas, is yielding to shallower, horizontal, and more integrated structures. Modern organizations must be flexible, be agile, be able to make swift decisions and take quick actions, and be customer-oriented. They must be capable of managing the process of technological innovation and able to deal with the social and environmental impacts of technology. All of this requires vision, teamwork, and overall sensitivity to people issues.

## THE VERTICAL ORGANIZATION

An organization is an arrangement that channels individual and group efforts toward achieving goals or satisfying needs. The most basic type of organization is the *line organization,* in which each person reports to a boss and everyone knows who the boss is. There is a clear division of authority and responsibility. An example of line organization is shown in Figure 13-1. This format is suitable for privately owned small enterprises where specialization is not an important factor for the success of the business. The head of the organization has complete knowledge of all aspects of the business and controls all decisions. In line organizations workers perform line functions, which implies that their jobs directly affect the work flow of the organization. Staff functions are support

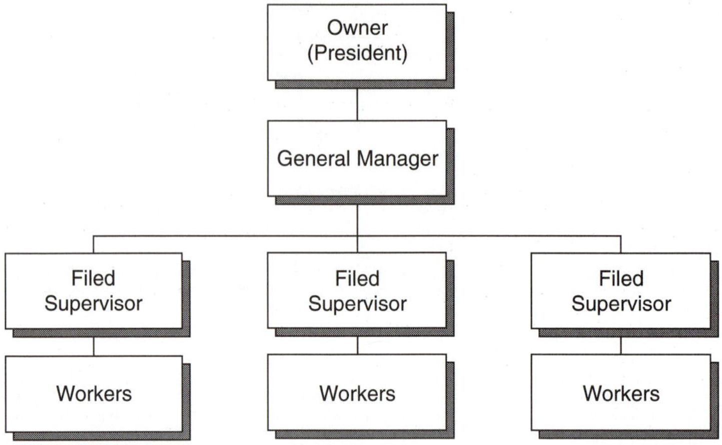


**FIGURE 13-1**
LINE ORGANIZATION

jobs for line departments. If staff are used, such as a secretary or an assistant to the president, their function is service-oriented or advisory in nature.

The complex nature of modern business requires the use of specialists, such as engineers, marketing people, accountants, lawyers, human resource specialists, and finance people, who are familiar with specialized facets of the business. They form the functional staff groups, divided by specialty (Vaughn, 1985). In this type of design staff members may be given authority and have responsibility in the line of their specialties. In the twentieth century, the *line-and-functional-staff organization* was the most widely used model by corporations. It served business well at a time when technology was relatively stable, technological change predictable, and competition limited.

An example of this model is shown in Figure 13-2. It combines line and functional-staff characteristics. Authority and responsibility are usually delegated to each functional specialty area, with the coordination of these functions handled at the upper-management level. The organization may have several vertical layers of management.

The type of organization shown in Figure 13-2 is challenged in effectively responding to the needs of the technology revolution. First, its multilayered management structure makes communicating ideas and taking initiatives very difficult. Ideas emanating either from the top or, more problematically, from the bottom will have to move several layers before they reach the other end. At each one of these layers a new initiative may encounter resistance to change. Multiple layers increase the chances that there will be a manager or a system that may find a conflict and a reason to delay or totally kill the idea. A second problem in functional-staff organizations is that functional specialty units tend to congregate and draw boundaries around their disciplines. Sometimes their loyalty is more to their discipline than to the organization as a whole. (This is also true in a university structure.)

A third drawback to this type of vertical organization is that multilayers are costly and respond slowly. Today, with competition based on new innovations and fast time to

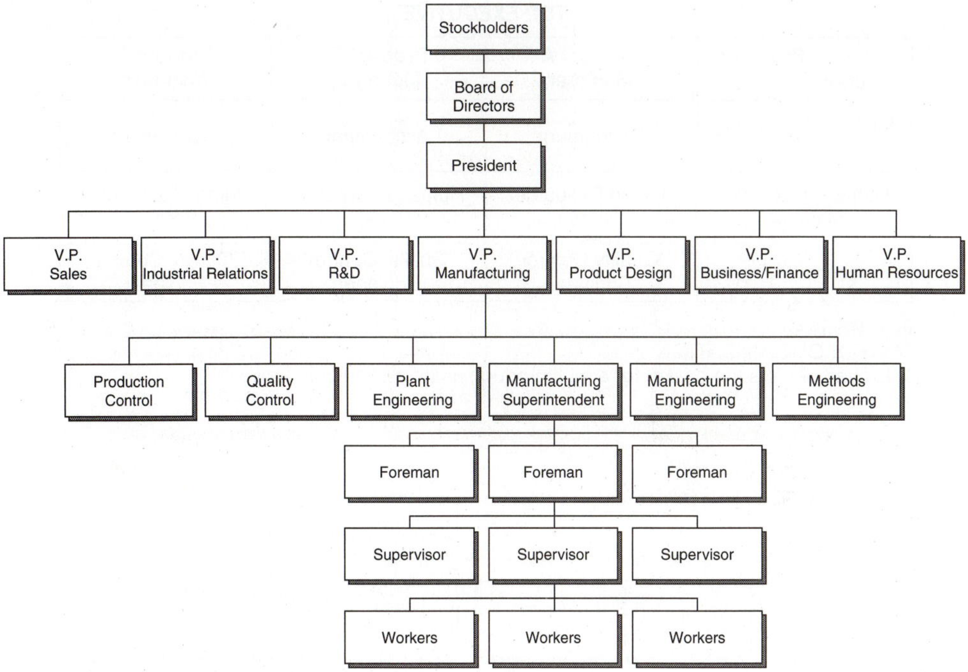


**FIGURE 13-2**
LINE-AND-FUNCTIONAL-STAFF ORGANIZATION, SHOWING THE MANUFACTURING HIERARCHY

market, the vertical organization style has reached its mature stage and is being retired by many companies, particularly high-tech companies. The interest in reengineering companies is one response of business enterprises that want to change or revitalize an aging organizational structure.

## THE MATRIX ORGANIZATION

A matrix structure may be used as a model for organizational design. In this arrangement, a supervisor may have authority in and a relationship with more than one system. An example of this type of organizational design is shown in Figure 13-3. A matrix design usually exists where there is a need for a combination of functional departments and different products. A company having multiple lines of products may appoint a product manager for each product line. The product manager will have a relationship with functional departments such as finance, accounting, personnel, and production. Both the product managers and the function managers may report to the company president. In matrix organizations there are some employees that report to two managers—a functional manager and a product manager. For example, an accountant responsible for a product might report to the product manager as well as to a functional corporate department of accounting. A quality control engineer responsible for a product line might report to both the product manager and a corporate production manager.

**TOP EXECUTIVE**

| Products / Function | Product A Manager | Product B Manager | Product C Manager |
|---|---|---|---|
| Finance Accounting (Manager) | Accountant | Accountant | Accountant |
| Human Resources (Manager) | Human Resources Person | Human Resources Person | Human Resources Person |
| Production (Manager) | Quality Control Engineer | Quality Control Engineer | Quality Control Engineer |

**FIGURE 13-3**
MATRIX ORGANIZATION
This chart depicts the design for a multiproduct, multifunction company.

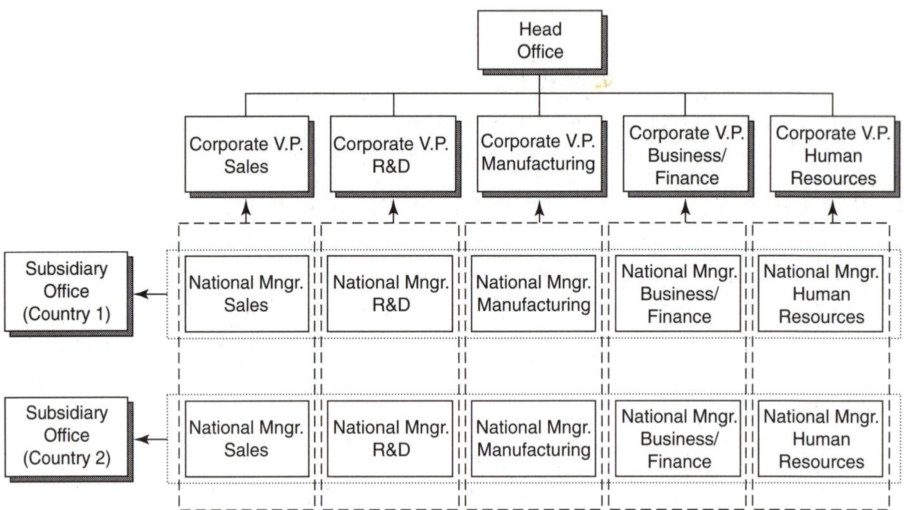


**FIGURE 13-4**
MATRIX ORGANIZATION FOR MULTINATIONAL COMPANY

The matrix type of organization is common among multinational organizations. Managers are appointed in each country, and they have to deal with corporate functional managers at the company's headquarters. The company may employ specialists who will report to specific country managers as well as function managers in the head office (Figure 13-4).

A matrix organization evolves in stages. It can start as a task force or as a venture team consisting of members assigned from various departments in the company, such as finance, accounting, marketing, and human resources. The task force will have a leader. Members of the team report to their functional department manager and are also accountable to the team leader.

As the task force progresses, its activity usually requires the creation of a permanent team—the second stage in matrix organization. This team leads the way to implementation of a full-blown project. At this late stage a manager may be appointed. The manager will need to negotiate acquiring the necessary human resources from the functional departments, and a new multiple-authority relationship develops. This relationship is the major distinguishing characteristic of matrix designs. It is contrasted to the hierarchical design of vertical organizations, where formal line reporting is clearly delineated. Managing in a matrix organization is challenging. It requires continuous negotiation, tolerance for peers, a cooperative attitude, and acceptance of shared power. A matrix organization design is more conducive to innovation than is a vertical design. It provides for better flexibility and team effort.

The matrix design is costly and is difficult to implement. However, it is appealing to organizations using complex technology, competing with innovation, requiring high degrees of interaction among employees, needing strong cooperation among projects and functions, and having to adopt to changing market conditions.

## THE HORIZONTAL ORGANIZATION

A new wave of thinking calls for changing the conventional vertical organizational structure by eliminating the functional or departmental boundaries and the organizational hierarchy. In the emerging forms of organizational structure, work is done by multidisciplinary teams organized around core processes, not around functions. For example, product development is a core process that requires a team of design engineers, market analysts, business strategists, and a process owner. Another core process is sales, which also has a process owner. It may consist of a team of salespeople and production, shipping, and pricing experts. Customer support, a third core process, requires a team of researchers, service personnel, and advertisers working with a process owner to provide the necessary support to guarantee customer satisfaction. The entire organization is flat, with few layers, and is led by a president and a skeleton group of senior executives who support functions such as finance and human resources.

The important characteristics unique to a horizontal corporation are:

- It is organized around processes as opposed to staff functions.
- It is composed of multidisciplinary teams as opposed to individuals working within the walls of a specialized department.
- It is a flat organization.
- It is lean, with few layers of management compared to traditional organizations.
- It empowers the employees. The process owner (manager) and the team are given the authority to move projects expeditiously.

Byrne (1993) conducted a review of new corporate models and depicted managing in horizontal corporations as shown in Figure 13-5. Corporations that have moved toward this model include:

- *AT&T Network Systems Division,* which organized its entire business around process and set its budget by process.

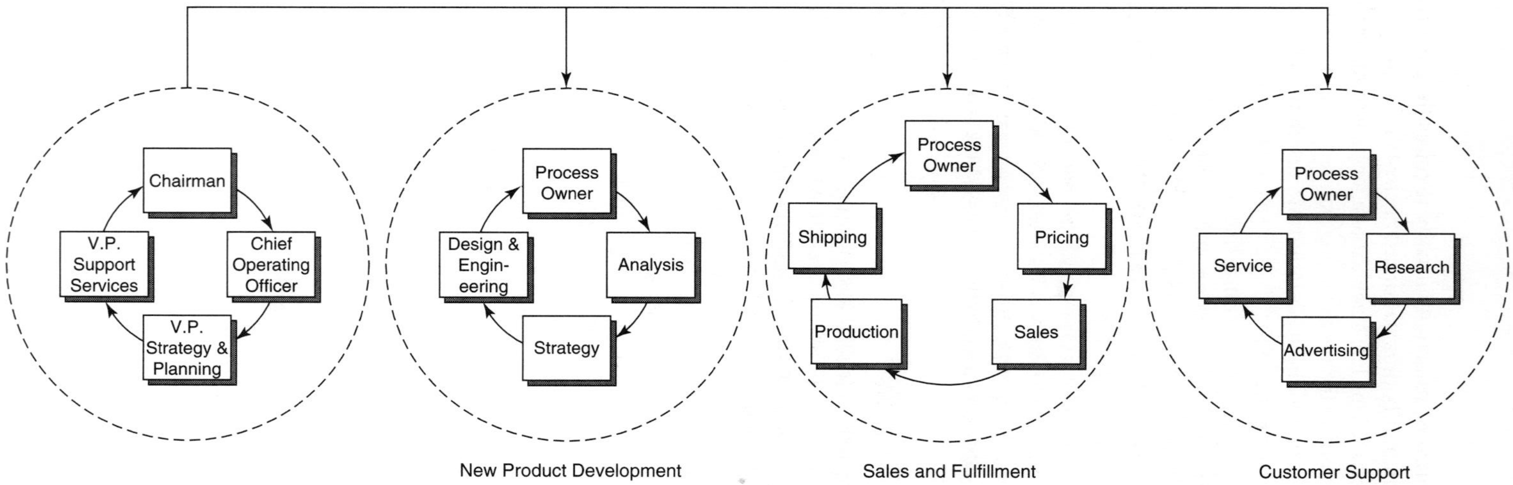


**FIGURE 13-5**
THE HORIZONTAL CORPORATION
*Source:* Adapted from Byrne, 1993.

• *Eastman Kodak,* which eliminated several vice president positions (R&D, manufacturing, and finance) and built 1,000 self-directed teams.

• *Xerox,* which now develops new products through multidisciplinary teams that work in a single process.

• *General Electric,* which moved from a vertical structure to a horizontal one with more than 100 processes and programs.

As this list indicates, organizational design is responding to the changing business environment (see Figure 13-6). Byrne (1993) reported on contemporary organizational charts that reflect the work practices of some leading American companies (Figure 13-7):

• Eastman Kodak utilizes a pepperoni-pizza-like chart: "Each pepperoni typically represents a cross-functional team responsible for managing a business, a geographic area, a function, or a core competence in a specific technology or area such as innovation" (Byrne, 1993). The surrounding areas provide space for interaction among the teams. The type of network design used by Eastman Kodak is intended to facilitate managing complex, diverse, and dynamic factors involving multiple units and many people (Hellriegel et al., 1995).

• Pepsico uses an inverted-pyramid organizational chart. Field representatives are put at the top to denote the importance of the customer to the organization. The inverted-pyramid design is intended to be more responsive to customer needs.

• The starbust model symbolizes a company that "splits off units like shooting stars."

• The shamrock model of three leaves connected to one stem symbolizes the importance of partnering between constituencies of the organization: (1) core competencies, (2) external contractors, and (3) part-time staffers (Handy, 1991).

**FIGURE 13-6**
ORGANIZATIONAL DESIGN AND CHANGING ENVIRONMENTS

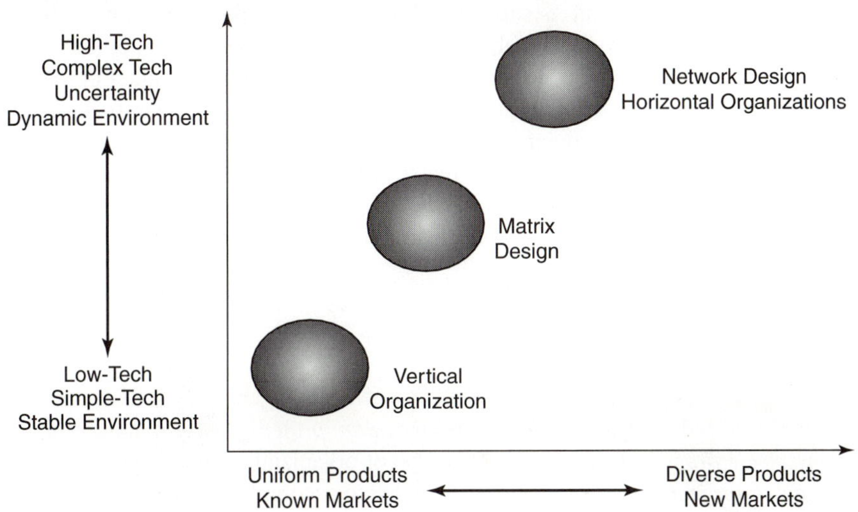

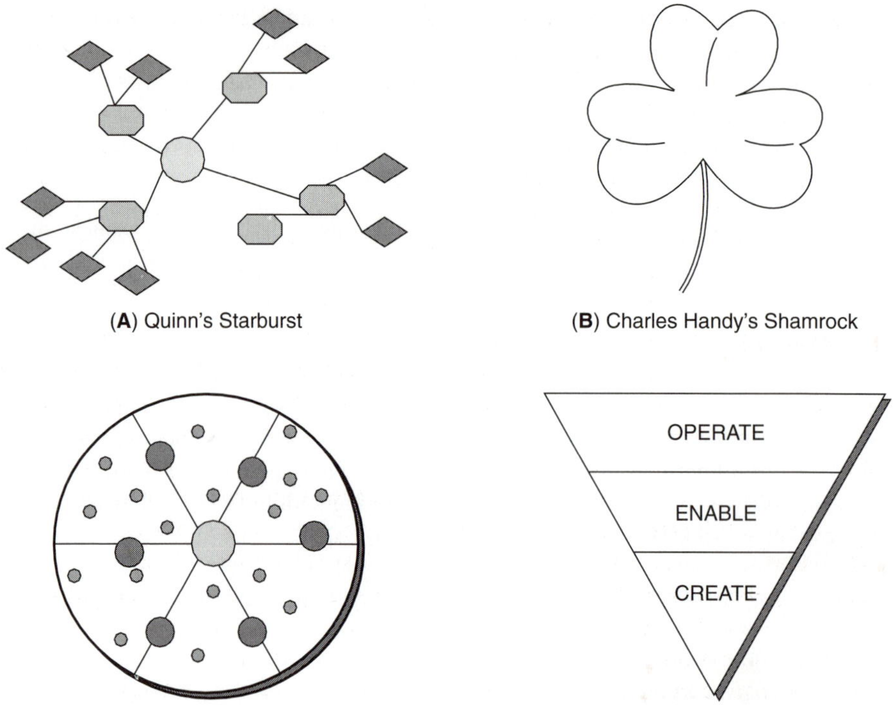


(**A**) Quinn's Starburst

(**B**) Charles Handy's Shamrock

(**C**) Eastman Kodak's Pizza

(**D**) Pepsico Inverted Pyramid

**FIGURE 13-7**
MODELS FOR THE MODERN CORPORATION
*Source: Business Week,* Dec. 20, 1993; as used in Byrne, 1993.

Dell computers embraces this model to connect customers' orders that come on the Internet with its factories and its suppliers. The wide use of the Internet permits plugging the customers into the supply chain via the company's web site. Suppliers and assemblers in a Dell factory can respond directly by providing the product on demand. Customers are able to track the progress of their order on-line. The shamrock model of organization permits better efficiency of operation through value chain integration. It reduces cost of production and provides greater perceived value to the customer. It also permits a company to concentrate on its core competencies. Many other new structures are being advanced by corporations as they struggle to be in the forefront of competition in the twenty-first century.

## PROJECT-BASED ORGANIZATIONS

A concept that is increasingly being adopted by firms is the *project-based organization,* sometimes referred to as a *team-based organization, project portfolio management,* or *enterprisewide project management.* This type of organization is built around projects and teams. A *project* is a set of activities intended to accomplish a specific end result. A special characteristic of a project is that when its objective is accomplished, the project ends. Projects are usually undertaken by teams of workers. Teams are assembled to tackle a problem or achieve a specific objective of a company. Examples are launching a new product, installing a new computer system, or constructing a building. The completion of a project may lead to the start of a new project or an ongoing operation. Ongoing operations usually have multiple objectives, while a project usually has a single objective. The use of teams and teaming mechanisms to integrate organizations laterally has increased dramatically in recent years (Mohrman et al., 1995). This is seen as an effective structure for implementing firm strategies formulated to deal with the business environment. Project-based organizations are viewed as organizational designs that enable firms to execute better, learn faster, and be more flexible.

Creating a project-based organization involves more than simply using teams to carry out projects. It requires a total redesign of the entire organization. It provides the organization with the ability to dynamically form and dismantle teams as needed to respond to performance challenges, to develop and support the teams, to design career paths and reward systems, and to design processes of information flow, communication, and feedback mechanisms. Figure 13-8 illustrates the difference between the activity-oriented, hierarchical organization and a project-based structure formed by combining tasks, people, and processes. In the traditional hierarchical structure of organizations, each department performs a series of discrete tasks. In the project-based structure, a team organizes to focus on a project and results. A team could also be assembled to focus on a business process to solve customers' problems. For example, at AT&T's credit corporation, employees used to perform individual tasks such as review application, check credit standing, notify the customer, and produce contracts. AT&T reorganized, switching from a division of labor to cross-functional teams that focus on performing all functions and concentrate their goal on extending credit to qualified customers. Here the focus shifted from compartmentalized activities of receive, review, check, and notify to a process of finding a quick solution for the customer. The net result is a much quicker turnaround time in processing applications (Montebello, 1994).

When IBM wanted to get quickly into the PC production and marketing business in the early 1980s, it assembled a new-venture project team to accomplish the task. When

**FIGURE 13-8**
FUNCTIONALLY ORIENTED VERSUS PROJECT-BASED ORGANIZATIONAL DESIGNS

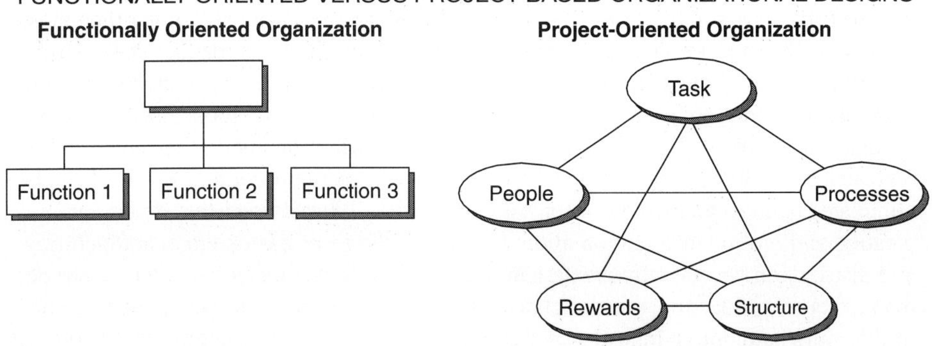

Rubbermaid wanted to increase revenues from the sale of new products, it formed cross-functional teams to conduct research on new products (Montebello, 1994).

Organizations have discovered that project-focused and team-based strategies produce better results, provide greater flexibility, and make them more responsive to changes in technology and markets.

## ORGANIZATIONS OF THE FUTURE

Organizations of the future will probably be significantly different from past or current organizations. Peter Drucker (1988) described a shift from the "command-and-control organization" to the "information-based organization." In fact, progress in information technology is changing the way organizations do business. It is forcing a new style of organizational structure based on communication and information flows. It is also transforming management thinking into new patterns totally different from the ideas that dominated nineteenth- and twentieth-century managerial philosophies.

Consider a few of the changes that took place during the last two decades of the twentieth century: The automotive industry experienced a fundamental restructuring of operations, a change in management philosophies, and a different attitude in labor-management relations. General Motors, which has been credited with establishing many of the management paradigms of the mid-twentieth century, mostly under the leadership of its chairman at that time, Alfred Sloan, found in the early seventies that it had become frozen in the paradigm it mastered (Fine and St. Clair, 1996). The corporation started a slow and arduous task of renewal. It established a joint venture with Toyota in Freemont, California—the NUMI plant. The plant uses Japanese lean methods of production and several Japanese labor practices of shared responsibilities and empowerment. GM also created a new type of organization with its Saturn Company in Spring Hill, Tennessee. Saturn introduced a new relationship between the company, its employees, dealers, and customers. This relationship is based on partnership, mutual trust, and concern for equality. The Saturn Corporation departed from the traditional conflict of labor-management relations by instituting a cooperative model of organizational governance.

Equally dramatic is GM's, and many other U.S. and international automotive manufacturers', switch from vertical integration to new supply-chain and distribution policies. The information age permits communications with a much looser collection of organizations that are not closely controlled by the automaker. Smaller organizations may be capable of producing products and delivering services at costs lower than those possible within the confines of larger organizations. Supply-chain management permits the effective use of supplies for development and production of components and subsystems. It is therefore important to establish a close, trusting relationship with suppliers and distributors. It is estimated that automotive distribution, marketing, and retailing represent 20 to 30 percent of the value of a new car. A large portion of this cost can be eliminated by introducing new, information technology-based practices. Better communication and on-line information allow for direct linkage of customers, manufacturing, and distribution and thereby create lean distribution. Better matching of customer demand, factory production, and direct distribution methods results in increased efficiency of the entire automotive industry's value chain.

Fine and St. Clair (1996) categorized the uses of information technology for post-manufacturing business as follows:

- Technologies that enhance the existing business structure.
- "Spec-ing" systems that allow salespeople to match factory options to customer needs more accurately and to create a more accurate demand forecast.
- In-house training and service management systems that connect factory-engineering expertise to all technicians within brand dealership.
- Customer outreach systems linked to factory-sponsored advertising and marketing.
- Technologies that will accelerate current trends toward nontraditional distribution structures.
- Customer relations systems (via the Internet or similar customer-oriented media) that allow brokers or order takers to become viable economic organizations (reducing significantly the role of traditional dealerships in many transactions).
- Analogs to shop-by-phone or catalog channels.
- Retail systems that allow mega-dealers to solidify their multibrand positions (similar to the systems that support mega-retailers such as Wal-Mart).

The new trends in organizational structure include:

- *Adaptive organizations:* The demands placed upon organizations by the changing business environment are driving them to respond by attempting to design more flexible, adaptive organizations. This implies eliminating the traditional, rigid organizational charts in favor of a more agile system that is capable of responding to demand. Managing change requires swift response in organizational structure. Thus a design that facilitates building interfirm and intrafirm teams and networks is put in place. It permits the building of teams, coalitions, and alliances to work on projects. Project leaders and employees are empowered to make decisions and to cut across boundaries of departments and functional units. They are encouraged to seek whatever support they may need to accomplish their task, whether such support comes from within or outside the boundaries of the organization.

  There are many types of adaptive organizations, including the matrix organization and horizontal designs. Automotive manufacturers, high-tech companies, and most progressive organizations are continuously changing their infrastructures and policies to adapt to the new environment. Many are restructuring to become adaptive organizations.
- *Shallow corporations:* Many organizations have reduced the number of managerial layers to create shallower organizational structures. Shallow organizations eliminate most of the middle-management positions. Studies have shown that successful corporations (based on long-term financial performance) use almost four fewer layers of management than do poor performers ("Middle Managers . . . ," 1988).
- *Hollow corporations:* These are organizations that are abandoning production activities and concentrating on marketing for other producers, usually foreign ones. Several U.S. companies have been formed to market products made in Asia and in the former eastern European bloc. They have no manufacturing facilities and no direct labor force. Some major multinationals are hollowing out. An article in *Global Competitor* ("Sun Rises . . . ," 1994) reports on the operation of NEC, a huge Japanese corporation,

in Hong Kong: "In Hong Kong NEC is a company without a factory. The NEC subsidiary in Hong Kong supervises the production of PC's, printers, and disk drives, but manufactures nothing. Some factories in China make the printer parts and others assemble them. PC parts come from manufacturers in Hong Kong and other Asian regions and the disk drives from the Philippines. NEC has no investment in any of these companies producing NEC products, but maintains a team of 26 roving production specialists to see that quality control and product reliability meet NEC standards."

- *Virtual corporations:* These are corporations that consist of a network of companies linked together by information technology. The relationships among the companies can be permanent or temporary in nature. They are formed to exploit a specific or an emerging opportunity. They share resources, skills, and access to markets. Their success depends on the ability of managers to spot opportunities, negotiate alliances with partners, provide solutions to customers, and design systems that optimize both resource-sharing and reward-sharing among partners.

## ORGANIZING FOR TECHNOLOGY PLANNING

In order to effectively plan the creation, acquisition, and use of technological resources, companies need to develop and use special organizational structures that facilitate attaining their objectives. The Battelle Institute's experience in technology planning led it to propose that companies use two parallel-track groups, one to plan for technology and one to plan for business. Neither plan should submerge the other. Each plan can proceed at its own pace, because business planning may require a shorter time cycle than technology planning. The technology structure may be formed under the leadership of a chief technology officer, and the business structure may be formed under the leadership of a chief business officer. The technology structure may have units responsible for technology audit, technology licensing, technology forecasting, R&D, and so on. The business structure may have its own financial, marketing, sales, and human resource units. Of paramount importance, however, is the mechanism by which the organization integrates and coordinates the strategies emerging from the two tracks. Both tracks have to focus on the organization's objectives and should continuously communicate across the organization. They should integrate their plans into a unified company plan. The unified plan should reconcile the time horizons of the technology and business cycles and address the issues of resource allocation.

Figure 13-9 shows the integrated technology-business plan structure proposed by Bhalla (1987). Using this model as a base, one can structure a technology group and a business group to perform the needed tasks. Teams may also be formed, with the team members spanning the horizon of tasks that need to be carried out. Technology gatekeepers can be of great value here. These are people who possess a high level of technical competence and can bring their knowledge to bear on the team's decisions. Integrating the technology planning with the business planning should result in the creation of new opportunities for the company, the development of new products and services, the improvement of profit margins, and a more competitive business. Organizational structure can facilitate attaining these results.

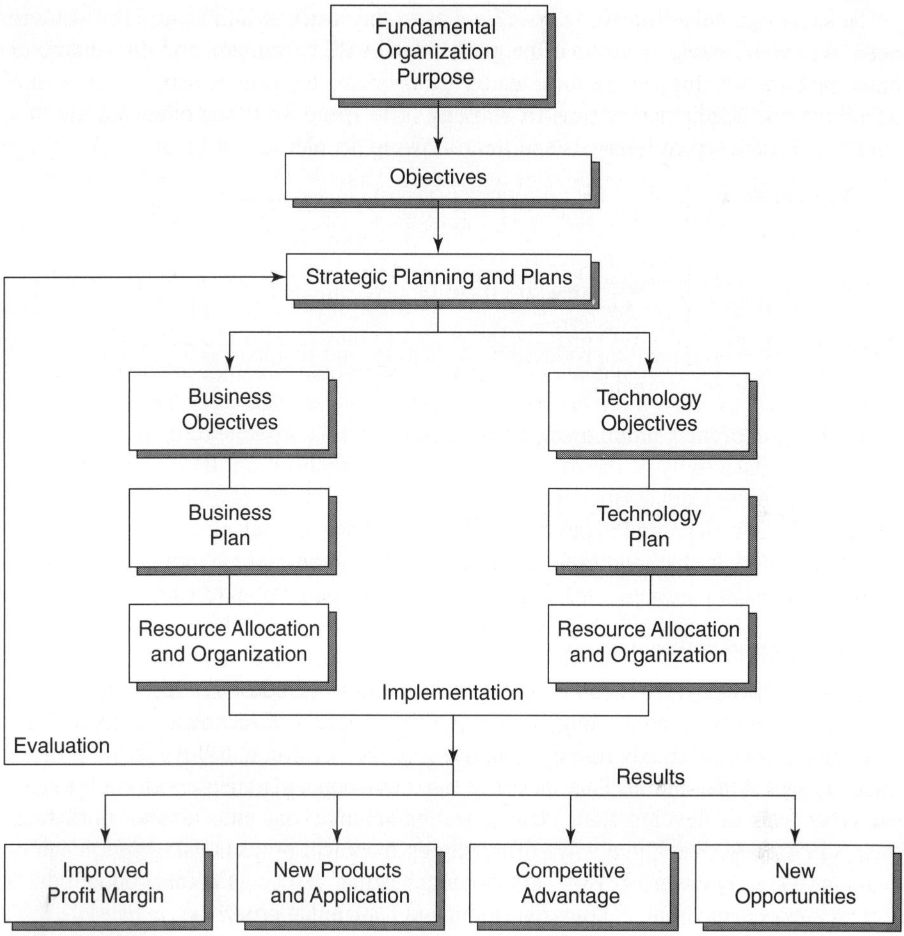


**FIGURE 13-9**
INTEGRATED TECHNOLOGY-BUSINESS PLAN MODEL
*Source:* S. Bhalla, *The Effective Management of Technology,* Battelle Press, 1987. Reprinted with permission.

## ORGANIZING FOR R&D AND NEW VENTURES

Most corporations have discovered that R&D groups cannot function effectively under a heavy bureaucratic structure. Creativity flourishes when the environment permits people to work on their own areas of interest, interact with like-minded people, take risks, and not be punished for failure. Mechanistic organizations tend to stifle individual creativity and restrict the autonomy needed within a research group. It is therefore more appropriate to keep R&D people away from the bureaucratic claws of large corporations. This is why Xerox moved its research group (PARC) to Palo Alto (see Reading 6.4). It is also why companies use "skunkwork" spaces away from the main operating units of the organization to let creative people concentrate on their work.

The same rationale discussed above applies to the structure and location of venture teams. A *venture team* is a group of people charged with carrying out a new venture or championing a new project. Venture teams are good mechanisms for companies to use to facilitate the acquisition of new technology or to speed up the implementation of a project. The team typically consists of the following people:

- A *champion,* who plays the entrepreneurial role in the venture. He or she must be technically competent, aggressive, and knowledgeable about the company and the market and must be a good communicator.
- *Technological gatekeepers,* who create or give guidance on the technology that needs to be developed or transferred.
- *Members,* who have the production, marketing, and financial skills that are necessary to move a project forward.

The location of the venture team, away from the main corporate unit, gives it autonomy and rids its activity of the routine of the parent organization. IBM was successful in using its venture team at Boca Raton, Florida, to enter the PC market. This helped the company come from behind in developing its PC technology and attain a position ahead of its competitors in that market for a good number of years.

### Removing Organizational Barriers

To improve efficiency, reduce the time to market, and be responsive to customer needs, organizations must eliminate impediments to the progress of technical projects. New products or services usually move through a sequence of events following the process of technological innovation. This sequence starts as a concept in the research laboratory and progresses to development, design, testing, production, and, finally, marketing. Potter (1990) describes three ways of managing technical projects using sports analogies introduced by Lorenz (1987). The approaches that can be used for product development are (1) sequential, (2) iterative-loop, and (3) team approaches.

The sequential functional management structure is analogous to a relay race. The functional areas of research and development, design and tooling, pilot production, and full production are compartmentalized. Each has its own specialists and carries the baton (i.e., the project) at different stages of the race. When its task is finished, it hands the baton over to the next functional unit, which continues the race (Figure 13-10). This sequential operational mode is not conducive to rapid product development. Each functional area has to wait for the previous one in the sequence to complete its activity before it can react or question a decision made by the previous group.

Functional groups are separated symbolically and virtually by walls; thus, they cope badly with the pressure involved in rapid product development. The walls between departments prevent effective communication, resulting in a wasteful iterative-loop structure.

The iterative-loop functional management style is analogous to a volleyball game. The project, after being passed sequentially between departments, is returned to the preceding department for correction. A volleyball game ensues, with the project being tossed back and forth over the wall between departments before problems are corrected (Figure 13-11). This back-and-forth tossing of a project is not an unusual occurrence in

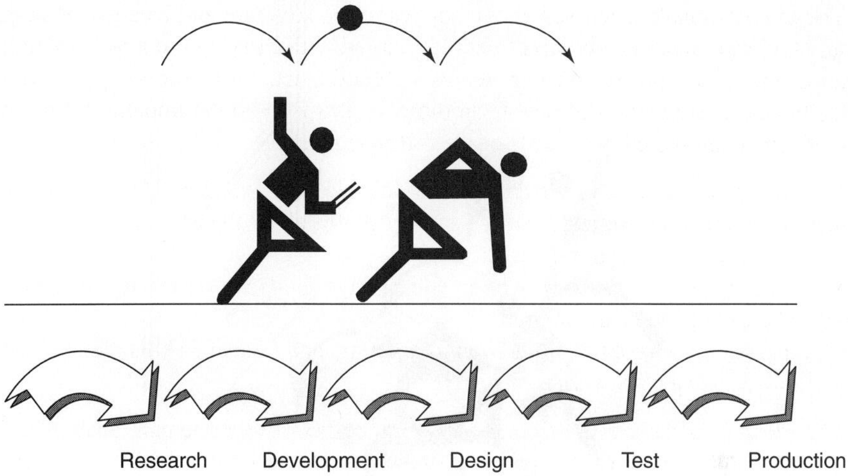


**FIGURE 13-10**
SEQUENTIAL FUNCTIONAL MANAGEMENT: THE RELAY RACE
*Source:* Potter, 1990; © 1990, Institute of Industrial Engineers.

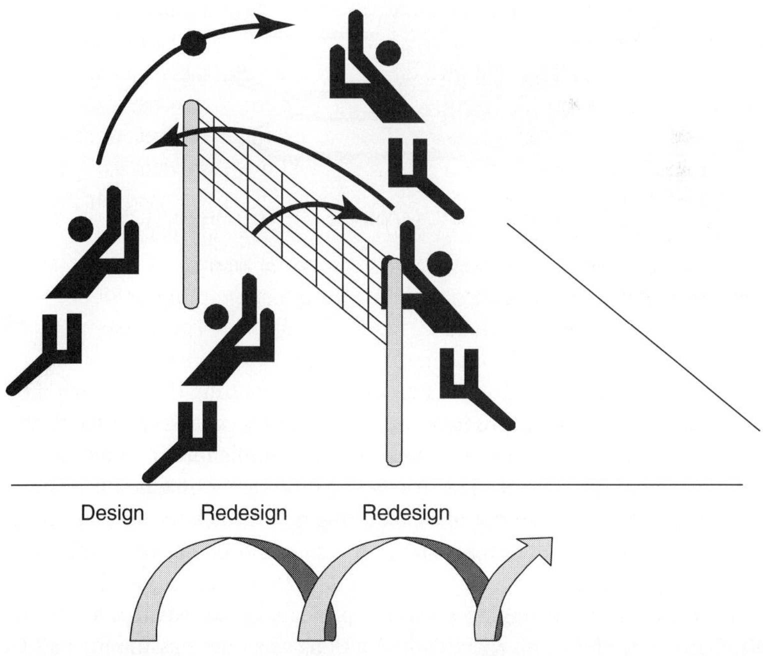


**FIGURE 13-11**
SEQUENTIAL FUNCTIONAL MANAGEMENT: THE VOLLEYBALL GAME
*Source:* Potter, 1990; © 1990, Institute of Industrial Engineers.

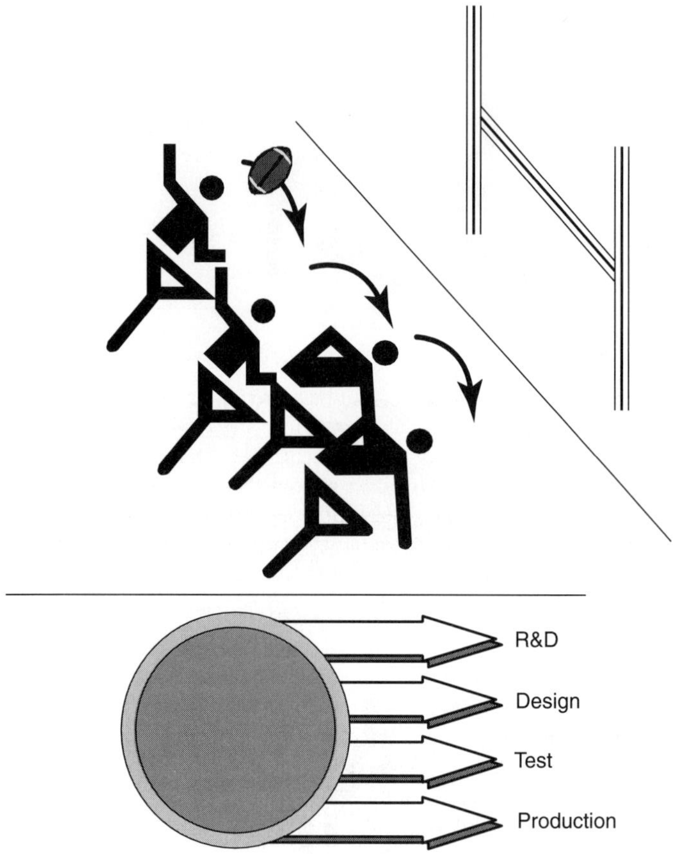


**FIGURE 13-12**
SEQUENTIAL FUNCTIONAL MANAGEMENT: THE RUGBY TEAM

*Source:* Potter, 1990; © 1990, Institute of Industrial Engineers.

many companies that have separate design and manufacturing departments. For example, an engineer may design a product and send it to the manufacturing floor only to have it returned with a note stating that it cannot be manufactured to specifications with the existing machines in the shop. This forces the designer to change the design. It may also lead designers to conclude that manufacturing people lack the skill to execute what they perceive to be an elegant design. This can create friction between departments and result in unnecessary delays.

The third approach to managing technical projects is the use of multidepartmental teams. Similar to a rugby team, each team functions as a unit, passing the ball back and forth in a quick, synchronized fashion. This shortens the time for iterative processes, and any problem can be resolved on the spot (Figure 13-12). Potter (1990) advocates the synchronized team approach, which was successful in designing the advanced passenger train and Inter-city 225 projects in the United Kingdom.

## DISCUSSION QUESTIONS

1 Look for the following articles in *Business Week:* "The Horizontal Corporation" (Dec. 20, 1993); "The Virtual Organization" (Feb. 8, 1993); "Congratulations. You Are Moving to a New Pepperoni" (Dec. 20, 1993). What are the "forces" driving the changes in organizational structure? Compare the previous articles with "No More Mr. Nice Guy at P&G—Not by a Long Shot" (Feb. 3, 1992). Look for more recent articles and compare trends.

2 Look for articles on companies and new ventures in Silicon Valley. What is the management style and organizational structure at each? Why?

## ADDITIONAL READINGS

Steven L. Goldman, Roger N. Nagel, & Kenneth Preiss. *Agile Competitors and Virtual Organizations.* Van Nostrand Reinhold, New York, NY, 1995.

This book presents a new paradigm for corporate structure and business practices. Organizations have to exist no longer as permanent institutions with a specific objective. *Virtual* refers to the ability to enter into temporary alliances with several partners only during the time an opportunity exists. This demands *agility,* the capacity to quickly adapt to an ever-changing environment.

Senge, Peter M. *The Fifth Discipline: The Art and Practice of the Learning Organization.* Doubleday/Currency, 1990.

Michael Hammer & James Champy. *Reengineering the Corporation: A Manifesto for Business.* Harper Business, New York, NY, 1993.

Hammer and Champy's book launched the term "reengineering" as a common buzzword in managerial practice and academia. The authors claim that competitiveness is achieved not by improving current processes but by totally redesigning them. This involves looking for totally new ways of achieving objectives by asking, "Why are we doing this in the first place?"

Michael Audet, Roger Blais, & Roger Miller. "Human Resource Management in Technology Based Firms." In T. Khalil & B. Bayraktar (eds.), *Management of Technology III, Proceedings of the Third International Conference on Management of Technology,* Vol. II, pp. 973–983. Industrial Engineering and Management Press, Atlanta/Norcross, GA.

This paper describes the Human Resource Management (HRM) practices that distinguish the most innovative and successful firms from the poor performers. The paper analyzes work design, staffing, compensation, training, performance assessment, and labor relations as key practices towards an effective organization.

Anthony D. Wilbon. "Organizational and Development Dynamics of Project Review Teams in Technology Environments." *The Qualitative Report,* vol. 3, no. 3, September 1997 (http://www.nova.edu/SSS/QR/QR3-3/Wilbon.html).

## SUGGESTED CASES

- "Business Teams at Rubbermaid, Inc." Harvard Business School, Case 9-897-165.
- "Chaparral Steel: Rapid Product and Process Development." Harvard Business School, Case 9-692-018.
- "Quantum Corp.: Business and Product Teams." Harvard Business School, Case 9-692-023.

## REFERENCES

Bhalla, Sushil K. 1987. *The Effective Management of Technology.* Battelle Press, Columbus, OH.

Byrne, J. A. 1993. "The Horizontal Corporation." *Business Week,* December 20.

Drucker, Peter. 1988. "The Coming of the New Organization." *Harvard Business Review,* January–February, pp. 45–53.

Fine, C. H., & St. Clair, R. 1996. *US Industry Faces the 21st Century: The Automobile Manufacturing Industry.* U.S. Department of Commerce, Office of Technology Policy, Washington, DC.

Handy, Charles. 1991. *The Age of Unreason,* 2nd ed. Random House, New York, NY.

Hellriegel, D., Slocum Jr., J., & Woodman, R. 1995. *Organizational Behavior,* 7th ed. West Publishing, Minneapolis/St. Paul, MN.

Lorenz, Cristopher. 1987. "Scrum and Scrabble—The Japanese Style." *Financial Times.* June 19.

"Middle Managers Face Extinction." 1988. *The Economist,* Jan. 23, p. 53.

Mohrman, S. A., Cohen, S. G., & Mohrman Jr., A. M. 1995. *Designing Team-Based Organizations.* Jossey-Bass, San Francisco.

Montebello, A. R. 1994. *Work Teams That Work: Skills for Managing across the Organization.* Best Sellers, Minneapolis, MN.

Potter, S. 1990. "Successfully Managing Research, Design and Development." In Khalil, T., & Bayraktar, B. (eds.), *Management of Technology II.* Industrial Engineering and Management Press, Atlanta/Norcross, GA.

"Sun Rises in the East, The." 1994. *Global Competitor,* pp. 27–31.

Vaughn, R. C. 1985. *Introduction to Industrial Engineering,* 3rd ed. Iowa State University, Ames, IA.

# 14

# THE CHANGING GAME OF MANAGEMENT

When Adam Smith published his famous book, *The Wealth of Nations,* in 1776, he laid the foundation of the capitalistic economic system of the West. He introduced many of the ideas that have been practiced by entrepreneurs, managers, and governments to create wealth. He advanced the concepts of private enterprise, specialization in tasks, and mass production, and he initiated the thought process of countless management ideas that have been practiced for the past two centuries.

Frederick Taylor, in the late nineteenth century, introduced the principles of scientific management, for exercising better managerial control of operations, and of work incentives, for boosting workers' productivity. Many management gurus followed Taylor with their own ideas of how to boost industrial efficiency and improve profitability. Management theories have proliferated, each with its own merits and drawbacks. Some are just fads that come and go without much influence on companies' performance, and others are techniques that are useful in enhancing quality or productivity. However, managers seem to continue to be fascinated with names that they can use as slogans for strategies to improve profitability in their organizations. Engineers and managers have applied the methodologies of time-and-motion study, process analysis, operation analysis, and optimization techniques to improve the productivity of their organizations. Management philosophies such as theory X, theory Y, theory Z, management by objectives (MBO), management by results, total quality management (TQM), continuous process improvement, business process reengineering, reengineering, and many more have been adopted by a large number of organizations (see Exhibit 14-1). All of these tools and strategies have been the focus of managers' attention during the past century. They have been applied, with results showing mixed levels of success and failure. The overall business environment may guide changes in the management style needed to effect optimal organizational performance. However, one thing has not changed

**EXHIBIT 14-1**
MANAGEMENT PHILOSOPHIES

*Scientific management:* A philosophy and methodology established by Frederick Taylor to manage work. It is based on scientific measurement that prescribes a division of work. Planning work is done by management. Execution is left up to the workers and their supervisors, who should control the work and asses workers' performance against management's established standards.

*MBO:* Management by objectives (MBO) is a method of associating people's positions with objectives and linking the objectives to the corporate plan. Peter Drucker first proposed this management tool in the early 1950s as a method of managing the complexity of growing organizations. The common elements in MBO are objectives that are established for positions; managers and subordinates negotiate to set the objectives; objectives are linked with corporate goals; and emphasis is placed on measurement and control and on establishment of a review and recycle system (Reddin, 1971). Generally, the manager is more involved than the subordinate in setting the objectives, commonly establishing and *selling* them. MBO can be perceived either as a way of linking evaluation to performance or as a planning aid. Reddin suggests the following as elements of success: flexibility to accept inputs from subordinates, emphasis on change, acceptance of the human side, group emphasis, emphasis on effectiveness, and perception of the situation. There are a few problems associated with MBO: People tend to play it safe and set low objectives, it requires too much paperwork, and it is a time-consuming activity.

*Theory X and Theory Y:* McGregor's paper "The Human Side of the Enterprise" (1957) constituted an important step toward modern organization theory. McGregor referred to the predominant managerial paradigm of his time as *Theory X*. This "conventional view" can be summarized as follows: (1) Management is responsible for organizing the elements of production. (2) Management is a process of controlling people. (3) Without management, people would be passive. This approach does not take into account the needs of individuals (i.e., psychological, safety, social, ego, and self-fulfillment needs), so McGregor proposed a new paradigm called *Theory Y,* with the following dimensions: (1) Management should pursue economic ends through the administration of the elements of production. (2) People have become passive as a result of bad experiences in organizations. They are not passive by nature. (3) Management must develop motivation in people through recognition of their needs. (4) "The essential task of management is to arrange organizational conditions so that people can achieve their best goals by directing their own efforts" (McGregor, 1957).

Theory Y is often associated with a liberal management style where delegation is the rule. This is misleading. Theory Y acknowledges managers' role as leaders, but suggests that managers should take into account the personal fulfillment of employees. Research in organizational behavior has found that fulfilled employees are able to make better contributions to a business.

*Theory Z:* This theory, developed by William Ouchi (1981), focuses on the Japanese management style. Big Japanese corporations have a managerial system deeply rooted in Japan's culture and tradition. Its main characteristics are lifetime employment, loyalty, subtle control (apparently nonexistent for an outsider), a participative approach in decision making, collective values, teamwork, collective effort, holistic concern about people (to the level of family involvement), slow evaluation and promotion, and nonspecialized career paths.

Ouchi applies the label *Z organizations* to American firms characterized by the Japanese management style. Examples of Z organizations are icons such as Eastman Kodak, GM, Hewlett-Packard, IBM, Procter & Gamble (P&G), and Xerox. As in Japan, these corporations base their management style on strong corporate culture. They expect employees to behave responsibly as corporate citizens and to "always do what is correct" (from the P&G mission statement).

Characteristics of Z organizations are interdependence, interpersonal skills, corporate communication, participation (although the manager is still responsible), informal relationships (open-door policies), trust, a certain autonomy level, hierarchy as a mode of control, and commitment. Since a strong culture has a tendency to reject novel ideas, the major problem for this type of organization is the possibility of becoming an industrial dinosaur.

*Benchmarking:* This is a technique for assessing a firm's performance against the performance of other firms. The firms selected for benchmarking can be from the same or a completely different industry. The objective of benchmarking is to find the best practice and to take necessary actions to improve the firm's own performance so that it meets or exceeds that of its competitors.

*TQM:* Total quality management (TQM) is a philosophy that was adopted first by Japanese companies and then implemented in the United States and around the world by companies such as Florida Power and Light, Xerox, and Motorola. TQM involves adopting quality as a culture in all areas of the company, not only for products but for administrative and managerial processes also. Quality is perceived as a big umbrella under which all activities are performed.

Kaoru Ishikawa proposes six quality principles: (1) Quality first, (2) consumer and not producer orientation, (3) customer as next process, (4) presentation with facts and data, (5) respect for humanity as management philosophy, and (6) cross-functional management. There are many techniques used under TQM; some of the better known are PDCA (plan-do-check-act) cycle, the seven tools (Pareto chart, cause-effect diagram, stratification, checksheet, histogram, scatter diagram, Shewhart control chart), quality circles (workers' involvement), *kaizen* (small and continuous improvements), and quality function deployment (QFD, a tool to design product in line with customer needs).

Many gurus have contributed to quality improvement efforts. Deming, Juran, and Crosby are among the best known.

"Total quality" is not merely a slogan to be used by managers. It is an organizational strategy that drives a continuous process of improvement. It may require radical changes in organizational design and daily procedures. Successful implementation of TQM in an organization depends greatly on top management's belief in its value and commitment to its implementation.

*Just in Time:* Developed by Toyota, just in time (JIT) could have also been called "the no-waste culture." The basic principle is that each activity must add value to the final product; otherwise, it should be eliminated. Even though JIT is commonly known as a manufacturing-related concept, it is not limited to the production floor. It should permeate all levels of the company. JIT is also known as a *zero-inventory* production system, but this term is misleading. Shingo, 1990, 1992 (production engineer at Toyota) perceives inventory as a "necessary evil"; thus it cannot be zero. JIT holds that inventory hides problems in the production process and, therefore, that in order to find and correct such problems, low inventory is required. Another term commonly associated with JIT is *kanban*. Kanban is a simple tool first used by Toyota to control what is called a *pull system,* a process in which production is not *pushed* by the process to the customers but is adapted to the customers' demand pace, fabricating only what is needed when it is needed.

JIT development was relatively straightforward for Toyota. The *kereitsu* (a group of financial and manufacturing organizations with strong ties to each other) facilitated the process. Since suppliers are close to customers (Toyota's plant is literally surrounded by suppliers in what is called "Toyota City"), JIT's logistic demands are reduced. This is more difficult to achieve in countries where customers and suppliers are hundreds or thousands of miles apart.

**EXHIBIT 14-2**
MANAGERIAL ROLE IN THE EFFECTIVE MANAGEMENT OF TECHNOLOGY

- Carrying out the essential elements of the management process (planning, organizing, leading, and controlling).
- Strategically integrating business and technology.
- Using proper accounting and financial methods to evaluate the health of a business (financial statement, ratio analysis).
- Allocating capital resources and cost of capital, including time value of money, net present value, rate of return, etc.
- Following contemporary advances and challenges in science and technology, forecasting direction and rate of technological advance, and evaluating the impact of technological development in the marketplace and the firm.
- Selecting of "appropriate technology."
- Exploring the characteristics of innovations and the decision processes that influence their rates of diffusion and adoption in the marketplace.
- Engaging in aggressive marketing: knowing how marketing resources are developed in different high-technology firms, strategies for communicating product attributes to potential customers, tools for competitive intelligence gathering and analysis.
- Promoting product development processes to bring new products to the market.
- Managing customer relations.
- Evaluating technology-based initiatives according to their economic and financial feasibility as well as strategic fit within the organization.
- Effectively organizing and managing the R&D efforts to maintain a competitive advantage.
- Allocating R&D resources among basic research, applied research, and development efforts.
- Following appropriate methods for evaluating the performance of R&D groups.
- Using strategic alliances as an alternative to vertical integration and the roles of various types of business alliances, including partnerships and joint ventures.
- Supplementing firm's own technology development capabilities with externally developed technology.
- Interacting with sources of external technology, including federal laboratories, universities, and other firms.
- Assessing candidate technology for compatibility with the acquirer's core technologies.
- Establishing appropriate mechanisms to facilitate the transfer of technology to and from the firm.
- Understanding contract law and negotiation techniques.

throughout history whether before or after the time of Adam Smith: Technology and the wise management of technological resources create wealth.

Managerial functions, responsibilities, and obligations for the effective management of technology are listed in Exhibit 14-2. Management fads and styles will come and go, but the factors in Exhibit 14-2 will remain the major management challenges influencing the competitiveness of organizations.

## REENGINEERING

As explained in previous chapters, companies today face explosive growth in knowledge, a fast pace of technological change, global competition, and a customer-driven economy. The need to adjust to this new environment has forced many executives to resort to the reengineering of processes or organizations. The idea itself is not new; it is

- Finding ways to reduce exposure, such as project phasing, risk sharing, insuring, and other techniques.
- Pursuing breakthroughs in manufacturing systems—new methods for transforming concepts into products.
- Using logistics, concurrent engineering, and integration of design, development, production, and organizational infrastructure.
- Knowing how management information systems and functions are used and structured to support decision-making in technology management.
- Recognizing that human resources are the firm's most valuable assets and technology developed by these resources represents the most important source of competitive advantage. Success of the organization depends on effective leverages of intellectual property.
- Creating progressive strategies for managing and leading engineering and other technical professionals and cross-cultural management.
- Understanding legal and organizational issues inherent in managing intellectual property.
- Promoting intra/entrepreneurs and approaches to stimulating creativeness and innovativeness.
- Motivating knowledge-based workers and optimal use of intellectual capital (IC).
- Measuring and improving productivity and effectiveness.
- Designing fair reward systems.
- Managing the learning process of the organization.
- Utilizing technology gatekeepers in the context of a learning organization.
- Establishing quality as a corporate value and strategy.
- Creating links between missions, the alignment of goals, and individual empowerment.
- Using written and oral communication effectively and persuasively.
- Presenting complex technical material in a manner that can be appreciated by nontechnical decision makers.
- Mastering information technologies (data processing, storage and retrieval, telecommunications, and multimedia).
- Acquiring knowledge and understanding of pertinent emerging technologies.
- Developing organizational structure and staffing requirements in response to a constantly changing environment.
- Protecting the environment, health, and safety.
- Demonstrating social responsibility toward employees, families, and society.

the competition in global environments that is new. The reengineering of processes has been practiced by industrial engineers for several decades. However, the urgency to initiate or implement changes did not exist in the relatively stable business environment of the mid-twentieth century. The emergence of the present dynamic environment—led by the knowledge and technology revolutions of the late twentieth century—combined with strong global competition is making changes in processes and organizational structure a management priority.

The idea of reengineering the corporation was very well argued by Hammer and Champy in 1993. The main premise of reengineering is that for companies to succeed in today's world, they cannot rely on incremental improvements but should engage in radical change, or "discontinuous thinking" as Hammer and Champy refer to it. The company should be ready to challenge and possibly abandon fundamental assumptions about how it does its work. Reengineering looks beyond tasks, jobs, and organizational structure to focus on processes. A *process* is defined as a set of activities that produce a result of value to customers. It has input, such as material, labor, or information, and produces outputs to meet customer needs. An example would be the development of a new product (Hammer and Champy, 1993).

*Reengineering* calls for the examination of every option available, including elimination, combination, or streamlining processes. The focus is on improving efficiency and limiting inputs while adding value to outputs. This implies doing more with less resources which sounds contradictory. This is one of the reasons why some managers have difficulty implementing reengineering.

In the early 1990s, a large number of U.S. companies rushed to adopt business process reengineering or to reengineer their entire companies in an effort to improve efficiency and restore competitiveness lost to foreign organizations. Many executives used the idea of reengineering to downsize their workforces and eliminate jobs rather than to restore the long-term effectiveness of their companies. "Reengineering" became synonymous with "downsizing," or, as some called it, "right-sizing." The practice resulted in an improved profit margin for many corporations. Yet, when executives got carried away with downsizing, it had a negative effect on employee morale and weakened the companies' long-term core strength. It is critical that when management attempts the reengineering of corporations it consider both the short- and the long-term effects of the change.

Reengineering, when done appropriately, can result in a lean and efficient organization. Teams usually carry out the reengineering efforts. The overall leader of the reengineering is the senior executive of the organization. *A process owner*—someone with knowledge of the concept—is chosen to assemble a team (or several teams). The leader of the team must have some clout in the organization and be a good motivator. The reengineering team itself may consist of five to ten people. Their task is to reexamine every aspect of the existing process, propose necessary changes, and oversee implementation. The team may include outside consultants to help expand its thinking horizon and give it extra capability. To be fully effective, each team should concentrate on reengineering one process at a time. A company may have several teams working simultaneously on different processes. In such a case, an overall coordinator for the teams may be appointed to work with the process owners.

## THE REVOLUTION AT GENERAL ELECTRIC

General Electric (GE) is one of America's largest and most prestigious companies. It has more than 220,000 employees operating in more than 100 countries. Its revenues in 1993 exceeded $60 billion and exceeded $90 billion in 1998. It manufactures many products, from light bulbs to locomotives to consumer products to jet engines, in more than 20 divisions. Each one of its divisions could be a *Fortune* 500 company in itself.

In 1981, GE hired Jack Welch to be its CEO. Welch is one of the icons of American industry executives. He has made more changes and contributions to the style and practice of industrial management than have most executives of modern times. Welch is a visionary and is extremely competitive. He will do whatever it takes to win. He made the decision that GE will be number one or number two in every market in which it competes, and he made the employees and the world aware of this. Welch has been a leader in management innovation for two decades, having predicted the growing trend in global competitiveness. When he took the helm at GE, Welch started a major restructuring of the company's operations. This was probably the first reengineering effort in U.S. industry, even before the term "reengineering" was coined.

In 1985 Welch hired Noel Tichey, a professor at the University of Michigan, to help him turn GE training centers into an engine for change. Tichey coauthored a book (Tichey and Stratford, 1993) about the changes instituted by Welch, which he considered to be revolutionary. The book presents the GE case in the form of a three-act drama, with protagonists, antagonists, and a deep plot. The protagonists are Welch and his people, who were determined to change the corporation, start the revolution, and see it through to a new order. The antagonists are the people who had a stake in keeping the company the way it was—those resisting the change.

When Welch took control in 1981, GE had a strong balance sheet but an average earnings growth. Productivity was increasing at a slow rate of 1 to 2 percent per year, and the company had persistent cash-flow problems because of high capital expenditures and working capital expansion. GE's technology was mature in several areas of core business, such as electrical and consumer products. These areas dominated the company, yet they were starting to feel the pressure of foreign competition.

Welch's philosophy at the time was to "automate, emigrate, or evaporate." He was ready for action, yet the organization was inwardly focused, bureaucratic, slow in decision making, and lacking in innovation. Act 1 of the drama was set to begin.

### Act 1: The Awakening

In this act, the need to change the organization is articulated. The leadership team is carefully selected and scrutinized for leadership qualities and conformity of thought with the ideals of the revolution.

At GE it was explained that change was necessitated by the need to:

- Meet the challenges of global competition.
- Take advantage of global markets.
- Improve product-development cycle time.
- Be a nimble organization.
- Improve responsiveness to customers.
- Take advantage of growth opportunities.

Welch had to deal with four types of resistance to change:

**1** *Political:* Resistance comes from leaders of existing units whose status is threatened. In the GE case, the power-system and lighting business, which was dominant in 1980, was slated to lose its grip on the power structure of GE. The fight for resources became another political problem, since Welch asked his divisions and employees to do more with fewer resources and held them accountable for productivity gains.

**2** *Cultural:* An organization must be able to convince its leaders and employees to buy into the necessary continued change. It was difficult for GE managers to perceive the competitive threat to their traditionally dominant products.

**3** *Technical:* When an organization asks people to change their way of doing things, employees become worried about losing their investment in training and doing things their way. Welch's employees considered their way to be the "GE way."

**4** *Emotional:* People have a fear of the unknown. GE managers were not used to dealing with global markets and Welch demanded that they go global. Welch had to overcome the habit and inertia that came with bureaucratic traditions.

### Act II: Envisioning

The vision is translated into a group effort to implement it. *Envisioning* addresses the three fundamental building blocks of the organization:

**1** *The political system:* Here, power, career opportunities, and rewards control the functioning and operation of the organization.
**2** *The cultural system:* Here, every member of the organization should share in the norms, beliefs, and values expressed by the vision.
**3** *The technical system:* In this system, the technological assets of the company—including technology, people, capital, and information—are organized to optimize the production of goods and services

In order to carry out a revolution, revolutionaries overturn an existing system by gaining control of the police, media, and education systems. Welch did that. The police at GE were the internal auditing staff, headed by a top finance executive who controlled the GE businesses. Welch forced them to change their focus and eliminate many financial measures, and he requested that they benchmark with competitors. Their objective became to serve GE's businesses, not to control them. Welch indirectly controlled the media by using them to spread his own ideas throughout the company.

Welch also took direct control of the GE training facility at Crotonville, New York. He abandoned GE's old training guides and promoted his own ideas. He appeared personally at the training facility every two weeks and demanded that everyone in the organization undergo training in the new curriculum.

### Act III: Re-Architecting

Revolutions destroy existing systems and replace them with new ones. Welch did that with his organization. To position GE for the twenty-first century, he removed the company's boundaries and nonproductive structures. His new organization was described as "boundaryless": Vertical boundaries were removed by delayering the hierarchy, reducing executives' perks, and broadening gain-sharing among employees. Introducing partnership concepts, using project teams, and permitting cross-functional activities removed horizontal boundaries. External boundaries between a business and its suppliers, customers, and competitors were restructured to create an alliance with competitors, track customer satisfaction, and build teams with customers and partnerships with suppliers.

This boundaryless organization permitted information to flow freely across functional business units and throughout the organization. It made the organization more responsive to changes in the marketplace. Flexibility and time to market with new products and services improved under this structure.

To complete the re-architecting phase of the revolution, the social architecture of how people relate to each other and how decisions are made also needed to be changed.

Detailed plans for physical locations and social networks were built. At GE the right people were selected for the appropriate posts. Guidelines for conducting strategic and operational review were laid out to complete the structure.

Welch knew that once everything was in place and the three-act process was complete, it would be time to start all over again because change is never finished. GE continued to thrive under his leadership for many years, and it remains one of America's most admired corporations.

## THE FPL STORY

Florida Power and Light Company (FPL) is the state of Florida's biggest utility company. It provides power to the South Florida region, the most populated region in the fourth most populated state in the United States. FPL has a near-monopoly in providing electricity in South Florida. The company owns the power distribution network, and it operates 13 power generation facilities including 2 nuclear power generation plants.

The utilities in Florida are regulated by the Public Service Commission (PSC), a statewide watchdog commission that protects the public interest in receiving quality service at a fair price.

In the mid-1980s FPL's operating budget was about $4.75 billion, and the company employed nearly 15,000 employees. In 1984 FPL's management conceived a vision of becoming the best-managed electric utility in the United States. The population growth in Florida was booming, and the company had to respond to the needs of this rapid growth. The business environment was changing at a fast pace. The company's management team felt that it needed a new management system that would allow it to cope with these problems and enable it to plan for future changes (Hudiburg, 1991).

FPL's top managers decided to implement a total quality management (TQM) system. This system had proved successful in many Japanese companies. TQM is based on four principles: (1) customer satisfaction, (2) the use of data and analysis, (3) the use of Deming's famous plan-do-check-act cycle of continuous improvement, and (4) respect for people. Japanese companies used TQM systems to improve the quality of products and services. They were impressed by W. Edward Deming, who helped introduce this system to Japan. To honor him, Japan introduced the "Deming Prize," an award given to companies in recognition of outstanding success in the practice of quality control. FPL management, convinced that the Japanese-style TQM would help it achieve its vision, decided to fully implement a TQM system and compete for the Deming Prize. John Hudiburg, chairman of the board and CEO of FPL, was fully convinced that FPL's survival, as well as the survival of most other companies in the twenty-first century, would depend on implementing TQM. In his 1991 book, Hudiberg stated, "In the highly competitive worldwide markets that will emerge in the twenty-first century, only those companies able to do an excellent job in providing quality and customer satisfaction are going to survive and prosper."

FPL elected to go "all the way with the Japanese style of quality management" (Hudiburg, 1991). Management sought to change FPL's company culture to a quality culture, and it did not spare any effort to make this happen. The effort paid off when, after four years of very intense work, FPL became the first company outside Japan to be awarded the Deming Prize. FPL's effort also paid off handsomely in terms of improving its quality of service. Customers' complaints were reduced by 70 percent, and the percentage of customers extremely satisfied increased from 41 to 62 percent. Dramatic quality improvements were achieved in service availability, reliability, safety, and employee satisfaction. FPL was also able to "reduce the price of electricity to its customers, reversing an inflationary trend" (Hudiburg, 1991).

The campaign that Hudiburg and his coworkers waged in the quest for quality seemed rather costly to FPL's board, so the board decided to change management. A new CEO was hired who immediately decided to reengineer the company. Many employees of FPL were laid off and instructed to reapply in the restructured organization. Several vice presidents were retired, and the workforce was reduced drastically in stages. Many cost-cutting measures were introduced, and several operations were streamlined. The restructuring effort met with a positive response on Wall Street, and FPL stock rose in value. The FPL board and the stockholders seemed to be more satisfied with the company's performance. The reduction in the workforce resulted in improved efficiency and increased profit margins. However, the TQM program instituted by the previous management team was virtually scrapped. Employee morale started to sink, and loyalty to the company suffered.

In a 1998 report, the Public Service Commission stated, "By virtually every measure examined FPL's distribution service quality has declined over the period 1992 through 1996." Some of the figures cited in the PSC report were:

- It took FPL an average of 81 minutes to respond to trouble calls in 1997, 20 minutes longer than in 1993.
- Power outages lasted an average of 165 minutes in 1996, up from 124 minutes in 1992.
- FPL customers filed 1,447 complaints in the first nine months of 1997, more than twice the number of complaints in all of 1993.
- The amount FPL spent on operating and maintaining its 27,000-mile power distribution grid dropped 20 percent between 1992 and 1996.

McNair (1998) cited an FPL worker and union officer statement: "You just can't cut that many people in a company and expect productivity and reliability to remain at a satisfactory level." A senior management analyst with PSC was also quoted: "Its been no secret that they've [FPL] been in an operating and maintenance reduction mode since 1991. . . . First they cut fat maybe, then they began to cut muscle."

### Lessons from FPL

Several lessons can be extracted from the events taking place at FPL over the last two decades of the twentieth century. These are:

**1** In the mid-1980s CEOs of major U.S. corporations realized that the business environment was undergoing major changes. A strong feeling emerged that competition in the twenty-first century will require drastic changes in management systems.

**2** The emergence of Japanese firms as strong global competitors attracted the attention of U.S. executives to the Japanese management style. TQM was one of the leading systems that U.S. executives were trying to emulate.

**3** Significant time, effort, and money are needed to transform the culture of corporations and implement a TQM program.

**4** Implementing a TQM system of management improves service quality, customer satisfaction, and employee self-respect.

**5** U.S. corporate boards and stockholders are strongly interested in cost-cutting measures and improved profitability. Wall Street rewards companies taking such measures through increased stock prices.

**6** When applied prudently, reengineering helps reduce cost and improve profitability.

**7** When employee layoffs and cost-cutting measures, instituted through a reengineering effort, exceed their reasonable limits, quality will suffer. The long-term competency of the corporation may also be adversely affected.

**8** A change in a corporation's top management may result in a total change of direction in the management system adopted. This can mean a reversal of fortune for many of the old employees, and it can affect morale and loyalty to the company.

**9** Many U.S. corporations shifted their focus in the 1990s from the Japanese TQM style of management to a management style based on reengineering. The latter seems to be more in line with U.S. corporate culture and the stockholders' interests.

**10** Managers should be warned not to follow new management fads without fully evaluating their long-term effect on the sustainability of the organization's competitive position in the marketplace.

## MANAGING WITH TECHNOLOGY

A concept that has recently gained popularity among managers is based on managing with technology. In the 1990s managers have been through reengineering, downsizing, and outsourcing. They have been driven to embrace manufacturing concepts and to design agile enterprises. They have installed packaged information-technology (IT) applications to effect "Enterprise Resource Planning" (ERP). IT serves as a platform that permits automation of internal processes and a way to manage the supply chain. Most managers embraced the flexibility and relatively low cost of PC-based client/server computing. These systems empower employees and customers through decentralization but distribute complexity to all users. The pace of technological change is fast, and now managers have to embrace business on the Internet. The Internet applications force a new philosophy different from ERP. The Internet is about communicating and connecting with the outside world while ERP is more inward looking, focusing on making the enterprise more efficient as an isolated entity. The Internet's management philosophy uses the power of communication to integrate the customers and suppliers with the company's major competencies. It provides as much of a competitive advantage as the ability to deploy the technology within the boundaries of the company. An example of this is Amazon.com competing with Barnes and Noble. Another is Price.line.com competing with major airlines and travel agents reservation systems.

The ability of managers to continuously embrace technological change and to manage with technology is essential for survival and growth.

## DISCUSSION QUESTIONS

1 *Harvard Business Review* (September–October 1997) published a time line of management ideas and practices. It shows the evolution of managerial practices since 1922. Do you think that history influences management or that it's the other way around? Support your answer.
2 It seems that the only constant is change. How can a manager cope with it?
3 Discuss the differences between the TQM and the reengineering management styles. Can these differences be reconciled in a new management style?

## ADDITIONAL READINGS

J. Collins & J. Porras. *Built to Last.* Harper Business, New York, NY, 1994.
The authors follow the example of Peters and Waterman (*In Search of Excellence*) in looking for the secrets of well-managed corporations. However, they focus on long-lasting successful enterprises. What are the practices that have allowed such firms to remain in business for several decades? Collins and Porras claim the secret is vision and goal setting.

James Swartz. *The Hunters and the Hunted.* Productivity Press, Portland, OR, 1994.
This book reviews many of the traditional managerial practices, focusing on improving operations in manufacturing facilities. Management techniques such as quality, empowerment, reengineering, and JIT are all components of an effort to enhance firms' performance. The book is written as a novel.

Tim Smart. "Jack Welch's Encore: How GE's Chairman is Remaking His Company Again." *Business Week,* Oct. 28, 1996, pp. 155–160.

## SUGGESTED CASE

- "Siemens Rolm Communications, Inc.: Integrated Logistics Core Process Redesign (ILCPR)." Harvard Business School, Case 9-195-214.

## REFERENCES

Hammer, Michael, & Champy, James. 1993. *Reengineering the Corporation.* Harper Business, New York.

Hudiburg, John J. 1991. *Winning with Quality—The FPL Story.* Quality Resources, White Plains, NY.

McGregor, J. 1957. "The Human Side of the Enterprise." *Management Review,* November.

McNair, James. 1998. "Regulators Zap FPL, Say Service Declined." *The Miami Herald,* Jan. 10.

Ouchi, William. 1981. *Theory Z.* Addison-Wesley, Reading, MA.

Reddin, W. J. 1971. *Effective Management by Objectives.* McGraw-Hill, New York.

Tichey, Noel & Stratford, Sherman. 1993. *Control Your Destiny or Someone Else Will: How Jack Welch Is Making General Electric the World's Most Competitive Corporation.* Doubleday, New York.

Shingo, Shingoe. 1992. *Production Management System: Improving Process Functions.* Productivity Press, Cambridge, MA.

Shingo, Shingoe. 1990. *Modern Approaches to Manufacturing Improving: The Shingo System.* Productivity Press, Cambridge, MA.

# 15

# HOW AMERICA DOES IT

This chapter provides examples of how corporations are managed, or should be managed, in the era of the technology revolution. It presents a number of case studies and technology management lessons for demonstration and discussion.

READING 15.1

## 3M: The Innovative Corporation

**J. Garcia-Arreola and T. Khalil**

For nearly a century, 3M's culture has fostered creativity and given employees the freedom to take risks and try new ideas. This culture has led to a steady stream of products. With no boundaries to imagination and no barriers to cooperation, one good idea swiftly leads to another. So far the company can claim credit to more than 60,000 innovative products.

### ORIGINS AND DEVELOPMENT

3M was founded in 1902 in the Lake Superior town of Two Harbors, Minnesota:

> Five investors (a doctor, a lawyer, two railroad executives and a meat-market manager) got together in 1902 to excavate what they thought was a mother lode of high-quality corundum, a mineral used in the abrasives industry. They expected to profit greatly by

> selling it to companies that made sandpaper. What they dug up was a low-grade, worthless mineral. They sold one order, $20 worth. (Allen, 1996)

Years of struggle ensued until the company could master quality production and develop a supply chain. New investors were attracted to 3M, such as Lucius Ordway, who moved the company to St. Paul in 1910. Early technical and marketing innovations began to produce successes; in 1916 the company paid its first dividend of six cents a share. William L. McKnight is today considered the spiritual father of the company. After he became chairman and CEO, he penned an innovation manifesto in which one of the principles of managing innovation was established: "If management is intolerant and destructively critical when mistakes are made, I think it kills initiative" (Mitchell, 1989).

The company's turnaround, a struggle since 3M was formed, was due in large part to a product that was an extension of the company's original interests in abrasives. Launched in 1914, the new abrasive cloth was made with aluminum oxide and branded Three-M-ite. It proved far superior to the natural mineral emery for cutting metal. The automotive and machine-tool industries were its chief consumers until the United States entered World War I, when staggering quantities of sandpaper were needed to help turn out automobiles and other vehicles used in the war effort.

The 3M product became even more successful after, quite by accident, plant superintendent Orson Hull drew a sheet of abrasive cloth over the sharp corner of an iron bar and broke down the adhesive backing in a way that made the sheet more flexible. This feature gave production workers access to otherwise inaccessible places on car parts they were sanding. Curved metal surfaces could be sanded with greater efficiency. Sales climbed to $1,386,383 by 1919, with a net income of $439,407.

Another major milestone occurred in 1925 when Richard G. Drew, a young lab assistant, invented masking tape, an innovative step toward diversification and the first of many Scotch-brand pressure-sensitive tapes. In the following years, technical progress resulted in Scotch Cellophane Tape for box sealing.

In the early 1940s, 3M was diverted into defense materials for World War II. This was followed by new ventures such as Scotchlite Reflective Sheeting for highway markers, magnetic sound-recording tape, filament adhesive tape, and the start of 3M's involvement in the graphic arts with offset printing plates.

In the 1950s, 3M introduced the Thermo-Fax copying process, Scotchgard Fabric Protector, videotape, Scotch-Brite Cleaning Pads, and several new electromechanical products.

In the 1960s, dry-silver microfilm was introduced, as well as photographic products, carbonless papers, overhead projection systems, and a rapidly growing health care business of medical and dental products.

Markets further expanded in the 1970s and 1980s into pharmaceuticals, radiology, energy control, and office products and spread globally to almost every country in the world. In spite of diversification, 3M encountered trouble in the early 1980s. Costs were out of control and quality required a boost. Through stepped-up research spending and some intelligent cost cutting, CEO Allen F. Jacobson revived 3M's financial ratios. His cost-reduction program, called J-35, had the goal of reducing labor and manufacturing costs by 35% in five years (from 1985 to 1990) as well as reducing product development cycles. Quality improvement was achieved by redesigning positions in the plants and putting more responsibilities on the shoulders of individual workers. For instance, in the tape plant at Bedford Park, a Chicago suburb, an operator is expected to identify quality problems immediately and have the production line stopped before more of the damaged product is produced.

In early 1989, Jacobson started to insist on the necessity of creating products able to generate bigger revenues. 3M's strategists were focusing on 45 new product areas, each with $50 million in annual sales potential within three to five years. The big question that must be answered is how to balance product development priorities based on customer needs with a climate of freedom. As Mitchell (1989) points out, prioritizing and freedom cannot be tradeoffs, for both are needed: "Contrary to what one might think, we have found that prioritizing not only enhances productivity and the flow of the products but also affords individual researchers more time for projects of their own choosing."

3M makes about 40% of its sales to the industrial sector, so it is sensitive to a recession. Thus, the company made efforts to diversify to health care. 3M business overseas accounts for 50% of the company's total (Kelly, 1991), so it is sensitive to the strong dollar as well.

During the first half of 1991, the company again experienced financial problems: Net income fell 11.5% to $599 million without "any recovery coming" soon. Jacobson's approach to this crisis was not the usual layoffs and desperate expense cuts. The company was still investing $1.3 billion to modernize 101 plants worldwide as part of the - J-35 plan. The project included increasing automation, reconfiguring plant layouts, and using just-in-time inventory control. R&D expenses were expected to be $890 million, or 6.6% of sales, about double that of U.S. industry (Kelly, 1991).

The company launched yet another five-year program, called Challenge '95, to cut unit manufacturing cost by 10% and manufacturing-cycle time by 50% from 1985 levels (Kelly, 1991). In the 1990s 3M set new sales records of over $15 billion annually, with about 30% of sales coming from products created within the past four years. 3M's growth has come through a desire to participate in many markets where the company's core technologies can make a significant contribution, rather than being dominant in just a few markets. The company's stock has split eight times and had a market value in 1996 of over $55 per share and by 1999 it was over $97 per share.

## 3M'S BUSINESS STRUCTURE

3M is a diverse company which produces more than 45 major product lines with thousands of individual items. About 12,000 of its 86,000 employees are based at 3M Center, the worldwide headquarters in St. Paul, Minnesota.

3M's business units are grouped into three major sectors: Industrial and Consumer Sector; Information, Imaging and Electronic Sector; and Life Sciences Sector. This structure includes about 50 international subsidiary companies and more than 50 divisions in the U.S.—the basic business units of the company that have their own technical, manufacturing, sales, and marketing functions. The three main sectors are subdivided as shown in Exhibit 15-1.

**EXHIBIT 15-1**
3M ORGANIZATIONAL STRUCTURE

| Industrial and Consumer Sector | Information, Imaging, and Electronic Sector | Life Sciences Sector |
|---|---|---|
| • Abrasive, Chemical and Film Products Group<br>• Automotive Systems Group<br>• Consumer Markets Business<br>• Office Market Business<br>• Tape Group | • Electro and Communications Systems Group<br>• Imaging Systems Group<br>• Memory Technologies Group | • Medical Products Group<br>• Pharmaceutical, Dental, and Personal Care Products Group<br>• Traffic and Personal Safety Products Group |

As a result of this organizational structure, 3M is big but is able to act small. Planning and executing strategies for its businesses varies considerably, depending on the business's size and range and regional issues (Hammerly, 1992). The divisional structure used in 3M creates empowerment in the employees because "you are never far from either the top or the bottom. So what you do really has an effect on the total success of a project. You feel responsible and empowered by the situation. You feel that you've got the ball, and you can run with it. This has a remarkable effect on people" (Fry, 1988).

Each 3M division may borrow on the learned experiences, knowledge, and other valuable assets of the global organization. While each unit operates with autonomy, it is also part of a broader strategic market cluster. It is this internal synergy that makes 3M unique, as it brings to market products that reflect the knowledge gained from its global operations.

3M also innovates through organization. R&D is conducted on three levels, with some projects active at one, two, or all three levels. Corporate research laboratories work on projects for 10 or more years. Sector laboratories work on midterm projects for 5 to 10 years. Divisional laboratories focus on current programs for one to five years. Divisional laboratories can go to the sector and corporate laboratories for special expertise, such as computational chemistry. Computational chemists simulate experiments on computers. They can speed up the development process significantly, because computer experiments go faster than test-tube experiments. 3M also has a dozen centers of technology excellence and five analytical resource centers where researchers can find specialized help in areas ranging from nonwoven microfibers to electronic hardware design. However, the lab to market is only one route to innovation. The company also assesses customer needs and anticipates market trends (Mitsch, 1990).

Research must be funded at an adequate level. In 1994, 3M spent more than $1 billion on R&D, or over 7 cents of every sales dollar. The company had some outstanding results from this R&D spending: Nearly 500 new products were introduced that year alone; $1 billion in sales resulted from products introduced in 1994; and 543 patents were received that year. In the first quarter of 1995, 3M had another 500 patent applications in the works.

Innovation also requires reviewing operations. For example, the company has determined that 25% of revenues comes from products developed within five years. The numbers are now 30% in four years and, because of global competition, probably will have to rise again soon. As Allen (1995) says,

> This 30% is a crucial yardstick at bonus time, so managers take it seriously. Within 3M, the operating units monitor their performance against established plans. And they have a Corporate Technical Planning and Coordination organization. It offers resources, such as computational modeling and literature searches, and also provides a structure for peer review of technical programs.

## 3M'S CULTURE

The following are central points in the 3M cultural system (based on Mitchell, 1989, and Mitsch, 1990):

- Corporate rules are scarce.
- Control results from constant peer review and feedback.
- Salaries and promotions are tied to the success of new products (from inception to commercialization).
- Product champions are allowed to manage new initiatives as if they are managing their own businesses.
- Divisions are kept relatively small ($200–$300 million in sales).
- Divisions are expected to share technology and human resources.
- Information flows informally among people.
- Promotions occur from within. The company rarely hires from the outside (except, of course, entry level), and never at the senior level.
- Innovations do not happen unless people know it is a top priority, management is committed to it, and resources are assigned to innovators.

What is the 3M system of innovation? Encouragement, recognition, and reward are keys to innovation. There are different programs to support initiative and intrapreneurship. A typical project is managed as follows: An employee comes up with a new idea that can be converted into a product. The employee forms a team by recruiting full-time members from different areas of the company (laboratories, manufacturing, marketing, finance, etc.). The team designs the product and figures out how to market it. As the product generates more revenues, the members of the team get promotions: "When the sales grow to $5 million, the product originator becomes a project manager, at $20 million to $30 million, a department manager, and at $75 million a division manager" (Mitchell, 1989). There is a dual-track career system, one for scientists who do not want to become managers. However, they still can advance their careers without having to rise to the management ranks.

It is important to note that the company does not require a particular market size to support a new venture. This allows apparently marketless products to become market winners, for example, Post-it Notes and Scotch Tape.

3M makes it easy for technical people to bounce ideas off each other and to exchange information informally. One way is through 3M's Technical Forum. All 3M scientists, engineers, and technicians belong to the Technical Forum. The forum provides formal and informal means of communication designed to foster the cross-fertilization of ideas. The forum's activities include mind-stretching seminars conducted by speakers such as space scientist Wernher Von Braun, physicist Edward Teller, and Nobel laureate Linus Pauling. The Technical Forum also includes various chapters devoted to a specific discipline, for example, polymer chemistry. In their chapter, polymer chemists can maintain their skills at state-of-the-art levels. And a materials scientist, for example, knows where to look for advice about polymeric problems such as how to bond an advanced aerospace composite to titanium.

A program named Genesis is a court of last resort for scientists looking for start-up funds. If they can't get the money elsewhere, they can petition 3M for seed money to jump-start their "orphan ideas." Each grant offers up to $85,000 to carry projects past the idea stage. All technical people are allocated 15% of their time to work on projects of their own choosing. And this 15% spawns many of the ideas that wind up as Genesis grants.

Personal recognition is important, too. 3M's highest recognition for scientists is election to the Carlton Society, which is named after Richard P. Carlton, 3M's first laboratory worker with a college technical degree and 3M's fifth president. In the 1920s, Carlton supplanted trial-and-error methods with more specific scientific standards. He also favored risk taking, pointing out, "You don't stumble, unless you're in motion." Carlton Society membership goes to scientists who have contributed to 3M in a massive way, such as Richard Drew, for his invention of masking tape and transparent tape. One of the Carlton Society members, Matthew Scholz, was only in his thirties and a 3M scientist for only about a decade when he was inducted. He had 14 patents and was named inventor on 19 additional patent applications. Matt Scholz is the principal inventor of 3M Scotchcast Orthopedic Casting Materials. They are the leading fiberglass replacements for plaster casts on broken arms and legs. Incidentally, there are 718 patent holders among about 8,400 scientists, technicians, and engineers.

Another highly coveted award, the Golden Step Award, goes to cross-functional teams that create and market products that reach at least $4 million of profitable sales in one year. There are similar award programs called Pathfinder and Pathfinder Merchant for product market or development overseas.

The third element is to reward innovators. 3M does this with a dual-ladder system for promotion. On the one hand, a scientist can advance into technical management or even general management. In fact, most of the top managers are technical people by training and experience. On the other hand, a scientist who loves the laboratory work of creation can continue working at the bench without an artificial limit on advancement. There are specialist ranks up to and including the high-level job of corporate scientist. Setting high expectations also helps. That's why it took less than two years to reach the new goal of obtaining 30% of worldwide sales from new products within four years.

## INNOVATION TALES

There are many examples of successful innovations throughout 3M's history, two of which are discussed in Boxes 15-1 and 15-2.

**Box 15-1**

MASKING TAPE AND SCOTCH TAPE

In the 1920s, everyone who bought a car wanted a two-tone finish, which the automobile manufacturers had just introduced. Because a straight, sharp demarcation between the two colors was essential when a car was being painted two colors, one area of the car had to be masked off while the other was being painted. The most satisfactory method of masking in the early twenties was to separate the two areas to be painted by old newspapers. But the homemade glue and cloth-backed surgical tapes used to hold the newspapers in place created trouble.

While visiting an auto-body shop in 1925, Richard G. Drew, a laboratory assistant, heard shouting, a particularly violent outburst from a worker who was disgusted with the mess the crude masking method had made of his paint job (removing the glued newspaper peeled away fresh paint, ruining a day's work).

Drew returned to 3M and explained an idea to management. What the auto industry needed was a masking tape that would stick tightly, yet pull off easily without leaving a gummy residue or taking the paint with it; a tape that wouldn't let solvents in lacquer seep through and ruin the paint job underneath; a tape strong enough to provide a sharp edge for two tones. Drew and his assistants cooked dozens of batches of sticky substances using linseed oil, various resins, gum chicle, and naphtha. The laboratory finally ended up with a formula containing a good grade of cabinetmaker's glue kept sticky with the addition of glycerin. It gave the painter a sharp, clean edge. This first tape was called simply, "3M Non-Drying Tape."

3M finally named this new product "Scotch" Brand masking tape and has been trying to explain why ever since. One story has become plausible legend. When the masking tape was first sold, the company, as an economy measure, applied adhesive only to the outer edges of the two-inch strips, leaving the center plain. One edge was to be taped to the paper, the other to the car to hold the masking paper in place. This turned out to be a new headache to car painters, for the partially coated tape didn't stick too well, and the story is that these painters growled at 3M salesmen, "Why be so Scotch with the adhesive?" Soon, the tape was officially trademarked "Scotch" Brand.

Yearly volume grew steadily from \$164,279 its first full year on the market to \$1,151,023 in 1935, and on up to a multimillion-dollar gross. From this basic masking tape, 3M branched out into the manufacture of hundreds of pressure-sensitive tapes for home and industry.

*Source:* Reproduced, with slight modifications, from the 3M web site: http://www.3M.com.

**Box 15-2**

POST-IT

Post-it Notes are a more recent example, involving an extensive exchange of technology within the company. Spence Silver created a novel adhesive that clung to its own molecules better than to other molecules. It would stick between two surfaces, but wouldn't bond firmly to either. For a decade, Silver took his adhesive around the company: a solution in search of a problem. In 1974, another scientist, Art Fry, decided to play with the novel adhesive, taking advantage of 3M's 15% rule, which allows scientists and technical personnel to spend up to 15% of their time on projects of their

(*continued*)

**Box 15-2**

POST-IT (*Continued*)

choosing. He sang in a church choir and wanted to create a hymnal marker that wouldn't fall out of the book but could be removed without tearing pages. After several experiments, Art Fry came up with just what he wanted, as well as an insight that led Silver and Fry to look for innovative ways to use this interesting material on other kinds of paper.

Silver and Fry were told it would be impossible to engineer equipment to assemble the adhesive-coated paper into the pads. Fry wouldn't take no for an answer, so he went home and built prototype hardware in his basement. The equipment worked, but Art Fry's new hardware wouldn't fit through his door. So he had to knock out a basement wall to remove the hardware to his laboratory.

The Post-it team added several members, but they had trouble getting 3M's attention for the novel product idea. They solved this in a unique way. The team sent sample pads of Post-it Notes to all the senior executives, who quickly reordered additional samples. When the executives were "addicted" to Post-it Notes, the development team cut off their supply. They referred all executive calls to the marketing organization. Executive complaints tend to concentrate the mind wonderfully. So, with new attention being paid to Post-it Notes, the team was able to move its product out of the laboratory and into the test market.

In 1978, the Post-it project was at the point of cancellation because of a near-zero response to the new product. The problem was, nobody had the foggiest idea about what to do with tiny pads of paper with sticky stuff on one edge. People had to be shown how to use the product. That led to the last chance to save the product. The sponsoring division took an expensive gamble. It flooded Boise, Idaho, with sales representatives to demonstrate Post-it Notes in banks, offices, stores, and everywhere paper clips and staplers might be used. The test-market turnaround was immediate and electrifying. Most people received a sample, played with it, and started to find applications for it in their own work (Fry, 1988). Customers loved Post-it Notes once they knew how to use them.

Post-it Notes were introduced nationally in 1980 and have grown into a family of products. There are many standard sizes, plus special shapes for file-folder tabs, phone messages, and so forth, as well as tape flags, glue sticks, and even a removable version of 3M Magic Transparent Tape.

One year after its introduction, Post-it Notes was named 3M's Outstanding New Product. Fry was promoted to division scientist in 1984 and, in 1986, to corporate scientist, the highest rung on the technical side of 3M.

*Source:* 3M web site: http://www.3M.com.

### Reading 15.1 References

Allen, George M. 1996. Speech given at the 3M South Africa Management Forum, February 20. Reported in http://www.3M.com, June.

Fry, Art. 1988. "Lessons from a Successful Intrapreneur." *The Journal of Business Strategy.* March/April, pp. 20–24.

Hammerly, Harry. 1992. "Matching Global Strategies with National Responses." *The Journal of Business Strategy.* March/April, pp. 8–12.

Kelly, Kevin. 1991. "3M Run Scared? Forget about It." *Business Week.* September 16, pp. 59–60.

Mitchell, Russell. 1989. "Masters of Innovation. How 3M Keeps Its New Products Coming." *Business Week.* April 10, pp. 58–63.

Mitsch, Ronald A. 1990. "Three Roads to Innovation." *The Journal of Business Strategy.* September/October, pp. 18–21.

### Lessons from 3M: The Innovative Corporation

A review of 3M's history provides technology managers with many lessons showing how a company can succeed in balancing innovation with marketing and with operations to achieve sustainable success and staying power in the global industrial scene. The following points summarize factors that contributed to 3M's success and can be generalized into generic lessons for others to follow:

**1** Innovation can be encouraged in a corporation through the appropriate policies.

**2** A successful firm must balance its desire for strict management control with the need to give employees freedom to create new things.

**3** R&D is an expensive activity. Even 3M has been careful in managing cost. However, in spite of financial pressures, management must ensure that cost-cutting programs do not adversely affect innovations.

**4** Communication among the different elements of the corporation is crucial for technology transfer.

**5** Market analyses are not always useful for the launching of new products. Post-it Notes could have been killed as an innovation if market analysis was to determine its fate.

**6** Although technology-science push is an important source of innovation, the market pull must be created, or at least carefully considered, by companies if they are going to introduce products that really satisfy market needs.

**7** The market pull is more effective when technical people visit a customer's facilities. They can link needs with competencies.

**8** Although R&D may be the source of innovation within the corporation, other value-added functions, such as manufacturing and distribution, are key for the market success of the products. Even the brightest idea will not be accepted in the market without a reasonable price and good quality.

**9** Technology on a shelf is worth nothing: The glue used in Post-it Notes was stored for 10 years before an application was found.

### Factors in 3M's Corporate Structure Permitting Innovation

The design of organizations can facilitate the process of technological innovation and improve competitiveness. 3M provides an excellent example of how one company has been successful structuring itself to permit innovation. Here is how it is done:

- 3M is a very big company with more than 50 divisions, and it can continue to grow, but it acts small. It allows employees' ideas to circulate and provides growth opportunity for innovators.
- 3M's success in taking core technologies and coming up with variations of products has kept it in the forefront of innovation.

• 3M initiated a rule requiring that one-quarter of any division's sales come from products introduced within the last five years. Known as the 25 percent rule, it highlights 3M's emphasis on renewal and competing with new products. The rule is now 30 percent from products created in the last four years.

• Meeting the company's goals, such as the introduction of new products under the 25 or 30 percent rule, is rewarded with bonuses.

• The company has another rule that permits employees to spend part of their time on any product-related activity of their choice. Known as the 15 percent rule, it gives an employee time to create and invent. Post-it Notes was developed as a result of this rule.

• 3M's philosophy of supporting innovation, whether or not it fits into the firm's business portfolio and regardless of its potential market size, makes 3M a progressive, caring corporation. If a product fits one of its divisions, the innovator is allowed to take it there. If it does not, 3M encourages the innovator to explore outside avenues for exploitation.

• The company provides financial support to innovators. Genesis grants, for example, allow researchers to carry their projects past the idea stage.

• A technology council made up of researchers from various divisions meets regularly to exchange information and provide technical support as needed.

• An employee with an idea is allowed to form an action team that draws on company expertise in technical areas, manufacturing, marketing, sales, and finance. Acting as a venture team, it shepherds the product from design to market. All team members are rewarded for their success by promotions and raises as the project moves forward. When sales grow to a certain level, the product originator is further rewarded according to the level of success and can eventually become a project manager, a department manager, or a division manager. If the originator prefers to stay in technical areas, a dual track exists for those who do not want to be managers.

3M's continued success is a fitting tribute to the encouragement of entrepreneurial spirit, to a reliance on innovation and being first to market, to a spirit of renewal, and to a structure permitting creativity and growth.

READING 15.2

## The Development of the PC Industry

The story of the personal computer industry and the companies involved in it superbly illustrates the thrill of victory and the agony of defeat for the major players in this industry. We are in the midst of a revolutionary change in information technology.

Reviewing the development of the PC industry from its inception in the mid-1970s to the present time offers a unique perspective on one of the most recent and far-reaching technology upheavals in modern history. Studying the individuals who launched this revolution and the companies they built offers valuable lessons for technology managers. To extract these lessons, we must go back in time and review attitudes, actions, and decisions which gave birth to this new industry—an industry that has grown to be the third-largest industry in the world and one of the great stories in creating wealth through technology.

## THE START OF THE PC INDUSTRY

In the 1950s, mainframe computers were huge and often took up entire rooms and sometimes entire buildings. They needed to be cooled to prevent the overheating of vacuum tubes, valves, and later transistors that made up the machines. It took a revolutionary invention—the transistor at Bell Laboratories in 1947 (Isaacson, 1998)—and the development of the semiconductor business at Schockley Semiconductor and Fairchild Semiconductor to build better and faster computers. IBM made a fortune on building, selling, and, later, leasing mainframe computers. IBM and other big computer companies, such as UNIVAC and Digital Equipment Corporation (DEC), possessed excellent technological and financial resources, but according to Cringely (1996), did not have the vision needed to invent the personal computer. IBM's CEO, Tom Watson, Jr., and president, Al Williams, were not technologists and placed more attention on marketing and finance issues. They chose to believe that American business executives would never be interested in a PC.

Schockley Semiconductor was started by William Schockley, one of the inventors of the transistor, who had just moved for personal reasons from the Bell labs in New Jersey to Palo Alto, California, his hometown. Schockley was a good scientist and tried to capitalize on the new technology of transistors, but he was not a good manager. He was unable to motivate and keep his excellent group of employees inside his company. Robert Noyce and seven other employees left his employment and started Fairchild Semiconductor. Noyce later cofounded Intel in 1968 with Gordon Moore.

Jack Kilby at Texas Instruments had developed the idea of connecting germanium resistors and capacitors together on a slice of germanium using gold wires as connectors. This is the idea of the integrated circuit board, but it had to be assembled by hand. Noyce sought to reduce the cost of assembling the many electronic components by connecting them together on a single piece of silicon, thus introducing the concept of silicon-chip integrated circuits. This innovation gave Silicon Valley its name.

Gordon Moore predicted that the number of transistors that could be built on a piece of silicon would double every 18 months. Moore's prediction still seems to hold true for the industry today. Intel memory chips held 1,024 bits of data in 1972 and 1,024,000 bits in 1990 (Cringely, 1996). In 1998 Intel's Pentium II chips can make 588 million calculations a second (Isaacson, 1998).

In 1971, Ted Hoff at Intel introduced the microprocessor, which is a chip that contains logic elements used in a computer. It is small and powerful and can be programmed to perform different tasks. Intel introduced the microprocessor to be used in the calculator market and in traffic lights. Intel did not adequately appreciate the brilliance of its own product. It had all the elements for inventing the PC business but missed the opportunity.

Major computer companies, such as IBM and DEC, also failed to capitalize on Intel's microprocessor in regard to building PCs. The idea that a computer could be built around a tiny microprocessor, as opposed to their huge mainframes, did not seem to capture their imagination or suit their image of building large, complex, and expensive machines. Management's attitude of protecting the status quo seemed to pervade. The business of mainframes was lucrative, so why change? This view proved to be a mistake in the years to follow. It took small companies and young entrepreneurs to change the computer industry landscape and start the PC industry.

It was in January 1975 that the world's first PC was developed, by Ed Roberts at a company called MITS in Albuquerque, New Mexico. "It was not even a real computer but rather an unassembled kit of electronic components sold for $350" (Cringely, 1996). You had to build it yourself and even then it usually did not work. There was no keyboard or computer language to go with it. The Altair 8800 microcomputer from MITS was featured on the cover of *Popular Electronics;* fortunately it was seen by Paul Allen. Allen contacted his friend Bill Gates, who was then a student at Harvard, and both decided to write a BASIC programming language for the Altair. Allen and Gates's effort was successful. Their effort helped launch the PC industry into a new era of progress and helped Gates and Allen build Microsoft and become multibillionaires.

The Altair computer itself was a hobbyist machine. It attracted a bunch of "nerds," engineers, programmers, computer scientists, and entrepreneurs who saw the potential of this new technology. Ed Roberts's machine, which looked like a box with some switches and a few lights, allowed many interested "hackers" to use it as a piece of test equipment. It could be programmed only in machine language, and many of its users could flip its switches and see the results of their code only through the light bulbs on the box (Cringely, 1996).

It has been said that the very uselessness of the Altair is what drove the hobbyists together to brainstorm and offer suggestions. Information was shared freely at local gatherings of enthusiasts who formed the Homebrew Computer Club, which met regularly on the Stanford University campus. Attendance at such gatherings included young entrepreneurs such as Steve Jobs and Steve Wozniak, who later introduced the Apple computer and established the Apple Computer Company at Cupertino, California.

### Reading 15.2 Reference

Cringely, Robert. 1996. *Accidental Empires.* 2nd edition. HarperCollins, New York.
Isaacson, Walter. 1998. "Driven by the Passion of Intel's," *TIME,* Dec. 29, 1997/ Jan. 5, 1998.

### Lessons from the Development of the PC Industry

Development of the PC industry offers many valuable lessons in management of technology. These include:

- Inventions such as the transistor created revolutionary changes in the computer industry.
- The rapid changes in semiconductor technology made it possible to introduce the microprocessor. This was a great triumph for miniaturization and lowered the cost of assembling electronic components. It made the power of the computer available to the public by facilitating the development of affordable PCs.

- Silicon Valley developed because of entrepreneurial fever and the presence of good technology resources that existed in that region of the world.
- In the infancy of the PC industry, everyone shared information, and everybody won. Sharing information freely at the Homebrew Computer Club created breakthroughs in PC technology.
- The presence of Stanford University and the availability of its technical resources and facilities spawned interest in the PC industry.
- Technologists (also known as "techies" or "technology wizards"), such as Steve Wozniak, have certain characteristics. They become immersed in the technology and enjoy developing it as a challenge and to have fun and impress their friends.
- CEOs, company presidents, and industry boards of directors without a technical background tend to concentrate on downstream activities, such as marketing and financial issues, and place less emphasis on upstream activities, such as technological innovation. The actions of IBM provide some evidence to support this point of view.
- Possession of the technology does not mean best use of the technology. Intel developed the microprocessor and marketed it for use as a controller and in calculators. The company did not envision or exploit it in the development of the PC.
- The technology may arrive before the application of the technology is evident. The Altair was a solution looking for a problem. IBM did not even think that it was needed or that it would be accepted by its clients.
- New technologies can be scuttled by large, bureaucratic organizations. Small companies and entrepreneurs can more successfully introduce them.
- Software awaits the developing hardware, but hardware relies on software to succeed. This was shown in the case of the Altair and, later, during the development of the IBM PC.

**READING 15.3**

## Microsoft: The Challenges of New Technology

Microsoft's history has paralleled the development of the PC: The founders have been able to take advantage of innovations in PC hardware and then adapt products to several generations of technology. They have been leaders in software technology from the early days' 8080 microprocessor, to the 32-bit technology included in the Pentium, to many of today's advances in software. Microsoft has been able to control the industry standards in software designed for the PC industry, especially in the corporate market. Microsoft leaders have shown a remarkable ability to embrace change in response to new market demands and technological advancements. In the second half of the 1990s, Microsoft has developed a corporate structure aimed at competing in the new PC era. This structure is characterized by network access and multimedia, fields in which the company has not had previous experience.

This case study presents two phases: The first corresponds to the early entrepreneurial days and the consolidation of DOS and Windows as the industry standards; the second describes Microsoft's efforts and strategies to continue its leadership during the new technological phase in the home computer market.

## I. THE PC ERA

### BASIC for the Altair

Microsoft's history started at a computer terminal in a Seattle school, where Bill Gates and Paul Allen developed a passion for computer software. Years later, although Allen went to study computer science at the University of Washington and Gates went to Harvard University, the old friends managed to keep in touch.

Allen had always been very enthusiastic about the advances in Intel's microprocessors, truly believing that a small computer using one of those chips would be designed soon. In December 1974, while working at Honeywell, Allen read about a PC developed by Ed Roberts in Albuquerque, New Mexico: the machine called Altair, which utilized Intel's 8080 technology. He contacted Gates and they put together a simple BASIC language program for the new computer. In the early PC days, one of the problems confronted by the hardware developers was the very limited amount of available software; without it, personal computers were not useful tools. Allen and Gates offered to give Roberts a demonstration. He agreed to have an interview with them despite the fact that Intel representatives had told him that the 8080 chip was not powerful enough to run a—relatively speaking—advanced language such as BASIC. The main challenge for the two friends was that they did not have an Altair computer, or even an 8080 chip. Overcoming these difficulties, Allen developed a simulator of the Altair's processor on a PDP-10 machine, as well as the assembler for programming the chip. Using Allen's simulator, Gates wrote the BASIC code in one of the mainframes in Harvard. Three months after Allen read the article, he was flying to meet Roberts in New Mexico. When the transfer of files was completed, the teletype connected to the Altair computer printed the word "Ready>", indicating that the machine was ready to receive instructions. This surprised Allen more than it did Roberts. Allen had to run a moon-landing simulation program to prove that BASIC worked properly. One small mistake, either in the simulator used to develop the program or in the BASIC code itself, would have led to the failure of the demonstration. But things went right for Allen and Gates. They foresaw the potential of the emerging technology and decided to fully exploit it.

Gates and Allen went down to Albuquerque to develop software for the microcomputer. They called their company Microsoft.

### Microsoft's DOS and IBM

Bill Gates's management style is characterized by vision and a profound technical knowledge that has allowed him to develop software for the latest hardware technology advancements. When Intel presented him with the 8086 chip in April 1978, Gates decided to go with the new technology and agreed to develop BASIC for the 8086. At that time there was no company manufacturing a computer based on the 8086 technology. His vision allowed him to lead the market using the technology. He expected the project to take three weeks. It actually took six months—not an unusual occurrence in R&D projects. Gates was committed to the technology and he persisted until he succeeded.

By 1979, Microsoft had moved to Seattle with yearly sales of $2.5 million, and Microsoft's BASIC became the industry standard (Ichbiah and Knepper, 1991).

In 1980, IBM realized that it was missing out on the expanding PC market. It contacted Microsoft to develop *Project Chess,* whose objective was to design and manufacture a personal computer based on the 8080 microprocessor. Gates suggested using the newer 8086 chip because it presented many advantages to programmers (Ichbiah and Knepper, 1991). The first step in the development of software for the IBM PC was to look for an operating system. IBM had not been involved in the development of the PC technology, nor did it have all the components to put a PC together. Microsoft did not own an operating system. When IBM approached Gates to develop an operating system for its new PC, he sent IBM to Gary Kildall at Digital Research, who had the technology at that time.

That year, the dominant operating system was CP/M (Control Program/Monitor for Microcomputer) which was developed by Kildall and sold by his company, Digital Research. Many companies had adopted it, and even Microsoft chose it for its FORTRAN and COBOL software. Given CP/M's characteristics and popularity, IBM approached the company to incorporate it in Project Chess. But Kildall was involved in other business deals and did not pay attention to the opportunities presented by signing a deal with IBM. His wife, Dorothy, who was a vice president of Digital Research, refused even to sign the IBM nondisclosure agreement. This seemed to upset the ego of the giant IBM (Cringely, 1996).

IBM went back to Microsoft, which would not let an opportunity of a business deal with IBM slip through its hands twice. Microsoft decided to look for a new operating system. Seattle Computer Products had developed an operating system called QDOS. Paul Allen contacted the company and negotiated an agreement to license the operating system for whatever uses Microsoft wanted. Microsoft obtained QDOS for $50,000 (Cringely, 1996). This must have been the deal of the century. It allowed Microsoft to offer the operating system to IBM and license BASIC, FORTRAN, and eventually COBOL to IBM.

The development of the IBM PC was completely in line with Gates and Allen's early dream of one PC in each home. Although late in coming to the market, it was supported by IBM's name and complimentary assets. IBM's decision to go with open architecture allowed software developers to create numerous applications for the machine, which made it very popular with users. The best part of the Microsoft contract with IBM was that it allowed the company to sell DOS to other hardware producers, anticipating the growth of IBM-compatible machines. IBM's mistake in this deal was not acquiring exclusive rights to this technology. It did not own or protect this very important technology used in their PCs. This decision came back to haunt IBM in the future.

IBM decided to use the 8088 microprocessor in the PC. Microsoft assigned 35 out of its 100 employees to the project. The development presented technical problems, accentuated cultural differences between Microsoft and IBM, and increased personnel stress. Finally, however, the IBM PC was launched in August 12, 1981, with prices ranging from $1,556 to $6,000 (Ichbiah and Knepper, 1991).

Two years after the IBM PC release, MS DOS and CP/M were fighting for the right to become the leader in operating systems. Gates convinced many software developers to design software for MS DOS, knowing that applications represent the actual value to the user: the broader the number of applications, the better the operating system's chance for survival. A factor that helped the dominance of MS DOS over its competitors was the launch in 1983 of Lotus 1-2-3, developed exclusively to run under MS DOS. By the end of the 1980s, 30 million MS DOS–programmed PCs were in the market (Ichbiah and Knepper, 1991). This gave Microsoft market dominance in operating systems. In a strategic move to break Microsoft's dominance, IBM acquired Lotus in mid-1995 in an effort to ensure the development of applications for its OS/2 operating system. OS/2 technology was developed by IBM in an attempt to recapture the leadership in technology in the PC industry by developing IBM proprietary technology. It proved to be a late attempt to compete with the already established de facto "industry standards" and market dominance of Microsoft systems.

### Windows Introduction

Although DOS was the standard operating system in the 1980s, it was not a user-friendly one. The system was text-based and monochrome. In contrast to DOS, Apple's system was graphically displayed and was much easier for nonexperts to use. Bill Gates would not allow Microsoft to lag behind in the graphic user interface (GUI) technology. His vision of a graphical environment consisted of an interface manager lying between DOS and the applications. The interface design would not affect the sales of the already successful DOS, nor would it affect the software applications developed for DOS. Microsoft started the project in September 1981.

In November 1983, Gates announced that the graphical interface would be operational by the end of 1984. This notice came a month after a close competitor, VisiCorp, announced the shipping of its own Windows environment. It seems that the announcement was made to urge customers to wait for Microsoft's product rather than rush to buy a competitor's product.

As proven in the DOS case, applications were fundamental for the success of Windows. Although Microsoft had the commitment to release technical information to developers in early 1984, the date was postponed to May and then to August. This represented a major predicament for the company, since many competitors were already trying to establish their respective systems. Had they been successful, it could have cost Microsoft its lead in the GUI market. Fortunately for Gates, all of them were unsuccessful (Ichbiah and Knepper, 1991).

At that time, instability was the best adjective to describe Microsoft's organizational structure: Bill Gates was in charge of all the development divisions, changing specifications continuously and moving programmers from one team to another. In August 1984 the company was reorganized to confront the challenges of an industry in a different stage of maturity. It had to move from an entrepreneurial style to a more steward-management-type style. Divisions in charge of operating systems and applications were developed, deploying responsibilities. The organizational changes helped to launch Windows 1.03 in November 1985. However, the product faced problems such as a lack of applications designed specifically for it and a lack of broad-based modern machines to better exploit the system's features (at that time it was the PC AT model).

By 1989, applications developed for Windows flooded the market. In 1990, Windows 3.0 was launched, selling 1 million copies during that year only (Ichbiah and Knepper, 1991). Windows 95 was launched in late 1995 to take advantage of the 32-bit technology included in Intel's Pentium microprocessor—the project had been due in the first half of 1994. It was delayed, and once again Microsoft was lucky that neither IBM nor any other competitor was able to come to the market with alternative technology during that period. Windows 95 arrived with the ability to fully integrate the graphical environment with the operating system and the ability to run many Windows 3.1 applications, therefore providing an excellent application base immediately. There are many experts who considered Windows 95 "as primitive as the early days of Macintosh" (Lewis, 1996). In spite of the critics, Microsoft's share for the office suites (Microsoft Office) was at 90% in 1995 (Berst, 1995). Apple was unable to capitalize on its superior GUI technology by licensing it to other PC manufacturers and diffusing it into the market. Again, Microsoft was able to penetrate the market and diffuse its software technology throughout the PC industry.

## II. INTO MULTIMEDIA AND NETWORKING SERVICES

The second half of the 1990s has proved to be of significant importance to the growth of Microsoft. As PCs are finding their way into everyday life, their functionality goes beyond work: Computer networks are designed to take computer capabilities to any place, making the hardware less important, and moving the value added to the owners of the networks and their content. Areas such as home education, entertainment, and banking are markets that Microsoft is pursuing by launching games, movie reviews, baseball statistic services, and on-line encyclopedias.

With the release of Windows 95, Microsoft included software to access the Microsoft Network. This network competes with other on-line service providers such as America Online and CompuServe. Connection with these services is one of the fastest-growing segments in the industry. The base for the network is a Windows NT server. Windows NT is a server operating system, used by Microsoft to exploit the 32-bit technology in an effort to compete against UNIX, the more recognized client/server operating system. Microsoft has been pushing software companies to develop compatibility between Windows 95 or 98 and Windows NT, creating two major difficulties for developers: (1) entirely different platforms for the two programs: Windows NT applications run in UNIX-based systems, while Windows is based on Intel's technology, and (2) very low demand for NT applications (Brandel and Scannell, 1994). The battle to dominate the market in this technology continues. The race toward better technology never ends.

### The Race for Technology Continues at Microsoft

Microsoft's strategy in 1994 seemed to focus around five main areas (Brandt, 1994):

**1** *Consumer software:* multimedia developments for the home-computer market, including programs designed for kids.

**2** *Advanced technology:* future applications such as interactive TV and electronic banking.

**3** *On-line services:* Microsoft Network (MSN).

**4** *Network computing:* the transformation of Windows NT into the base for a host of communications programs.

**5** *Office equipment:* PCs with copiers, faxes, and phones.

Microsoft pursued network computing. "As Gates's ambitions have grown, so has his bet on NT. NT serves as a foundation for a slew of advanced systems, including Tiger video servers, and Cairo, an 'intelligent' operating system that will, on its own, fetch information on a network. NT is also the base technology for server applications, programs for electronic messaging, groupware, and network management. In short, NT was considered fundamental to Gates's plan for making Microsoft a communication giant" (Cortese and Brandt, 1994).

Another important area in which the company attempted to create new capabilities is database management: services such as the distribution of digitized movies through a network demand sophisticated database software for delivering signals to thousands of users and for billing purposes. Nonetheless, Microsoft's experience in database management is limited only to desktop systems. A database of such magnitude would be a challenge even for a database software leader such as Oracle.

Microsoft is facing the technology challenges with new plans and changes in organizational structure (Janison, Swope, and Tirso, 1996; Markoff, 1996). Examples of some of these changes are:

- Internet Platform and Tools Division (client/server-oriented browsers and application development products).
- Desktop and Business Systems (support and maintenance of Windows 95 and Windows NT, the BackOffice family of NT server-based applications, and object linking and embedding [OLE] technology which allows interaction among different applications).
- Software Strategy (definition of future architectures and technologies).
- Consumer Systems (development and marketing of noncomputer devices and software for interactive television).
- Online Services (Microsoft Network, Explorer browser).
- Office Business Unit (design and development of Microsoft Office).
- Long-Term Research (natural language, speech recognition, and artificial intelligence).
- Strategic Relations (alliances and suppliers).

Microsoft is aggressively pursuing new goals and visions, as demonstrated in its broad structure. To do so, Gates has taken steps to move Microsoft into a competitive position: Budgets for research and development areas had grown to more that $600 million by 1994, and since 1987 the company has been hiring every talent from physicists and computer experts to marketing experts, sociologists, and linguists (Brandt, 1994).

Cusumano and Selby (1996) identified seven strategies that Microsoft is using to cope with the challenges of the frantic pace of technological change it is facing:

**1** Find smart people who know the technology and the business.

**2** Organize small teams of overlapping functional specialists.

**3** Pioneer and orchestrate evolving mass markets.

**4** Focus creativity by evolving features and concentrating resources.
**5** Do everything in parallel, with frequent synchronization.
**6** Improve with continuous self-critiquing, feedback, and sharing.
**7** Attack the future!

In addition to the technology challenges Microsoft faces, industry observers detect organizational deficiencies within the company. Cusumano and Selby (1995) report: "Microsoft is highly dependent on its leader, Bill Gates. He plans to run the company for another decade, but a successor has not been chosen and there are no obvious choices for the job. Microsoft has weaknesses in its middle management and its overdiversification, as well as a declining market focus. The company tends to produce products with many more features than the consumers really need."

In addition, the company is also incorporating marketing tactics that may create problems, as Rebello et al. (1993) remark: "Competitors have spurred a two and a half year investigation by the Federal Trade Commission into Microsoft's marketing tactics. Competitors accuse Microsoft of a variety of unscrupulous tactics, including using inside knowledge of its operating system to write better application packages, announcing products years before they are ready to ship in order to kill off promising competitor's products, and offering DOS licensing agreements that make it unrealistic for PC vendors to sell any other operating system." Bill Gates has responded to these accusations, maintaining "his company's right to not share operating system data with competitors by stating that firms doing innovative work do not have an obligation to share the work before putting out products" (Johnston, 1995).

Microsoft is expanding operations into virtually every possible technology and market available in the information/computer industry. Although the company has signed partnership contracts (Mermigas, 1996)—such as licensing Sun Microsystems' Java, a language used in developing homepages on the Internet, and including access to America Online in Windows 95 in exchange for software distribution—the new businesses are launching Microsoft into a fierce competition with rivals, such as AT&T, Oracle, Novell, Sun Microsystems, and Netscape (Rebello, 1996).

Microsoft has proved its resiliency as a successful technology-based company. Bill Gates* has shown superb capabilities in managing technology; his wealth doubled from $18.5 billion in 1996 to $36.5 billion in 1997, making him the wealthiest person in the United States (*Forbes,* 1997). Paul Allen's worth has been estimated at more than $14 billion. Many of the company employees have become instant millionaires. These figures keep multiplying as the company's stock price continues to surge in the stock market.

### Reading 15.3 References

Berst, J. 1995. "Earth to Lotus . . . Earth to Novell . . . Come in Please." *PC Week.* June 12, p. 63.

Brandel, William, and Scannell, Ed. 1994. "Developing for the Next Generation of Windows May Mean Running on NT." *Computerworld.* November, p. 4.

*Bill Gates wealth was estimated to be in excess of $90 billion in 1999.

Brandt, R. 1994. "Bill Gates's Vision." *Business Week.* June 27, pp. 56–62.
Cortese, A., and Brandt, R. 1994. "Microsoft's Network Wares Still Aren't Connected." *Business Week.* June 27, pp. 60–61.
Cringely, Robert. 1996. *Accidental Empires.* 2nd edition. HarperCollins, New York.
Cusumano, Michael, and Selby, Richard. 1995. "What? Microsoft Weak?" *Computerworld.* October 2, p. 105.
Cusumano, Michael, and Selby, Richard. 1996. "How Microsoft Competes." *Research Technology Management.* Jan.–Feb., pp. 26–31.
Ichbiah, D., and Knepper, S. 1991. *The Making of Microsoft.* Prima Publishing, Rocklin, CA.
Janison, M., Swope, R., and Tirso, J. 1996. "Microsoft. Lessons in the Management of Technology." Nonpublished report prepared for IEN 699, University of Miami.
Johnston, Stuart J. 1995. "Defensive Gates Uses Keynote to Rebut Critics." *Computerworld.* March 27, p. 2.
Lewis, Ted. 1996. "Is It Too Late for Apple Computer?" *Computer.* May (IEEE).
Markoff, John. 1996. "Microsoft Shifts Focus of Software: Revamps Key Units in Bow to Internet." *The New York Times.* Feb. 21, pp. C3, D4.
Mermigas, Diane. 1996. "Bill Gates Mines Internet Gold." *Advertising Edge.* April, p. 55.
Rebello, Kathy, Schwartz, Evan, Verity, John, Lewyn, Mark, and Levine, Jonathan. 1993. "Is Microsoft Too Powerful?" *Business Week.* Mar. 1, pp. 82–90.
Rebello, Kathy. 1996. "Inside Microsoft." *Business Week,* July 15, pp. 56–68.

### Lessons from the Alliance of IBM and Microsoft

The business relationship that developed between IBM and Microsoft in the early days of the PC development offer many valuable lessons that deserve attention.

- In management of technology, timing is critical. It looked like time might be running out for IBM to get into the PC market. If IBM had waited another year, the PC industry would have been too big even for IBM to take on. IBM needed to produce its PC in a year! Instead of a radical innovation, IBM went the easy route—it bought components off the shelf and assembled them. IBM's executives went against their tradition. They went with open architecture, non-IBM technology, non-IBM software, non-IBM sales, and non-IBM service. This was done in order to gain fast access to technology and quick entry into markets.
- For the business world in the 1970s, everything was still mainframe computers. IBM was huge in the technology market. Its focus was on making mainframe computers, not personal computers, for large companies.
- Microsoft understood early that for the PC to be taken seriously by big business, it needed a company like IBM to get behind it. Gates and his colleagues were ready to do anything to ride on the coattails of IBM's entry into the PC market.
- IBM needed an operating system for its PC. Digital Research blew the opportunity to get the IBM business. So you can be good in technology but if you don't have business savvy, you can lose opportunities. Bill Gates seized the opportunity of a lifetime. Microsoft licensed IBM BASIC, FORTRAN, COBOL Assembler, Typing Tutor, and Venture. Every product the company had was committed to IBM in a very short time.

- IBM's entry into the PC market legitimized the industry. For corporate America, it signaled the computer giant's approval to start buying and using PCs. Public interest followed—if it's okay for corporate America, it's got to be okay for everybody.
- By the late 1980s, IBM had created a PC that anyone could copy. IBM always thought its inside track would keep it ahead. IBM's slow pace and high overhead put it at a disadvantage with the leaner clone makers. IBM eventually lost its dominant market share.
- Microsoft kept the right to license its operating system to any company. IBM did not buy exclusive use from Microsoft. Companies started making clones, and buying their operating systems from Microsoft. IBM never imagined Bill Gates would sell DOS to anyone else, but he did.
- Alliances and sharing of technology can play a major role in the success of a company. There are many companies that have benefited from combining technologies. Even huge companies like IBM can benefit from alliances with small companies.
- The prize does not always go to the inventor but often goes to the exploiter of an invention. Gary Kildall of Digital Research invented CP/M. Tim Patterson of Seattle Computer Products invented DOS, but it was Paul Allen and Bill Gates of Microsoft who succeeded in exploiting the technology.
- No complacency in technological development is allowed in high-tech industry. Continuing success is dependent on being at the forefront of innovation. This is true for IBM, for Digital Research, and for Microsoft. A company's size or its early record of success is not adequate to guarantee continued success or even survival.
- Microsoft was able to establish market dominance and the industry standard around its DOS operating system and later its Windows series. A saying attributed to Gates is, "Money is made by setting de facto standards" (Cringely, 1996).
- Incremental innovations need to be managed with the same vigor as are radical innovations. Once a radical innovation such as the PC hits the market, several incremental innovations are needed for it to succeed and survive.

### Lessons from Microsoft: The Challenges of New Technology

Microsoft's history and strategies provide the perfect textbook example for managing high technology. Important lessons are extracted below:

- Gates's early entrepreneurial personality, vision, ability to forecast the future of the technology and influence it, fierce competitive spirit, and excellent marketing skills, combined with his technological skills, are some of the most important factors that influenced the success of Microsoft.
- Paul Allen persevered in his dream of the PC. He foresaw the potential of the technology and pursued its development with vigor.
- Gates and Allen took risks, yet they were calculated risks.
- The technical capability of both founders and Gates's business ability were major factors in Microsoft's success.
- Luck was a significant factor. Nevertheless, the risks taken were based on technical confidence and market need. It may be that sometimes people can create their own luck.

• The tendency to be late in the introduction of technology has its perils. Microsoft fell into this situation a few times. Maybe Gates underestimates the difficulty of the projects, or maybe he is confident that with his resources he still can be in a position to kill off competition.

• Announcements of new technology must be made at the proper time. Gates announced the development of Windows shortly after VisiCorp announced its own product; he was trying to stop VisiCorp's sales. That may have been a smart move. However, problems in the development of Windows caused Microsoft's credibility to be low.

• Any new successful technology requires a proper environment to prosper. Software technology depends on the available hardware. It is interesting to note that Microsoft has kept pace with the development of the PC hardware technology. Gates's entrepreneurial and champion spirit always pushed development of his software to take advantage of the latest hardware technology available.

• At the beginning of each subtechnology cycle there are problems such as lack of applications or a very reduced base of advanced hardware users. Gates had the ability to forecast technology trends and designed products to take advantage of these trends.

• In 1996, the PC technology was a mature technology. Innovations were only incremental, focusing more on multimedia features such as sound and video. Intel's microprocessor—the Pentium—uses a 32-bit technology and Microsoft was there with Windows 95. But the technology may take a different path. Microsoft has made adjustments and tried to cope with the challenges imposed by new technologies. Gates and Allen were present since the development of early chips such as the 8008. Yet the information era of the second half of the 1990s was completely new to the company: Internet connections and massive data management were not part of Microsoft's know-how. Microsoft was not the leader in those markets. Competition against the leaders of those emerging technologies, such as Netscape, Sun Microsystems, or Novell, represented a true challenge. Gates made a monumental change in the company's strategic direction. He switched course and decided to compete in the Internet technology. He has shown, once more, that he has the vision and the courage to orient—even drastically change—his company's direction of R&D toward the emerging areas of information technology. This ability to embrace technological change is the mark of great corporate leaders and the secret to the continued survival and prosperity of leading companies.

• As proven in the development of Windows, competitiveness can sometimes be tracked to managerial style and organizational structure. This combination was specifically important at Microsoft. The founders were technical people with the business spirit embodied by Bill Gates. He was able to make decisions and set directions during the early years of the company. The development of DOS was in part possible due to the centralization of the decision making and the relatively small size of the company. In 1984 the position of the PC software industry was very peculiar: On the one hand, the PC was in the growth stage of the technology cycle and DOS had been established as the standard operating system. Microsoft's organization was requiring a more stable managerial style in order to consolidate its leadership position and make better use of the resources available. On the other hand, graphical environments for the PC were in the launching phase, requiring entrepreneurial spirit and a less rigid managerial style. Microsoft was able to reconcile these two types of management styles and remain competitive.

- Microsoft has been successful in changing from one technology to the next, always introducing incremental innovations into its products. Microsoft takes into account the fact that applications and a broad base of hardware are necessary for the new platforms to succeed. The change from DOS to a graphical environment was not drastic, allowing time for software developers to react and for users to acquire new computers. On top of this, Gates's timing of announcements of new releases causes consumers to delay purchasing other options, thus blocking competitors' efforts to obtain market share from Microsoft.
- Microsoft created vast wealth for the nation, the region, its founders, stockholders, and employees.

READING 15.4

## Apple: A Vision to Change the World

The creation, growth, and recent decline of Apple Computer, Inc., is a storybook of success and failure in technology management. Apple brought the PC to people's homes in the late 1970s and early 1980s and helped change the world forever. Yet today Apple struggles to survive in the competitive technological environment it helped to create.

Two young computer enthusiasts, Steve Jobs and Stephen Wozniak, established Apple in a garage in Silicon Valley, California. Wozniak was the technical wizard and Jobs was the entrepreneur. Wozniak had quit the engineering program he attended to follow his passion to build computers and started to work at Hewlett-Packard. He was a regular attendee of the Homebrew Computer Club in California. Wozniak enjoyed working with computer components and programming logic so much that he would go home from the Computer Club meetings with a problem and return with a solution to share with his friends by the next meeting. By 1976, Wozniak had developed a Basic programming language for the 6502 microprocessor. He passed out copies to his Homebrew friends and helped them build their own machines. Jobs shared Wozniak's passion for computers. He lacked Wozniak's technical ability but had a vision to make his mark on the world. The two made a perfect team. They had collaborated before on several projects, including developing games for Atari. Jobs persuaded Wozniak to join him in packaging his design in a kit and selling it under the name Apple I. The garage of Jobs's parents was the first site of Apple Computer. Jobs had a vision for Apple that went beyond building kits for the hobbyist market. He saw it as a business opportunity. He met and solicited the help of Mike Markkula, who at age 33 retired from a lucrative finance and marketing career at Intel. Markkula provided the marketing expertise needed. Jobs also sought venture capital and was successful in persuading Arthur Rock, the venture capitalist who helped start Intel, to fund Apple.

In 1978, Wozniak was able to package several components of the PC together, and the Apple II was presented like a consumer electronic product. It was an instant success at that year's computer show. Jobs commissioned a local factory to build 1,000 machines. Markkula signed up dealers across the country and placed advertisements. Apple moved out of Jobs's garage and into a building in Cupertino, California. Apple needed a business application to take the Apple II from a hobbyist market to the business world. This application came in 1979 with the development of VisiCalc, a spreadsheet program created by Dan Bricklin while studying at Harvard Business School. Bricklin and his professor Bob Frankston developed the program for their own convenience, but later it proved to be the major element needed to take the Apple II computer to a new level of application in the business world. Finally, the personal computer was capable of doing useful functions.

The growth of Apple was impressive in sales and profits. Jobs started Apple when he was 21 years old. At age 23, he was worth $1 million; at age 24, $10 million; and at age 25, $100 million. Apple also made Wozniak and many other people millions of dollars in return for their investment in developing and marketing the new technology.

## THE GROWTH PERIOD

As the 1980s began, the world witnessed phenomenal growth in the PC industry. An avalanche of software programs that turned computers into special-purpose tools kindled the boom. The computer entered into people's everyday lives. It could be used to perform word processing—replacing the typewriter—analyze financial data, file information, sort files, and do whatever else the user wanted—even play games. By 1980, an estimated 25,000 Apple II's were bought specifically for their ability to run VisiCalc. That year, Apple went public with the largest stock offering since Ford Motor Company in 1956.

In 1982, a huge infrastructure of retail dealers sprang up worldwide. Apple's revenues grew to $583 million for fiscal year 1982, the fifth year of company operation. Soon, Apple became the first PC company to reach the milestone of $1 billion in revenues. Many companies watched Apple's success in awe and decided to enter the market with their own PC. IBM, DEC, and AT&T are some of the big-name companies that decided to join the fray and compete. Software and accessory developers rushed to create add-on products and software for the IBM machine. Compaq and other foreign manufacturers in Taiwan and Korea started developing IBM-compatible clones. As a result, the dominance of the Apple line dwindled.

Apple used a marketing strategy that worked for it well at the time. The company prohibited mail-order sales and relied on dealers whom Apple trained and qualified. People at Apple believed that a local dealer who was trained in service and support and was armed with a good technical manual was essential to customer satisfaction. For a new or emerging technology the customers need advice and technical support. Apple provided both by giving extensive training to its sales and service people. Apple also started diffusing its technology by targeting primary and secondary schools. It launched a program called Kids Can't Wait, which donated computers to 10,000 schools throughout California. Apple also awarded many financial grants to individuals and schools for the development of software.

Steve Jobs and his coworkers did not lose sight of the importance of continued innovation. They knew that they had formidable competition from IBM and wanted to stay ahead of the pack with innovation. They introduced new computer models—the IIe and the Lisa models. Lisa was not a commercial success, but it formed the technological foundation for the Macintosh, a highly successful computer released in 1984. The Lisa idea was conceived by Jobs after a visit to Xerox PARC in Silicon Valley. Steve became convinced that the graphical user interface (GUI) that Xerox engineers were developing was the technological breakthrough needed to make the interface between people and the machine more friendly. Apple invested $100 million over two years to develop software that was simple and easy for the average user to understand.

## CHANGES AT THE HELM

Steve Jobs demands perfection, and some technical difficulties forced delays and cost overruns for the Lisa project. At the same time, Apple's growing pains started to appear. The company was in the growth phase; revenues reached $1 billion in 1983. Apple was growing invincibly and production demands were high. Steve Jobs needed help managing the company, and it was time to move to a stewardship management style. Apple seemed to have found the perfect man—John Sculley, a PepsiCo executive who was initially reluctant to make the move. It was reported that Jobs persuaded him by saying, "Do you want to spend the rest of your life selling sugar water, or do you want to join me and change the world" ("Revenge of the Nerds," PBS television documentary, 1996). Sculley signed on and was elected president and CEO of Apple in 1983. Sculley wanted tighter management controls. He thought that R&D expenses were excessive and that Jobs's ideas and zeal for technological innovation on the Lisa project needed to be harnessed. Friction between Jobs, the entrepreneur, and Sculley, the steward, emerged and resulted in the ousting of Jobs by Apple's board in 1985.

Even though the Lisa project was expensive and did not have the commercial success desired, it resulted in a milestone product that reestablished Apple as a technological pioneer in GUI. Elements of the user interface included icons, windows, mouse pointing devices, and pull-down menus. Xerox researchers at PARC initially conceived many of these features, but Xerox's management failed to see their potential for the industry. Apple was able to develop the technology and bring it to the marketplace in its products. These unique PC features set new standards for personal computing and captured the attention of users. They also drew the attention of other software developers, such as Microsoft, to the need for developing software that would make the use of the computer easier. Eventually, Microsoft introduced Windows, its version of user-friendly software.

## APPLE UNDER SCULLEY

In 1984, Apple introduced the Macintosh, a faster and less expensive computer than the Lisa. However, 1985 proved to be a difficult year for Apple. The departure of Jobs and Wozniak, the founders of Apple, took away its entrepreneurial spirit and its soul. Sales were below expectations, IBM had made a forceful entry into the PC market, and the demand for Apple computers was depressed. Sculley started reengineering the organization. Three plants were closed and 1,200 jobs were cut. Costs associated with these measures resulted in the first loss for the company. Sculley was able to reorganize Apple and engineer a turnaround in 1986. His vision was to target office computing and become a real contender in business applications. New software and accessories were released, and the company released high-end Macintosh computers suitable for the office environment. This wave of new technology and Sculley's business strategy were extremely successful and were responsible for an increase in revenues from \$1.96 billion in 1986 to \$5.3 billion in 1989.

Apple managed to get a foothold in the business market and expanded in that area, continuing to push the high-end products until 1989–1990. The premium pricing of these products and proliferation of clones began to slow the company's overall market-share growth. By that time, the PC industry had reached the mature phase. The IBM-clone competition was fierce. Microsoft introduced its Windows 3.0 in the spring of 1990, made it available to all IBM clones, and subsequently steamrolled the industry.

In an attempt to control costs and reverse the company's slow downtrend in 1988, Sculley decentralized Apple. Separate sales divisions were established in the United States, Europe, and the Pacific. A separate division was established for product development and manufacturing. The European and Pacific marketing divisions performed well, but sales in the U.S. division were poor, mainly due to the IBM-clone competition. The change in the organizational structure also caused problems between marketing and R&D labs. New product development was not moving fast enough, and marketers and product developers started blaming each other for the company's woes.

Sculley's attitude and management style focused on changing Apple from Jobs's culture of "we are changing the world" to a down-to-earth image of sustainable, steady, incremental change. He included process improvement initiatives, such as smart manufacturing, and geared the company more toward market-pull than technology-push strategies. His strategies worked well when there was steady growth of the technology. However, when the competition from clones intensified, he was unable to keep the edge with continued innovation.

### DECLINE OF APPLE

Apple had a strong edge with its GUI technology. The company refused to license its operating system or architecture to other computer manufacturers. It elected to keep this technology proprietary and exploit it only in its own products. This superior technology gave Apple a price advantage for its products. Then Microsoft came up with its own Windows GUI software and made it available to clone producers. This broke Apple's hold on the technology. Microsoft diffused its Windows technology fast by putting it on each IBM PC and clone. Apple finally realized its mistake of not exploiting its own superior technology through licensing. Apple announced that it would license its technology in 1993, but the move came too late for Apple to recover. Microsoft had already established the market dominance and de facto industry standard based on its own product.

In 1991, Apple felt the pressure from competitors and made the decision to participate in an IBM/Motorola/Apple consortium. This alliance consists of five distinct initiatives: better integration with IBM compatibles, a new family of Reduced Instruction Set Computers (RISCs), a new open-system environment called AIX derived from IBM's UNIX standard, a multimedia joint venture, and the next generation of operating-system software based on object-oriented programming. The alliance was aimed at breaking Intel's and Microsoft's newly found strongholds on the PC market. Intel and Microsoft were not about to stand still and relinquish their positions. They fought back with innovation and new technology. The game continues. This process is the essence of technology management.

In 1993, Michael Spindler took over as CEO of Apple. He faced a drop in sales, loss in profits, and decrease in market share from 14 to 10%. He took many aggressive actions to deal with the problems. He immediately started cost-cutting measures by laying off 2,500 workers, freezing executive salaries, canning some projects, and reducing R&D by more than $100 million a year. He also reorganized the company and made some changes in marketing strategy. Such changes are inadequate, however, if the competition has a superior technology or a hold on the market.

Another change of Apple's leadership was announced in February 1996. Gilbert Amelio replaced Spindler as CEO. Another wave of layoffs, cost-cutting measures, and change of strategy took place. Apple's second-quarter loss in 1996 was $740 million. The company sales and market share continued to decline. In 1997, Apple stock fell to a 12-year low. Amelio resigned in 1997, as did Ellen Hancock, Executive Vice President of Technology. Steve Jobs was brought back as an advisor to the company. It was indicated that he would be tapped as the company's new CEO, but he declined. Jobs remains as CEO at Pixar, his new, successful high-tech venture. Apple continues to struggle to find the right formula to regain its eminent position in the PC industry. The struggle for survival and for restoring competitive edge at Apple continues.

The story does not end here; it has just begun anew. Who is going to win in the next round of play? The answer depends on who will better manage the ever-changing technology and markets.

---

### Lessons from Apple: A Vision to Change the World

MOT lessons extracted from Apple's case are:

**1** The free flow of information and the sharing of ideas permit innovations. An example is the Homebrew Computer Club, which fueled the development of industry in Silicon Valley.

**2** Venture capital is a strong catalyst in the initiation of start-up companies.

**3** Synergy between an inventor and an entrepreneur leads to successful innovation.

**4** The development of application software was a primary factor in making the PC a success in the marketplace. The development of the VisiCalc spreadsheet fueled the demand for personal computers. Application is the key for technology growth.

**5** The packaging of a product contributes to the success of the product. The packaging and presentation of Apple as a consumer product contributed to Apple's initial success and the broader market acceptance of the PC.

**6** The prize does not always go to the inventor but sometimes goes to the exploiter of an invention. Dan Briklin and Bob Frankston of Harvard designed and wrote the actual code for VisiCalc. They did not protect their technology. VisiCalc was vital for Apple's success. Apple used it to expand its sales, and the inventors reaped limited monetary reward for their invention.

**7** Scientists and engineers at Xerox PARC were the first to introduce many of the features of the modern PC, including GUI. Xerox failed to commercialize the technology. It was Jobs's vision and drive that brought GUI to Apple and made it a commercial success.

**8** Apple was an innovative leader. It introduced the first PC to the average consumer market. At that time, the market was not asking for personal computers for use in any of the major areas in which PCs exist today—education, business, and home. Apple led again in innovation with GUI, which dispelled the public's fear of computers and led the industry with its innovation.

**9** Apple's innovators, Jobs and Wozniak, had a vision to change the world and worked feverishly to project this philosophy. Innovation, whether incremental or radical, was their credo. They were highly successful.

**10** Incremental innovations need to be managed with the same vigor as radical innovations. Once a radical innovation hits the market, it needs to be followed by a series of incremental innovations to keep the product vigor. Apple developed a broad spectrum of products to support the system; monitors, software, add-on cards, and networking products offered the user a highly integrated product line to ensure flexibility. Innovation, whether incremental or radical, is necessary at all times during the life cycle of any industry.

**11** There comes a time in the life cycle of an industry when an entrepreneurial management style needs to yield to a stewardship management style. Apple hired John Sculley, who made good organizational changes to reduce cost, emphasized process improvement, and established a highly automated production line to produce the Macintosh. Each of these measures was necessary for a maturing industry in which the competition had shifted to price and quality.

**12** Sculley realized the importance of forming alliances with competitors to change the rules of competition in technology. He signed a technology accord with IBM and Motorola to develop the PowerPC RISC microprocessor in 1991. This was a way to distribute the cost of developing the processor while taking advantage of core competencies available within the other two technology giants.

**13** Apple decided to license its operating system and architecture in an attempt to increase the presence of Apple's platform in the industry. Unfortunately for Apple, that decision was not made soon enough. Microsoft and the PC clones were able to enter into the GUI technology, diffuse their products, and break Apple's supremacy in the technology before it could take full advantage of it.

**14** Following the departure of Jobs the entrepreneur, Apple attempted to reorganize its R&D activities several times. It also attempted to regain the lead in innovation by introducing new products such as the Newton Personal Digital Assistant. It has not been successful, and Apple continues to struggle to regain its position in the PC industry.

### Some Thoughts about the PC Industry

Timing is the most critical factor in success. Probably the worst mistake Apple made was not diffusing its technology by licensing its operating system and architecture soon enough. Apple protected its technology, and doing so forced the company to support all aspects of the technology and thus kept costs high. Licensing of one's technology can increase market share, reduce cost, and take steam out of competitors attempting to develop similar technology.

IBM did not develop the PC technology. When IBM introduced its PC, the only protected piece of technology in it was the ROMBIOS. Compaq was able to reverse-engineer it and developed a clone. The clones proliferated. IBM initially had an opportunity to get exclusive rights to Microsoft's DOS operating system but passed up this opportunity. Later on, Microsoft refused to give IBM exclusive rights to the software. IBM took a tumble in the PC market, due to bureaucracy and poor management of technology, not clones or Microsoft.

READING 15.5

## Intel: Creating Market Pull

Along with the IBM PC launching in 1981, two companies emerged that would eventually eclipse IBM and become the leaders and creators of industry standards: Microsoft and Intel. Both have shown such dominance that the word "Wintel" (in reference to Windows and Intel) has been used to designate the non-Apple PC industry (Lewis, 1996; Moore, 1996). Another sign of supremacy is that the microprocessor is used to describe the type of computer. For example, a computer owner will say, "I have a 486 computer." This type of product dominance has the power to change the language in the same way that the word "xerox" does when used as a verb. Intel has developed a particular strategy for bringing technology to market early through innovation and de facto dominance of the market. Lewis (1996) reports that there are 100 million Intel computers in 90% of the corporate United States.

### EARLY DAYS OF INTEL

Jack Kilby at Texas Instruments (TI) received credit as the inventor of the integrated circuit. Intel was founded in 1968 by Robert N. Noyce, who is credited with developing industrial applications for the integrated circuit on a silicon chip. During those early times, the company was dedicated to the production of memory chips.

In 1969, while working for a client at Intel, Ted Hoff, a Stanford engineer, conceived the idea of a general-purpose microprocessor that could be programmed to perform different tasks. By 1971, Intel produced its first microprocessor chip, labeled the 4004.

In 1972, Computer Terminals Corporation approached Intel with a request for a more powerful microprocessor, giving birth to the next-generation microprocessor, 8008, which included eight-bit technology. Things seemed to go well, except the client wanted a second supplier for the chip and took the design to TI. With more than one supplier, the competition shifted to price and diffusion of application. Intel's chip was smaller and thus cheaper, because the cost of a chip is a function of its size (Rogers and Larsen, 1984). Intel moved to apply the microprocessor in a wide variety of products, contrasting with TI's focus on the calculator market. Gordon Moore, an Intel cofounder with Noyce, forecasted that technology would permit the number of transistors built on a silicon chip to double every 18 months. This became known as "Moore's law" and it drove the competition in the microprocessor industry.

In 1973, the 8080—twenty times faster that the 4004—came out, opening the door to personal computer developers such as Ed Roberts, who developed the Altair in Albuquerque, initiating the PC era.

## MANAGING INNOVATION: THE X86 SERIES

In the early 1980s, Intel followed an aggressive licensing policy, giving more than 20 licenses in order to diffuse its technology and establish the industry standard. Intel retrenched somewhat from licensing its 286 microprocessor, with only six sources receiving licenses for the 286. By 1985, Intel commanded a monopoly in the 386 until cheaper and faster clones were developed by competitors such as Advanced Micro Devices and Cytrix, causing Intel's market share to drop from 90% to less than 30% (Chang, 1994).

Managers at Intel decided that the best strategy to fight against 386 competitors was to aggressively compete with innovation and time to market—to introduce new generations of microprocessors every two or three years, given the fact that competitors took several months or even years after the launch of a chip to respond with clones. This proved to be a successful strategy. To accomplish the company's goals, different teams at Intel had to work in parallel, a system that increased R&D costs but maintained Intel's leadership: the company spent $2.9 billion in capital during 1993 and another $3.5 billion in 1994 (Moore, 1996). In 1993, less than a year after the clones had absorbed the 386 market, Intel convinced computer manufacturers to switch to the superior 486 to meet the speed requirements of Windows, making the 386 obsolete. During that year, the company achieved $5 billion in revenues and over $1 billion in net profits, mainly due to the 75% profit margin in the 486 (Chang, 1994). In an effort to avoid conceding any niche to competitors, Intel marketed 25 variations of the 486.

## NEW BATTLES

During the early 1990s, the company faced several important challenges. According to Moore (1996), the challenges were as follows:

- Although most PCs had an Intel microprocessor, the final consumer did not realize this. To promote its products, Intel emphasized its central attributes—high-technology chips, first in the market, compatibility, and reliability—with the "Intel Inside" campaign.

• Many of the popular applications, such as word processing, did not require the computational power that Intel could offer. However, Intel's strategy against clones was continuous innovation, keeping competitors off balance as they tried to keep up with the speed and power of Intel's chips that provided a customer's perceived added value.

• The growth of the PC market had overwhelmed the capacity of Intel's manufacturing system. The microprocessors that the company produced were allocated among different computer manufacturers, provoking companies such as Compaq and Dell to be receptive to buying from other suppliers. This allowed competitors to share Intel's market.

• The technological progress made possible by Intel's chips did not fully result in better performance of the PC because the standard computer architecture embedded in the PC was creating bottlenecks in the system.

• Since each new microprocessor required changes in computer design, some manufacturers started to resist adopting more powerful chips.

To overcome these difficulties, Intel had to play a more active role in directing the PC industry toward more advanced computational needs, such as multimedia. With this in mind, Intel created the Intel Architecture Laboratories, whose mission was doing computer architectural research to serve the industry and facilitate the adoption of new developments.

Other initiatives were aimed at stimulating demand for new products: The marketing strategy consisted of seeding machines with the latest microprocessor developments through distribution companies, eventually forcing computer manufacturers to include them in their designs. One additional marketing approach was to refocus the so-called Intel System Business. The purpose of this unit is to facilitate the adoption of new chip generations by selling chip sets, mother boards, or even complete systems to manufacturers during the launching of a new microprocessor.

Intel also shifted the industry toward Pentium, the chip introduced in the spring of 1993. This microprocessor contains over 3 million transistors and doubles the speed of the 486 chip. The Pentium chip technology was capable of being networked with other chips. Networked PCs with Pentiums are able to compete in the workstation market, dominated by companies such as Sun Microsystems, DEC, and Silicon Graphics, at a fraction of the cost and to offer the possibility of running Microsoft software (Chang, 1994).

The Pentium is based on RISC, which is a faster technology than that of the X86 family. Although not 100% RISC, Pentium is the first RISC-design chip able to profit from the PC market due to its ability to run the broad spectrum of software developed for the previous generation.

Intel is not alone in the RISC technology. Many other companies, such as Sun Microsystems, Digital Equipment Corporation, Silicon Graphics, and Hewlett-Packard, are pursuing this technology. The consortium among Apple, IBM, and Motorola to produce the PowerPC was an attempt to counter Intel's dominant position in the microprocessor business. However, it failed to break Intel's hold on the leading position in the market.

In 1997, Intel slashed the price of the Pentium microprocessor by 50% to reduce its inventory and open the road for its new generation of microprocessor. The competition is regrouping to once again break Intel's dominance of the PC microprocessor market. Mergers among less dominant companies and the introduction of a replacement technology would be the strategy to pursue in this effort. As Chang concludes, "Intel has also taken a hedging strategy by diversifying into related products and business, working on new technical standards and products for fax and modem services, video-conferencing, and massively parallel supercomputing" (Chang, 1994).

#### Reading 15.5 References

Chang, Ike. 1994. *The Economics of Dominant Technical Architectures: The Case of the Personal Computer Industry.* RAND, Santa Monica, CA.

Lewis, Ted. 1996. "Is It Too Late for Apple Computer?" *Computer,* May (IEEE).

Moore, James. 1996. *The Death of Competition.* Harper Business, New York.

Rogers, Everett, and Larsen, Judith. 1984. *Silicon Valley Fever.* Basic Books, New York.

### Lessons from Intel: Creating Market Pull

Intel's history and strategies provide another excellent case for managing high-tech companies. Lessons learned are:

**1** Intel used a strategy of applying the microprocessor to a more diverse market than its competitors did. This strategy was decisive in acquiring economies of scale that allow the company to compete not only in technology but also in price.

**2** The exploitation of technology through licensing allows a company to diffuse its technology and establish a dominant design—a de facto industry standard. Intel's success of, and profit from, a new product such as the 386 chip brought about clones. However, Intel was prepared to confront the competition by heavily investing in R&D.

**3** Intel competed with being first to market with innovation. This proved to be a winning strategy.

**4** Compatibility is an important factor in Intel's dominance. Even with the introduction of new and more advanced technology in each product, the company has managed to maintain compatibility with the basic hardware and software of the time, allowing a smoother transition for users from one generation of technology to the next generation.

**5** Intel depends on its core competencies to create new business. Although today it is primarily a microprocessor company, it is diversifying into new businesses that require integrated circuitry.

**6** Technological advancement cannot flourish in isolation. For a technology to be a commercial success, it requires complementary technologies around it. (A classic example of this is the design of portable radios: Even though the transistor was invented in the United States, the Japanese were able to miniaturize other electronic components, making possible the portable radio.) Intel had to confront and overcome technological obstacles outside the microprocessor in order to exploit its own technology.

**7** Being successful in a technological business depends not only on advanced R&D programs but on planned strategies and their perfect execution as well. The plans and strategies must include smart marketing. Intel not only had to use market pull but had to create it.

## THE 15 COMMANDMENTS OF PROPER MOT

Studying companies that successfully manage technology indicates that they share some common characteristics. The key elements in their MOT philosophies are summarized here:

**1** Set your vision and long-term strategy.
**2** Design an aggressive technology plan.
**3** Develop a sound business plan.
**4** Set up organizational procedures to permit the integration of the business and technology plans and allocate resources to support strategic objectives.
**5** Focus on core competencies.
**6** Release workers' creativity.
**7** Focus on customers' needs and demands.
**8** Drive both radical and incremental innovations.
**9** Strive for operational efficiency and adopt a continuous improvement policy.
**10** Build partnerships with suppliers and distributors and take advantage of strategic alliances.
**11** Adjust to market needs and search for new markets.
**12** Educate, train, then educate, train, then educate, train.
**13** Motivate and reward.
**14** Build an organizational structure to support the strategy.
**15** Enforce organizational vigilance to changes in the environment.

## DISCUSSION QUESTIONS

**1** Identify the factors that led to the success of the 3M Company. Class discussion of the 3M case study provides a good interactive forum for discussing the identified factors.
**2** Discuss the Microsoft case, focusing on technology strategy.
**3** Read a recent magazine article on Microsoft and analyze its findings. Compare these findings with what you have read in this chapter about Microsoft's strategies.
**4** The cases included in this chapter end around 1997. However, information technology changes at a fast rate. What is the latest news on the PC-server-network computer? Who seems to be winning the game?
**5** Read recent updates on Microsoft. Is there any apparent weakness in the company? What main alliances does the company have?
**6** On July 1, 1996, a spinoff from 3M created a new company called Imation. Look for information on it. Why did 3M decide to create Imation instead of keeping the business within 3M?
**7** Analyze the development of the IBM PC. What was IBM's major flaw? Compare IBM's PC innovation strategy with the strategies of Intel and Microsoft.
**8** Analyze Internet technologies (i.e., browsers, networking, servers, etc.). Which company do you think will have the advantage in the future?

**9** Obtain recent information on Apple Computer Company. Analyze its recent decisions from a management-of-technology point of view, and discuss their effect on the company's performance.

## ADDITIONAL READINGS

Catherine Arnst and Peter Burrow. 1993. "Showdown in Silicon Valley." *Business Week,* Nov. 1, p. 146.

Lee B. Burgunder. *Legal Aspects of Management of Technology.* South-Western, Cincinnati, Ohio, 1995.

This book discusses all legal issues facing technology managers. It covers intellectual property protection, patent and copyright policies, tort liability, and key contract issues of high-technology companies.

## SUGGESTED ASSIGNMENT

Working in groups of three to five students, prepare a comprehensive case study for a major corporation or a new start-up company. Each group should present a written report and make a class presentation about their project. The report should contain:

1. A description of the company, its business, technology, and management structure. A narrative describing the company's history and major events would be appropriate.
2. Delineation of the issues from a management-of-technology point of view. Both strategic and operational considerations should be examined.
3. A summary of the technology management lessons learned, preferably in list form.
4. Personal opinion as to what the company's strategy should be in the future.

This assignment is suggested for students interested in advanced analysis. It is suitable as a capstone project for MOT students.

# INDEX